# Genetic Modification of Hematopoietic Stem Cells

*Series Editor*
John M. Walker
School of Life Sciences
University of Hertfordshire
Hatfield, Hertfordshire, AL10 9AB, UK

For other titles published in this series, go to
www.springer.com/series/7651

METHODS IN MOLECULAR BIOLOGY™

# Genetic Modification of Hematopoietic Stem Cells

## Methods and Protocols

Edited by

## Christopher Baum

*Department of Experimental Hematology, Hannover Medical School, Hannover, Germany*

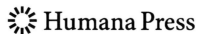 Humana Press

*Editor*
Christopher Baum
Department of Experimental Hematology
Hannover Medical School
Carl-Neuberg-Straße 1
30625 Hannover
Germany

ISBN 978-1-58829-980-2          ISBN 978-1-59745-409-4 (eBook)
ISSN: 1064-3745                e-ISSN: 1940-6029
DOI: 10.1007/978-1-59745-409-4

Library of Congress Control Number: 2008939908

Printed on acid-free paper

springer.com

# Preface

## Gene Transfer into Hematopoietic Cells: From Basic Science to Clinical Application

**Christopher Baum**

### 1. The Potential of Gene Transfer into Hematopoietic Cells

Vectors with the potential for stable transgene integration are widely used in basic hematology and clinical trials of gene medicine. In basic research, both gain-of-function and loss-of-function situations of individual genes can be created by gene transfer, leading to a wide range of applications in developmental biology, stem cell biology, immunology, leukemia research, and human genetics. With the first evidence of successful modification of murine hematopoietic cells using retroviral gene vectors (*1, 2*), researchers have also explored the therapeutic potential of this approach. To date, the emerging discipline of gene therapy is a highly diversified field that offers entirely novel approaches to treat a great variety of human diseases (*3*). All hematopoietic cell types are of major interest in this context, since the modification of the hematopoietic stem cell population may potentially give rise to a completely transgenic hematopoiesis with the potential to cure genetic disorders or fight severe chronic infections, and the targeting of mature cells such as lymphocytes or antigen-presenting dendritic cells offers all types of transient and semipermanent modifications of the immune system.

The unifying principle of gene medicine is the need to transfer complex nucleic acids cells that do not contribute to the germline (somatic cells). "Complex nucleic acids" refers to genetic sequences arranged as expression units with the capacity to encode proteins or RNAs. The rich resource of ~25,000 human protein-coding genes with multiple splice variants, additional genes that express a variety of noncoding RNAs, sequences derived from the remaining living world, and engineered variants of all of these different sources of genetic information creates a tremendous repertoire of options for therapeutic interventions. This great spectrum is further expanded by the availability of many different gene vector systems and the potential choice of different target cell types and various routes and modes of application (ex vivo or in vivo, single or repeated use) (*4*). Clinical considerations addressing biosafety and efficiency of gene medicine need to reflect this enormous diversity and yet try to develop common rules for efficient procedures to evaluation in nonclinical models and patient monitoring. An important emerging field is a specific toxicology for gene medicine. As in other medical disciplines, it can be predicted that the future of gene medicine will heavily rely on the strength and predictive value of "gene toxicology."

Both, evidence for a major clinical benefit (*5–10*) and observations of dose-limiting adverse events in pioneering clinical trials (*10, 11*), have triggered major international activities to obtain insight into underlying mechanisms and develop further improved strategies and technologies for gene transfer and monitoring of gene-modified hematopoiesis. The flexibility in the design of gene vectors can generate relatively straightforward solutions to overcome present hurdles in the development of gene medicine.

Insertional mutagenesis currently represents the most intensely discussed side effects of approaches that use stable integration of transgene sequences into long-lived hematopoietic cells to cure genetic disorders, fight cancer, or improve antiviral immunity. Clinical trials have provided evidence that deregulation of cellular proto-oncogenes after quasirandom transgene insertion in the neighboring chromatin may initiate cell transformation. Potentially premalignant cell expansion of transplanted autologous hematopoietic cells triggered by insertional mutagenesis has been observed in a clinical trial to treat patients suffering from chronic granulomatous disease (CGD) (*10*). Induction of overt lymphoproliferative disorders was observed several years after treating children suffering from an X-linked form of severe combined immunodeficiency (SCID-X1) with autologous hematopoietic cells in which retroviral vectors were introduced to establish a functional copy of the defective gene (*11, 12*). Other clinical trials have been confronted with a lower frequency of severe adverse events, despite molecular evidence of integrations close to proto-oncogenes (*13–15*). This observation has triggered a discussion of potential cofactors that contribute to malignant transformation after insertional mutagenesis. These may include culture conditions of the cells, the nature of the transgene product, and patient-specific factors. Of note, the great majority of related studies performed in large animals and patients, including approaches targeting hematopoietic stem cells or T cells, have not provided evidence for clinically relevant side effects triggered by insertional mutagenesis (*13, 15–17*).

Active research in this field is important to develop new clinical trials using novel, potentially safer generations of integrating gene vectors, such as those based on lentiviruses or transposons. Accumulating evidence convincingly demonstrates that the nature of the vector used and the type of cells treated are crucial variables in the risk profile of insertional mutagenesis *(18–20)*. New technologies are on the horizon that mediate nonviral genome modification in somatic cells *(21)*, including site-specific approaches *(22)*. All these efforts have one common goal: introduce transgenes into cells in a way that maintains the functional integrity of the surrounding genomic sequences while also guaranteeing physiological levels of transgene expression. We would thus expect that the therapeutic index for future clinical trials can be improved. However, these expectations will have to rely on experimental evidence, for which exact and reliable protocols such as those provided in this book are required.

## 2. Organization of Chapters in This Book

This book has been produced to fill an important gap. Previously, no compendium of protocols was available that covers the various aspects of genetic modification of hematopoietic cells: from the purification and culture of various types of hematopoietic cells for subsequent genetic modification over the basic science of vector development and technical issues of vector production in small and large scale, to the complex issue of monitoring and biosafety studies related to gene-modified hematopoiesis. Here, the reader finds a rich resource of protocols that cover these topics, written by leading scientists in their area of expertise. An overview is provided in Fig. 1 .

Having introduced protocols for immunomagnetic (Chap. 1) and flow sorting of hematopoietic cells (Chap. 2), we continue with protocols for gammaretroviral gene transfer into murine (Chap. 3) and human hematopoietic cells (Chap. 4), followed by protocols for lentiviral vector-mediated gene transfer into these cells (Chap. 5). Gammaretroviral, lentiviral, and nonviral transduction protocols are then presented for murine and human T lymphocytes (Chaps. 6–9), natural killer cells (Chap. 10), and dendritic cells (Chap. 11). Protocols for the emerging fields of in vivo application of lentiviral vectors into bone-marrow resident cells (Chap. 12) or thymocytes (Chap. 13) conclude the topic of gene delivery. Of note, these protocols cover the three most widely used gene delivery principles: lentiviral vectors with their potential for genetic modification of nonmitotic cells *(23)*, gammaretroviral vectors that can only target replicating cells *(24)*, and the emerging DNA-based transposons as a nonviral alternative for integrating gene delivery *(21)*.

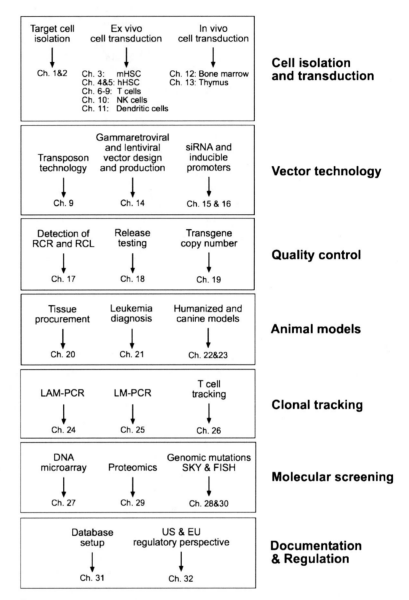

Fig. 1. Organization of chapters in this book.

For those interested in details of vector design and production, an intermezzo of protocols is provided that introduce principles for the design and production of gammaretroviral and lentiviral vectors (Chap. 14), the design and use of vectors expressing small inhibitory RNAs (Chap. 15), and vectors for conditional transgene expression (Chap. 16). Transposon technology is covered in Chap. 9.

The following section of Monitoring and Biosafety Studies contains 15 additional highly valuable chapters that offer protocols applicable for both nonclinical and clinical studies. This includes the detection of replication-competent integrating retrovirus or lentivirus (Chap. 17), procedures for release testing of viral vectors and genetically modified hematopoietic cells (Chap. 18), and copy number determination of gene-modified cells (Chap. 19).

Chapters on tissue procurement for molecular studies using laser microdissection (Chap. 20) and leukemia diagnosis in the most commonly used murine model (Chap. 21) will also be of great interest. These are complemented by protocols for the monitoring of human hematopoiesis in immunodeficient mice (Chap. 22) and canine models of gene-modified hematopoiesis (Chap. 23).

Considering the risk of insertional mutagenesis and the impact of cell culture conditions on cell engraftment, clonal tracking studies are of key importance for monitoring of biodistribution and clonal dynamics of gene-modified hematopoietic cells. In this book, you will find a protocol for LAM-PCR, the most widely used method that allows highly sensitive detection of multiple clones from very small cell samples (Chap. 24), a more simple clonal tracking procedure designed to detect only the dominant clones, including guidelines for bioinformatical data processing (Chap. 25), and a protocol for clonal tracking of T cells (Chap. 26).

Four further chapters cover modern methods of molecular screening approaches for toxicology: DNA microarrays (Chap. 27), proteomics (Chap. 29) a novel approach to the quantification of genomic mutations (Chap. 28) and spectral karyotyping of murine cells (Chap. 30).

The second to last chapter introduces a concept for database setup (Chap. 31), a topic that is of increasing importance also for researchers engaged in the preclinical field. To round up the need for exact protocols and documentation in nonclinical and clinical studies addressing the genetic modification of hematopoietic cells, a most important final chapter (Chap. 32) discusses the regulatory perspective in both the USA and the EU.

In summary, this book provides a unique and comprehensive resource of protocols for the genetic modification of various hematopoietic cell types and up-to-date procedures for molecular and systemic monitoring. The regulatory perspective discussed here covers both the EU and the USA, and may thus also be of value for other regions of the world. It was a great pleasure to be able to work with all the authors who devoted a significant amount of their precious time to realize this compendium. We all hope that this book will be a great help for the design of excellent research in basic hematology, oncology, genetics, and immunology, and also promote the implementation of investigator-driven clinical studies using gene-modified hematopoietic cells.

## Acknowledgements

Work in the laboratory of the editor was supported by the EU-funded network of excellence CLINIGENE, the integrated project CONSERT, the Research Priority Program "Mechanisms of Gene Vector Entry and Persistence" of the Deutsche Forschungsgemeinschaft, the German Ministry for Research and Education, and the National Cancer Institute of the USA.

## References

1. Williams, D. A., Lemischka, I. R., Nathan, D. G., and Mulligan, R. C. (1984) Introduction of new genetic material into pluripotent haematopoietic stem cells of the mouse. *Nature* **310**, 476–80

2. Dick, J. E., Magli, M. C., Huszar, D., Phillips, R. A., and Bernstein, A. (1985) Introduction of a selectable gene into primitive stem cells capable of long-term reconstitution of the hemopoietic system of W/Wv mice. *Cell* **42**, 71–9

3. Alton, E. (2007) Progress and prospects: gene therapy clinical trials (part 1). Gene Ther **14**, 1439–47

4. Thomas, C. E., Ehrhardt, A., and Kay, M. A. (2003) Progress and problems with the use of viral vectors for gene therapy. *Nat Rev Genet* **4**, 346–58

5. Hacein-Bey-Abina, S., Le Deist, F., Carlier, F., Bouneaud, C., Hue, C., De Villartay, J. P., Thrasher, A. J., Wulffraat, N., Sorensen, R., Dupuis-Girod, S., Fischer, A., Davies, E. G., Kuis, W., Leiva, L., and Cavazzana-Calvo, M. (2002) Sustained correction of X-linked severe combined immunodeficiency by ex vivo gene therapy. *N Engl J Med* **346**, 1185–93

6. Aiuti, A., Slavin, S., Aker, M., Ficara, F., Deola, S., Mortellaro, A., Morecki, S., Andolfi, G., Tabucchi, A., Carlucci, F., Marinello, E., Cattaneo, F., Vai, S., Servida, P., Miniero, R., Roncarolo, M. G., and Bordignon, C. (2002) Correction of ADA-SCID by stem cell gene therapy combined with nonmyeloablative conditioning. *Science* **296**, 2410–3

7. Gaspar, H. B., Parsley, K. L., Howe, S., King, D., Gilmour, K. C., Sinclair, J., Brouns, G., Schmidt, M., Von Kalle, C., Barington, T., Jakobsen, M. A., Christensen, H. O., Al Ghonaium, A., White, H. N., Smith, J. L., Levinsky, R. J., Ali, R. R., Kinnon, C., and Thrasher, A. J. (2004) Gene therapy of X-linked severe combined immunodeficiency by use of a pseudotyped gammaretroviral vector. *Lancet* **364**, 2181–7

8. Morgan, R. A., Dudley, M. E., Wunderlich, J. R., Hughes, M. S., Yang, J. C., Sherry, R. M., Royal, R. E., Topalian, S. L., Kammula, U. S., Restifo, N. P., Zheng, Z., Nahvi, A., de Vries, C. R., Rogers-Freezer, L. J., Mavroukakis, S. A., and Rosenberg, S. A. (2006) Cancer regression in patients after transfer of genetically engineered lymphocytes. *Science* **314**, 126–9

9. Bonini, C., Ferrari, G., Verzeletti, S., Servida, P., Zappone, E., Ruggieri, L., Ponzoni, M., Rossini, S., Mavilio, F., Traversari, C., and Bordignon, C. (1997) HSV-TK gene transfer into donor lymphocytes for control of allogeneic graft-versus-leukemia. *Science* **276**, 1719–24

10. Ott, M. G., Schmidt, M., Schwarzwaelder, K., Stein, S., Siler, U., Koehl, U., Glimm, H., Kuhlcke, K., Schilz, A., Kunkel, H., Naundorf, S., Brinkmann, A., Deichmann, A., Fischer, M., Ball, C., Pilz, I., Dunbar, C., Du, Y., Jenkins, N. A., Copeland, N. G., Luthi, U., Hassan, M., Thrasher, A. J., Hoelzer, D., von Kalle, C., Seger, R., and Grez, M. (2006) Correction of X-linked chronic granulomatous disease by gene therapy, augmented by insertional activation of MDS1-EVI1, PRDM16 or SETBP1. *Nat Med* **12**, 401–9

11. Hacein-Bey-Abina, S., Von Kalle, C., Schmidt, M., McCormack, M. P., Wulffraat, N., Leboulch, P., Lim, A., Osborne, C. S., Pawliuk, R., Morillon, E., Sorensen, R., Forster, A., Fraser, P., Cohen, J. I., De Saint-Basile, G., Alexander, I., Wintergerst, U., Frebourg, T., Aurias, A., Stoppa-Lyonnet, D., Romana, S., Radford-Weiss, I., Gross, F., Valensi, F., Delabesse, E., Macintyre, E., Sigaux, F., Soulier, J., Leiva, L. E., Wissler, M., Prinz, C., Rabbitts, T. H., Le Deist, F., Fischer, A., and Cavazzana-Calvo, M. (2003) LMO2-associated clonal T cell proliferation in two patients after gene therapy for SCID-X1. *Science* **302**, 415–9

12. Cavazzana-Calvo, M., and Fischer, A. (2007) Gene therapy for severe combined immunodeficiency: are we there yet? *J Clin Invest* **117**, 1456–65

13. Aiuti, A., Cassani, B., Andolfi, G., Mirolo, M., Biasco, L., Recchia, A., Urbinati, F., Valacca, C., Scaramuzza, S., Aker, M., Slavin, S., Cazzola, M., Sartori, D., Ambrosi, A., Di Serio, C., Roncarolo, M. G., Mavilio, F., and Bordignon, C. (2007) Multilineage hematopoietic reconstitution without clonal selection in ADA-SCID patients treated with stem cell gene therapy. *J Clin Invest* **117**, 2233–40

14. Deichmann, A., Hacein-Bey-Abina, S., Schmidt, M., Garrigue, A., Brugman, M. H., Hu, J., Glimm, H., Gyapay, G., Prum, B., Fraser, C. C., Fischer, N., Schwarzwaelder, K., Siegler, M. L., de Ridder, D., Pike-Overzet, K., Howe, S. J., Thrasher, A. J., Wagemaker, G., Abel, U., Staal, F. J., Delabesse, E., Villeval, J. L., Aronow, B., Hue, C., Prinz, C., Wissler, M., Klanke, C., Weissenbach, J., Alexander, I., Fischer, A., von Kalle, C., and Cavazzana-Calvo, M. (2007) Vector integration is nonrandom and clustered and influences the fate of lymphopoiesis in SCID-X1 gene therapy. *J Clin Invest* **117**, 2225–32

15. Schwarzwaelder, K., Howe, S. J., Schmidt, M., Brugman, M. H., Deichmann, A., Glimm, H., Schmidt, S., Prinz, C., Wissler, M., King, D. J., Zhang, F., Parsley, K. L., Gilmour, K. C., Sinclair, J., Bayford, J., Peraj, R., Pike-Overzet, K., Staal, F. J., de Ridder, D., Kinnon, C., Abel, U., Wagemaker, G., Gaspar, H. B., Thrasher, A. J., and von Kalle, C. (2007) Gammaretrovirus-mediated correction of SCID-X1 is associated with skewed vector integration site distribution in vivo. *J Clin Invest* **117**, 2241–49

16. Nienhuis, A. W., Dunbar, C. E., and Sorrentino, B. P. (2006) Genotoxicity of retroviral integration in hematopoietic cells. *Mol Ther* **13**, 1031–49

17. Recchia, A., Bonini, C., Magnani, Z., Urbinati, F., Sartori, D., Muraro, S., Tagliafico, E., Bondanza, A., Stanghellini, M. T., Bernardi, M., Pescarollo, A., Ciceri, F., Bordignon, C., and Mavilio, F. (2006) Retroviral vector integration deregulates gene expression but has no consequence on the biology and function of transplanted T cells. *Proc Natl Acad Sci U S A* **103**, 1457–62

18. Modlich, U., Bohne, J., Schmidt, M., von Kalle, C., Knoss, S., Schambach, A., and Baum, C. (2006) Cell-culture assays reveal the importance of retroviral vector design for insertional genotoxicity. *Blood* **108**, 2545–53

19. Montini, E., Cesana, D., Schmidt, M., Sanvito, F., Ponzoni, M., Bartholomae, C., Sergi, L. S., Benedicenti, F., Ambrosi, A., Di Serio, C., Doglioni, C., von Kalle, C., and Naldini, L. (2006) Hematopoietic stem cell gene transfer in a tumor-prone mouse model uncovers low genotoxicity of lentiviral vector integration. *Nat Biotechnol* **24**, 687–96

20. Evans-Galea, M. V., Wielgosz, M. M., Hanawa, H., Srivastava, D. K., and Nienhuis, A. W. (2007) Suppression of clonal dominance in cultured human lymphoid cells by addition of the cHS4 insulator to a lentiviral vector. *Mol Ther* **15**, 801–9

21. Izsvak, Z. and Ivics, Z. (2004) Sleeping beauty transposition: biology and applications for molecular therapy. *Mol Ther* **9**, 147–56

22. Urnov, F. D., Miller, J. C., Lee, Y. L., Beausejour, C. M., Rock, J. M., Augustus, S., Jamieson, A. C., Porteus, M. H., Gregory, P. D., and Holmes, M. C. (2005) Highly efficient endogenous human gene correction using designed zinc-finger nucleases. *Nature* **435**, 646–51

23. Ailles, L. E. and Naldini, L. (2002) HIV-1-derived lentiviral vectors. *Curr Topics Microbiol Immunol* **261**, 31–52

24. Baum, C., Schambach, A., Bohne, J., and Galla, M. (2006) Retrovirus vectors: toward the plentivirus? *Mol Ther* **13**, 1050–63

# Contents

# Contributors

OUMEYA ADJALI • *Institut de Génétique Moléculaire de Montpellier, CNRS UMR 5535, IFR122, Montpellier, France*

EVREN ALICI • *Department of Medicine, Karolinska Institutet, Karolinska University Hospital Huddinge, Stockholm, Sweden*

MIRANDA R.M. BAERT • *Department of Immunology, Erasmus University Medical Center, Rotterdam, The Netherlands*

BRENDEN BALCIK • *Division of Experimental Hematology, Cincinnati Children's Hospital Medical Center, Cincinnati, OH, USA*

CYNTHIA BARTHOLOMAE • *Department of Translational Oncology, National Center of Tumor Diseases, Heidelberg, Germany*

BRIAN C. BEARD • *Fred Hutchinson Cancer Research Center, Seattle, WA, USA*

DOMINIQUE BONNET • *Haematopoietic Stem Cell Laboratory, Cancer Research UK, London Research Institute, London, UK*

GUNDA BRANDENBURG • *Preclinical & Clinical Development, Ganymed Pharmaceuticals AG, Mainz, Germany*

RENIER BRENTJENS • *Department of Medicine and Immunology Program, Memorial Sloan-Kettering Cancer Center, New York, NY, USA*

TULIN BUDAK-ALPDOGAN • *Department of Medicine, The Cancer Institute of New Jersey, Robert Wood Johnson Medical School, University of Medicine and Dentistry of New Jersey, New Brunswick, NJ, USA*

JUAN A. BUEREN • *Hematopoiesis and Gene Therapy Division, Centro de Investigaciones Energéticas, Medioambientales y Tecnológicas (CIEMAT) and Centro de Investigación en Red sobre Enfermedades Raras (CIBER-ER), Madrid, Spain*

KLAUS CICHUTEK • *Paul-Ehrlich-Institut, Langen, Germany*

KENNETH CORNETTA • *Department of Medical and Molecular Genetics, Indiana University School of Medicine, Indianapolis, IN, USA*

FRANCOIS-LOIC COSSET • *Ecole Normale Superior Lyon, Lyon, France*

CAROLINE COSTA • *Ecole Normale Superior Lyon, Lyon, France*

EDWIN F.E. DE HAAS • *Department of Immunology, Erasmus University Medical Center, Rotterdam, The Netherlands*

DICK DE RIDDER • *Department of Immunology, Erasmus University Medical Center, Rotterdam, The Netherlands*

M. SIRAC DILBER • *Department of Medicine, Karolinska Institutet, Karolinska University Hospital Huddinge, Stockholm, Sweden*

MATTHIAS EDER • *Department of Hematology, Oncology, Hemostaseology, and Stem Cell Transplantation, Hannover Medical School, Hannover, Germany*

BORIS FEHSE • *Clinic for Stem Cell Transplantation, University Medical Centre Hamburg-Eppendorf, Hamburg, Germany, Experimental Paediatric Oncology and Haematology, University Hospital of the Johann Wolfgang Goethe-University, Frankfurt am Main, Germany*

ARNOLD GANSER • *Department of Hematology, Hemostasis, and Oncology, Hannover Medical School, Hannover, Germany*

HARTMUT GEIGER • *Division of Experimental Hematology, Cincinnati Children's Hospital Medical Center and Department of Medicine, University of Cincinnati, Cincinnati, OH, USA*

HANNO GLIMM • *Department of Translational Oncology, National Center of Tumor Diseases, Heidelberg, Germany*

ELKE GRASSMAN • *Division of Experimental Hematology, Cincinnati Children's Hospital Medical Center, Cincinnati, OH, USA*

GUILLERMO GUENECHEA • *Hematopoiesis and Gene Therapy Division, Centro de Investigaciones Energéticas, Medioambientales y Tecnológicas (CIEMAT) and Centro de Investigación en Red sobre Enfermedades Raras (CIBER-ER), Madrid, Spain*

XIN HUANG • *The Division of Blood and Marrow Transplantation, Department of Pediatrics, The Cancer Center, University of Minnesota, Minneapolis, MN, USA*

CHANTAL JACQUET • *Institut de Génétique Moléculaire de Montpellier, CNRS UMR 5535, IFR 122, Montpellier, France*

ANNE KAISER • *Division of Experimental Hematology, Cincinnati Children's Hospital Medical Center, Cincinnati, OH, USA*

HANS-PETER KIEM • *Fred Hutchinson Cancer Research Center, Seattle, WA, USA University of Washington School of Medicine, Seattle, WA, USA*

HANS KREIPE • *Institute of Pathology, Medizinische Hochschule Hannover, Hannover, Germany*

OLGA S. KUSTIKOVA • *Department of Experimental Haematology, Hannover Medical School, Hannover, Germany, Engelhardt Institute of Molecular Biology, Russian Academy of Sciences, Moscow, Russia*

JAMES LEE • *Department of Medicine and Immunology Program, Memorial Sloan-Kettering Cancer Center, New York, NY, USA*

ULRICH LEHMANN • *Institute of Pathology, Medizinische Hochschule Hannover, Hannover, Germany*

ZHIXIONG LI • *Department of Experimental Hematology, Hannover Medical School, Hannover, Germany*

RAINER LÖW • *EUFETS AG, Idar-Oberstein, Germany*

SOPHIE MARTY • *Institut de Génétique Moléculaire de Montpellier, CNRS UMR 5535, IFR 122, Montpellier, France*

FULVIO MAVILIO • *Department of Biomedical Sciences, University of Modena and Reggio Emilia, Modena, Italy*

R. SCOTT MCIVOR • *The Cancer Center, Department of Genetics, Cell Biology and Development, University of Minnesota, Minneapolis, MN, USA*

ANJALI MISHRA • *Division of Experimental Hematology, Cincinnati Children's Research Foundation, Cincinnati, OH, USA*

UTE MODLICH • *Department of Experimental Hematology, Hannover Medical School, Hannover, Germany*

CEDRIC MONGELLAZ • *Institut de Génétique Moléculaire de Montpellier, CNRS UMR 5535, IFR122, Montpellier, France*

AMÉLIE MONTEL-HAGEN • *Institut de Génétique Moléculaire de Montpellier, CNRS UMR 5535, IFR122, Montpellier, France*

LUIGI NALIDINI • *The San Raffaele Telethon Institute for Gene Therapy HSR-TIGET, Fondazione San Raffaele del Monte Tabor, Milano, Italy*

KALPANA J. NATTAMAI • *Division of Experimental Hematology, Cincinnati Children's Hospital Medical Center and Department of Medicine, University of Cincinnati, Cincinnati, OH, USA*

SEBASTIAN NEWRZELA • *Institute for Biomedical Research, Georg-Speyer-Haus, Frankfurt am Main, Germany*

YUK YIN NG • *Department of Immunology, Erasmus University Medical Center, Rotterdam, The Netherlands*

DIANA NORDLING • *Division of Experimental Hematology, Cincinnati Children's Hospital Medical Center, Cincinnati, OH, USA*

DAO PAN • *Division of Experimental Hematology, Cincinnati Children's Research Foundation, Cincinnati, OH, USA*

KARIN PIKE-OVERZET • *Department of Immunology, Erasmus University Medical Center, Rotterdam, The Netherlands*

ALESSANDRA RECCHIA • *Department of Biomedical Sciences, University of Modena and Reggio Emilia, Modena, Italy*

LILITH REEVES • *Division of Experimental Hematology, Cincinnati Children's Hospital Medical Center, Cincinnati, OH, USA*

ISABELLE RIVIÈRE • *The Gene Transfer and Somatic Cell Engineering Facility, Department of Medicine and Immunology Program, Memorial Sloan-Kettering Cancer Center, New York, NY, USA*

CORNELIA RUDOLPH • *Institute of Cell and Molecular Pathology, Hannover Medical School, Hannover, Germany*

MICHEL SADELAIN • *Department of Medicine and Immunology Program, Memorial Sloan-Kettering Cancer Center, New York, NY, USA*

FRANCESCA SANTONI DI SIO • *The San Raffaele Telethon Institute for Gene Therapy HSR-TIGET, Fondazione San Raffaele del Monte Tabor, Milano, Italy*

LAKSHMI SASTRY • *Department of Medical and Molecular Genetics, Indiana University School of Medicine, Indianapolis, IN, USA*

AXEL SCHAMBACH • *Department of Experimental Hematology, Hannover Medical School, Hannover, Germany*

MICHAELA SCHERR • *Department of Hematology, Oncology, Hemostaseology, and Stem Cell Transplantation, Hannover Medical School, Hannover, Germany*

BERNHARD SCHIEDLMEIER • *Department of Experimental Hematology, Hannover Medical School, Hannover, Germany*

BRIGITTE SCHLEGELBERGER • *Institute of Cell and Molecular Pathology, Hannover Medical School, Hannover, Germany*

DAVID SCHLEIMER • *Division of Experimental Hematology, Cincinnati Children's Hospital Medical Center and Department of Medicine, University of Cincinnati, Cincinnati, OH, USA*

MANFRED SCHMIDT • *Department of Translational Oncology, National Center of Tumor Diseases, Heidelberg, Germany*

TOM SCHONEWILLE • *Department of Immunology, Erasmus University Medical Center, Rotterdam, The Netherlands*

TODD SCHUESLER • *Division of Experimental Hematology, Cincinnati Children's Hospital Medical Center, Cincinnati, OH, USA*

KERSTIN SCHWARZWAELDER • *Department of Translational Oncology, National Center of Tumor Diseases, Heidelberg, Germany*

JOSE C. SEGOVIA • *Hematopoiesis and Gene Therapy Division, Centro de Investigaciones Energéticas, Medioambientales y Tecnológicas (CIEMAT) and Centro de Investigación en Red sobre Enfermedades Raras (CIBER-ER), Madrid, Spain*

FRANK J.T. STAAL • *Department of Immunology, Erasmus University Medical Center, Rotterdam, The Netherlands*

RENATA STRIPECKE • *Department of Hematology, Oncology, Hemostaseology, and Stem Cell Transplantation, Hannover Medical School, Hannover, Germany*

TOLGA SUTLU • *Department of Medicine, Karolinska Institutet, Karolinska University Hospital Huddinge, Stockholm, Sweden*

LOUISE SWAINSON • *Institut de Génétique Moléculaire de Montpellier, CNRS UMR 5535, IFR 122, Montpellier, France*

WILLIAM S. SWANEY • *Division of Experimental Hematology, Cincinnati Children's Research Foundation, Cincinnati, OH, USA*

NAOMI TAYLOR • *Institut de Génétique Moléculaire de Montpellier, CNRS UMR 5535, IFR 122, Montpellier, France*

JOHANNES C.M. VAN DER LOO • *Division of Experimental Hematology, Cincinnati Children's Research Foundation, Cincinnati, OH, USA*

LETIZIA VENTURINI • *Department of Hematology, Oncology, Hemostaseology, and Stem Cell Transplantation, Hannover Medical School, Hannover, Germany*

ELS VERHOEYEN • *Ecole Normale Superior Lyon, Lyon, France*

RITA VICENTE • *Institut de Génétique Moléculaire de Montpellier, CNRS UMR 5535, IFR 122, Montpellier, France*

JAN VIJG • *Buck Institute for Age Research, Novato, CA, USA*

CHRISTOF VON KALLE • *Department of Translational Oncology, National Center of Tumor Diseases, Heidelberg, Germany, and Division of Experimental Hematology, Cincinnati Children's Hospital Medical Center, Cincinnati, OH, USA*

DOROTHEE VON LAER • *Institute for Biomedical Research, Georg-Speyer-Haus, Frankfurt am Main, Germany*

EVA M. WEISSINGER • *Department of Hematology, Hemostasis, and Oncology, Hannover Medical School, Hannover, Germany*

ANDREW C. WILBER • *Department of Genetics, Cell Biology and Development, University of Minnesota, Minneapolis, MN, USA*

CAROLYN A. WILSON • *Division of Cellular and Gene Therapies, Office of Cellular, Tissues, and Gene Therapies, Center for Biologics Evaluation and Research, U.S. FDA, Bethesda, MD, USA*

XIANZHENG ZHOU • *The Division of Blood and Marrow Transplantation, Department of Pediatrics, The Cancer Center, University of Minnesota, Minneapolis, MN, USA*

VALÉRIE ZIMMERMANN • *Institut de Génétique Moléculaire de Montpellier, CNRS UMR 5535, IFR122, Montpellier, France*

PETRA ZÜRBIG • *Mosaiques Diagnostics GmbH, Hannover, Germany*

# Chapter 1

## Immunomagnetic Enrichment of Human and Mouse Hematopoietic Stem Cells for Gene Therapy Applications

### Guillermo Guenechea, Jose C. Segovia, and Juan A. Bueren

## Summary

The hematopoietic stem cells (HSCs) constitute an ideal target for the gene therapy of inherited diseases affecting the hematopoietic system. HSCs, however, constitute a very rare population of progenitor cells, most of which are out of cycle in normal bone marrow. To facilitate their transduction with gammaretroviral or lentiviral vectors, HSCs are generally enriched using physical or pharmacologic methods. In this chapter we describe efficient procedures which are frequently used to enrich human and mouse HSCs, aiming at the transduction of these cells with adequate gene therapy vectors or the subsequent purification of particular HSCs by fluorescence-activated cell sorting.

**Key words:** Immunomagnetic cell sorting, Stem cells, Gene therapy, Flow cytometry, 5-fluorouracil.

## 1. Introduction

One of the main advantages of the HSCs as targets of gene therapy relies on the fact that this population is capable of undergoing self-renewing divisions and also of differentiating into mature cells of different lineages. Therefore, transducing the true HSCs with integrative vectors facilitates the generation of genetically corrected T- and B-lymphocytes, and also of myeloid and erythroid cells. In contrast to other stem cells, HSCs can be readily harvested from accessible hematopoietic tissues and then concentrated using procedures routinely used in clinical settings (1), thus facilitating their genetic modification.

Nowadays, a variety of immunomagnetic cell sorting methods facilitate the efficient concentration of hematopoietic progenitors

Christopher Baum (ed.), *Methods in Molecular Biology, Methods and Protocols, vol. 506*
© Humana Press, a part of Springer Science+Business Media, LLC 2009
DOI: 10.1007/978-1-59745-409-4_1

and HSCs in a few hours. Although, in general, these methods are not directed toward the purification of specific subpopulations of HSCs, they are extremely efficient in terms of cell yield. Consequently, immunomagnetic cell sorting is widely used as a preliminary fractionation procedure prior to fluorescence activated cell sorting (FACS). Additionally, since most gene therapy applications do not require the use of highly purified populations of HSCs *(2–5)*, samples moderately enriched in HSCs (i.e., by immunomagnetic sorting procedures) constitute adequate targets of these gene therapy protocols.

In this chapter we describe in detail how to purify human CD34+ progenitor cells, and also how to purify mouse hematopoietic progenitors and HSCs based on the negative selection of differentiating cells. Additionally, the preparation of mouse bone marrow grafts enriched in HSCs after a single treatment with 5-fluorouracil treatment is described.

## 2. Materials

### 2.1. Enrichment of Human CD34+ Cells

1. As a source of human CD34+ cells, umbilical cord blood (CB) cells can be used. Samples are obtained from umbilical cord, after previous written informed consent of the mother (*see* **Note 1**).

2. Ficoll-Paque (1.077 g/mL, Pharmacia Biotech), stored at 4°C.

3. 15-mL and 50-mL conical tubes.

4. Magnetic cell separator MiniMACS, MidiMACS, VarioMACS, or SuperMACS columns, and column adapters (Miltenyi Biotec GmbH; **Fig. 1a–c**).

5. 30-μm nylon mesh, Pre-separation Filter #130-041-407 (Miltenyi Biotec GmbH; **Fig. 1a–c**).

6. Direct CD34 Progenitor Cell Isolation Kit (Miltenyi Biotec GmbH). It contains MACS CD34 MicroBeads (Colloidal superparamagnetic MicroBeads directly conjugated to monoclonal mouse antihuman CD34 antibody) and FcR Blocking Reagent (human IgG). Order No. 130-046-703 (*see* **Note 2**).

7. *Working solution (PBE)*. 0.5% BSA, 0.05 mM EDTA in Phosphate Buffer Saline (PBS), filtered by 0.22 μm and degasified (*see* **Note 3**).

8. DMSO (Dimethyl Sulfoxide, Sigma, St Louise, MO).

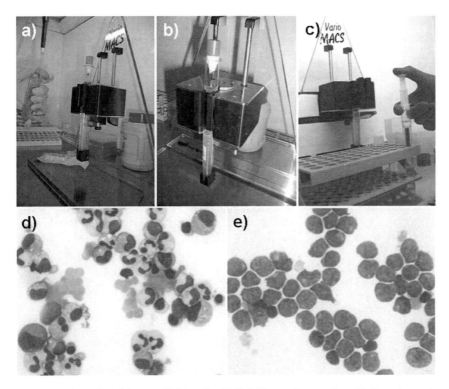

Fig. 1. Immunomagnetic sorting of human CD34+ cells. (a) VarioMacs cell separator with the column placed in the magnet. The picture shows the moment of applying the cell suspension onto the filter and allowing the cells to pass through the column. (b) The picture shows the collection of the CD34− cells contaminated with erythrocytes. (c) The picture shows the collection of the CD34+ fraction by washing the cells retained in the column (already removed from the magnet) with the plunge. (d) May-Grunwald Giemsa staining of the human hematopoietic cell population before enrichment and (e) after immunoselection of CD34+ cells.

9. Human albumin (Behring) and dextran-40 (Rheomacrodex 10%, Pharmacia Biotech).

10. *PBA.* PBS 1×, 0.1% BSA (w/v), 0.02% NaN$_3$ (w/v).

11. Fetal bovine serum (FBS) (GIBCO Laboratories).

12. Iscove's modified Dulbecco's medium (IMDM) (GIBCO Laboratories).

13. Flow cytometer EPICS XL (Coulter, Hialeah, FL).

14. *Monoclonal antibodies.* antihuman-CD34-PE (Anti-HPCA-2, Becton Dickinson Immunocytometry).

15. Propidium iodide (PI) at 2 mg/mL (Sigma).

16. *Turk solution.* 2% Acetic Acid, 0.01% Methylene Blue in water.

17. Trypan Blue (Sigma).

18. Hemocytometer (Neubauer chamber).

**2.2. Enrichment of Mouse HSCs by Depletion of Differentiating Cells**

1. *Mice.* We routinely use B6D2 mice F1 (C57BL/6J x DBA/2), although the procedure can be used with other mouse strains.

2. Iscove's modified Dulbecco's medium (IMDM) (GIBCO Laboratories).

3. Fetal bovine serum (FBS) (GIBCO Laboratories).

4. *Turk solution.* 2% Acetic Acid, 0.01% Methylene Blue in water.

5. Trypan Blue (Sigma).

6. Lineage depletion kit; order #140-090-858 (Miltenyi Biotec).

7. *Working solution (PBE).* 0.5% BSA, 0.05 mM EDTA in Phosphate Buffer Saline (PBS), filtered and degasified.

8. Magnetic cell separator MiniMACS, MidiMACS, VarioMACS, or SuperMACS columns, and column adapters (Miltenyi Biotec GmbH).

**2.3. Enrichment of Mouse HSCs by 5-FU**

1. 5-Fluorouracil (5-FU; F. Hoffmann-La Roche Ltd, Switzerland; Stock solution at 50 mg/mL).

2. *Mice.* We routinely use B6D2 mice F1 (C57BL/6J x DBA/2), although other mouse strains can be used (*see* **Note 4**).

# 3. Methods

**3.1. Methods for the Enrichment of Human CD34+ Cells**

1. Add the heparinized CB cells (2 volumes) onto Ficoll-Paque (1 volume) and centrifuge at $400 \times g$ for 30 min without break, at 20°C in a swinging-bucket rotor. Generally, the blood from one CB (approximately 90 mL) can be distributed in six conical tubes of 50 mL.

2. Aspirate the upper layer leaving the MNC layer undisturbed at the interphase.

3. Carefully collect the MNC layer in the interface.

4. Add the MNCs from the six tubes into one 50-mL conical tube. Dilute the samples 1:5 with PBE. Centrifuge at $400 \times g$ for 15 min.

5. Discard the supernatant, resuspend the cells, and dilute in 25 mL PBE. Before centrifugation, take a small aliquot to count the number of mononuclear cells (MNCs) in a hemocytometer with Turk solution.

6. Resuspend the cells in PBE at a final volume of 300 μL per $10^8$ MNCs. If the cell number is lower than $10^8$ cells, also resuspend the sample in 300 μL PBE.

7. The labeling of the cell suspension with anti-CD34+ MoAbs directly coupled to the magnetic beads requires one single

labeling step: Add 100 µL of FcR Blocking Reagent per 300 µL of the cell suspension ($10^8$ MNCs). Vortex to inhibit unspecific or Fc-receptor-mediated binding of CD34 MicroBeads to nontarget cells. Immediately, add another 100 µL of CD34 MicroBeads per 300 µL of initial cell suspension and vortex (*see* **Note 5**).

8. Incubate 30 min at 4°C in the dark.

9. Wash the cells by adding 20 mL PBE and centrifuge at 400 × *g* for 10 min. Discard the supernatant completely.

10. Resuspend the cells in 3 mL of PBE (maximum of $2 \times 10^8$ cells/mL).

11. Choose an appropriate magnetic column and magnetic separator according to the number of cells to be labeled. For up to $2 \times 10^8$ cells use an MS column. For up to $2 \times 10^9$ total cells, use an LS column (Miltenyi Biotec) (*see* **Note 6**).

12. Place the column in the appropriate magnet and place the 30-µm nylon mesh on top of the column to remove clumps. Wash the filter and the column with 3 mL PBE. Allow the buffer to flow through until stop dropping (**Fig. 1a**).

13. Apply the cell suspension onto the filter and allow the cells to pass through the filter and the column. Collect effluent, which will be enriched in CD34⁻ cells (**Fig. 1b**).

14. Separate the column from the magnet and recover the CD34⁺ cells with 1 mL PBE and the help of the plunger provided with the column (**Fig. 1c**).

15. To improve the purity of the CD34⁺ cell fraction, we recommend conducting a second purification of the CD34⁺ cell fraction: Add the CD34⁺ cell fraction from the first column (1 mL) into a second MS column (without the cell filter) placed in the magnet. Do not allow the column to stop dropping. Allow the cells to pass through the column and collect the effluent in a tube.

16. Wash the column with 1 mL PBE for three times, and collect the eluent in the same tube, which will contain a second fraction of CD34⁻ cells.

17. Separate the column from the magnet and recover the second fraction of CD34⁺ cells with 0.5 mL PBE and the help of the plunger provided with the column. At least 90% of these cells will be CD34⁺ cells, which could be directly used for vector transduction, or subjected to further purification (*see* **Note 7**). Cell numbers and controls of cell viability must be determined in a hemocytometer, using the Trypan blue solution.

18. To determine the purity of CD34⁺ cells, incubate an aliquot of the CD34⁺ fraction (around $10^4$ cells are enough) with antihuman-CD34-PE for 30 min at 4°C in the dark. Wash

the cells with PBA, resuspend in PBA plus 2 mg/mL PI stain to exclude death cells, and analyze in a flow cytometer, as described later.

19. Perform flow cytometry analyses by a serial gating strategy (**Fig.2**). Briefly, a forward size scatter (FS) vs. side size scatter (SSC) dot plot is set up and a gate is drawn to select the MNCs cells. This gate is applied to a second FL4 (675 nm ± 10 band pass filter) vs. FL3 (610 nm ± 10 band pass filter) dot plot in which a new gate is drawn avoiding cells that appear as a diagonal, which are considered dead cells. Finally, both gates are applied to an FL2 (575 nm ± 10) dot plot to analyze PE fluorescence. Collect at least 5,000 events within the two consecutive gates per sample analysis. Off-line analysis can be done with the WinMDI free software package http://www.cyto.purdue.edu/flowcyt/software/Winmdi.htm.

***3.2. Methods for the Enrichment of Mouse HSCs by Depletion of Differentiating Cells***

1. Obtain mouse BM from femora and tibiae by flushing IMDM throughout the shaft of the bones using a 25–27G needle attached to a 1-mL syringe. Use 0.5–1 mL of IMDM for each bone.

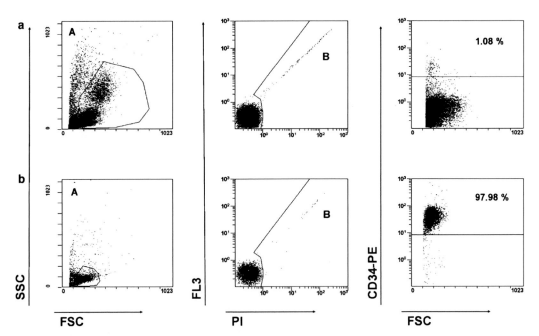

Fig. 2. FACS analysis of the CD34+ cell content in human cord blood before and after immunomagnetic separation. Panels (a) and (b) show *dot plots* corresponding to unpurified and CD34+ cord blood purified samples, respectively. In the first column, the forward size scatter (FSC) vs. side size scatter (SSC) is represented. These *dot plots* facilitate to select the mononuclear cells (gate A). Gate A is applied to a second IP vs. FL3 *dot plot* (*dot plots* in the middle of the figure) where Gate B is drawn. Cells that appear as a diagonal are dead cells and must not be considered. Finally, Gates A and B are applied to the *right dot plots*, to estimate the percentage of cells expressing the CD34 marker.

2. Make a single cell suspension by flushing the BM cells through the same needle, into a new tube. Alternatively, the cell suspension can be filtered with a 30-μm mesh (preseparation filters #130-041-407, Miltenyi Biotec).

3. Wash the cells by adding 8 mL of PBE and centrifuge at 300 × $g$ for 10 min at 4–8°C. Discard the supernatant completely.

4. Resuspend the cells in PBE buffer and determine the cell concentration by diluting a sample with Turk's solution and counting in a hemocytometer. Resuspend the cells in 40 μL of PBE buffer per every $10^7$ total cells. If less than $10^7$ cells are used, also use 40 μL of the PBE buffer. If higher cell numbers are required, scale up all reagent volumes, accordingly.

5. Add 10 μL of the Biotin-labeled-antibodies cocktail (containing antimouse-CD3, -B220, -TER119, -GR1, and -CD11b MoAbs) per aliquot containing $10^7$ BM cells.

6. Mix well and incubate for 10 min in the refrigerator (4–8°C).

7. Add 30 μL of PBE and 20 μL of antiBiotin Microbeads per $10^7$ BM cells.

8. Mix well and incubate for additional 15 min in the refrigerator.

9. Wash the cells by adding 2 mL of PBE per $10^7$ cells and centrifuge at 300 × $g$ for 10 min. Discard supernatant completely.

10. Resuspend up to $10^8$ cells in 500 μL of PBE.

11. Choose the appropriate magnetic column and magnetic separator according to the number of cells labeled. For up to 2 × $10^8$ BM cells use an MS column. If numbers are in the range of 2 × $10^8$–2 × $10^9$ cells, use an LS column (*see* **Note 8**).

12. Place the column in the appropriate magnet and place a cell filter on the top of the column. Wash the filter and the column by rinsing with the appropriate volume of buffer (500 μL for MS or 3 mL for LS columns). Allow the buffer to flow through until stop dropping.

13. Apply the cell suspension onto the filter and allow the cells to flow through the filter and the column. With a pipette tip collect the cell suspension retained in the bottom of the filter and apply it onto the column.

14. Collect the effluent as the fraction enriched in lineage negative (Lin⁻) cells.

15. Wash the column with the appropriate volume (500 μL for MS or 3 mL for LS column) and collect the effluent into the same tube containing the Lin⁻ cells (*see* **Note 9**).

16. Take a small aliquot for cell counting using Turk solution and evaluate the purity of the Lin population by staining a small aliquot (around $10^4$ cells are sufficient) with Tricolor™-labeled

biotin. Incubate in the dark for 30 min at 4°C. Wash the cells with PBA, resuspend in PBA plus 2 mg/mL propidium Iodide to exclude stained death cells, and analyze in a flow cytometer.

17. Perform flow cytometric analyses by a serial gating strategy (**Fig. 3**). Briefly, a forward size scatter (FSC) vs. side size scatter (SSC) dot plot is set up, and a gate is drawn to select only the mononuclear cells. This gate is applied to a second FL4 (675 nm ± 10 band pass filter) vs. FL3 (610 nm ± 10 band pass filter) dot plot in which a new gate is drawn avoiding cells that appear as a diagonal, which are considered dead cells. Finally, both gates are applied to an FL4 (675 nm ± 10) histogram to estimate Tricolor™ fluorescence. Collect at least 5,000 events within the two consecutive gates per sample analysis. Off-line analysis can be done with the Win-MDI free software package http://www.cyto.purdue.edu/flowcyt/software/Winmdi.htm.

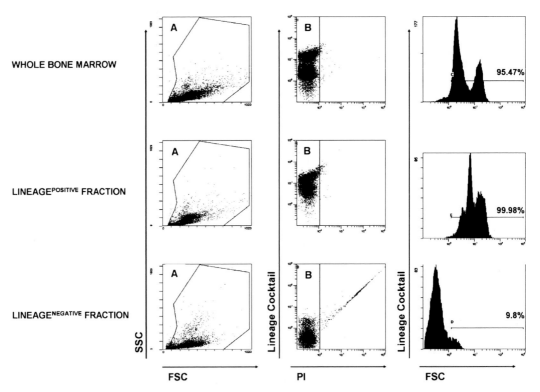

Fig. 3. Flow cytometry analysis of mouse bone marrow enriched in hematopoietic stem cells by depletion of cells expressing lineage markers. Gate A from the *left dot plots* (FSC vs. SSC) is applied to *dot plots* represented in the middle of the figure (Lin cocktail vs. PI). Gate B is drawn to exclude dead cells. Finally, gates A and B are applied to the *right dot plots*, to estimate the percentage of cells expressing the lineage markers. Region C is drawn by labeling cells with isotype antibodies. FSC forward size scatter; SSC side size scatter; PI propidium iodide.

18. Wash the Lin⁻ cell suspension with IMDM plus 20% FBS and centrifuge at $300 \times g$ for 10 min. Discard supernatant and resuspend cells in IMDM + 20%FBS. Cells can now either be applied to further HSC purification by FACS or directly subjected to transduction.

***3.3. Methods for the Enrichment of Mouse HSCs Using 5-Fluorouracil***

1. Weight the animals to be used as BM donors and determine the concentration of 5-FU to be i.v. injected in each animal. The final concentration of 5-FU must be 150 mg/kg body weight (*see* **Note 10**).

2. Inject the 5-FU solution through the tail vein.

3. Harvest the BM from the donors 2–5 days after 5-FU administration (*see* **Note 11**).

4. Bone marrow cells can be subjected to further HSC purification or directly used for vectors transduction.

# 4. Notes

1. Other human hematopoietic sources can be used instead of CB. However, for practical reasons CB samples generally constitute the preferential source of human HSCs in experimental studies. Compared to mobilized hHSCs or bone marrow-derived hHSCs, CB samples have improved properties for the engraftment of immunodeficient mice *(6)*. Samples scheduled for discard and processed during the next 12 h postpartum must be collected in heparin.

2. Other immunomagnetic separation methods for human hematopoietic progenitors are available like EasySep or Robosep (StemCell Technologies, Vancouver, Canada).

3. The PBE solution must be degasified in order to avoid microbubbles that could interfere with the flow during the cell sorting procedure. To degasify the PBE, filter the solution with a vacuum filter applied into a bottle and then close the container with the cup and leave it with the vacuum connected for at least 15 min.

4. The mouse strain to be used in these experiments must be carefully considered since the kinetics of BM damage after 5-FU treatment is mouse strain dependent. Higher HSCs enrichments are associated with lower numbers of BM/femur *(7–9)*.

5. For the magnetic cell labeling use precooled solutions, work fast, and keep cells cold.

6. It is very important to respect these threshold numbers since an overload of cells in the column results in clotting, flowing stop and losing cells. We recommend using one LS column as a first step, followed by an MS column.

7. In many times it is useful to cryopreserve CD34⁺ cells in liquid nitrogen. In these cases we recommend to use IMDM supplemented with 10% of DMSO and 20% of FBS as a cryoprotector. We recommend thawing the CD34⁺ cells according to an optimized protocol *(10)*. Briefly, dilute the cells 1:1 with IMDM containing 2.5% human albumin and 5% dextran-40, and maintain at room temperature for 10 min. Dilute the cells with IMDM plus 10% FBS and centrifuge at $400 \times g$ for 15 min. Finally, adjust the cell density by counting the cell suspension in a hemocytometer.

8. It is very important to maintain these concentrations because an overload of the column will result in clotting, flowing stop and losing cells.

9. This step can be avoided if a more enriched lineage negative fraction is needed. In these cases, the cell yield will be, however, reduced.

10. We recommend injecting each mouse with 100–200 µL of the 5-FU solution. Therefore, we recommend to dilute 2.5 times the stock solution of 5-FU (generally at 50 mg/mL).

11. The BM preparation will be highly contaminated with red blood cells and markedly depleted on dividing cells. The number of nucleated cells will be 5–10 times lower than the expected number in normal BM. If BM is harvested 4–5 days after 5-FU treatment, the primitive HSCs will be actively cycling, and therefore will not need a prestimulation period before gammaretroviral transduction.

## Acknowledgments

The authors especially thank M.E. Alonso, A. Gonzalez-Murillo, and M.L. Lozano for helpful collaboration; and I. Orman, J. Martínez-Palacio, and S. García for technical work. This work supported by grants from the European Program "Life Sciences, Genomics and Biotechnology for Health" (CONSERT; Ref 005242), the Ministerio de Educación y Ciencia (SAF2005-02381; SAF2005-00058), and the Fondo de Investigaciones Sanitarias, Instituto de Salud Carlos III (RD06/0010/0015) awarded to J.C.S and J.A.B.

# References

1. Gaipa, G., Dassi, M., Perseghin, P., Venturi, N., Corti, P., Bonanomi, S., Balduzzi, A., Longoni, D., Uderzo, C., Biondi, A., Masera, G., Parini, R., Bertagnolio, B., Uziel, G., Peters, C., and Rovelli, A. (2003) Allogeneic bone marrow stem cell transplantation following CD34 + immunomagnetic enrichment in patients with inherited metabolic storage diseases. *Bone Marrow Transplant* 31, 857–860

2. Cavazzana-Calvo, M., Hacein-Bey, S., de Saint Basile, G., Gross, F., Yvon, E., Nusbaum, P., Selz, F., Hue, C., Certain, S., Casanova, J.L., Bousso, P., Deist, F.L., and Fischer, A. (2000) Gene therapy of human severe combined immunodeficiency (SCID)-X1 disease. *Science* 288, 669–672

3. Gaspar, H.B., Parsley, K.L., Howe, S., King, D., Gilmour, K.C., Sinclair, J., Brouns, G., Schmidt, M., Von Kalle, C., Barington, T., Jakobsen, M.A., Christensen, H.O., Al Ghonaium, A., White, H.N., Smith, J.L., Levinsky, R.J., Ali, R.R., Kinnon, C., and Thrasher, A.J. (2004) Gene therapy of X-linked severe combined immunodeficiency by use of a pseudotyped gammaretroviral vector. *Lancet* 364, 2181–2187

4. Ott, M.G., Schmidt, M., Schwarzwaelder, K., Stein, S., Siler, U., Koehl, U., Glimm, H., Kuhlcke, K., Schilz, A., Kunkel, H., Naundorf, S., Brinkmann, A., Deichmann, A., Fischer, M., Ball, C., Pilz, I., Dunbar, C., Du, Y., Jenkins, N.A., Copeland, N.G., Luthi, U., Hassan, M., Thrasher, A.J., Hoelzer, D., von Kalle, C., Seger, R., and Grez, M. (2006) Correction of X-linked chronic granulomatous disease by gene therapy, augmented by insertional activation of MDS1-EVI1, PRDM16 or SETBP1. *Nature Medicine* 12, 401–409

5. Aiuti, A., Slavin, S., Aker, M., Ficara, F., Deola, S., Mortellaro, A., Morecki, S., Andolfi, G., Tabucchi, A., Carlucci, F., Marinello, E., Cattaneo, F., Vai, S., Servida, P., Miniero, R., Roncarolo, M.G., and Bordignon, C. (2002) Correction of ADA-SCID by stem cell gene therapy combined with. *Science* 296, 2410–2413

6. Wang, J.C., Lapidot, T., Cashman, J.D., Doedens, M., Addy, L., Sutherland, D.R., Nayar, R., Laraya, P., Minden, M., Keating, A., Eaves, A.C., Eaves, C.J., and Dick, J.E. (1998) High level engraftment of NOD/SCID mice by primitive normal and leukemic hematopoietic cells from patients with chronic myeloid leukemia in chronic phase. *Blood* 91, 2406–2414

7. Ogawa, M., Shih, J.P., and Katayama, N. (1994) Enrichment for primitive hemopoietic progenitors of marrow cells from 5-fluorouracil-treated mice and normal mice. *Blood Cells* 20, 7–11; discussion 11–13

8. Szilvassy, S.J., Lansdorp, P.M., Humphries, R.K., Eaves, A.C., and Eaves, C.J. (1989) Isolation in a single step of a highly enriched murine hematopoietic stem cell population with competitive long-term repopulating ability. *Blood* 74, 930–939

9. Barker, J.E., Wolfe, J.H., Rowe, L.B., and Birkenmeier, E.H. (1993) Advantages of gradient vs. 5-fluorouracil enrichment of stem cells for retroviral-mediated gene transfer. *Experimental Hematology* 21, 47–54

10. Rubinstein, P., Dobrila, L., Rosenfield, R.E., Adamson, J.W., Migliaccio, G., Migliaccio, A.R., Taylor, P.E., and Stevens, C.E. (1995) Processing and cryopreservation of placental/umbilical cord blood for unrelated bone marrow reconstitution. *Proceedings of the National Academy of Sciences of the U S A* 92, 10119–10122

# Chapter 2

# Isolation of Human and Mouse Hematopoietic Stem Cells

## Yuk Yin Ng, Miranda R.M. Baert, Edwin F.E. de Haas, Karin Pike-Overzet, and Frank J.T. Staal

## Summary

Hematopoietic stem cells (HSC) are rare with estimated frequencies of 1 in 10,000 bone marrow cells and 1 in every 100,000 blood cells. The most important characteristic of HSC is their capacity to provide complete restoration of all blood cell lineages after bone marrow ablation. Therefore they are considered as the ideal targets for various clinical applications including stem cell transplantation and gene therapy. In adult mice and men, the main stem cell source is the bone marrow. For clinical applications HSC derived from umbilical cord blood (UCB) and G-CSF mobilized peripheral blood (PB) have been demonstrated to have several advantages compared to bone marrow; therefore, they are slowly replacing BM as alternative source of stem cells. The mouse is the model organism of choice for immunological and hematological research; therefore, studies of murine HSC are an important research topic. Here we described the most often used protocols and methods to isolate human and mouse HSC to high purity.

**Key words:** Hematopoiesis, Stem cell, CD34, Cord blood, Magnetic labeling, Cell sorter.

## 1. Introduction

In humans the mostly used marker for the hematopoietic stem cells is the sialomucin CD34 antigen, while in mouse, hematopoietic stem cells are enriched in a cell population often referred to lineage negative, Sca-1[+] and c-kit[+] (LSK) cells, where stem cell antigen-1 (Sca-1) plays a similar role (1, 2). CD34 expression has been found in immature stem cells and committed progenitors, but as cells mature the level of CD34 expression declines (3, 4). Thus, immature stem cells express the highest level. CD34 expression is present on 1–3% of bone marrow mononucleated cells, while in steady state peripheral blood CD34[+] cells are extremely

Christopher Baum (ed.), *Methods in Molecular Biology, Methods and Protocols, vol. 506*
© Humana Press, a part of Springer Science + Business Media, LLC 2009
DOI: 10.1007/978-1-59745-409-4_2

rare (< 0.1%), but this can be increased by various mobilization protocols of bone marrow stem cells into the circulation. In umbilical cord blood, the percentage of CD34+ stem cells is about 1% of the mononucleated cells. CD34+ cells are heterogeneous; using multiparameter flow cytometry analysis, subsets of CD34+ cells were identified and have been shown to possess different stem cell and committed progenitor activity. Several groups including ours have demonstrated that CD34+CD38- cells contain stem cell including the multilineage potential of CD34+CD38- cells at clonal level by using single cells sorting (5–7).

Hematopoietic stem cells, which are enriched in CD34+ cell population in human and LSK cells in mouse, can be efficiently isolated from various stem cell sources using different methods (8). Here we describe two widely used methods to isolate stem cells from umbilical cord blood and mouse bone marrow. Human CD34+ cells are enriched by magnetic separation using an automated magnetic purification system, while mouse LSK cells are sorted by high-speed cell sorting. Human HSC can be further purified into CD34+CD38-lin- cells by similar FACS cell-sorting experiments as described for murine LSK cells, using the appropriate markers (i.e., CD34, CD38, lineage markers).

## 2. Materials

### 2.1. Mononuclear Cell Purification by Density Centrifugation

1. Umbilical cord blood or other source of CD34+ cells in a heterogeneous mixture (bone marrow, peripheral blood, human thymus) (*see* **Note 1**).
2. *Complete medium.* RPMI+ 1640 with 25mM Hepes and Ultraglutamine 1 (BioWhittaker/Cambrex, Belgium) supplemented with 2.5% heat-inactivated fetal calf serum, 100U/mL penicillin, and 100µg/mL streptomycin.
3. Ficoll-Paque™ PLUS solution (GE Healthcare Bio-sciences AB, Uppsala, Sweden).
4. Phosphate-buffered saline (PBS), pH 7.4.
5. Centrifuge with swinging bucket rotor.

### 2.2. Magnetic Separation by AutoMACS™

1. CD34 Microbead kit (Miltenyi Biotec Inc., Bergisch-Gladbach, Germany).
2. AutoMACS™ separator (Miltenyi).
3. *Rinsing buffer.* Phosphate-buffered saline supplemented with 2mM EDTA (*see* **Note 2**).
4. *Running buffer.* Phosphate-buffered saline supplemented with 2mM EDTA and 0.5% bovine serum albumin (BSA).

5. 30μm pre-separation filters (Miltenyi).

6. Fluorochrome-conjugated antibody for flow cytometric analysis, e.g., CD45 FITC and CD34 APC (clone HPCA-2) (*see* **Note 3**).

7. 5-mL and 15-mL collection tubes.

### 2.3. High-speed Cell Sorter of Lin⁻ Sca-1⁺c-kit⁺ Cells

1. *FACS buffer.* Phosphate-buffered saline supplemented with 2mM EDTA and 0.5% bovine serum albumin (BSA).

2. *Antibody mix.* Biotinylated-B220, CD3, NK1.1, Mac1/CD11b, Gr1, Ter-119, Sca1 FITC, and c-kit APC (*see* **Note 5**).

3. Streptavidin-PE and 7-AAD (FL3).

4. Filtered heat-inactivated fetal calf serum.

5. 15-mL collection tubes.

6. High-speed cell sorter, i.e. BD FACSAria.

---

# 3. Methods

### 3.1. Human CD34 Cell Enrichment Using an Automated Magnetic Purification System (AutoMACS)

*3.1.1 Purification of Mono-nuclear Cells by Density Centrifugation*

1. Determine the volume of cord blood samples.

2. Dilute the cord blood four times with PBS pH 7.4.

3. Pipette 12.5mL of Ficoll-Paque solution into a 50-mL tube.

4. Carefully layer 25mL of diluted cord blood over the Ficoll-paque solution.

5. Centrifuge at 900×$g$ for 15min in a swinging bucket rotor *at room temperature without break.*

6. Carefully transfer the interphase cells to a new 50-mL tube.

7. Add complete medium, mix, and centrifuge at 920×$g$ for 10min at 4°C. Discard the supernatant completely.

8. Resuspend the pellet in 20mL complete medium and count the cells.

9. Centrifuge at 520×$g$ for 5min at 4°C.

10. Cells can be frozen and stored in liquid nitrogen or use for CD34⁺ cell enrichment for further experiments.

*3.1.2 Magnetic Cell Labeling of CD34⁺ Cells*

1. Resuspend the cell pellet in 300μL running buffer per $10^8$ cells.

2. Add 100μL FcR Blocking Reagent per $10^8$ total cells to the cell suspension. Mix well.

3. Add 100μL anti-CD34 MicroBeads per $10^8$ total cells, mix well, and incubate for 30min at 4°C.

4. Wash cells with 1mL running buffer per $10^8$ cells, centrifuge at 520×$g$ for 10min at room temperature, and carefully remove the supernatant.

5. Resuspend cells in 500μL running buffer per $1 \times 10^8$ cells.

6. Filter through a 30-μm pre-separation filter and proceed to enrichment with the autoMACS™ separator (*See* **Note 4**).

*3.1.3 Separation with the AutoMACS Separator*

1. Clean the area around the autoMACS separator with 70% ethanol.

2. Make sure the fluid bottles are filled with the appropriate buffers and installed correctly. Ensure that the waste bottle is empty.

3. Turn the power on and the separator is ready for priming.

4. Perform the Program Clean before the separation.

5. Place appropriate collection tubes under the outlet ports.

6. *Choose program.* "posseld2" for the enrichment.

7. Remove carefully the positive fraction from the POS2 outlet port, centrifuge at $520 \times g$ for 5min at room temperature, and carefully remove the supernatant completely.

8. Resuspend cells in 1–2mL running buffer and take 20μL aliquot for cell counting.

9. To determine the purity of the positive fraction, transfer 50,000–100,000 cells into a microtube and add 25μL anti-CD34 APC/Anti-CD45 FITC mixture (*see* **Note 3**). Incubate 10min at room temperature.

10. Add 1mL running buffer and centrifuge $520 \times g$ for 5min. Resuspend cells with 250μL running buffer and analyze by flow cytometry (**Fig. 1**).

11. CD34+ cells can further be used in different applications including in vitro expansion, in vivo transplantation, or viral transduction experiments. If not used immediately CD34+ cells can be frozen and stored in liquid nitrogen.

*3.2 Purification of Mouse Lin⁻ Sca1⁺c-kit⁺ Cells by High-Speed Cell Sorting*

*3.2.1 Sample Preparation*

1. Collect mouse bone marrow cell form tibias and femurs. Flush the bones with 10mL complete medium using a syringe and a 25G needle in a medium-size sterile culture dish.

2. Resuspend cells by gently pipetting them several times using the syringe and a 25G needle.

3. Add 500 μL complete medium to wet the pre-separation filters. Pass the cell suspension through the filter to remove cell clumps.

4. Centrifuge at $520 \times g$ for 5min at 4°C. Remove the supernatant completely.

5. Resuspend cells with 2mL of FACS buffer and take 20μL aliquot for cell counting.

6. Centrifuge at $520 \times g$ for 5min at 4°C.

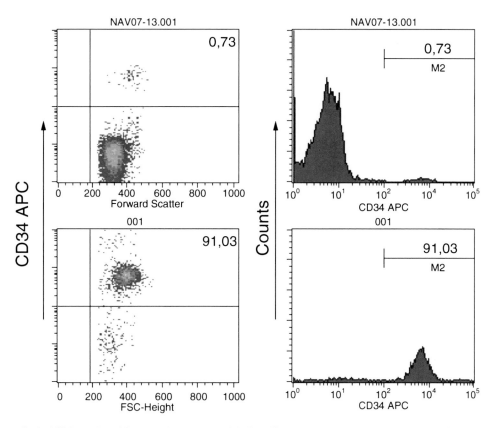

Fig. 1. Typical CD34+ cell enrichment using magnetic labeling. After enrichment, 50.000–100.000 cells from positive fraction were stained with antibody against CD34 to determine the purity.

7. Resuspend cells in 100μL of antibody mix per $10^7$ total cells.

8. Mix gently with vortex and incubate for 30min on ice.

9. Wash cells by adding 1mL FACS buffer per $10^7$ cells.

10. Centrifuge at 520×$g$ for 5min at 4°C. Remove the supernatant completely.

11. Add 100μL of Streptavidin-PE per $10^7$ cells and 50μL of 7-AAD per $10^7$ cells. Mix gently and incubate for 30min on ice.

12. Wash cells by adding 1mL FACS buffer per $10^7$ cells.

13. Centrifuge at 520×$g$ for 5min at 4°C. Remove the supernatant completely.

14. Resuspend cells in 1mL running buffer per $2.5×10^7$ cells.

15. Filter through a 30-μm pre-separation filter. Continue with the high-speed cell sorting with BD FACSAria.

1. *Power and fluidics.* Turn on the main power switch and the cooling device. Set the temperature of the cooling device at 4°C. Ensure that the sheath tank is filled with the sheath fluid and the waste tank is empty. Also make sure that the filters are free of air bubbles.

2. *Computer and software.* Turn on the computer and start up the FACSDiVa software.

3. *Fluidics.* Perform the start-up protocol by choosing "Instrument/Fluidics Startup,"

4. *Stream.* When the startup is completed turn on the Stream. Select the Sort Setup Menu and select one of the three choices (*see* **Note 7**). The software will adjust all necessary parameters.

5. *Prepare the instrument.* Clean carefully the sort plates. Insert the plates into the chamber and adjust the angle. To prevent sparking, ensure the plates are not too close together. The distance must be around 7mm. Turn the plate voltage on. Choose 6,000V for High and Medium sort setup and 5,000V for Low sort setup. Ensure the center stream hits the waste aspirator, adjust if necessary.

6. *Stability check.* Let the system run for at least 30min. Check: Break off point, Image of the stream. Make sure that the value of "Drop1" is somewhere between 150 and 250 and the gap is 6. Make sure that there are only a few, fast moving satellite drops (2–4 satellites are acceptable). This will benefit the stability of the sorting procedure. Check the side stream quality and adjust the frequency and amplitude, if necessary. Also check Center stream quality, adjust if necessary with second, third, and fourth Drops. Turn on the "SweetSpot."

7. *Setup the collection devices.* Select the two-way holder with 15-mL tubes and place it into the chamber. Turn the Plate Voltage on. Make sure that the side stream are separated and hit inside the tubes.

8. *Determine Drop delay.* Generate a dot plot of FSC against SSC with a gate "P1" Make sure that this gate is free from any event; thereafter invert P1 in the Population Hierarchy. Select "INITIAL" as sort mode in the Sort Layout and start to sort with "not P1" in the left or right tube. Ensure that the AccuDrop laser illuminates the streams maximally. Select "optical filter" and adjust the drop delay until nearly all the beads are deflected and the center stream is virtually empty. Repeat this procedure with the sort mode FINE TUNE. A correct drop delay is the most important step in the machine setup. With an incorrect drop delay no cells or wrong cells will be sorted in your collection tube.

9. Check the Area scaling factors for the FSC and different lasers by comparing the mean of FSC-H with the mean of FSC-A values. Adjust these Area scaling factors if the FSC-H differs from the FSC-A value.

10. Check the performance of the Aria by running fluorescent beads (e.g., rainbow beads, BD) which their emmision signals are detected in all detectors with a fixed voltage. Plot the mean values and CVs in time.

11. The cell sorter is ready for sorting.

12. *Start of a sort experiment.* Start a new experiment or copy an old experiment protocol. Make sure that you run the sample with a correct instrument setting. If this is not available yet measure an unlabeled sample together with the single stained samples (with different fluorochrome) in order to set up the instrument setting. Run the autocompensation command for the correct instrument setting. Set up four dotplots with the following parameters: SSC-A/FSC-A (A), FSC-W/7-AAD (B), c-kit/lineage (C), and Sca1/c-kit (D) (*see* **Fig. 2**).

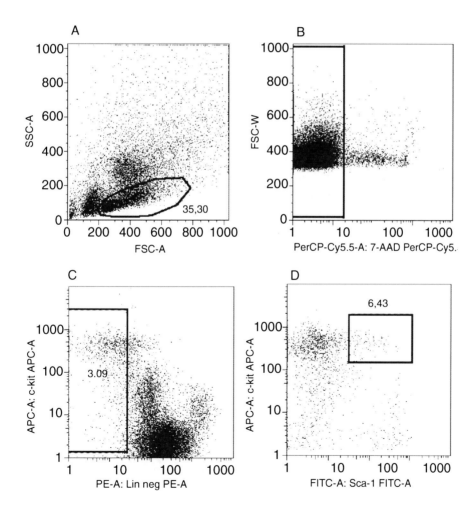

Fig. 2. Mouse LSK cells sorted with high-speed cell sorter. (a) scatter parameters, (b) the live gate, (c) lineage markers versus c-kit (d) the boxed LSK population (Lin⁻Sca-1⁺c-kit⁺) is indicated.

13. Place a lymphogate in plot A and a second live gate in plot B to exclude 7-AAD positive cells (apoptotic cells).

14. The third lineage gate is placed to select c-kit$^+$/lineage$^-$ cells (C), and finally, a sorting gate (D) was placed to sort Sca-1$^+$c-kit$^+$cells.

15. First run the Yield Mode to enrich the Lin$^-$Sca-1$^+$c-kit$^+$ cells. Collect the positive fraction in 15-mL tube filled with 1mL heat-inactivated FCS. Centrifuge 520×$g$ for 5min at 4°C.

16. Resuspend cells in 1mL FACS buffer.

17. Change the value of the Precision Mode (Purity) the Yield Mask and the Purity Mask from 32 to 16. Run the Purity Mode for further purify of the Lin$^-$Sca-1$^+$c-kit$^+$ cells.

18. After sorting, aliquot 30μL for reanalysis of the positive fraction.

19. Lin$^-$Sca-1$^+$c-kit$^+$ cells can be used for further experiments or stored in liquid nitrogen if not used immediately.

## 4. Notes

1. The cord blood samples should not be older than 24h.

2. Unless indicated otherwise, all solutions should be prepared freshly in MilliQ water and sterilized or filtered through a 0.2μm filters.

3. It is important to note that for FACS staining after magnetically labeling, an antibody is needed that requires a monoclonal antibodies recognizes an epitope other than used for magnetic labeling, e.g., for anti-CD34 (My10 or HPCA-2).

4. Prefilter of cell suspension before separation and cell sorting prevents clogging of the column of the separator and the cell sorter.

5. It is important that the antibodies used for staining are well titrated.

6. In most institutions, a well-trained, specialized cell-sorting operator performs the cell-sorting procedure.

7. *Cell sorter setup.* For setting up the Stream, the choice of setup mode is dependent on the cell size (large/small) of the samples. For cell sorting of Lin$^-$Sca-1$^+$c-kit$^+$ cells, choose setup mode HIGH.

## Acknowledgments

The authors would like to thank Tiago Luis and Sjanneke Heuvelmans for techincal assistance.

## References

1. Kondo, M., Wagers, A.J., Manz, M.G., Prohaska, S.S., Scherer, D.C., Beilhack, G.F., Shizuru, J.A., Weissman, I.L. (2003) Biology of hematopoietic stem cells and progenitors: implications for clinical application. *Annu Rev Immunol.* 21, 759–806

2. Coulombel, L. (2004) Identification of hematopoietic stem/progenitor cells: strength and drawbacks of functional assays. *Oncogene.* **23**, 7210–7222

3. Gangenahalli, G.U., Singh, V.K., Verma, Y.K., Gupta, P., Sharma, R.K., Chandra, R., Luthra, P.M. (2006) Hematopoietic stem cell antigen CD34: role in adhesion or homing. *Stem Cells Dev.* 15, 305–313

4. Buhring, H.J., Asenbauer, B., Katrilaka, K., Hummel, G., and Busch, F.W. (1989) Sequential expression of CD34 and CD33 antigens on myeloid colony-forming cells. *Eur J Haematol.* 42, 143–149

5. Ng, Y.Y., Bloem, A.C., van Kessel, B., Lokhorst, H., Logtenberg, T., Staal, F.J. (2002) Selective in vitro expansion and efficient retroviral transduction of human CD34+ CD38 haematopoietic stem cells. *Br J Haematol.* 117, 226–237

6. Ng, Y.Y., van Kessel, B., Lokhorst, H.M., Baert, M.R., van den Burg, C.M., Bloem, A.C., Staal, F.J. (2004) Gene-expression profiling of CD34+ cells from various hematopoietic stem-cell sources reveals functional differences in stem-cell activity. *J Leukoc Biol.* 75, 314–323

7. Pike-Overzet, K., de Ridder, D., Weerkamp, F., Baert, M.R., Verstegen, M.M., Brugman, M.H., Howe, S.J., Reinders, M.J., Thrasher, A.J., Wagemaker, G., van Dongen, J.J., Staal, F.J. (2007) Ectopic retroviral expression of LMO2, but not IL2Rgamma, blocks human T-cell development from CD34+ cells: implications for leukemogenesis in gene therapy. *Leukemia.* 21, 754–763

8. Lagasse E., Connors H., Al-Dhalimy M., Reitsma M., Dohse, M., Osborne, L., Wang, L., Finegold, M., Weissman, I.L., Grompe, M. (2000) Purified hematopoietic stem cells can differentiate into hepatocytes in vivo. *Nat Med.* 6, 1229–1234

9. BD Biosciences (2006). BD FACSAria User's Guide

# Chapter 3

# Murine Hematopoietic Stem Cell Transduction Using Retroviral Vectors

## Ute Modlich, Axel Schambach, Zhixiong Li, and Bernhard Schiedlmeier

## Summary

Hematopoietic stem cells (HSCs) represent an important target cell population in bone marrow transplantation and gene therapy applications. Their progeny cells carry the genetic information of the HSCs and replenish the blood and immune system. Therefore, in the setting of inherited diseases, transduction of HSCs with retroviral vectors (including gammaretro- and lentiviral vectors) offers the possibility to correct the phenotype in all blood lineages as demonstrated in clinical trials for immuno-deficiencies (e.g., X-SCID). In the process of developing gene therapy strategies for patient applications, suitable mouse models for the human gene therapy are important to validate the concept. Stem-cell-enriched populations such as lineage negative cells as the functional equivalent of human CD34+ cells can be isolated from murine bone marrow and efficiently transduced using retroviral vectors. This chapter provides a step-by-step protocol for retroviral transduction of murine lineage negative cells.

**Key words:** Hematopoietic stem cells, Retroviral vector, Lentiviral vector, Gene therapy, Transduction, Cell culture, Serum-free medium, Cytokines.

## 1. Introduction

Hematopoietic stem cells (HSCs) represent important target cells for gene therapy of inherited disorders. They have the capability to self-renew and differentiate into all mature cells of the blood and immune system. Thus, their remarkable potential to repopulate the blood is used in clinical HSC transplantation to treat leukemias or inborn errors of metabolism.

Aiming for gene therapy, viral gene delivery into HSCs using gammaretroviral and lentiviral vectors is efficient and can be used

Christopher Baum (ed.), *Methods in Molecular Biology, Methods and Protocols, vol. 506*
© Humana Press, a part of Springer Science+Business Media, LLC 2009
DOI: 10.1007/978-1-59745-409-4_3

to cure a genetic defect *(1)*. The recently published successful gene therapy trials for immunodeficiencies (X-SCID1, ADA, CGD) *(2–5)* provide proof of concept for gene therapy of HSCs as a therapeutic option.

On the long track to bring a gene therapy application into the clinic, suitable mouse models and murine primary cell culture systems are mandatory and a *conditio sine qua non* to validate the concept. Murine HSCs can be harvested from the bone marrow (BM) and require further purification using FACS or magnetic cell separation (MACS) technologies. Lineage (marker) depleted cells (lin⁻; frequency ~1% of adult mouse BM cells) and LSK cells (lin⁻ sca⁺ ckit⁺; 0.1% of adult mouse BM cells) are frequently used as enriched HSC cell populations which harbor one competitive repopulating unit (CRU) per ~1,000 or ~100 cells, respectively *(6–8)*. Moreover, CD34⁻/low lin– sca⁺ ckit⁺ (CD34⁻LSK; 0.005% of adult mouse BM cells) can be used in single-cell colony assays and transplantation procedures *(9)*. In order to maintain as much HSCs as possible *in vitro*, the use of serum-free media compositions and optimal cytokine cocktails *(10)* is important in ensuring engraftment and repopulation ability *(11, 12)*. Since serum per se can contain cytokines or unwanted mediators, serum-free culture conditions have been developed for ex vivo cultivation of HSCs. The correct interplay of cytokines via activating signaling cascades determines whether an HSC divides and stays an HSC (self-renewal) or differentiates into precursor and lineage-committed mature cells. Here, we provide a protocol allowing expansion and transduction of HSCs-enriched fraction (lineage negative cells) as needed in preclinical gene therapy studies.

## 2. Materials

All materials will be used for cell culture work and have to be sterile!

**2.1. Prestimulation of Purified Lineage Negative Bone Marrow Cells**

1. StemSpan medium (StemCell Technologies, Grenoble, France), long-term storage in aliquots at –20°C.
2. 200 mM glutamine (Biochrom, Berlin, Germany), long-term storage in aliquots at –20°C.
3. Penicillin/streptomycin 100× (PAA, Pasching, Austria) long-term storage in aliquots at –20°C, for usage store at 4°C.
4. Recombinant murine stem cell factor (mSCF), (PeproTech, Heidelberg, Germany), store at –20°C.
5. Recombinant human FLT3-ligand (hFlt3L), (PeproTech), store at –20°C.

6. Recombinant human interleukin 11 (hIL-11), (PeproTech), store at –20°C.

7. Recombinant murine interleukin 3 (mIL-3) (PeproTech), store at –20°C.

8. Cell culture plastic ware: 6-well, 12-well, and 24-well plates (Sarstedt, Nümbrecht, Germany).

9. Serological pipettes at 2, 5, 10 mL (Sarstedt, Nümbrecht, Germany).

10. Sterile conical tubes at 15 and 50 mL (TPP, Trasadingen, Switzerland).

**2.2. Retroviral Transduction of Lineage Negative Bone Marrow Cells**

1. RetroNectin® recombinant human fibronectin fragment (48 μg/mL, TaKaRa, Saint-Germain-en-Laye, France), store at –20°C.

2. Phosphate-buffered saline (PBS) (PAA, Pasching, Austria).

3. Bovine serum albumin (BSA) (USBiological, Swampscott, MA, USA).

4. Hanks balanced salt solution (HBSS; Invitrogen Corporation, Karlsruhe, Germany).

5. 1 M Hepes buffer (PAA, Pasching, Austria).

6. Cell Dissociation Buffer (enzyme-free; Invitrogen Corporation, Karlsruhe, Germany).

7. Cooling centrifuge (Multifuge 3 S-R, Heraeus, Berlin, Germany) with rotatable tissue culture plate holders.

8. Cell-free retroviral or lentiviral vector preparations with known titers from frozen stocks kept at –80°C. Ecotropic or VSVg-pseudotyped particles are preferred.

# 3. Methods

The transduction protocol for murine HSCs described here is based on the transduction of a stem-cell-enriched fraction of bone marrow cells (lineage negative cells; cf. introduction). The purification of bone marrow cells that leads to enrichment of HSCs before retroviral transduction is of advantage for several reasons. The complete bone marrow cell pool contains end-differentiated cells that divide only poorly. They will absorb retroviral particles but will not contribute to the expansion culture or the repopulation of recipient mice in case the transduced cells are transplanted. Therefore higher amounts of retroviral particles would be needed, and the calculation of the virus particle to cell ratio (multiplicity of infection, MOI) becomes more difficult. In addition, mature cells secrete inhibitory cytokines (13) and deplete the medium of stimulatory cytokines.

For transduction of lineage negative BM cells, RetroNectin (a recombinant fibronectin fragment) offers several advantages over conventional exposure to virus-containing medium. Retro-Nectin colocalizes retrovirus and target cells through binding of Retronectin domains to cellular integrin receptors and thereby increases productive transduction *(14, 15)*. In addition, preloading with multiple administrations of viral supernatants can be used to concentrate the amount of viral particles bound to RetroNectin. Moreover, preloading of (serum-containing) viral supernatants ensures that lineage negative cells can be cultured in serum-free media avoiding exposure to serum components, which prevents unwanted differentiation and impairment of HSC self renewal.

**3.1. Prestimulation of Purified Lineage Bone Marrow Cells**

The proliferation of the target cells is a prerequisite of the gammaretroviral transduction, as gammaretroviral vectors can only efficiently enter dividing cells *(16)*. Also lentiviral vectors, which are thought to access the nucleus via transport along the nuclear pore and can infect nondividing cells, transduce proliferating hematopoietic cells more efficiently than nonproliferating cells *(17)*. However, prestimulation for transduction by lentiviral vectors can be reduced to less than 12 h *(18, 19)*.

1. Dilute your lineage negative cells (lin-) into culture at a density of $5 \times 10^5$ cells/mL in serum-free medium (StemSpan), 2 mM glutamine, 1% penicillin/streptomycin containing 50 ng/mL mSCF, 100 ng/mL hFlt3L, 100 ng/mL hIL-11, 10 ng/mL mIL-3. Freshly isolated cells should be prestimulated 2 days prior to gammaretroviral transduction (*see* **Notes 1** and **2**).

2. Incubate the cells at 37°C, 5% $CO_2$, 100% humidity in a cell culture incubator. During the 2 days' culture time the cell number should double indicative of cell proliferation.

**3.2. Retroviral Transduction of Lineage Negative Bone Marrow Cells**

1. Precoat the tissue culture well of a 6-well plate (12-well, 24-well plate, respectively) with 2 mL (0.8 mL, 420 μL) of RetroNectin (10 μg/cm², *see* **Table 1**) and incubate for 2 h at room temperature (RT) or overnight at 4°C. The size of culture plate used for transduction is defined by the cell number that should get transduced. The density of cells should not drop below $2 \times 10^5$/mL.

2. Remove the RetroNectin solution using a micropipette and block the well with 2 mL (0.8 mL, 420 μL) 2% BSA in PBS for 30 min at RT (*see* **Note 3**).

3. Remove the blocking solution and wash the well with 3 mL (1.2 mL, 630 μL) of HBSS/Hepes buffer (2.5%, v/v) once.

4. Remove the washing solution and add your viral supernatant to the tissue culture well. Add 2 mL (0.8 mL, 420 μL) of viral

**Table. 1**
**Amounts of RetroNectin (RN) needed for plate precoating**

|  | RN stock for 20 µg/cm² | RN stock for 10 µg/cm² | RN stock for 5 µg/cm² |
|---|---|---|---|
| 6-well plate (9.62 cm²) | 4 mL | 2 mL | 1 mL |
| 12-well plate (3.8 cm²) | 1.6 mL | 0.8 mL | 400 µL |
| 24-well plate (2.0 cm²) | 840 µL | 420 µL | 210 µL |

supernatant (diluted in ice-cold PBS). A further increase of the supernatant volume does not increase the efficiency of the preloading of viral particles. An MOI (multiplicity of infection) of three (transducing units/cell, as determined on SC-1 cells, murine fibroblast line) transduces approx. 30% lin– cells when using particles with an ecotropic envelope (*see* **Note 4**).

5. Centrifuge the plate at 2,000 rpm (700 × *g*) and 4°C for 30 min in a Heraeus Multifuge *(20)*. The preloading can be repeated to increase viral particles on the plate.

6. While the plate with preloaded viruses is centrifuging, carefully remove the cultured bone marrow cells using gentle pipetting and a 1-mL micropipette, and count the cells (*see* **Notes 5** and **6**). Alternatively, cells can be detached from the tissue culture plate by using the Cell Dissociation Buffer (2 mL in a 6-well plate).

7. Remove the viral supernatants from wells and add 2 mL [0.8 mL, 420 µL] of the cell suspension (5 × 10⁵ lineage negative cells/mL) in StemSpan medium (including cytokines supplements, *see* 3.1.1). It is advisable to prewarm the preloaded culture plates to RT shortly before the cells are seeded into the wells.

8. Incubate the cells at 37°C, 5% $CO_2$, 100% humidity in a cell culture incubator (*see* **Note 7**).

9. The same transduction procedure can be repeated the following day, if necessary.

*3.3. Detection of Retrovirally Transduced Cells*

Typically, 24–48 h after retroviral transduction (dependent on expression strength of the promoter) the expression of the transferred transgene can be detected in transduced primary bone marrow cells. Gently mix the cells in your tissue culture well with a micropipette and take out a small aliquot of cells. For FACS analysis 5 × 10⁴ cells are sufficient (**Figs. 1** and **2**).

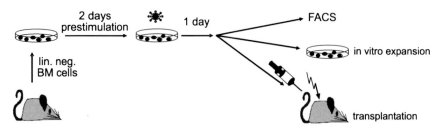

Fig. 1. Flow sheet of transduction and experimental procedures. Bone marrow cells are harvested from femurs of C57BL/6 mice and a stem-cell-enriched fraction (lineage negative cells) is isolated using MACS purification. Lineage negative cells are cultivated using serum-free media (StemSpan) and prestimulated in presence of cytokines (mSCF, hFlt3L, mIL-3, hIL-11) for 2 days. Transduction with gammaretroviral or lentiviral vectors is performed using RetroNectin precoated dishes. After additional cultivation for 1 day, cells can be analyzed by FACS, further expanded for in vitro experiments, or transplanted into lethally irradiated C57BL/6 mice for bone marrow transplantation.

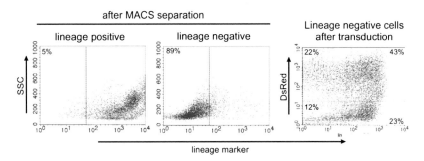

Fig. 2. MACS purification and analysis of lineage negative cells before and after cultivation. After MACS separation the lineage positive (*left*) and lineage negative cell fractions (*middle*) are given in FACS dot plots. The lineage negative cells were cultivated for 2 days, transduced with a gammaretroviral vector encoding DsRed and cultured for another day. The dot plots on the right shows the percentage of lineage negative cells and transduction efficiency (DsRed). Percentages of positive cells in the respective quadrants are given.

### 3.4. Expansion of Transduced Bone Marrow Cells

Prolonged expansion of primary bone marrow cells may reduce the frequency of stem cells. However, for gammaretroviral transduction the proliferation is essential for transduction efficiency, and therefore a prestimulation step is included. Expansion of primary bone marrow cells after retroviral transduction may result in loss of engraftment potential and clonal drift. Therefore, for transplantation studies it is important to keep the ex vivo cultivation time as short as possible to maintain as much stem cells with engraftment potential as possible. **Figure 1** shows a flow sheet embedding the murine HSC transduction in further ex vivo analysis or transplantation settings (*see* **Notes 8** and **9**).

Elucidating the exact molecular mechanisms which control HSC self-renewal and differentiation will help to identify and improve cell culture conditions to amplify HSCs in vitro *(21, 22)*.

## 4. Notes

1. Lineage negative cells can be frozen in FCS with 10% DMSO and stored in liquid nitrogen. If frozen cell aliquots are thawed, they should be taken into culture for 2–3 days before retroviral transduction.

2. Different cytokine conditions for the in vitro stimulation of murine HSC have been described. Most of them prefer serum-free conditions to exclude any influences of unknown components of the serum. The positive influence of IL-11 and IL-3 in the absence of serum was shown by Bryder and Jacobsen (23). Some other protocols favor a combination of SCF, Flt3L, and TPO (24) or SCF, Flt3L, and IL-11 (25). Furthermore, fibroblast growth factor-1 (FGF-1) expands murine HSCs and has a positive effect on transduction rates (26). In combination with IGF-2, TPO, and SCF, FGF-1 expands HSC in vitro over prolonged culture times (27).

3. RetroNectin can be reused after initial coating for a second pre-coating procedure. To do this, transfer the used RetroNectin into a new well and proceed as described in the protocol. Store the new tissue culture plate at 4°C for use on the next day.

4. Detachment of cultured primary cells from plates by pipetting is a critical step. Alternatively, it is recommended to use a nonenzyme-based Cell Dissociation Buffer (Invitrogen, Karlsruhe, Germany) to gently resuspend the cells.

5. Primary cells are very sensitive and prefer growing in conditioned media. Therefore, if splitting primary cells it is advisable to include half fresh and half conditioned media.

6. For the calculation of the multiplicity of infection (MOI) when using ecotropic supernatants, take into account that only half of the particles are capable of binding to RetroNectin, the remaining will be removed during washing.

7. For optimal transduction and expansion conditions of primary BM cells, it is advisable to use a separate cell culture incubator. The repeated opening of the cell culture incubator leads to fluctuations in $CO_2$ levels resulting in pH changes of the medium.

8. The protocol described here can also be used for HSC that were further purified, such as LSK cells. In addition, this protocol can be used for transduction of human CD34+ cells, if cytokines are adapted accordingly (28).

9. For the transplantation of transduced lin– cells at least $1 \times 10^5$ cells should be reinfused into one recipient. This corresponds to a maximum of 100 stem cells. Due to the differentiation of the lineage negative cells during culture time only 20–30% of

the cells are still lineage negative at the day of transplantation. This should be taken into consideration when calculating the amount of cells needed for a transplantation experiment.

## Acknowledgments

The authors thank Christopher Baum for supporting this work and Sabine Knoess for comments on the methodology. This work was supported by grants by the DFG, the BMBF, and the EU. A.S. is financed by a stipend of the Else-Kröner-Fresenius foundation.

## References

1. Karlsson, S., Ooka, A. & Woods, N. B. (2002). Development of gene therapy for blood disorders by gene transfer into haematopoietic stem cells. *Haemophilia* 8, 255–60

2. Hacein-Bey-Abina, S., Le Deist, F., Carlier, F., Bouneaud, C., Hue, C., De Villartay, J. P., Thrasher, A. J., Wulffraat, N., Sorensen, R., Dupuis-Girod, S., Fischer, A., Davies, E. G., Kuis, W., Leiva, L. & Cavazzana-Calvo, M. (2002). Sustained correction of X-linked severe combined immunodeficiency by ex vivo gene therapy. *N Engl J Med* 346, 1185–93

3. Gaspar, H. B., Parsley, K. L., Howe, S., King, D., Gilmour, K. C., Sinclair, J., Brouns, G., Schmidt, M., Von Kalle, C., Barington, T., Jakobsen, M. A., Christensen, H. O., Al Ghonaium, A., White, H. N., Smith, J. L., Levinsky, R. J., Ali, R. R., Kinnon, C. & Thrasher, A. J. (2004). Gene therapy of X-linked severe combined immunodeficiency by use of a pseudotyped gammaretroviral vector. *Lancet* 364, 2181–7

4. Aiuti, A., Slavin, S., Aker, M., Ficara, F., Deola, S., Mortellaro, A., Morecki, S., Andolfi, G., Tabucchi, A., Carlucci, F., Marinello, E., Cattaneo, F., Vai, S., Servida, P., Miniero, R., Roncarolo, M. G. & Bordignon, C. (2002). Correction of ADA-SCID by stem cell gene therapy combined with nonmyeloablative conditioning. *Science* 296, 2410–3

5. Ott, M. G., Schmidt, M., Schwarzwaelder, K., Stein, S., Siler, U., Koehl, U., Glimm, H., Kuhlcke, K., Schilz, A., Kunkel, H., Naundorf, S., Brinkmann, A., Deichmann, A., Fischer, M., Ball, C., Pilz, I., Dunbar, C., Du, Y., Jenkins, N. A., Copeland, N. G., Luthi, U., Hassan, M., Thrasher, A. J., Hoelzer, D., von Kalle, C., Seger, R. & Grez, M. (2006). Correction of X-linked chronic granulomatous disease by gene therapy, augmented by insertional activation of MDS1-EVI1, PRDM16 or SETBP1. *Nat Med* 12, 401–9

6. Ikuta, K. & Weissman, I. L. (1992). Evidence that hematopoietic stem cells express mouse c-kit but do not depend on steel factor for their generation. *Proc Natl Acad Sci U S A* 89, 1502–6

7. Harrison, D. E. (1980). Competitive repopulation: a new assay for long-term stem cell functional capacity. *Blood* 55, 77–81

8. Szilvassy, S. J., Humphries, R. K., Lansdorp, P. M., Eaves, A. C. & Eaves, C. J. (1990). Quantitative assay for totipotent reconstituting hematopoietic stem cells by a competitive repopulation strategy. *Proc Natl Acad Sci U S A* 87, 8736–40

9. Ema, H., Morita, Y., Yamazaki, S., Matsubara, A., Seita, J., Tadokoro, Y., Kondo, H., Takano, H. & Nakauchi, H. (2006). Adult mouse hematopoietic stem cells: purification and single-cell assays. *Nat Protoc* 1, 2979–87

10. Heike, T. & Nakahata, T. (2002). Ex vivo expansion of hematopoietic stem cells by cytokines. *Biochim Biophys Acta* 1592, 313–21

11. Nakauchi, H., Sudo, K. & Ema, H. (2001). Quantitative assessment of the stem cell self-renewal capacity. *Ann N Y Acad Sci* 938, 18–24; discussion 24–5

12. Li, Z., Schwieger, M., Lange, C., Kraunus, J., Sun, H., van den Akker, E., Modlich, U., Serinsöz, E., Will, E., von Laer, D., Stocking, C., Fehse, B., Schiedlmeier, B. & Baum, C. (2003). Predictable and efficient retroviral gene transfer into murine bone marrow

repopulating cells using a defined vector dose. *Exp Hematol* 31, 1206–1214

13. Manfredini, R., Zini, R., Salati, S., Siena, M., Tenedini, E., Tagliafico, E., Montanari, M., Zanocco-Marani, T., Gemelli, C., Vignudelli, T., Grande, A., Fogli, M., Rossi, L., Fagioli, M. E., Catani, L., Lemoli, R. M. & Ferrari, S. (2005). The kinetic status of hematopoietic stem cell subpopulations underlies a differential expression of genes involved in self-renewal, commitment, and engraftment. *Stem Cells* 23, 496–506

14. Hanenberg, H., Xiao, X. L., Dilloo, D., Hashino, K., Kato, I. & Williams, D. A. (1996). Colocalization of retrovirus and target cells on specific fibronectin fragments increases genetic transduction of mammalian cells. *Nat Med* **2**, 876–82

15. Hanenberg, H., Hashino, K., Konishi, H., Hock, R. A., Kato, I. & Williams, D. A. (1997). Optimization of fibronectin-assisted retroviral gene transfer into human CD34+ hematopoietic cells. *Hum Gene Ther* 8, 2193–206

16. Suzuki, Y. & Craigie, R. (2007). The road to chromatin - nuclear entry of retroviruses. *Nat Rev Microbiol* 5, 187–96

17. Manganini, M., Serafini, M., Bambacioni, F., Casati, C., Erba, E., Follenzi, A., Naldini, L., Bernasconi, S., Gaipa, G., Rambaldi, A., Biondi, A., Golay, J. & Introna, M. (2002). A human immunodeficiency virus type 1 pol gene-derived sequence (cPPT/CTS) increases the efficiency of transduction of human nondividing monocytes and T lymphocytes by lentiviral vectors. *Hum Gene Ther* 13, 1793–807

18. Kurre, P., Anandakumar, P. & Kiem, H. P. (2006). Rapid 1-hour transduction of whole bone marrow leads to long-term repopulation of murine recipients with lentivirus-modified hematopoietic stem cells. *Gene Ther* 13, 369–73

19. Mostoslavsky, G., Kotton, D. N., Fabian, A. J., Gray, J. T., Lee, J. S. & Mulligan, R. C. (2005). Efficiency of transduction of highly purified murine hematopoietic stem cells by lentiviral and oncoretroviral vectors under conditions of minimal in vitro manipulation. *Mol Ther* 11, 932–40

20. Kuhlcke, K., Fehse, B., Schilz, A., Loges, S., Lindemann, C., Ayuk, F., Lehmann, F., Stute, N., Fauser, A. A., Zander, A. R. & Eckert, H. G. (2002). Highly efficient retroviral gene transfer based on centrifugation-mediated vector preloading of tissue culture vessels. *Mol Ther* 5, 473–8

21. Eaves, C., Miller, C., Conneally, E., Audet, J., Oostendorp, R., Cashman, J., Zandstra, P., Rose-John, S., Piret, J. & Eaves, A. (1999). Introduction to stem cell biology in vitro. Threshold to the future. *Ann N Y Acad Sci* 872, 1–8

22. Sorrentino, B. P. (2004). Clinical strategies for expansion of haematopoietic stem cells. *Nat Rev Immunol* 4, 878–88

23. Bryder, D. & Jacobsen, S. E. (2000). Interleukin-3 supports expansion of long-term multilineage repopulating activity after multiple stem cell divisions in vitro. *Blood* 96, 1748–55

24. Ramsfjell, V., Bryder, D., Bjorgvinsdottir, H., Kornfalt, S., Nilsson, L., Borge, O. J. & Jacobsen, S. E. (1999). Distinct requirements for optimal growth and In vitro expansion of human CD34(+)CD38(-) bone marrow long-term culture-initiating cells (LTC-IC), extended LTC-IC, and murine in vivo long-term reconstituting stem cells. *Blood* 94, 4093–102

25. Miller, C. L. & Eaves, C. J. (1997). Expansion in vitro of adult murine hematopoietic stem cells with transplantable lympho-myeloid reconstituting ability. *Proc Natl Acad Sci U S A* 94, 13648–53

26. Crcareva, A., Saito, T., Kunisato, A., Kumano, K., Suzuki, T., Sakata-Yanagimoto, M., Kawazu, M., Stojanovic, A., Kurokawa, M., Ogawa, S., Hirai, H. & Chiba, S. (2005). Hematopoietic stem cells expanded by fibroblast growth factor-1 are excellent targets for retrovirus-mediated gene delivery. *Exp Hematol* 33, 1459–69

27. Zhang, C. C. & Lodish, H. F. (2005). Murine hematopoietic stem cells change their surface phenotype during ex vivo expansion. *Blood* 105, 4314–20

28. Schiedlmeier, B., Klump, H., Will, E., Arman_Kalcek, G., Li, Z., Wang, Z., Rimek, A., Friel, J., Baum, C. & Ostertag, W. (2003). High-level ectopic HOXB4 expression confers a profound in vivo competitive growth advantage on human cord blood CD34+ cells, but impairs lymphomyeloid differentiation. *Blood* 101, 1759–68

# Chapter 4

# Genetic Modification of Human Hematopoietic Cells: Preclinical Optimization of Oncoretroviral-mediated Gene Transfer for Clinical Trials

## Tulin Budak-Alpdogan and Isabelle Rivière

## Summary

This chapter provides information about the oncoretroviral transduction of human hematopoietic stem/progenitor cells under clinically applicable conditions. We describe in detail a short –60 h transduction protocol which consistently yields transduction efficiencies in the range of 30–50% with five different oncoretroviral vectors. We discuss a number of parameters that affect transduction efficiency, including the oncoretroviral vector characteristics, the vector stock collection, the source of CD34+ cells and transduction conditions.

**Key words:** Retroviral gene transfer, CD34+ cells, Hematopoietic stem cells, Transduction, Oncoretroviral vector, Vector production, RetroNectin, Tissue culture bags, Preclinical optimization.

## 1. Introduction

Human hematopoietic stem cells (HSCs) can be genetically modified with oncoretroviral vectors to express therapeutic genes for the treatment of either inherited *(1–12)* or acquired disorders *(1–4, 6, 7, 11)*. Oncoretroviral vectors have the ability to integrate permanently into the chromosomes of mammalian cells. HSCs have both self-renewal and multilineage differentiation capacities. Thus, stable engraftment of genetically modified HSCs potentially leads to the stable expression of transgene in stem and/or progenitor cells, depending on the promoter/enhancer combination that controls transgene expression.

Christopher Baum (ed.), *Methods in Molecular Biology, Methods and Protocols, vol. 506*
© Humana Press, a part of Springer Science + Business Media, LLC 2009
DOI: 10.1007/978-1-59745-409-4_4

Active target cell cycling is required for integration of oncoretroviral vectors into the host cell's chromosomal DNA *(12)*. Primitive HSCs which provide long-term engraftment are mostly quiescent *(9)* and thus transduced at a lower frequency than cycling progenitors. Prestimulation with early acting cytokines is therefore required for efficient gene transfer into HSCs with oncoretroviral vectors. Compared with several cytokine combinations that contain IL-3, and/or IL-6 *(5, 10)*, a mix of stem cell factor, FLT3 ligand, and thrombopoietin induces better, more synchronous ex vivo CD133+ *(8)* or CD34+/Thy.1+ *(13)* cell expansion, yields higher transgene marking *(14)*, improves survival *(13)*, and maintains the multipotential engraftment ability of HSCs in NOD/SCID mice *(5)*. Ex vivo stimulation with hematopoietic growth factors increases the expression of the envelope receptors on HSCs *(15–19)* and the number of cells that are actively cycling, and also maintains cell viability by inhibiting HSC apoptosis *(20)*. Prolonged ex vivo manipulation of HSCs may result in either differentiation of HSCs and/or loss of their engraftment potential *(21–24)*. Prestimulation must therefore allow the quiescent stem cells to enter the cell cycle while preserving their engraftment potential.

GaLV and amphotropic envelope receptors, Pit-1 and Pit-2, respectively, are inducible phosphate receptors. Their density on the target cells defines the rate of transduction efficiency *(19, 25)*. Human CD34+ cells enter active cycling after 36–48 h of cytokine stimulation *(26)*, while the expression of Pit-1 is concomitantly induced *(15, 25)*. The length of prestimulation, the number of transduction cycles, and time interval between two transductions varies among various published retroviral gene transfer protocols *(14, 26–34)*, so that total ex vivo cell manipulation time fluctuates between 60 and 120 h. In this chapter, we describe in detail our short transduction protocol of 60 h *(27)* which consistently yields transduction efficiencies in the range of 30–50% with five different oncoretroviral vectors, i.e., SFG$^{mpsv}$-eGFP, SFG$^{mpsv}$-NTP4, SFG$^{mpsv}$-DHFR/CD, SFG$^{mpsv}$-huTyr, and SFG$^{mpsv}$-huTyr-ires-eGFP produced under serum-free conditions. We have previously demonstrated that this protocol does not alter the engraftment potential of the CD34+ transduced cells and that the transgene expression in the progeny of the repopulating CD34+ cells in NOD/SCID mice in vivo is at least as good as that of cells transduced in the presence of serum or using longer, more conventional transduction protocols *(27)*.

Physical parameters such as the number of viral particles per cell (multiplicity of infection-MOI) and the virus concentration also affect retroviral transduction efficiency *(27)*. Increased human CD34+ cell transduction has been demonstrated by colocalizing the vector particles with the target cells, using the RetroNectin®-CH-296 domain of fibronectin to coat tissue culture vessels *(29, 35–38)*, adding polycations into the transduction media *(39–43)*, or spinoculation *(44–50)*. We and others have reported that polycations *(27, 35)*, i.e., polybrene or protamine sulfate, and centrifugation *(14, 27)* are not required for efficient transduction of HSCs in the presence of RetroNectin®. Vector preloading might improve gene transfer efficiency by both increasing viral particle-cell interaction and decreasing exposure to inhibitory factors contained in the vector stocks *(51)*. Systematic prescreening of vector stocks titers on human CD34+ cells might eliminate producer cell clones that secrete inhibitory factors. In our hands, preloading with prescreened vector stocks does not result in higher CD34+ cell transduction efficiency *(27)*.

Keeping the CD34+ transduction efficiency around 30–50% is also advisable to reduce the probability of multiple integrations in single cells *(52)* and limit the risk of insertional mutagenesis previously observed in four patients who developed T-cell leukemia after integration of a Mo-MuLV-derived, LTR-driven vector. Indeed, it has been shown that oncoretroviruses preferentially integrate in open chromatin regions at transcriptionally active sites in the vicinity of cellular promoters. High vector copy numbers per cell increases the risk of insertional mutagenesis *(53)*, and there is a quantitative correlation between overall gene transfer rate and integration frequency in single cells *(52)*. There is usually less than one vector copy per cell when gene transfer rates are less than 30%, and an average of three vector copy per cell for gene transfer rates around 60% *(52)*. Evaluation of retroviral insertion sites with linear amplification PCR (LAM-PCR) has been considered for monitoring clonal integration sites in patient cells. LAM-PCR is an informative test more than a diagnostic one, but data accumulated through systematic monitoring of integration sites might help to further understand the mechanism of this potential adverse effect.

Retroviral transduction protocol for human CD34+ cells under clinically relevant serum-free conditions is depicted in **Fig. 1**. Briefly, isolated human CD34+ cells are prestimulated at a cell density of $2–8 \times 10^5$ cells/mL for 36 h in X-Vivo ten serum-free media supplemented with recombinant human stem cell factor, thrombopoietin, and FLT3-ligand. Following prestimulation cells are washed and transferred into RetroNectin®-coated bags with serum-free vector stock supplemented with fresh cytokines. Approximately 12 h after, cells are washed and resuspended in a new batch of vector stock with fresh cytokines. Cytokines are maintained throughout ex vivo transduction at a concentration of 100 ng/mL for each cytokine. Twelve hours following the end of the second transduction, cells are then wash and concentrated in a transfusion bag. Samples are collected for FACS or colony-forming assays, or to transplant into NOD/SCID mice. Additionally an aliquot of transduced cells, and DNA and RNA isolated from genetically modified cells are archived.

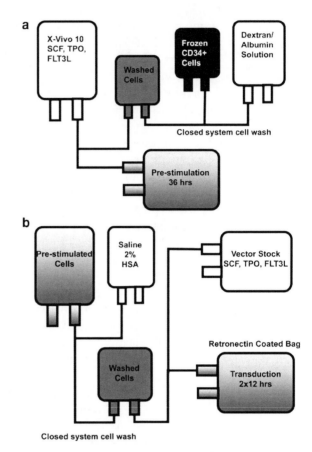

Fig. 1. Closed system retroviral transduction of human CD34+ cells; (**a**) Prestimulation, and (**b**) Transduction in RetroNectin-Coated Bags.

Serum-free ex vivo cell manipulations do not compromise either the transduction efficiency or the engraftment potential of human CD34+ cells. Prestimulation for 36 h *(27, 35)* and two cycles of transduction 12 h apart are sufficient to maximize the transduction efficiency. Adding a third transduction cycle or performing transduction cycles 24 h apart does not result in better gene transfer. Coating of tissue culture vessels with Retro-Nectin® improves both human CD34+ cell transduction by eight to tenfold and increases cell viability by protecting the cells from apoptosis. These effects are observed with doses of RetroNectin® as low as 2 µg/cm² *(27)* (**Fig. 2**).

The important factors to achieve adequate levels of gene transfer are the vector titer, the physical proximity of the target cells and viral particles, the cell cycle status of the cells, and the level of expression and density of the envelope receptor on the target cells. The vector concentration becomes limiting when it is lower than 4–5 × 10⁵ tu/mL, and this limitation cannot be compensated by using higher transduction volumes to increase

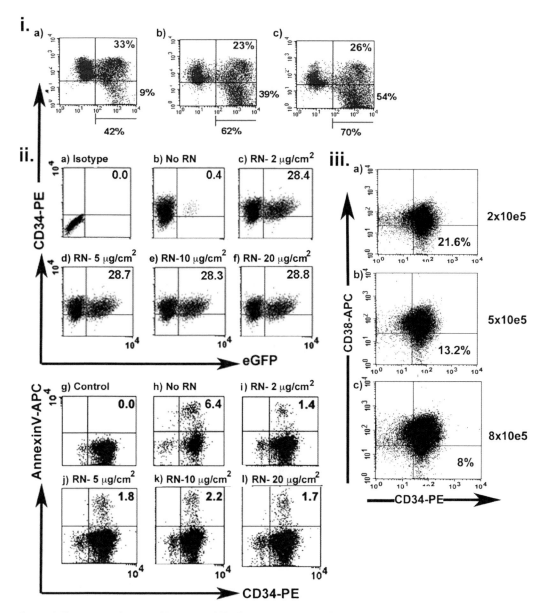

Fig. 2. (i) Expression of human CD34 and eGFP after transduction of CD34+ cells transduced under three different conditions; (**a**) prestimulation in serum-free media and vector stock for transduction in serum-free X-Vivo 10 media (**b**) prestimulation in serum-free X-Vivo 10 media, and transduction with serum-free vector stock supplemented with 10% FBS, (**c**) all steps are in serum-containing conditions. (ii) Influence of RetroNectin dose on both transduction efficiency and cell apoptosis; (**a–f**) RetroNectin-coating increases the transduction, but this effect is independent of the amount of Retronectin (between 2 and 20 µg/cm$^2$). (**g–l**) Apoptotic cells are stained with Annexin V-APC and 7-AAD staining. The percentage of Annexin-V+7-AAD- is higher in the absence of Retronectin and lower in presence of Retronectin at all concentrations. (iii) Cell density during the transduction protocol might influence CD34+/CD38-cell population intensity. After 60 h of ex vivo manipulation, cultures plated at starting cell densities of $2 \times 10^5$, $5 \times 10^5$, and $8 \times 10^5$ CD34+ cells/mL contain CD34+/CD38- populations of $18.9 \pm 1.7\%$, $12 \pm 0.6\%$, and $7.5 \pm 0.6\%$, respectively. Under low cell density conditions CD34+/CD38- immature hematopoietic progenitor/stem cell population is preserved better than high-density cultures.

the multiplicity of infection (MOI) *(27)*. Spinoculation might increase gene transfer in human CD34+ cells *(44–50)*; however, under serum-free conditions combining polycations with spinoculation significantly decreases cell viability without any gain in gene transfer *(27)*.

## 2. Materials

*2.1. Maintenance of Vector-Producing Cell Line and Serum-Free Vector Production*

1. PG13 producer cell line (*see* **Note 1**).
2. Dulbecco's Modified Eagle Medium (DMEM) high glucose (1×), liquid, with L-glutamine and sodium pyruvate, but no phenol red (DMEM-HG), 500 mL.
3. Fetal Bovine Serum (FBS), prescreened for producer cell growth and vector production, 500 mL.
4. Gentamicin 50 mg/mL, 10 mL storage at room temperature. Final concentration in media, 50 µg/mL.
5. Trypsin-EDTA.
6. Dulbecco's Phosphate-Buffered Saline (D-PBS), without calcium and magnesium, 500 mL.
7. X-Vivo 10 Media without phenol red, 1 L (BioWhittaker Cambrex Bioscience Walkersville, Inc.) (*see* **Note 1**).
8. T75, T150 Vented Tissue Culture Flasks.
9. Ten tray-Cell Factories, cell culture surface area: 6,320 cm$^2$ (Nalgene Nunc, Inc.).
10. Cell Factory Accessories; CF HDPE Connectors, White Tyvek Cover caps, Blue sealing caps, Bacterial Air Vent filter, silicone tubing, and a tubing clamp. (Nalgene Nunc, Inc.).
11. Kimax Aspirator bottle, 2 L (Kimble Glass, Inc.).
12. Sterile 2- and 4-Spike to Membrane Port Adapter Sets (Baxter Healthcare).
13. Sterile Plasma Transfer Sets (Baxter Healthcare).
14. Autoclavable Septum Closure (Nalgene Nunc, Inc.).
15. Sterile Square Media Bottles, 125 mL, 250 mL, 500 mL, 1 L, and 2 L (Nalgene Nunc, Inc.).
16. Dual inline filter (No. 4C8030, Baxter Healthcare).
17. 20 µm filter (No. 4C7704, Baxter Healthcare).
18. Sepa-cell filter (No. 4C2481, Baxter Healthcare).
19. Vented Spike Adapter (Nalgene Nunc, Inc.).
20. Bioprocessing Bag 5, 10 L (TC Tech).

21. Cryocyte Freezing Containers, 250, 500, 1,000 mL (Baxter Healthcare Corporation).

22. Cryocyte Cassettes (Custom Biogenics).

**2.2. Cell Preparation and Prestimulation**

1. Human bone marrow (BM), mobilized peripheral blood (MPB), or umbilical cord blood (UCB)-derived human CD34+ cells, either fresh or frozen, will be used for gene transfer. Purification of human CD34+ cells is beyond the scope of this chapter.

2. Human Serum Albumin Solution 5%, 250 mL (Buminate; Baxter Healthcare Corporation).

3. 10% Dextran 40 in 0.9% Sodium Chloride IV, 500 mL (Gentran Viaflex, Baxter Healthcare Corporation).

4. Transfer Packs with Spike, 300 mL (Baxter Healthcare Corporation).

5. VueLife FEP bag, 300–500 mL (American Fluoroseal Corporation). The average fill volume limits the solution height approximately to 1 cm allowing optimal gas transfer.

6. Interlink System Blood Bag Injection Site (Baxter Healthcare Corporation).

7. X-Vivo 10 Media without phenol red, 1 L, storage at +4°C.

8. Gentamicin 50 mg/mL, 10 mL storage at room temperature. Final concentration in media, 50 µg/mL.

9. Sterile 2- and 4-Spike to Membrane Port Adapter Sets.

10. Sterile Plasma Transfer Sets.

11. Autoclavable Septum Closure.

12. Recombinant Human Stem Cell Factor (SCF), lyophilized 10 µg vial (PeproTech, Inc.) Both the lyophilized protein and the reconstituted solution are stored at -20°C. The lyophilized protein is diluted in 1 mL injectable $dH_2O$ and filtered to generate a stock solution at 100 ng/µL; 1 µL should be added for each mL of media for a final concentration of 100 ng/mL.

13. Recombinant Human Fms-related tyrosine kinase 3 ligand (FLT3L), lyophilized 10 µg vial (PeproTech, Inc.) Both the lyophilized protein and the reconstituted solution are stored at -20°C. The lyophilized protein is diluted in 1 mL injectable $dH_2O$ and filtered, to generate a stock solution at 100 ng/µL; 1 µL should be added for each mL of media for a final concentration of 100 ng/mL.

14. Recombinant Human Thrombopoietin (TPO), lyophilized 10-µg vial (PeproTech, Inc.) Both the lyophilized protein and the reconstituted solution are stored at -20°C. The

lyophilized protein is diluted in 1 mL injectable dH$_2$O and filtered to generate a stock solution at 100 ng/μL; 1 μL should be added for each mL of media for a final concentration of 100 ng/mL.

15. Sterile Water Injection, USP, 10-mL vial.

16. Millex-GV PVDF 0.2-μm, syringe-driven filter unit (Millipore Corporation).

### 2.3. Coating of Cell Culture Bags with RetroNectin®

1. RetroNectin, lyophilized 2.5 mg protein (Takara Bio Inc.). The lyophilized protein should be stored at +4°C and the reconstituted solution (1 mg protein/mL) should be stored in aliquots at -20°C. For reconstitution, add sterilized 2.5 mL distilled water to obtain a 1-mg protein/mL solution, filtrate through 0.22-μm filter and store in aliquots at –20°C. To keep the protein integrity, avoid vigorous mixing (no vortex) and repeated freeze-thaw cycles.

2. Human Serum Albumin Solution 5%, 250 mL.

3. 0.9% Sodium Chloride Injection, USP, 250 mL.

4. Transfer Packs with Spike, 300 mL (Baxter Healthcare Corporation).

5. VueLife FEP bag, Average fill 225 or 255 mL, and surface areas are 500 and 561 cm$^2$, respectively (American Fluoroseal Corporation). For RetroNectin® coating, bags with improved cell adherence should be used.

6. Interlink System Blood Bag Injection Site.

7. Sterile Water Injection, USP, 10 mL vial.

8. Millex-GV PVDF 0.2 μm, syringe-driven filter unit.

### 2.4. Oncoretroviral Transduction in RetroNectin®-Coated Bags (Fig. 1b)

1. Frozen vector stock in Cryobags from **Subheading 4.2.1**.

2. Prestimulated human CD34+ cells in VueLife FEP bag from **Subheading 4.2.2**.

3. RetroNectin®-coated adherence improved VueLife FEP bag from **Subheading 4.2.3**.

4. 0.9% Sodium Chloride Injection, USP, 250 mL.

5. Transfer Packs with Spike, 300 mL.

6. Interlink System Blood Bag Injection Site.

7. X-Vivo 10 Media without phenol red, 1 L, storage at +4°C.

8. Gentamicin 50 mg/mL, 10 mL, storage at room temperature. Final concentration in media, 50 μg/mL.

9. Sterile 2- and 4-Spike to Membrane Port Adapter Sets.

10. Sterile Plasma Transfer Sets.

11. Autoclavable Septum Closure.

12. Recombinant Human SCF, stock concentration 100 ng/μL, 1 μL should be added for each mL of media, final concentration 100 ng/mL (*see* **Subheading 4.2.2**).

13. Recombinant Human FLT3 L, stock concentration 100 ng/μL, 1 μL should be added for each mL of media, final concentration 100 ng/mL (*see* **Subheading 4.2.2**).

14. Recombinant Human TPO, stock concentration 100 ng/μL, 1 μL should be added for each mL of media, final concentration 100 ng/mL (*see* **Subheading 4.2.2**).

15. Sterile Water Injection, USP, 10 mL vial.

*2.5. Single-Colony PCR for Determining Transduction Efficiency*

1. Colony-forming units that are growing in methylcellulose semisolid media.

2. DNA Lysis Buffer- Tris 5 mM, 0.45% Tween 20 at pH 8.0. Store at room temperature.

3. Proteinase K, Solution (Roche) in 10 mM Tris–HCL, pH 7.5, 15.1 mg/mL. Store at 2–8°C.

4. Dulbecco's Phosphate-buffered Saline (D-PBS), without calcium and magnesium, 500 mL.

5. Phenol/$CHCl_3$/isoamyl alcohol (25:24:1) mixture.

6. Glycogen, 20 mg/mL solution (Boehringer–Mannheim).

7. Ammonium Acetate, 7.5 M Solution.

8. Absolute Ethanol.

9. AmpliTaq Gold-DNA Polymerase (Applied Biosystems).

10. GeneAmp 10× PCR Buffer II (Applied Biosystems).

11. 25 mM $MgCl_2$ solution (Applied Biosystems).

12. dNTP GeneAmp Blend, 10 mM (Applied Biosystems).

13. β-Actin forward and reverse primers.

14. Transgene specific forward and reverse primers.

# 3. Methods

*3.1. Maintenance of Vector-Producing Cell Line and Serum-Free Vector Production*

1. After thawing the vial(s) at 37°C, dilute the cells at 1:10 (v:v) ratio with complete DMEM media supplemented with 10% FBS containing 50 μg/mL Gentamicin and pellet the cells by centrifugation. Aspirate the DMSO-containing supernatant. Plate the retroviral packaging cells in complete media in T150 flasks and expand the cells into four *(4)* 10-tray Cell Factories (**Note 2**).

2. Once the cells are plated in ten tray-Cell Factories in 1,000 mL complete media, let them grow for approximately 24 h

at 37°C and 5% $CO_2$. The following day, wash the cells two to three times with D-PBS, and replenish the cell factories with 625 mL of X-Vivo 10 media. After incubation at 32°C (or 37°C) and 5% $CO_2$ for 24 h, collect the vector stocks from cell factories in a 5-L bioprocessing bag by constructing a four-way manifold system using sterile 4-spike to Membrane Port Adapter sets and a peristaltic pump, at a flow rate of 500 mL/min. Each harvest should be step-filtered on the day of collection through a serially connected filtration system using a dual inline filter, a 20-μ filter, and a Sepacell filter and transferred into a second sterile bioprocessing bag *(54, 55)*. Two additional harvests can be performed on each of the following 2 days without significant vector titer loss *(27, 54, 56)*.

3. Store the first two harvests at 4°C until pooling. After filtration of the third harvest, pool the three filtered harvests into a sterile 10-L bioprocessing bag, and aliquot as needed into sets of eight sterile Cryocyte Freezing Containers using a peristaltic pump *(54)*. After filling, load the vector stocks aliquots into Cryocyte cassettes and store at -80°C (*see* **Note 3**).

4. Biosafety testing and vector titration (*see* **Note 4**) should be performed on each batch of clinical-grade vector stocks.

*3.2. Cell Preparation and Prestimulation (see Note 5) (Fig. 1a)*

1. Transfer 125 mL of Dextran 40 in 0.9% Sodium Chloride and 125 mL of 5% Human Serum Albumin solution to a transfer bag by using the fluid pump mode of the Cytomate™ Cell Washer (*see* **Note 6**). This solution should be kept at +4°C with an expiration time of 24 h.

2. Transfer X-Vivo 10 Media to a transfer bag through a septum closure using a 2-spike transfer set. Add gentamicin and cytokines to the bag through an Interlink System Blood Bag Injection Site. The volume of prestimulation media should be calculated according to the total viable cell number using a final cell concentration of $2–8 \times 10^5$ cells/mL.

3. Remove the cryopreserved CD34+ cells from the vapor phase of liquid nitrogen, place into a sterile zipper-locked bag, and thaw the cells at 37°C in a waterbath until total conversion to liquid phase.

4. Wash the cells on the Cytomate™. Set up the Cytomate disposable cell wash set on the Cytomate™ as per the manufacturer's instructions. Connect two bags to the wash/buffer line with a Y connector through spikes. The first wash buffer contains 500 mL D-PBS with 2% HSA, and the second bag should have ice-cold Dextran/Albumin solution. Close the line on the first wash buffer with Roberts clamp. Attach the bag containing the CD34+ cells to the cell source line on the Cytomate disposable cell wash set; dilute the cells twice with

the Dextran/Albumin solution (v:v) at the rate of 20 mL/min, and then with the same volume at the rate of 60 mL/min. Total volume of Dextran/Albumin solution transferred to the source bag will be at least equal to the volume of source bag and should not be in excess of two volumes of the source bag. Gently shake the bags during dilution steps in order to avoid cell sedimentation. Close the Roberts clamp on the line of the Dextan/Albumin solution and open the clamp to D-PBS with 2% HSA solution. The CD34+ cells are washed and concentrated by flowing through the spinning membrane, and subsequently transferred into a collection bag. Rinse and wash the bag containing the CD34+ cells and tubing with an additional 100 mL of D-PBS with 2% HSA solution in order to minimize the cell loss.

5. Detach the cell collection bag and with an Interlink System Blood Bag Injection Site take a sample for cell count, viability and flow cytometry analysis. Calculate the amount of prestimulation media needed and store in a transfer bag as described in **Subheading 4.2.2**.

*Example*. plating density – $5 \times 10^5$ cells/mL, total number of cells – $150 \times 10^6$ cells

Volume of Prestimulation Media = Total number of cells/ Plating density

Volume of Prestimulation Media = $150 \times 10^6$ cells/$5 \times 10^5$ cells/mL = 300 mL

An additional 10% volume should be accounted for loss in tubing; therefore, 330 mL (300 + 30 mL) of X-Vivo 10 media supplemented with cytokines (330 µL from the main 100 ng/µL stock) and Gentamicin (330 µL from the main 50 mg/mL stock) should be prepared.

An additional bag containing only X-Vivo 10 without cytokines and without Gentamicin should be transferred to a transfer bag. The volume should be equal to two volumes of cell suspension.

Install a new cell washer set on the Cytomate™ as per the manufacturer's instructions. Connect the bag containing the prestimulation media and the bag containing the X-Vivo 10 wash media through a Y-connection and attach to the wash buffer line. The cell collection bag containing the washed CD34+ cells from **step 5** will now be the source bag. Close the Roberts clamp on the prestimulation media until the cells are washed and concentrated through the spinning chamber with the X-Vivo 10 wash buffer. Subsequently transfer the CD34+ cells to a Vuelife bag in the prestimulation media. Clamp the line to X-Vivo wash media and unclamp the Roberts clamp on the prestimulation media line.

6. Detach the VueLife bag from the set and incubate at 37°C, in 5% $CO_2$ incubator for 36 h.

### 3.3. Coating of Culture Bags with Retro-Nectin®

1. Calculate the amount of RetroNectin® according to the surface of the bags; the surface area for a 225-mL volume VueLife FEP bag is 500 cm², and the bags will be coated with RetroNectin® at a dose of 2 µg/cm².

   Total amount of RetroNectin® = Surface Area × 2 µg/cm² = 1,000 µg = 1 mg

   Dilute 1 mL of 1 mg/mL stock solution (*see* **Subheading 4.2.3.1**) in 49 mL of either 0.9% saline or D-PBS (final concentration of 20 µg/mL; for effective coating RetroNectin® concentration should be kept between 20 and 100 µg/mL) and transfer into VueLife bag through interlink injection site. Incubate either at room temperature for 2 h or overnight at +4°C.

2. Prepare 100 mL blocking solution in a transfer bag as 2% Human Serum Albumin (40 mL 5% human serum albumin solution) and 0.9% NaCl (60 mL). After draining the RetroNectin®-containing solution to a waste bag through a 2-spike connector, transfer the blocking solution to the VueLife bag.

3. After 1/2 h incubation at room temperature drain the solution to the waste bag and gently wash the bag with another 100 mL of normal saline. If the RetroNectin®-coated bag is not used immediately, it can be kept at +4°C for approximately 1 week.

4. Depending on the target CD34+ cell dose, it may be necessary to coat two or more bags with RetroNectin® (*see* **Note 7**).

### 3.4. Retroviral Transduction in RetroNectin®-Coated Bags (Fig. 1b)

1. Install a new cell washer set on the Cytomate™ as per the manufacturer's instructions. Connect the bag containing the prestimulated cells to the source line, the VueLife FEP–RetroNectin®-coated bag/s to the wash/collection line. Connect two bags to the wash/buffer line with a Y connector through spikes. The first wash buffer contains 300 mL D-PBS with 2% HSA or 0.9% NaCl, and the second bag will contain the transduction media including the vector stocks, the fresh cytokines, i.e., SCF, FL3L, TPO, and gentamicin. During the cell wash process, the Roberts clamp on the second bag line should be closed.

2. Transfer the cells washed with the first wash buffer back to the source bag in 50 mL volume, then give a pause interval and through an Interlink Injection Site collect a sample. Determine the cell count and viability. The viable cell count defines the amount of vector stocks that need to be thawed.

3. Thaw the cryobag(s) containing the frozen vector stocks quickly by placing the bag in an overwrap pouch in a 37°C water bath until disappearance of the ice crystals. For vector stocks with high concentration (above 5 × 10⁵ tu/mL), vector titer should be adjusted by adding X-Vivo 10 Media with Gentamicin to prevent multiple vector copy integration per cell.

Cytokines, SCF, FLT3L, TPO, should be added to the transduction media at a final dose of 100 ng/mL through an Interlink Injection Site. While attached to the cell washer for fluid transfer, the transduction media is roughly at room temperature.

4. For the second "cell wash step" on the Cytomate™, use the transduction media prepared in **step 3** and transfer the cells and transduction media to the RetroNectin®-coated bag(s). By using a Y connector, two RetroNectin®-coated bags can be connected to the wash/collection line.

5. Detach the RetroNectin®-Coated bags from the wash set and spinoculate at $1,000 \times g$ for 2 h at 10°C, then incubate at 37°C, in a 5% $CO_2$ incubator for 10 h.

6. For the second transduction cycle, repeat **steps 1–5** again.

7. At the end of the second transduction cycle (2 h spinoculation and 10 h of incubation), wash and resuspend the cells in 2% HSA-containing normal saline. Concentrate the cells at a final volume of 50 mL. Take samples for cell count and viability, flow cytometry analysis for CD34 and other relevant markers, hematopoietic progenitor colony assay (*see* **Notes 8** and **9**), sterility, Mycoplasma, Endotoxin, RCR, and for archives, i.e., DNA/RNA and frozen cell samples.

8. Infuse the CD34+ transduced cells or freeze as needed. Freezing will likely be required if complete biosafety testing and vector copy number are required prior to infusion.

*3.5. Single-Colony PCR for Determining Transduction Efficiency (see Note 9)*

1. Fill microcentrifuge tubes with 1 mL D-PBS. Aspirate single colonies from the methylcellulose medium in a volume of 20–50 µL with individual plugged pipette tips (P200), under visualization with a phase-contrast inverted microscope and add to the D-PBS-containing tube. Place each colony into an individual tube.

2. Leave the pipette tip in the tube for about an hour at room temperature to allow the methylcellulose to dissolve. Then pipette D-PBS in and out to thoroughly resuspend the cells in D-PBS.

3. Centrifuge the tubes at 2,000 rpm for 10 min in a microcentrifuge. Aspirate carefully the supernatant using individual pipette tips without disturbing the tiny cell pellet. It is acceptable to leave up to 10 µL D-PBS in the tube to avoid cell loss.

4. Prepare the appropriate volume of DNA lysis buffer as per **Subheading 4.2.5** for the number of picked colonies (Total volume = Number of colonies picked × 50 µL + 10% of volume). Add proteinase K to DNA lysis buffer so that the final concentration is 0.1 mg/mL.

5. Transfer 50 µL of proteinase K-containing lysis buffer to each tube and resuspend cell pellet by pipetting in and out.

6. Incubate the samples at 56°C for 90 min.

7. Inactivate proteinase K at 95°C for 5 min.

8. Add 150 μL of D-PBS to each tube.

9. Mix 200 μL phenol/CHCl$_3$/isoamyalcohol with DNA-containing solution from **steps** 7. Leave at room temperature for 10 min and centrifuge at 14,000 rpm on microcentrifuge for 10 min.

10. Collect the aqueous upper phase (approximately 150–180 μL) and precipitate by adding 2 μg glycogen, 18 μL of 7.5 M Ammonium Acetate, 500 μL absolute ethanol. Incubate overnight at –20°C.

11. Centrifuge the DNA at 14,000 rpm on microcentrifuge for 10 min. Wash the pellet with 70% ethanol. Dry the DNA pellet on the bench top; do not use vacuum drying.

12. Resuspend the DNA in 50 μL of either TE Buffer or dH$_2$O. 3–5 μL of this solution should be adequate to detect the presence of most transgene by PCR.

13. Run the PCR reaction with the transgene specific primers. A PCR reaction with β-Actin primers should also be run to confirm the presence of DNA in the sample.

## 4. Notes

1. The PG13 parental packaging cell line (ATCC Catalog number; CRL-10686) is derived from the murine NIH 3T3 TK⁻ embryo fibroblasts *(57)* and contains a bipartite retroviral packaging system. The Moloney murine leukemia virus gag-pol expression construct was introduced by cotransfection using the herpes simplex virus thymidine kinase gene, and the gibbon ape leukemia virus (GaLV) envelope was introduced by cotransfection with a mutant methotrexate dihydrofolate reductase gene (DHFR*). These selection markers confer resistance to amethopterin. Selection against loss of the plasmid DNAs conferring the packaging functions can be performed by growing the cells in medium containing dialyzed FBS and 100 nM amethopterin for 5 days followed by cultivation in medium containing HAT and untreated FBS for an additional 5 days. Resting the cells in hypoxantine-containing media for at least 2 days is required for eliminating amethopterin from the environment (*see* ATCC recommendations).

Cross-infection of the packaging cells with vector stocks originally obtained by transfection increases the number of high-titer packaging clones relative to direct transfection *(43, 58)*. It is possible to generate stable high-titer PG13 clones by cross-infection with vector stocks derived from either Phoenix-Eco (http://www.stanford.edu/group/nolan/publications/publications.htmL). or VSV-G pseudotyped 293GPG cells *(59)*. Generation and selection of high-titer PG13 producer cell lines are summarized elsewhere *(60)*.

2. Production of vector stocks under serum-free conditions increases the biosafety of retroviral transduction of human CD34+ cells for clinical applications. The risk of transmitting spongiform encephalitis by exposure of target cells to poorly defined bovine products, the demonstration that better cell expansion can be obtained in the absence of serum *(61)*, and the fact that some fetal bovine serum proteins have been shown to be immunogenic *(62, 63)* are prompting the development of gene transfer protocols under serum-free conditions. Although the PG13 packaging cells are serum-dependent for proliferation, they adapt to short-term culture in serum-free medium, allowing serum-free vector stock collection *(64)*. Vector stocks produced under serum-free conditions display similar *(65)* or lower *(27, 66)* titers on indicator Hela cells. However, the transduction efficiency in human CD34+ cells is comparable, the differentiation of CD34+ cells is less *(27)* (**Fig. 2i**), and in vivo gene marking levels in NOD/SCID mouse are at least as good under serum-free conditions *(27)* in comparison to those obtained in presence of serum. Among the tested serum-free media (X-Vivo 10, X-Vivo 15, Stem-Pro 34 SFM, IMDM, QBSF60), X-Vivo 10 media provided the highest titers after either 16 or 24 h of incubation *(27, 56)*.

The stability of Mo-MuLV-derived retroviral vectors can be augmented by increasing the medium osmotic pressure from 335 up to 410–450 mOsm/kg, which decreases the cholesterol content of both virus particles and producer cells *(67)*. The vector stocks produced under high osmolar conditions have been shown to yield three to fourfold higher vector titers. However, these conditions have not been tested yet on large-scale vector stocks production *(67)*.

Adding sodium butyrate to the media during vector production has been shown to enhance expression of the vector and packaging construct, leading to a 10–1,000-fold increase in viral production *(68)*. However, in our experience, adding sodium butyrate to X-Vivo 10 media did not increase vector titers (unpublished data).

Vector stocks with titers below $5 \times 10^5$ tu/mL may be concentrated, but the type of envelope should be considered. The vector stocks produced by the packaging cell lines that encode either Eco-, Ampho-, or GaLV-envelope proteins cannot be concentrated by ultracentrifugation. Those envelope proteins are composed of two domains: an extracellular domain and a transmembrane domain which are linked by disulfide bonds only. The stress of centrifugation and filtration often causes the surface domain to be shed which results in soluble, free-floating surface domain peptides that can block infection by saturating receptors on the target cells. On the other hand, VSV-G pseudotyped viral vectors infect cells via membrane fusion, and, are therefore not dependent on receptor recognition. VSV-G is a strong glycoprotein that can withstand ultracentrifugation at $50,000 \times g$ for 90 min at 4°C and allows viral particle concentration up to 100–200 fold. Sheer-force sensitive viral particles pseudotyped with Eco-, Ampho-, or GaLV- envelopes can only be concentrated up to tenfold by either low-speed centrifugation (9,500 rpm in Beckman rotor JA-14 at 4°C for 12 h) (41), or centrifugation and filtering for 35 min at $3,000 \times g$, 15°C through centrifugal filter devices with a 100,000 molecular weight cut-off (27).

When the cells reach confluence, each T150 flask contains approximately $4–5 \times 10^7$ cells. Six T150 flasks 70–90% confluent are used to seed a single 10-tray Cell Factory which is itself used, once it reaches 70–90% confluence, to seed four (4) 10-tray-Cell Factories (27). By reducing the volume of harvest medium to 0.1 mL per $cm^2$, titers could be increased up to fourfold (66), and repetitive harvests of vector stocks are feasible over three to four consecutive days (27, 54, 56, 66) after an incubation period of 16 or 24 h at $5\%CO_2$ as previously published (56, 69, 70). The optimal incubation temperature for retroviral vector production is somewhat controversial. Some studies show greater retroviral vector inactivation and/or lower vector titers at 37°C when compared to 32°C (27, 54, 70, 71). On the other hand, the viral particles produced at 37°C are less rigid than those produced at 32°C. They are therefore more stable and thus provide higher transduction efficiency (56, 72). At this time, we recommend to compare side by side the titers of vector stocks harvested at 32°C and 37°C for the selected high-titer packaging cell clone.

3. Frozen vector stocks collected in serum-free conditions and stored at -80°C in Cryocyte bags are stable over a period of at least 12 months, and adding either human serum albumin or FBS does not change the stability of the frozen vector stocks (27). The half-life of frozen vector stocks is suggested to be biphasic. Following 25–30% loss of vector titer upon the first freeze/thaw cycle, the decay of viral particles is slow, and half-life of this stage varies from 18 to 41 months (54, 73). Interestingly, testing the frozen vector stocks only on target/ indicator cell lines may be misleading, as transduction efficiency in primary cells, i.e., T lymphocytes or CD34+ cells may not show the same decrease (27, 54). This different outcome between cell types may be due to differential cell surface receptor

saturation of envelope receptors on indicator cell lines such as Hela and primary cells.

It is a common research practice to filter vector stocks through a 0.45-µm filter to remove cellular debris. Do not use filters that contain detergents, as wetting agents will affect the integrity of the viral particles as well as the viability of the target cells (unpublished observation).

4. Vector particle counts can be determined either by direct particle count using an electron microscope, or by indirect methods such as quantitative real-time PCR *(74, 75)* or transduction of target/indicator cells. Vector stocks contain infectious viral particles as well as noninfectious particles. The vector titers defined by transduction of target cells depend in part on the level of expression of the envelope receptors such as PiT-1, PiT-2, or ecotropic receptor *(19, 76)* which varies among target cells *(27)*. When serial dilutions of the vector stocks are used to transduce CEM, HeLa, and human CD34+ cells, the transduction rates obtained on CD34+ cells are comparable to the transduction rates obtained on HeLa cells *(27)*. We recommend to initially test the transduction efficiency on both the type of primary cells that will be used for clinical application and various target/indicator cell lines, and for subsequent vector titrations, to select the cell line that gives transduction rates comparable to that obtained on primary cells. The methods for determining vector stock titers are beyond the scope of this chapter.

5. The source of hematopoietic stem cells (HSCs) used for retroviral transduction has been shown to affect the transduction efficiency and engraftment potential of the cells. UCB-derived CD34+ cells display the highest transduction efficiency and engraftment potential *(77–79)* as UCB contains larger numbers of primitive cells that are capable of forming secondary multipotential colonies upon replating in vitro *(80–82)*, NOD/SCID repopulation potential of UCB-derived CD34+ cells in G0 or G1 phase are similar *(83–85)*, while MPB-derived CD34+ cells SCID repopulating cells (SRCs) mainly reside in the G0 population *(22)*. Adult HSCs are more dormant and less responsive to cytokine stimulation when compared with UCB. They require prolonged cytokine stimulation to enter the cell cycle, to upregulate envelope receptor expression, and to be efficiently infected by oncoretroviral vectors *(35, 86)*. Higher levels of envelope receptor mRNA are present in UCB-derived HSCs when compared to HSCs derived from bone marrow *(17, 18)*. The expression of the amphotropic receptor on MPB-derived CD34+ cells, as assessed by indirect immunofluorescence assay, was approximately one log higher than that of steady-state BM or PB CD34+ cells

*(87)*. G-CSF-mobilized peripheral blood CD34+ cells are the
favored source of autologous HSCs in clinical transplantation;
however, G-CSF + stem cell factor (SCF)-mobilized peripheral blood CD34+ cells have been shown to engraft better
than G-CSF-mobilized peripheral blood or steady-state bone
marrow-derived CD34+ cells in nonhuman primates *(88–90)*.
Additionally, engraftment after transplantation using GzSCF/
SCF primed BM cells is more rapid than that using steady-
state bone marrow in nonhuman primates *(88)*. Similarly in
a dog model, G-CSF + SCF primed marrow provided the
highest in vivo gene marking, in comparison to steady-state
marrow, and MPB *(91)*. Resting the transduced cells for an
additional 2 days in presence of SCF on RetroNectin®-coated
support was suggested to decrease active cycling and proliferation, and consequently resulted in a significantly higher in
vivo engraftment when compared to nonrested cells in nonhuman primates *(92–94)*.

In planning a clinical trial, the source of the human CD34+ cells
should be carefully considered. Though it is possible to immunose-
lect CD34+ cells from frozen samples, the variability of the resulting CD34+ purity and yield *(95, 96)* makes it nearly impossible to
reliably predict how many cells should be collected and stored for
transduction. CD34+ cells from pooled leukapheresis MPB products or fresh bone marrow samples can be selected within 24–48 h
upon collection. Both negative and positive fractions should subsequently be frozen for future use. Thawing the frozen purified human
CD34+ cells in Dextran/Albumin solution increases the post-thaw
viability *(97)*. This method can be applied to a closed system with
a reasonable cell viability and recovery by using an automated cell
washer *(98)*. Frozen/thawed UCB-derived CD34+ cells have been
shown to express higher levels of amphotropic receptor mRNA, but
not of GaLV receptor (Pit 1) mRNA *(16)*. The ability of purified
CD34+ cells to respond to cytokine stimulation does not change
upon freezing and thawing *(99)*.

6. The fully automated cell washer system, Cytomate™ (Baxter
   Healthcare, Deerfield, IL), is a versatile instrument that can
   be programmed step by step for fluid transfer, media/solu-
   tion preparation, cell washing and concentration. The spinning membrane is connected to a filtered wash bag, a buffer
   bag, and a waste bag. Four weight scales and probes monitor
   the fluid transfer between bags. The sterile disposable sets that
   are used for either fluid transfer or cell wash create closed fluid
   path, and allow cell processing under current Good Manufacturing Practice (cGMP). Cryobags, transfer bags, etc. can be
   docked to sterile disposable sets either using spikes or sterile connecting device, so that closed system requirements can
   be established throughout all stages of the ex vivo retroviral
   transduction process.

According to our experience, the mean recovery of viable CD34+ cells using the closed system cell washer is usually in the range of 97 ± 5%, with a mean cell viability of 87 ± 6.5% (our unpublished data). In our hands, this constitutes a better recovery rate than conventional open system cell washing with centrifugation. The cell wash process takes approximately 1 h.

7. For prestimulation, CD34+ cells can be plated at cell densities of $1 \times 10^5$–$1 \times 10^6$ cells/mL without altering their viability during a 36-h incubation period. The plating of human CD34+ cells at different cell concentrations – $2 \times 10^5$, $5 \times 10^5$, $8 \times 10^5$ cells/mL – does not influence the ex vivo cell expansion rate – 1.7 ± 0.2, 2.1 ± 0.27, 1.85 ± 0.17 fold, respectively (unpublished data). At low cell density ($2 \times 10^5$ cells/mL), the percentage of CD34+/CD38- cells is statistically higher than the percentage obtained at higher cell densities (**Fig. 2iii**), but increasing the cell density from $2 \times 10^5$, $5 \times 10^5$ to $\times 10^5$ (MOI values range from 0.5 to 2) do not change the transduction efficiencies (45.1 ± 13, 54 ± 10.2, 49.3 ± 8.6%, respectively) ($p > 0.05$) (unpublished data). The culture of cells at low density requires more cytokines and bags. To our knowledge, the RC3 centrifuge (Sorvall) is the only centrifuge suitable for spinoculation. It can accommodate two bags with a maximum volume of $2 \times 250$ mL per cycle, which limits the volume of transduction to 500 mL per run. Starting with at least $2 \times 10^6$ viable CD34+ cells/kg for a 70-kg adult requires a total of at least $140 \times 10^6$ viable cells to be transduced. Using these calculations, we recommend a cell plating density in the range of $3$–$5 \times 10^5$ cells/mL. We recently observed that spinoculation at 1,000 $\times g$ for 2 h at 10°C increases the transduction rate of the human CD34+ cells approximately twofold (28.5 ± 9.7% vs. 55 ± 7.8%, $p < 0.05$), while the viability of the cells was similar among the groups (unpublished validation data).

After prestimulation our hypothetical $140 \times 10^6$ viable cells become approximately $210 \times 10^6$ viable cells with a mean 1.5 fold expansion. Transduction volume will be around 420 mL when the cells are plated at a density of $5 \times 10^5$ cells/mL. The only available centrifuge option that can hold the bags horizontally is so far Sorvall RC3, and the custom-made holder/rotor system can only hold two bags of 240 mL at the same centrifugation cycle. The maximum number of the cells that can be spinoculated at once is $240 \times 10^6$ viable cells (adequate for an 80-kg adult patient using a cell dose of $3 \times 106$ cells/kg). We prefer not to exceed $5 \times 10^5$ cells/mL ex vivo culture cell density because of the significant loss of CD34+/CD38- population in high-density cultures.

8. One can always produce one's own methylcellulose media for colony assay, but the variability of the ingredients from batch to batch necessitates the use of commercially available products.

We utilize serum-free MethoCult (Stem Cell Technologies, Vancouver, Canada), and plating 500–750 cells per mL. Each sample should be plated at least in triplicate and well mixed. Following 14 days of growth, colony-forming units should be enumerated. Enumeration of the colonies requires appropriate training as there can be substantial variability among operators. It is worth testing the proficiency of the operators annually. The details of plating and enumerating hematopoietic colonies in semisolid media are beyond the scope of this chapter.

9. There are actually no golden standard for determining the transduction rate of long-term hematopoietic stem cells. Calculating SCID-repopulating potential of transduced cells in NOD/SCID mice is an indirect estimation. This approach is useful when comparing different protocols, vector backbones, cell sources, etc. but highly impractical and costly when measuring gene transfer efficiency in patient cells enrolled in a clinical trial. It is more customary to report the transduction efficiency using single-colony PCR to detect the transgene, which actually defines the transduction rate in the hematopoietic progenitor pool. In this protocol, the DNA is extracted from individual hematopoietic colonies. The colonies growing in semisolid media usually contain less than 200 cells, and any DNA isolation method needs at least a couple of thousand cells to start with. The amount of DNA extracted varies from picograms to a couple of nanograms. When the isolated material is used to detect the transgene of interest by PCR, a second control PCR using $\beta$-actin primers has to be run to prove the presence of DNA in the sample. The failure rate of isolating DNA from the samples depends on the size of the colony but is usually less than 15%.

The number of colonies to pick depends on the expected transduction efficiency. There should be at least one colony among the picked colonies that will give a positive PCR product. Example: If the transduction efficiency in the hematopoietic progenitors is around 2%, you need to pick at least 58 colonies (1 out of 50 will be positive, and there can be a 15% failure in extracting DNA, so $50 + 15\% \times 50 = 58$).

## Acknowledgments

The authors wish to thank Michel Sadelain for critical review of the manuscript. This work is supported by PO1 CA-033049, P30 CA-008748, PO1 CA-059350, by Lymphoma Research Foundation MCLI-05-020, and by Mr. William H. Goodwin and Mrs. Alice Goodwin, and the Commonwealth Cancer Foundation for Research & the Experimental Therapeutics Center of MSKCC.

## References

1. Abonour, R., Williams, D.A., Einhorn, L., Hall, K.M., Chen, J., Coffman, J., et al. (2000) Efficient retrovirus-mediated transfer of the multidrug resistance 1 gene into autologous human long-term repopulating hematopoietic stem cells. *Nat. Med.* 6, 652–658

2. Cornetta, K., Croop, J., Dropcho, E., Abonour, R., Kieran, M.W., Kreissman, S., et al. (2006) A pilot study of dose-intensified procarbazine, CCNU, vincristine for poor prognosis brain tumors utilizing fibronectin-assisted, retroviral-mediated modification of CD34+ peripheral blood cells with O6-methylguanine DNA methyltransferase. *Cancer Gene Ther.* 13, 886–895

3. Cowan, K.H., Moscow, J.A., Huang, H., Zujewski, J.A., O'Shaughnessy, J., Sorrentino, B., et al. (1999) Paclitaxel chemotherapy after autologous stem-cell transplantation and engraftment of hematopoietic cells transduced with a retrovirus containing the multidrug resistance complementary DNA (MDR1) in metastatic breast cancer patients. *Clin. Cancer Res.* 5, 1619–1628

4. Hanania, E.G., Giles, R.E., Kavanagh, J., Fu, S.Q., Ellerson, D., Zu, Z., et al. (1996) Results of MDR-1 vector modification trial indicate that granulocyte/macrophage colony-forming unit cells do not contribute to posttransplant hematopoietic recovery following intensive systemic therapy. *Proc. Natl. Acad. Sci. U S A* 93, 15346–15351

5. Herrera, C., Sanchez, J., Torres, A., Bellido, C., Rueda, A., Alvarez, M.A. (2001) Early-acting cytokine-driven ex vivo expansion of mobilized peripheral blood CD34+ cells generates post-mitotic offspring with preserved engraftment ability in non-obese diabetic/severe combined immunodeficient mice. *Br. J. Haematol.* 114, 920–930

6. Hesdorffer, C., Ayello, J., Ward, M., Kaubisch, A., Vahdat, L., Balmaceda, C., et al. (1998) Phase I trial of retroviral-mediated transfer of the human MDR1 gene as marrow chemoprotection in patients undergoing high-dose chemotherapy and autologous stem-cell transplantation. *J. Clin. Oncol.* 16, 165–172

7. Kohn, D.B., Bauer, G., Rice, C.R., Rothschild, J.C., Carbonaro, D.A., Valdez, P., et al. (1999) A clinical trial of retroviral-mediated transfer of a rev-responsive element decoy gene into CD34(+) cells from the bone marrow of human immunodeficiency virus-1-infected children. *Blood* 94, 368–371

8. McGuckin, C.P., Forraz, N., Pettengell, R., Thompson, A. (2004) Thrombopoietin, flt3-ligand and c-kit-ligand modulate HOX gene expression in expanding cord blood CD133 cells. *Cell Prolif.* 37, 295–306

9. Passegue, E., Wagers, A.J., Giuriato, S., Anderson, W.C., Weissman, I.L. (2005) Global analysis of proliferation and cell cycle gene expression in the regulation of hematopoietic stem and progenitor cell fates. *J. Exp. Med.* 202, 1599–1611

10. Piacibello, W., Gammaitoni, L., Bruno, S., Gunetti, M., Fagioli, F., Cavalloni, G., et al. (2000) Negative influence of IL3 on the expansion of human cord blood in vivo long-term repopulating stem cells. *J. Hematother. Stem Cell Res.* 9, 945–956

11. Rahman, Z., Kavanagh, J., Champlin, R., Giles, R., Hanania, E., Fu, S., et al. (1998) Chemotherapy immediately following autologous stem-cell transplantation in patients with advanced breast cancer. *Clin. Cancer Res.* 4, 2717–2721

12. Roe, T., Reynolds, T.C., Yu, G., Brown, P.O. (1993) Integration of murine leukemia virus DNA depends on mitosis. *Embo J.* 12, 2099–2108

13. Murray, L.J., Young, J.C., Osborne, L.J., Luens, K.M., Scollay, R., Hill, B.L. (1999) Thrombopoietin, flt3, and kit ligands together suppress apoptosis of human mobilized CD34+ cells and recruit primitive CD34+ Thy-1+ cells into rapid division. *Exp. Hematol.* 27, 1019–1028

14. Murray, L., Luens, K., Tushinski, R., Jin, L., Burton, M., Chen, J., et al. (1999) Optimization of retroviral gene transduction of mobilized primitive hematopoietic progenitors by using thrombopoietin, Flt3, and Kit ligands and RetroNectin culture. *Hum. Gene Ther.* 10, 1743–1752

15. Barrette, S., Douglas, J., Orlic, D., Anderson, S.M., Seidel, N.E., Miller, A.D., et al. (2000) Superior transduction of mouse hematopoietic stem cells with 10A1 and VSV-G pseudotyped retrovirus vectors. *Mol. Ther.* 1, 330–338

16. Orlic, D., Girard, L.J., Anderson, S.M., Barrette, S., Broxmeyer, H.E., Bodine, D.M. (1999) Amphotropic retrovirus transduction of hematopoietic stem cells. *Ann. N. Y. Acad. Sci.* 872, 115–123

17. Orlic, D., Girard, L.J., Anderson, S.M., Pyle, L.C., Yoder, M.C., Broxmeyer, H.E., et al. (1998) Identification of human and mouse hematopoietic stem cell populations expressing high levels of mRNA encoding retrovirus receptors. *Blood* 91, 3247–3254

18. Orlic, D., Girard, L.J., Jordan, C.T., Anderson, S.M., Cline, A.P., Bodine, D.M. (1996) The level of mRNA encoding the amphotropic retrovirus receptor in mouse and human hematopoietic stem cells is low and correlates with the efficiency of retrovirus transduction. *Proc. Natl. Acad. Sci. U S A* 93, 11097–11102

19. Sabatino, D.E., Do, B.Q., Pyle, L.C., Seidel, N.E., Girard, L.J., Spratt, S.K., et al. (1997) Amphotropic or gibbon ape leukemia virus retrovirus binding and transduction correlates with the level of receptor mRNA in human hematopoietic cell lines. *Blood Cells Mol. Dis.* 23, 422–433

20. Zielske, S.P., Gerson, S.L. (2003) Cytokines, including stem cell factor alone, enhance lentiviral transduction in nondividing human LTCIC and NOD/SCID repopulating cells. *Mol. Ther.* 7, 325–333

21. Dorrell, C., Gan, O.I., Pereira, D.S., Hawley, R.G., Dick, J.E. (2000) Expansion of human cord blood CD34(+)CD38(-) cells in ex vivo culture during retroviral transduction without a corresponding increase in SCID repopulating cell (SRC) frequency: dissociation of SRC phenotype and function. *Blood* 95, 102–110

22. Gothot, A., van der Loo, J.C., Clapp, D.W., Srour, E.F. (1998) Cell cycle-related changes in repopulating capacity of human mobilized peripheral blood CD34(+) cells in non-obese diabetic/severe combined immune-deficient mice. *Blood* 92, 2641–2649

23. Mazurier, F., Gan, O.I., McKenzie, J.L., Doedens, M., Dick, J.E. (2004) Lentivector-mediated clonal tracking reveals intrinsic heterogeneity in the human hematopoietic stem cell compartment and culture-induced stem cell impairment. *Blood* 103, 545–552

24. Tisdale, J.F., Hanazono, Y., Sellers, S.E., Agricola, B.A., Metzger, M.E., Donahue, R.E., et al. (1998) Ex vivo expansion of genetically marked rhesus peripheral blood progenitor cells results in diminished long-term repopulating ability. *Blood* 92, 1131–1141

25. Kurre, P., Morris, J., Miller, A.D., Kiem, H.P. (2001) Envelope fusion protein binding studies in an inducible model of retrovirus receptor expression and in CD34(+) cells emphasize limited transduction at low receptor levels. *Gene Ther.* 8, 593–599

26. Hennemann, B., Conneally, E., Pawliuk, R., Leboulch, P., Rose-John, S., Reid, D., et al. (1999) Optimization of retroviral-mediated gene transfer to human NOD/SCID mouse repopulating cord blood cells through a systematic analysis of protocol variables. *Exp. Hematol.* 27, 817–825

27. Budak-Alpdogan, T., Przybylowski, M., Gonen, M., Sadelain, M., Bertino, J., Riviere, I. (2006) Functional assessment of the engraftment potential of gammaretrovirus-modified CD34+ cells, using a short serum-free transduction protocol. *Hum. Gene Ther.* 17, 780–794

28. Demaison, C., Brouns, G., Blundell, M.P., Goldman, J.P., Levinsky, R.J., Grez, M., et al. (2000) A defined window for efficient gene marking of severe combined immunodeficient-repopulating cells using a gibbon ape leukemia virus-pseudotyped retroviral vector. *Hum. Gene Ther.* 11, 91–100

29. Deola, S., Scaramuzza, S., Birolo, R.S., Carballido-Perrig, N., Ficara, F., Mocchetti, C., et al. (2004) Mobilized blood CD34+ cells transduced and selected with a clinically applicable protocol reconstitute lymphopoiesis in SCID-Hu mice. *Hum. Gene Ther.* 15, 305–311

30. Gaspar, H.B., Parsley, K.L., Howe, S., King, D., Gilmour, K.C., Sinclair, J., et al. (2004) Gene therapy of X-linked severe combined immunodeficiency by use of a pseudotyped gammaretroviral vector. *Lancet* 364, 2181–2187

31. Ott, M.G., Merget-Millitzer, H., Ottmann, O.G., Martin, H., Bruggenolte, N., Bialek, H., et al. (2002) Mobilization and transduction of CD34(+) peripheral blood stem cells in patients with X-linked chronic granulomatous disease. *J. Hematother. Stem Cell Res.* 11, 683–694

32. Relander, T., Karlsson, S., Richter, J. (2002) Oncoretroviral gene transfer to NOD/SCID repopulating cells using three different viral envelopes. *J. Gene Med.* 4, 122–132

33. Trarbach, T., Greifenberg, S., Bardenheuer, W., Elmaagacli, A., Hirche, H., Flasshove, M., et al. (2000) Optimized retroviral transduction protocol for human progenitor cells utilizing fibronectin fragments. *Cytotherapy* 2, 429–438

34. van der Loo, J.C., Liu, B.L., Goldman, A.I., Buckley, S.M., Chrudimsky, K.S. (2002) Optimization of gene transfer into primitive human hematopoietic cells of granulocyte-colony stimulating factor-mobilized peripheral blood using low-dose cytokines and comparison of a gibbon ape leukemia virus versus an RD114-pseudotyped retroviral vector. *Hum. Gene Ther.* 13, 1317–1330

35. Hanenberg, H., Hashino, K., Konishi, H., Hock, R.A., Kato, I., Williams, D.A. (1997) Optimization of fibronectin-assisted retroviral gene transfer into human CD34+ hematopoietic cells. *Hum. Gene Ther.* 8, 2193–2206

36. Moritz, T., Dutt, P., Xiao, X., Carstanjen, D., Vik, T., Hanenberg, H., et al. (1996) Fibronectin improves transduction of reconstituting hematopoietic stem cells by retroviral vectors: evidence of direct viral binding to chymotryptic carboxy-terminal fragments. *Blood* 88, 855–862

37. Moritz, T., Patel, V.P., Williams, D.A. (1994) Bone marrow extracellular matrix molecules improve gene transfer into human hematopoietic cells via retroviral vectors. *J. Clin. Invest.* 93, 1451–1457

38. Pollok, K.E., Williams, D.A. (1999) Facilitation of retrovirus-mediated gene transfer into hematopoietic stem and progenitor cells and peripheral blood T-lymphocytes utilizing recombinant fibronectin fragments. *Curr. Opin. Mol. Ther.* 1, 595–604

39. Davis, H.E., Morgan, J.R., Yarmush, M.L. (2002) Polybrene increases retrovirus gene transfer efficiency by enhancing receptor-independent virus adsorption on target cell membranes. *Biophys. Chem.* 97, 159–172

40. Davis, H.E., Rosinski, M., Morgan, J.R., Yarmush, M.L. (2004) Charged polymers modulate retrovirus transduction via membrane charge neutralization and virus aggregation. *Biophys. J.* 86, 1234–1242

41. Kwon, Y.J., Hung, G., Anderson, W.F., Peng, C.A., Yu, H. (2003) Determination of infectious retrovirus concentration from colony-forming assay with quantitative analysis. *J. Virol.* 77, 5712–5720

42. Kwon, Y.J., Peng, C.A. (2002) Transduction rate constant as more reliable index quantifying efficiency of retroviral gene delivery. *Biotechnol. Bioeng.* 77, 668–677

43. Persons, D.A., Mehaffey, M.G., Kaleko, M., Nienhuis, A.W., Vanin, E.F. (1998) An improved method for generating retroviral producer clones for vectors lacking a selectable marker gene. *Blood Cells Mol. Dis.* 24, 167–182

44. Bahnson, A.B., Dunigan, J.T., Baysal, B.E., Mohney, T., Atchison, R.W., Nimgaonkar, M.T., et al. (1995) Centrifugal enhancement of retroviral mediated gene transfer. *J. Virol. Methods* 54, 131–143

45. Campain, J.A., Terrell, K.L., Tomczak, J.A., Shpall, E.J., Hami, L.S., Harrison, G.S. (1997) Comparison of retroviral-mediated gene transfer into cultured human CD34+ hematopoietic progenitor cells derived from peripheral blood, bone marrow, and fetal umbilical cord blood. *Biol. Blood Marrow Transplant.* 3, 273–281

46. Kuhlcke, K., Fehse, B., Schilz, A., Loges, S., Lindemann, C., Ayuk, F., et al. (2002) Highly efficient retroviral gene transfer based on centrifugation-mediated vector preloading of tissue culture vessels. *Mol. Ther.* 5, 473–478

47. Movassagh, M., Desmyter, C., Baillou, C., Chapel-Fernandes, S., Guigon, M., Klatzmann, D., et al. (1998) High-level gene transfer to cord blood progenitors using gibbon ape leukemia virus pseudotype retroviral vectors and an improved clinically applicable protocol. *Hum. Gene Ther.* 9, 225–234

48. Sanyal, A., Schuening, F.G. (1999) Increased gene transfer into human cord blood cells by centrifugation-enhanced transduction in fibronectin fragment-coated tubes. *Hum. Gene Ther.* 10, 2859–2868

49. Takiyama, N., Mohney, T., Swaney, W., Bahnson, A.B., Rice, E., Beeler, M., et al. (1998) Comparison of methods for retroviral mediated transfer of glucocerebrosidase gene to CD34+ hematopoietic progenitor cells. *Eur. J. Haematol.* 61, 1–6

50. Zielske, S.P., Gerson, S.L. (2002) Lentiviral transduction of P140K MGMT into human CD34(+) hematopoietic progenitors at low multiplicity of infection confers significant resistance to BG/BCNU and allows selection in vitro. *Mol. Ther.* 5, 381–387

51. Chono, H., Yoshioka, H., Ueno, M., Kato, I. (2001) Removal of inhibitory substances with recombinant fibronectin-CH-296 plates enhances the retroviral transduction efficiency of CD34(+)CD38(-) bone marrow cells. *J. Biochem.* 130, 331–334

52. Kustikova, O.S., Wahlers, A., Kuhlcke, K., Stahle, B., Zander, A.R., Baum, C., et al. (2003) Dose finding with retroviral vectors: correlation of retroviral vector copy numbers in single cells with gene transfer efficiency in a cell population. *Blood* 102, 3934–3937

53. Suzuki, T., Shen, H., Akagi, K., Morse, H.C., Malley, J.D., Naiman, D.Q., et al. (2002) New genes involved in cancer identified by retroviral tagging. *Nat. Genet.* 32, 166–174

54. Przybylowski, M., Hakakha, A., Stefanski, J., Hodges, J., Sadelain, M., Riviere, I. (2006) Production scale-up and validation of packaging cell clearance of clinical-grade retroviral vector stocks produced in Cell Factories. *Gene Ther.* 13, 95–100

55. Reeves, L., Cornetta, K. (2000) Clinical retroviral vector production: step filtration using clinically approved filters improves titers. *Gene Ther.* 7, 1993–1998

56. Reeves, L., Smucker, P., Cornetta, K. (2000) Packaging cell line characteristics and optimizing retroviral vector titer: the National Gene Vector Laboratory experience. *Hum. Gene Ther.* 11, 2093–2103

57. Miller, A.D., Garcia, J.V., von Suhr, N., Lynch, C.M., Wilson, C., Eiden, M.V. (1991) Construction and properties of retrovirus packaging cells based on gibbon ape leukemia virus. *J. Virol.* 65, 2220–2224

58. Yu, S.S., Kim, J.M., Kim, S. (2000) The 17 nucleotides downstream from the env gene stop codon are important for murine leukemia virus packaging. *J. Virol.* 74, 8775–8780

59. Ory, D.S., Neugeboren, B.A., Mulligan, R.C. (1996) A stable human-derived packaging cell line for production of high titer retrovirus/vesicular stomatitis virus G pseudotypes. *Proc. Natl. Acad. Sci. U S A* 93, 11400–11406

60. Riviere, I., Sadelain, M.W.J. (1997) Methods for the construction of retroviral vectors and the generation of high titer producers. In Robbins, P. (ed.) *Methods in Molecular Medicine.* Totowa: Humana, pp. 59–78

61. Petzer, A.L., Hogge, D.E., Landsdorp, P.M., Reid, D.S., Eaves, C.J. (1996) Self-renewal of primitive human hematopoietic cells (long-term-culture-initiating cells) in vitro and their expansion in defined medium. *Proc. Natl. Acad. Sci. U S A* 93, 1470–1474

62. Schnell, S., Young, J.W., Houghton, A.N., Sadelain, M. (2000) Retrovirally transduced mouse dendritic cells require CD4+ T cell help to elicit antitumor immunity: implications for the clinical use of dendritic cells. *J. Immunol.* 164, 1243–1250

63. Tuschong, L., Soenen, S.L., Blaese, R.M., Candotti, F., Muul, L.M. (2002) Immune response to fetal calf serum by two adenosine deaminase-deficient patients after T cell gene therapy. *Hum. Gene Ther.* 13, 1605–1610

64. Seppen, J., Kimmel, R.J., Osborne, W.R. (1997) Serum-free production, concentration and purification of recombinant retroviruses. *Biotechniques* 23, 788–790

65. Kluge, K.A., Bonifacino, A.C., Sellers, S., Agricola, B.A., Donahue, R.E., Dunbar, C.E. (2002) Retroviral transduction and engraftment ability of primate hematopoietic progenitor and stem cells transduced under serum-free versus serum-containing conditions. *Mol. Ther.* 5, 316–322

66. Schilz, A.J., Kuhlcke, K., Fauser, A.A., Eckert, H.G. (2001) Optimization of retroviral vector generation for clinical application. *J. Gene Med.* 3, 427–436

67. Coroadinha, A.S., Silva, A.C., Pires, E., Coelho, A., Alves, P.M., Carrondo, M.J. (2006) Effect of osmotic pressure on the production of retroviral vectors: Enhancement in vector stability. *Biotechnol. Bioeng.* 94, 322–329

68. Pages, J.C., Loux, N., Farge, D., Briand, P., Weber, A. (1995) Activation of Moloney murine leukemia virus LTR enhances the titer of recombinant retrovirus in psi CRIP packaging cells. *Gene Ther.* 2, 547–551

69. Forestell, S.P., Bohnlein, E., Rigg, R.J. (1995) Retroviral end-point titer is not predictive of gene transfer efficiency: implications for vector production. *Gene Ther.* 2, 723–730

70. Kotani, H., Newton, P.B., 3rd, Zhang, S., Chiang, Y.L., Otto, E., Weaver, L., et al. (1994) Improved methods of retroviral vector transduction and production for gene therapy. *Hum. Gene Ther.* 5, 19–28

71. Eckert, H.G., Kuhlcke, K., Schilz, A.J., Lindemann, C., Basara, N., Fauser, A.A., et al. (2000) Clinical scale production of an improved retroviral vector expressing the human multidrug resistance 1 gene (MDR1). *Bone Marrow Transplant.* 25(2), S114–117

72. Carmo, M., Faria, T.Q., Falk, H., Coroadinha, A.S., Teixeira, M., Merten, O.W., et al. (2006) Relationship between retroviral vector membrane and vector stability. *J. Gen. Virol.* 87, 1349–1356

73. Wikstrom, K., Blomberg, P., Islam, K.B. (2004) Clinical grade vector production: analysis of yield, stability, and storage of gmp-produced retroviral vectors for gene therapy. *Biotechnol. Prog.* 20, 1198–1203

74. Carmo, M., Peixoto, C., Coroadinha, A.S., Alves, P.M., Cruz, P.E., Carrondo, M.J. (2004) Quantitation of MLV-based retroviral vectors using real-time RT-PCR. *J. Virol. Methods* 119, 115–119

75. Sanburn, N., Cornetta, K. (1999) Rapid titer determination using quantitative real-time PCR. *Gene Ther.* 6, 1340–1345

76. Scott-Taylor, T.H., Gallardo, H.F., Gansbacher, B., Sadelain, M. (1998) Adenovirus facilitated infection of human cells with ecotropic retrovirus. *Gene Ther.* 5, 621–629

77. Abe, T., Ito, M., Okamoto, Y., Kim, H.J., Takaue, Y., Yasutomo, K., et al. (1997) Transduction of retrovirus-mediated NeoR gene into CD34+ cells purified from granulocyte colony-stimulating factor (G-CSF)-mobilized infant and cord blood. *Exp. Hematol.* 25, 966–971

78. Lu, L., Xiao, M., Clapp, D.W., Li, Z.H., Broxmeyer, H.E. (1993) High efficiency retroviral mediated gene transduction into single isolated immature and replatable CD34(3+) hematopoietic stem/progenitor cells from human umbilical cord blood. *J. Exp. Med.* 178, 2089–2096

79. Pollok, K.E., van Der Loo, J.C., Cooper, R.J., Hartwell, J.R., Miles, K.R., Breese, R., et al. (2001) Differential transduction efficiency of SCID-repopulating cells derived

from umbilical cord blood and granulocyte colony-stimulating factor-mobilized peripheral blood. *Hum. Gene Ther.* 12, 2095–2108

80. Carow, C.E., Hangoc, G., Broxmeyer, H.E. (1993) Human multipotential progenitor cells (CFU-GEMM) have extensive replating capacity for secondary CFU-GEMM: an effect enhanced by cord blood plasma. *Blood* 81, 942–949

81. Lu, L., Xiao, M., Grigsby, S., Wang, W.X., Wu, B., Shen, R.N., et al. (1993) Comparative effects of suppressive cytokines on isolated single CD34(3+) stem/progenitor cells from human bone marrow and umbilical cord blood plated with and without serum. *Exp. Hematol.* 21, 1442–1446

82. Lu, L., Xiao, M., Shen, R.N., Grigsby, S., Broxmeyer, H.E. (1993) Enrichment, characterization, and responsiveness of single primitive CD34 human umbilical cord blood hematopoietic progenitors with high proliferative and replating potential. *Blood* 81, 41–48

83. Noort, W.A., Wilpshaar, J., Hertogh, C.D., Rad, M., Lurvink, E.G., van Luxemburg-Heijs, S.A., et al. (2001) Similar myeloid recovery despite superior overall engraftment in NOD/SCID mice after transplantation of human CD34(+) cells from umbilical cord blood as compared to adult sources. *Bone Marrow Transplant.* 28, 163–171

84. Wilpshaar, J., Bhatia, M., Kanhai, H.H., Breese, R., Heilman, D.K., Johnson, C.S., et al. (2002) Engraftment potential of human fetal hematopoietic cells in NOD/SCID mice is not restricted to mitotically quiescent cells. *Blood* 100, 120–127

85. Wilpshaar, J., Falkenburg, J.H., Tong, X., Noort, W.A., Breese, R., Heilman, D., et al. (2000) Similar repopulating capacity of mitotically active and resting umbilical cord blood CD34(+) cells in NOD/SCID mice. *Blood* 96, 2100–2107

86. Veena, P., Traycoff, C.M., Williams, D.A., McMahel, J., Rice, S., Cornetta, K., et al. (1998) Delayed targeting of cytokine-nonresponsive human bone marrow CD34(+) cells with retrovirus-mediated gene transfer enhances transduction efficiency and long-term expression of transduced genes. *Blood* 91, 3693–3701

87. Bregni, M., Di Nicola, M., Siena, S., Belli, N., Milanesi, M., Shammah, S., et al. (1998) Mobilized peripheral blood CD34+ cells express more amphotropic retrovirus receptor than bone marrow CD34+ cells. *Haematologica* 83, 204–208

88. Dunbar, C.E., Seidel, N.E., Doren, S., Sellers, S., Cline, A.P., Metzger, M.E., et al.

(1996) Improved retroviral gene transfer into murine and Rhesus peripheral blood or bone marrow repopulating cells primed in vivo with stem cell factor and granulocyte colony-stimulating factor. *Proc. Natl. Acad. Sci. U S A* 93, 11871–11876

89. Hematti, P., Sellers, S.E., Agricola, B.A., Metzger, M.E., Donahue, R.E., Dunbar, C.E. (2003) Retroviral transduction efficiency of G-CSF+SCF-mobilized peripheral blood CD34+ cells is superior to G-CSF or G-CSF+Flt3-L-mobilized cells in nonhuman primates. *Blood* 101, 2199–2205

90. Hematti, P., Tuchman, S., Larochelle, A., Metzger, M.E., Donahue, R.E., Tisdale, J.F. (2004) Comparison of retroviral transduction efficiency in CD34+ cells derived from bone marrow versus G-CSF-mobilized or G-CSF plus stem cell factor-mobilized peripheral blood in nonhuman primates. *Stem Cells* 22, 1062–1069

91. Thomasson, B., Peterson, L., Thompson, J., Goerner, M., Kiem, H.P. (2003) Direct comparison of steady-state marrow, primed marrow, and mobilized peripheral blood for transduction of hematopoietic stem cells in dogs. *Hum. Gene Ther.* 14, 1683–1686

92. Dunbar, C.E., Takatoku, M., Donahue, R.E. (2001) The impact of ex vivo cytokine stimulation on engraftment of primitive hematopoietic cells in a non-human primate model. *Ann. N. Y. Acad. Sci.* 938, 236–244

94. Sellers, S.E., Tisdale, J.F., Agricola, B.A., Donahue, R.E., Dunbar, C.E. (2004) The presence of the carboxy-terminal fragment of fibronectin allows maintenance of nonhuman primate long-term hematopoietic repopulating cells during extended ex vivo culture and transduction. *Exp. Hematol.* 32, 163–170

95. Koizumi, K., Nishio, M., Endo, T., Takashima, H., Haseyama, Y., Fujimoto, K., et al. (2000) Large scale purification of human blood CD34+ cells from cryopreserved peripheral blood stem cells, using a nylon-fiber syringe system and immunomagnetic microspheres. *Bone Marrow Transplant.* 26, 787–793

96. McNiece, I.K., Stoney, G.B., Kern, B.P., Briddell, R.A. (1998) CD34+ cell selection from frozen cord blood products using the Isolex 300i and CliniMACS CD34 selection devices. *J. Hematother.* 7, 457–461

97. Rubinstein, P., Dobrila, L., Rosenfield, R.E., Adamson, J.W., Migliaccio, G., Migliaccio, A.R., et al. (1995) Processing and cryopreservation of placental/umbilical cord blood for unrelated bone marrow reconstitution. *Proc. Natl. Acad. Sci. U S A* 92, 10119–10122

98. Perotti, C.G., Del Fante, C., Viarengo, G., Papa, P., Rocchi, L., Bergamaschi, P., et al. (2004) A new automated cell washer device for thawed cord blood units. *Transfusion* 44, 900–906

99. Martinson, J.A., Loudovaris, M., Smith, S.L., Bender, J.G., Vachula, M., van Epps, D.E., et al. (1997) Ex vivo expansion of frozen/thawed CD34+ cells isolated from frozen human apheresis products. *J. Hematother.* 6, 69–75

# Chapter 5

# Short-Term Culture of Human CD34+ Cells for Lentiviral Gene Transfer

## Francesca Santoni de Sio and Luigi Naldini

## Summary

Haematopoietic Stem Cells (HSCs) are attractive targets for the gene therapy. Upon ex vivo gene transfer and transplant, they may generate a progeny of gene-corrected cells potentially for a lifespan. The viral vectors most often used for HSC gene transfer are gamma-retroviral vectors (RVs) and HIV-derived lentiviral vectors (LVs). LVs have been proposed as improved tools for this task because they are able to transduce non-proliferating cells, while RVs are not. This implies that HSCs, which are mainly quiescent cells, need to be induced to proliferate in order to be transduced by RVs, whereas a prolonged stimulation is not needed for transduction by LVs. A short in vitro manipulation should reduce the risk of altering the characteristic biological properties of HSCs.

We describe here methods for short-term ex vivo culture and gene transfer into human HSCs. Cord blood-derived HSC gene transfer can be tuned to limit the average level of vector integration or instead to maximize the frequency of transduction and extent of transgene expression, according to the absence or presence of cytokines. Mobilized peripheral blood-derived HSCs need cytokine stimulation to maintain viability and obtain adequate levels of gene transfer. Although HSCs can be transduced by LVs in short ex vivo culture, they display low permissiveness to the vector. Recently, we have demonstrated that this is because proteasome activity restricts LV gene transfer in HSCs. We developed and describe here strategies that effectively overcome this restriction. Finally, we also provide methods for the assessment of the transduction efficiency.

**Key words:** Human Haematopoietic Stem Cells, Lentiviral Vectors, Cytokine, Proteasome Inhibitor, Flow Cytometry, Real-Time Quantitative-PCR.

## 1. Introduction

Haematopoietic Stem Cells (HSCs) are attractive targets for the gene therapy of a variety of inherited and acquired genetic diseases. Because of self-renewing ability, multipotency and clonogenic

Christopher Baum (ed.), *Methods in Molecular Biology, Methods and Protocols, vol. 506*
© Humana Press, a part of Springer Science+Business Media, LLC 2009
DOI: 10.1007/978-1-59745-409-4_5

potential, they can reconstitute all haematopoietic lineages in a transplanted host. Upon ex vivo gene transfer, they may generate a progeny of gene-corrected cells potentially for a lifespan. To be efficacious, the genetic material must be inserted into the target cell chromatin in order to be maintained during HSC proliferation and to be propagated to the progeny upon HSC differentiation. Moreover, efficient gene transfer should occur in primitive HSCs, without inducing detrimental effects on their biological properties.

The viral vectors most often used for HSC transduction (abortive infection) are gamma-retroviral vectors (RVs). More recently, HIV-derived lentiviral vectors (LVs) have been proposed as improved tools for this task. Both RVs and LVs integrate into the genome of host cells, but, while RV integration is dependent on target cell mitosis, LVs transduce both proliferating and non-proliferating cells (1). Since HSCs are considered to be mostly quiescent cells, they require stimulation with cytokines that trigger them into the cell cycle, in order to be transduced by RVs. Consequently, all the protocols developed for RV-mediated HSC gene transfer involve prolonged ex vivo manipulation and/or induction of cell cycle entry and proliferation (2–6), conditions that in most cases were shown to decrease the frequency of HSCs in the culture, and compromise their long-term repopulating ability (6–9). Nevertheless, recent clinical trials have demonstrated that RVs can be successfully used for HSC-based SCID gene therapy (10–12). However, the in vivo selective growth advantage of gene-corrected cells, which enabled amplification of a small input of transduced stem/progenitor cells, was a key factor in the success of these trials.

To broaden significantly the scope of HSC-based gene therapy, preservation and transduction of the majority of HSCs harvested for a transplant would be required. LVs are good candidates to this aim. Indeed, early studies by several groups showed efficient transduction of Cord Blood-derived NOD/SCID mouse-Repopulating Cells, considered to be closely related to HSCs, after a short incubation with LV, in the absence of cytokine stimulation (13–15). Surprisingly, and in apparent contrast with the lack of LV requirement for target cell proliferation, we later found a significant enhancement in gene transfer in the presence of a combination of early-acting cytokines (16). More recently, we demonstrated that the short-term cytokine stimulation did not impair HSC repopulation capacity and that the transduction enhancement by cytokines was not dependent on cell cycle progression (17). Although HSCs can be transduced by LVs in short ex vivo culture, they display low permissiveness to the vector, requiring cytokine stimulation and high vector doses to reach high-frequency transduction (16). We have demonstrated that this is because proteasome activity restricts LV gene transfer in HSCs. We are currently investigating the molecular mechanism of this restriction, which could be either direct (i.e., by degrading the entering particle) or indirect (i.e., by degrading a cellular factor required by the vector). Although it is still unclear how proteasome activity limits LV transduction, we developed new strategies that effectively overcome this restriction *(17)*.

## 2. Materials

1. Iscove Modified Dulbecco's Medium (IMDM).
2. Foetal Bovine Serum (FBS).
3. StemSpan serum-free medium (StemCell Technologies).
4. Cytokines (Peprotech): recombinant human-Interleukin 6 (rh-IL6), rh-IL3, rh-Stem Cell Factor (rh-SCF), rh-fms-like tyrosine kinase 3 Ligand (rh-FLT3L), and rh-thrombopoietin (rh-TPO).
5. MG132 (Calbiochem, EMD Biosciences).
6. ExVivo 15 (Biowhittaker).
7. CH296 fibronectin fragment (Retronectin – Takara Biomedicals).
8. Non-tissue culture-treated dishes.
9. 7-Amino-Actinomycin D (7AAD).
10. Blood and Cell Culture DNA kit (Qiagen).
11. Tris–EDTA (TE): 10 mM Tris–HCl, 1 mM EDTA, pH 8.0.
12. TaqMan Universal Master Mix (4304437; Applied Biosystem).

## 3. Methods

The viral vectors most often used for HSC gene transfer are gamma-retroviral vectors (RVs) and HIV-derived lentiviral vectors (LVs). LVs have been proposed as improved tools for this task because they are able to transduce non-proliferating cells, while RVs are not *(1)*. This implies that HSCs, which are mainly quiescent cells, need to be induced to proliferate in order to be transduced by RVs, whereas a prolonged stimulation is not needed for transduction by LVs. A short in vitro manipulation should reduce the risk of altering the characteristic biological properties of HSCs.

The tropism of retroviruses and retrovirus-derived vectors is dictated by the envelope glycoproteins. Thanks to the modality of the viral assembling process and to the strategy used to generate these vectors, it is possible to exchange the wild-type envelope proteins with other envelope proteins (a process termed pseudotyping) and thus modify the tropism of the vector. Different envelopes have been shown to efficiently pseudotype LVs and target haematopoietic cells; among them Vesicular Stomatitis Virus-G glycoprotein (VSV-G) and the modified form of the feline endogenous retrovirus RD114 (RD/TR) envelopes are the more promising *(16, 18)*. Thus, the methods described later outline the short-term ex vivo culture of human (h) CD34+ cells for gene transfer by VSV-G pseudotyped LVs in the **Subheading 3.1** and the short-term ex vivo culture of hCD34 cells for gene transfer by RD/TR pseudotyped LVs in the **Subheading 3.2**. Finally, we also describe the methods for the assessment of the transduction efficiency in the **Subheading 3.3**.

### 3.1. Ex Vivo Culture of hCD34+ Cells for Gene Transfer by VSV-G Pseudotyped LVs

Extracellular stimuli have an important role in regulating HSC features. Thus, it is commonly believed that HSCs derived from different sources possess different characteristics. Indeed, several studies demonstrated different frequency of progenitors, cell cycle status, gene expression profiles and engraftment ability of HSCs derived from Cord Blood (CB), Bone Marrow (BM), or Mobilized Peripheral Blood (MPB) *(19–23)*. HSCs derived from different sources also display different susceptibilities to gene transfer, thus requiring adjustments in the experimental protocols in order to obtain efficient transduction. The protocols for gene transfer into CB- and MPB- derived CD34+ cells are described now.

#### 3.1.1. CB-Derived CD34+ Cells

CD34+ cells can be used both fresh, i.e., right after the isolation, and frozen. Freeze CD34+ cells in Iscove Modified Dulbecco's Medium (IMDM) 10% dimethylsulphoxide, 50% Foetal Bovine Serum (FBS), and store them in liquid nitrogen. Thaw CD34+ cells by rapidly incubating them at 37°C and then adding a 10× volume of IMDM 5% FBS drop by drop. We usually try to avoid exposure of CD34+ cells to serum components; however, serum is required for freezing and thawing the cells. Thus, as a compromise we decreased the amount of serum during this procedure with respect to standard methods (50% instead of 90% for freezing and 5% instead of 10–20% for thawing). Let the cells recover for 5–10 min. Then, centrifuge them at $200 \times g$ for 10 min and resuspend them in an appropriate volume of transduction medium.

As transduction medium, use StemSpan serum-free medium (StemCell Technologies), in order to avoid any induction towards differentiation possibly induced by the serum components. Transduction can be carried out either without any supplement or supplementing the medium with a combination of early-acting cytokines. Indeed, our studies indicated that, according to the absence or presence of cytokines, HSC gene transfer can be tuned to limit the average level of vector integration, and thus reduce the risk of insertional mutagenesis, or instead to maximize the frequency of transduction and extent of transgene expression *(16, 17)* (**Fig. 1**). Seed CD34+ cells at a concentration of 5 × $10^5$–1 × $10^6$ cells/mL in StemSpan (*see* **Note 1**) with or without the following combination of cytokines: 20 ng/mL recombinant human-Interleukin 6 (rh-IL6), 100 ng/mL rh-Stem Cell Factor (rh-SCF), 100 ng/mL rh-fms-like tyrosine kinase 3 Ligand (rh-FLT3L), and 20 ng/mL rh-thrombopoietin (rh-TPO) (all from Peprotech). Add to the medium $10^8$ HeLa Transducing Units (TU)/mL of VSV-pseudotyped LV (*see* **Note 2** and **3**). Incubate for 20–24 h and then wash cells (see later).

Although HSCs can be transduced by LVs in this short ex vivo culture, they display low permissiveness to the vector, requiring cytokine stimulation and high vector doses to reach high-frequency transduction. Recently, we demonstrated that this is because LVs are effectively restricted at a post-entry step by the activity of the proteasome and developed a protocol to overcome this restriction and obtain very high level of gene transfer *(17)*. This protocol is based on the use of the proteasome inhibitor MG132. Thus, when very high frequency of transduction and integrated vector copies *per* cell are needed, transduce CD34+ as described earlier in presence of the cytokine cocktail and add MG132 (Calbiochem, EMD Biosciences) to the transduction medium at concentrations that range between 0.5 and 1 μM (*see* **Note 4**) (**Fig. 1**).

After the transduction period (20–24 h) wash CD34+ cells to remove vector (and MG132) by resuspending them in 10x

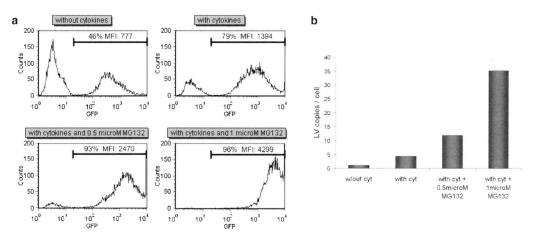

Fig. 1. LV-mediated gene transfer into CB-derived CD34+ cells. (**a**) Representative plots of flow cytometric analysis of CD34+ cells transduced by VSV-pseudotyped Green Fluorescent Protein (GFP)-expressing LV using different protocols, as indicated. Percentage of GFP positive cells and Mean Fluorescence Intensity (MFI) of the GFP positive cells are indicated for each condition. (**b**) Representative Real-time quantitative-PCR analysis of the same cells of panel A.

volume of IMDM medium and centrifuging them at $200 \times g$ for 10 min. Then, resuspend and maintain them in IMDM medium 10% FBS containing 20 ng/mL rh-IL6, 100 ng/mL rh-SCF, and 20 ng/mL rh-Interleukin 3 (rh-IL3) (all from Peprotech) for suspension culture (see **Note 5**) and/or plate for Colony-Forming Cell Assay and/or transplant into NOD/SCID mice. Analyse cells grown in suspension by Flow Cytometry or Real-Time Quantitative-PCR 2–3 weeks after transduction.

*3.1.2. MPB-Derived CD34+ Cells*

Like CB-derived CD34+ cells, MPB-derived cells can be used fresh or frozen. For freezing and thawing procedures see the preceding paragraph.

MPB-derived cells are less susceptible to gene transfer and more prone to apoptosis when cultured in vitro, as compared to CB-derived cells. Thus, a prolonged time in culture and more anti-apoptotic stimuli are needed, with respect to the protocol used for CB-derived cells, in order to increase the gene transfer level and reduce the mortality rate. Thus, incubate CD34+ cells for 24 h at a concentration of $10^6$ cells/mL in ExVivo 15 serum-free clinical grade medium (Biowhittaker) containing 300 ng/mL rhSCF, 100 ng/mL rhTPO, 300 ng/mL rhFlt3l, 60 ng/mL rhIL3 on non-tissue culture-treated dishes coated with CH296 fibronectin fragments (Retronectin; Takara Biomedicals) following manufacturer instructions (see **Note 6**). After this pre-stimulation period, expose the cells to $10^8$ TU/mL of VSV-pseudotyped LV for 12 h.

Also in this case, the proteasome inhibitor can be used in order to further increase the level of transduction. When using the proteasome inhibitor MG132, add it to the transduction media at concentrations that range between 0.5 and 1 µM only during the 12 h of exposure to the vector (see **Note 4**). After the transduction period, wash cells by resuspending them in 10x volume of IMDM medium and centrifuging them at $200 \times g$ for 10 min. Then, resuspend and maintain them in IMDM medium 10% FBS containing 20 ng/mL rh-IL6, 100 ng/mL rh-SCF, and 20 ng/mL rh-Interleukin 3 (rh-IL3) (all from Peprotech) for suspension culture (see **Note 5**) and/or plate for Colony-Forming Cell Assay and/or transplant into NOD/SCID mice. Analyse cells grown in suspension by Flow Cytometry or Real-Time Quantitative-PCR 2–3 weeks after transduction.

*3.2. Ex Vivo Culture of hCD34+ Cells for Gene Transfer by RD/TR Pseudotyped LVs*

Higher levels of transduction are obtained in presence of early-acting cytokines also when using RD/TR pseudotyped LVs. Moreover, these vectors need to be used on human Fibronectin Fragment (CH296 – Takara BIO)-coated dishes. Indeed, CH-296 fragments have been shown to help co-localizing haematopoietic stem/progenitor cells and vectors pseudotyped with specific envelopes and consequently increase transduction (24).

Thus, coat non-tissue culture-treated dishes with CH-296 following manufacturer instructions. Use RD/TR pseudotyped LVs at a concentration of $3 \times 10^7$–$1 \times 10^8$ TU/mL. Preload half dose of vector on CH-296-coated dishes by using a volume of Dulbecco's Phosphate-Buffered Saline (PBS) that allows covering all the dish area. After 2 h, seed CD34+ cells on the pre-loaded dishes at a concentration of $5 \times 10^5$–$1 \times 10^6$ cells/mL (*see* **Note 1**) in StemSpan (Stem Cell Technologies) containing 20 ng/mL rh-IL6, 100 ng/mL rh-SCF, 100 ng/mL rh-FLT3L, and 20 ng/mL rh-TPO (all from Peprotech) and expose them to the rest of the vector for 20–24 h.

In order to further increase the transduction efficiency, add MG132 at concentrations that range between 0.5 and 1 µM during the transduction period (*see* **Note 4**).

After the transduction period, wash cells to remove vector (and MG132) by resuspending them in 10x volume of IMDM medium and centrifuging them at $200 \times g$ for 10 min. Resuspend and maintain them in IMDM medium 10% FBS containing 20 ng/mL rh-IL6, 100 ng/ml rh-SCF, and 20 ng/mL rh-Interleukin 3 (rh-IL3) (all from Peprotech) for suspension culture (*see* **Note 5**) and/or plate for Colony-Forming Cell Assay and/or transplant into NOD/SCID mice. Analyse cells grown in suspension by Flow Cytometry or Real-Time Quantitative-PCR 2–3 weeks after transduction.

### 3.3. Assessment of Transduction Efficiency in In Vitro Cultures

Depending on the protein encoded by the vector used the transduction level can be assessed either by Flow Cytometry or by Real-Time Quantitative-PCR (Q-PCR) analysis. If the vector encodes for a reporter fluorescent protein or a protein that can be detected by staining with antibodies, Flow Cytometry is the optimal choice. Indeed, Flow Cytometry allows an analysis at single cell level and is the less time-consuming option. If the vector encodes for a protein that cannot be detected, it is possible to analyse the cells at a molecular level performing Q-PCR.

### 3.3.1. Analysis by Flow Cytometry

Perform the analysis on cells grown for 2 weeks. This time is required to reach steady-state protein expression and rule out pseudotransduction, which is detection of the protein encoded by the vector associated with the virus particles during vector production and transferred to the target cells during infection (the protein is diluted out of the cells as they divide).

Harvest $2$–$5 \times 10^5$ cells, wash in PBS containing 2% FBS, and centrifuge at $200 \times g$ for 10 min. If the vector-encoded protein is fluorescent, resuspend cells in 0.3–0.5 mL of PBS 2% FBS containing 10 µg/mL 7-Amino-Actinomycin D (7AAD; Sigma-Aldrich). If the vector-encoded protein requires an immunostaining to be detected, perform staining following antibody manufacturer instructions. After staining, centrifuge at $200 \times g$

for 10 min and resuspend cells in 0.3–0.5 mL of PBS 2% FBS containing 10 µg/mL 7AAD. 7AAD is used as staining for non-viable cells. Indeed, live cell membrane is not permeable to this compound, while dying/died cell membrane, which is damaged, allows entering of 7AAD. When analysing cells at the cytometer, use untransduced cells as negative control to set parameters and gates. Analyse only viable, 7AAD-negative cells for transgene expression (*see* **Note 7**). Score at least 10,000–20,000 events to have reliable statistics.

*3.3.2. Analysis by Real-Time Quantitative-PCR*

When performing this assay cells must be grown for at least 2 weeks in order to exclude non-integrated vector forms from the analysis (non-integrated forms are diluted out of the cells as they divide).

Perform genomic DNA extraction by using the Blood and Cell Culture DNA kit (Qiagen) (*see* **Note 8**). Follow manufacturer instructions, wash the recovered genomic DNA in 70% ethanol, by centrifuging at 3,300 × *g* for 30 min and, after removal of supernatant, air dry the pellet for 15 min. Resuspend in 100–200 µL TE (10 mM Tris–HCl, 1 mM EDTA, pH 8.0) and dissolve it on a shaker at 55°C for 2 h. Quantify genomic DNA by spectrophotometer analysis and run it on a 0.8% agarose gel to check for possible RNA contaminations and degradation of the sample (*see* **Note 9**). Store genomic DNA at –20°C.

Run each sample in triplicate in a total volume of 25 µL/reaction, containing 12.5 µL TaqMan Universal Master Mix (4304437; Applied Biosystem), 100 ng of sample DNA and primers and probe. To detect vector sequence it is possible to design primers and probe complementary to a specific sequence contained in the vector of interest, such as the transgene. Alternatively, it is possible to use primers and probe complementary to a sequence contained in all LV vectors, such as the encapsidation signal sequence. Useful sequences and final concentrations of primers and probe complementary to the encapsidation signal sequence are the following: forward primer: 5′ – TGAAAGCGAAAGGGAAAC CA – 3′ (750 nM); reverse primer: 5′ – CCGTGCGCGCTTCAG – 3′ (200 nM); probe: 5′ – VIC – AGCTCTCTCGACGCAG-GACTCGG – TAMRA – 3′ (200 nM).

Normalization for the DNA content is needed to determine the LV copy number *per* cell. Assessment of the DNA content by Q-PCR is usually more reliable than assessment by spectrophotometer. To determine human DNA content by Q-PCR, run samples as described earlier using primers and probe complementary to the human Telomerase Reverse Transcriptase (hTERT) sequence *(25)*. Sequences and final concentrations of oligonucleotides and probe are the following: forward primer: 5′ – GGCACACGTGGCTTTTCG – 3′(200 nM); reverse primer: 5′ – GGTGAACCTCGTAAGTTTATGCAA – 3′ (600 nM);

probe: 5′ – 6-FAM – TCAGGACGTCGAGTGGACACGGTG – TAMRA –3′ (200 nM).

Perform Q-PCR by using the following thermal cycling conditions: one cycle at 50°C for 2 min, one cycle at 95°C for 10 min, 40 cycles at 95°C for 15 s and 60°C for 1 min. For each sample, the software provides an amplification curve constructed by relating the fluorescence signal intensity (ΔRn) to the cycle number. Define Cycle threshold (Ct) as the cycle number at which the fluorescence signal is more than 10 SD of the mean background noise collected from the 3rd to the 15th cycle. Determine vector copy number by using as standard curve serial dilutions (200 ng, 100 ng, 50 ng, 25 ng, and 12,5 ng) of genomic DNA extracted from a human cell line clone stably containing a known number of LV integrations (copy number of the standard curve needs to be previously defined by Southern Blot analysis) (*see* **Note 10**). To determine vector copy number *per* cell, normalize LV copy numbers to the human TERT copy number. In order to avoid any possible variability due to the construction of two different standard curves, quantify human DNA content by using the same standard curve used for vector copy number assessment (*see* **Note 11**).

---

## 4. Notes

1. The cell concentration during the transduction period is critical for maintaining the viability of the cells themselves. Indeed, decreasing the number of cells *per* mL will not increase transduction efficiency (*see* **Note 2**) and will decrease cell recovery.

2. When the vector titer is calculated on a permissive cell line, refractory target cells may need to be transduced employing a high number of TU per mL. The fact that the expression TU/cell, which is equivalent to MOI (multiplicity of infection), is an arbitrary definition should be taken into account, because it does not consider the volume in which the transduction is performed, and therefore the particle concentration. Within this context, it is well acknowledged that vector concentration in the transduction medium is more important than the absolute number of particles available for each cell, since only a fraction of them comes into contact with the target. For this reason, it is more appropriate to express the vector dose as TU/mL rather than TU/cell (MOI).

3. When transducing low-permissive cells, such as HSCs, it is necessary to use a high-quality vector preparation. Infectivity is a reliable parameter to evaluate the quality of the vector

stock and can be defined as the transducing activity *per* unit of physical particles, where the first parameter is expressed as TU/mL, as obtained by end-point titration on a standard indicator cell line and the second as ng of HIV Gag p24/ mL, as determined by immunoenzymatic assay. A good vector preparation should range between $10^4$ and $10^5$ TU/ ng p24; using a vector stock with a lower infectivity may significantly decrease the cell viability and strongly reduce transduction efficiency, as the empty/defective particles will compete with infective particles for cell entry.

4. Proteasome inhibitor is toxic and can reduce cell recovery after the period of transduction. Concentrations higher that 1 μM strongly affect cell viability. Although the rate of mortality varies among cell donors, doses ranging between 0.5 and 1 μM should have a minor effect on the cell recovery. However, the use of a low infectivity vectors, and thus of a high amount of particles, in presence of MG132 can result in significant cell loss (*see* **Note 3**).

5. A medium supplemented with serum and cytokines is required to allow cells to recover from the stress induced by transduction. This is particularly relevant when using MG132 during transduction.

6. As mentioned in the text, one of the major issues in culturing and transducing MPB-derived CD34+ cells is their tendency to undergo apoptosis. Ex vivo 15 medium seems to maintain viability of the cells better than StemSpan. CH-296, and high cytokine concentrations have been used as anti-apoptotic stimuli. Another key factor in maintaining these cells is not only the cell concentration (*see* **Note 1**), but also the total number of cells seeded. Indeed, seeding less than $5 \times 10^5$ cells *per* experimental point may strongly reduce cell survival.

7. When performing flow cytometric analysis on primary cells it is important to use a staining for viable cells. Indeed, dead cells are auto-fluorescent and can be erroneously scored as positive cells if not excluded by the analysis.

8. The use of this specific kit is not strictly required. However, when using other methods for genomic DNA extraction it is necessary to verify that the recovered genomic DNA is not contaminated by chemicals that could inhibit the polymerase reaction.

9. The quality of the genomic DNA extracted is important for a reliable Q-PCR analysis. Indeed, degraded DNA or RNA contamination may lead to misleading results in the quantification of the human DNA and thus in the quantification of the vector copy number *per* cell.

10. The use of a standard curve based on serial dilutions of LV plasmid is not applicable when performing Q-PCR, since

the efficiency of amplification for the plasmid and the vector integrated into the host chromatin is different.

11. In order to have an appropriate quantification of the DNA content, the use of a cell line in the standard curve with a known ploidy (preferably euploid) for the gene of interest (in this case TERT) is needed.

## References

1. Vigna, E., and Naldini, L. (2000) Lentiviral vectors: excellent tools for experimental gene transfer and promising candidates for gene therapy. *J Gene Med* 2, 308–16

2. Larochelle, A., Vormoor, J., Hanenberg, H., Wang, J. C., Bhatia, M., Lapidot, T., Moritz, T., Murdoch, B., Xiao, X. L., Kato, I., Williams, D. A., and Dick, J. E. (1996) Identification of primitive human hematopoietic cells capable of repopulating NOD/SCID mouse bone marrow: implications for gene therapy. *Nat Med* 2, 1329–37

3. Conneally, E., Eaves, C. J., and Humphries, R. K. (1998) Efficient retroviral-mediated gene transfer to human cord blood stem cells with in vivo repopulating potential. *Blood* 91, 3487–93

4. Dao, M. A., Taylor, N., and Nolta, J. A. (1998) Reduction in levels of the cyclin-dependent kinase inhibitor p27(kip-1) coupled with transforming growth factor beta neutralization induces cell-cycle entry and increases retroviral transduction of primitive human hematopoietic cells. *Proc Natl Acad Sci U S A* 95, 13006–11

5. Barquinero, J., Segovia, J. C., Ramirez, M., Limon, A., Guenechea, G., Puig, T., Briones, J., Garcia, J., and Bueren, J. A. (2000) Efficient transduction of human hematopoietic repopulating cells generating stable engraftment of transgene-expressing cells in NOD/SCID mice. *Blood* 95, 3085–93

6. Guenechea, G., Gan, O. I., Dorrell, C., and Dick, J. E. (2001) Distinct classes of human stem cells that differ in proliferative and self-renewal potential. *Nat Immunol* 2, 75–82

7. Glimm, H., Oh, I. H., and Eaves, C. J. (2000) Human hematopoietic stem cells stimulated to proliferate in vitro lose engraftment potential during their S/G(2)/M transit and do not reenter G(0). *Blood* 96, 4185–93

8. Mazurier, F., Doedens, M., Gan, O. I., and Dick, J. E. (2003) Characterization of cord blood hematopoietic stem cells. *Ann N Y Acad Sci* 996, 67–71

9. Mazurier, F., Gan, O. I., McKenzie, J. L., Doedens, M., and Dick, J. E. (2004) Lentivector-mediated clonal tracking reveals intrinsic heterogeneity in the human hematopoietic stem cell compartment and culture-induced stem cell impairment. *Blood* 103, 545–52

10. Cavazzana-Calvo, M., Hacein-Bey, S., de Saint Basile, G., Gross, F., Yvon, E., Nusbaum, P., Selz, F., Hue, C., Certain, S., Casanova, J. L., Bousso, P., Deist, F. L., and Fischer, A. (2000) Gene therapy of human severe combined immunodeficiency (SCID)-X1 disease. *Science* 288, 669–72

11. Aiuti, A., Slavin, S., Aker, M., Ficara, F., Deola, S., Mortellaro, A., Morecki, S., Andolfi, G., Tabucchi, A., Carlucci, F., Marinello, E., Cattaneo, F., Vai, S., Servida, P., Miniero, R., Roncarolo, M. G., and Bordignon, C. (2002) Correction of ADA-SCID by stem cell gene therapy combined with nonmyeloablative conditioning. *Science* 296, 2410–3

12. Gaspar, H. B., Parsley, K. L., Howe, S., King, D., Gilmour, K. C., Sinclair, J., Brouns, G., Schmidt, M., Von Kalle, C., Barington, T., Jakobsen, M. A., Christensen, H. O., Al Ghonaium, A., White, H. N., Smith, J. L., Levinsky, R. J., Ali, R. R., Kinnon, C., and Thrasher, A. J. (2004) Gene therapy of X-linked severe combined immunodeficiency by use of a pseudotyped gammaretroviral vector. *Lancet* 364, 2181–7

13. Miyoshi, H., Smith, K. A., Mosier, D. E., Verma, I. M., and Torbett, B. E. (1999) Transduction of human CD34+ cells that mediate long-term engraftment of NOD/SCID mice by HIV vectors. *Science* 283, 682–6

14. Guenechea, G., Gan, O. I., Inamitsu, T., Dorrell, C., Pereira, D. S., Kelly, M., Naldini, L., and Dick, J. E. (2000) Transduction of human CD34 + CD38- bone marrow and cord blood-derived SCID-repopulating cells with third-generation lentiviral vectors. *Mol Ther* 1, 566–73

15. Uchida, N., Sutton, R. E., Friera, A. M., He, D., Reitsma, M. J., Chang, W. C., Veres, G.,

Scollay, R., and Weissman, I. L. (1998) HIV, but not murine leukemia virus, vectors mediate high efficiency gene transfer into freshly isolated G0/G1 human hematopoietic stem cells. *Proc Natl Acad Sci U S A* 95, 11939–44

16. Ailles, L., Schmidt, M., Santoni de Sio, F. R., Glimm, H., Cavalieri, S., Bruno, S., Piacibello, W., Von Kalle, C., and Naldini, L. (2002) Molecular evidence of lentiviral vector-mediated gene transfer into human self-renewing, multi-potent, long-term NOD/SCID repopulating hematopoietic cells. *Mol Ther* 6, 615–26

17. Santoni de Sio, F. R., Cascio, P., Zingale, A., Gasparini, M., and Naldini, L. (2006) Proteasome activity restricts lentiviral gene transfer into hematopoietic stem cells and is down-regulated by cytokines that enhance transduction. *Blood* 107, 4257–65

18. Sandrin, V., Boson, B., Salmon, P., Gay, W., Negre, D., Le Grand, R., Trono, D., and Cosset, F. L. (2002) Lentiviral vectors pseudotyped with a modified RD114 envelope glycoprotein show increased stability in sera and augmented transduction of primary lymphocytes and CD34+ cells derived from human and nonhuman primates. *Blood* 100, 823–32

19. Uchida, N., He, D., Friera, A. M., Reitsma, M., Sasaki, D., Chen, B., and Tsukamoto, A. (1997) The unexpected G0/G1 cell cycle status of mobilized hematopoietic stem cells from peripheral blood. *Blood* 89, 465–72

20. Steidl, U., Kronenwett, R., Rohr, U. P., Fenk, R., Kliszewski, S., Maercker, C., Neubert, P., Aivado, M., Koch, J., Modlich, O., Bojar, H., Gattermann, N., and Haas, R.

(2002) Gene expression profiling identifies significant differences between the molecular phenotypes of bone marrow-derived and circulating human CD34+ hematopoietic stem cells. *Blood* 99, 2037–44

21. Ueda, T., Yoshida, M., Yoshino, H., Kobayashi, K., Kawahata, M., Ebihara, Y., Ito, M., Asano, S., Nakahata, T., and Tsuji, K. (2001) Hematopoietic capability of CD34+ cord blood cells: a comparison with CD34+ adult bone marrow cells. *Int J Hematol* 73, 457–62

22. Ng, Y. Y., van Kessel, B., Lokhorst, H. M., Baert, M. R., van den Burg, C. M., Bloem, A. C., and Staal, F. J. (2004) Gene-expression profiling of CD34+ cells from various hematopoietic stem-cell sources reveals functional differences in stem-cell activity. *J Leukoc Biol* 75, 314–23

23. Theunissen, K., and Verfaillie, C. M. (2005) A multifactorial analysis of umbilical cord blood, adult bone marrow and mobilized peripheral blood progenitors using the improved ML-IC assay. *Exp Hematol* 33, 165–72

24. Hanenberg, H., Xiao, X. L., Dilloo, D., Hashino, K., Kato, I., and Williams, D. A. (1996) Colocalization of retrovirus and target cells on specific fibronectin fragments increases genetic transduction of mammalian cells. *Nat Med* 2, 876–82

25. Sozzi, G., Conte, D., Leon, M., Ciricione, R., Roz, L., Ratcliffe, C., Roz, E., Cirenei, N., Bellomi, M., Pelosi, G., Pierotti, M. A., and Pastorino, U. (2003) Quantification of free circulating DNA as a diagnostic marker in lung cancer. *J Clin Oncol* 21, 3902–8

# Chapter 6

# T Cell Culture for Gammaretroviral Transfer

## Sebastian Newrzela, Gunda Brandenburg, and Dorothee von Laer

## Summary

Gene transfer into mature T cells with gammaretroviral vectors requires prestimulation, as only mitotic cells are susceptible to integration of the gammaretroviral proviral genome. Costimulation via the CD3/TCR complex and a second costimulatory molecule, such as CD28 was found to better preserve functionality of the T lymphocytes during ex vivo expansion than stimulation with anti-CD3 alone. The protocols described here for prestimulation and transduction of human and murine T cells with gammaretroviral vectors were optimized for high-level gene transfer and maximum yield of functional T lymphocytes.

**Key words:** T cells, Immunotherapy, Retroviral vectors, Anti-CD3/anti-CD28 beads, Ex vivo T cell expansion.

## 1. Introduction

Activation and expansion of T lymphocytes for the treatment of cancer and infectious diseases require stimulation protocols that preserve T cell functionality. First strategies for T cell activation included the use of mitogenic agents like phytohemagglutinin (PHA) and concanavalin A (Con A) *(1–3)*. Proliferation and cytokine production of resting T cells are triggered by two major signals. The first signal is delivered via ligation of the CD3/TCR complex, naturally with the MHC-peptide complex. However, crosslinking anti-CD3 antibodies can replace the MHC-1 complex. The second signal is provided by interaction with costimulatory ligands on antigen-presenting cells (APCs) *(4)*. Several different

Christopher Baum (ed.), *Methods in Molecular Biology, Methods and Protocols, vol. 506*
© Humana Press, a part of Springer Science + Business Media, LLC 2009
DOI: 10.1007/978-1-59745-409-4_6

molecules like B7, LFA-3, and ICAM-1 *(5)* have been identified that can costimulate T cells. Two important receptors are CD28 and CTLA-4. While CTLA-4 is only expressed after T cell activation and has a negative regulatory function, CD28 is expressed on most T cells, triggers T cell activation, and is thus the ideal target for costimulation *(6)*.

There is a broad range of different applications for ex vivo expanded and genetically modified T cells in adoptive immunotherapy and genetically modified T lymphocytes have been used in several clinical trials for diseases such as ADA deficiency in children *(7)*, cancer *(8)*, and HIV infection *(9)*. Gene therapy approaches based on gammaretroviral gene transfer are highly dependent on optimal cell stimulation, as these vectors can only efficiently transduce dividing cells *(10)*. Thus, inappropriate stimulation of T lymphocytes will not allow efficient gene transfer and reduce the potential therapeutic benefit of a gene therapeutic strategy *(11)*.

In preceding studies by other groups and by the authors, costimulation with monoclonal antibodies (mAb) CD3 and mAb CD28-coated beads was found to support higher levels of gene transfer and a stronger expansion of T cells than stimulation with anti-CD3 alone *(12, 13)*.

On the other hand, it must be kept in mind that prolonged cultivation periods may lead to a loss of fitness and functionality of the expanded T cells. Telomere length e.g., was found to correlate with in vivo persistence of transferred T cells *(14)*.

Therapeutic efficacies of gene-modified T cells will critically depend on ex vivo expansion and gene transfer protocols that guarantee high transduction levels and large cell numbers, but furthermore preserve T cell function.

## 2. Materials

### 2.1. Isolation of Human Peripheral Blood Mononuclear Cells (PBMCs)

1. Fresh blood or buffy coats from healthy donors or patients.
2. Dulbecco's Phosphate-Buffered Saline (1×) (PBS) without calcium and magnesium (PAA Laboratories GmbH, Pasching, Austria).
3. Nylon 70- or 100-μm cell strainer (BD Falcon, San Jose, CA).
4. Ficoll lymphocyte separation medium (density 1.077) (PAA Laboratories).

### 2.2. Medium Supplements for Human T Cell Cultivation/Stimulation and Transduction

1. Complete medium: X-Vivo 15 with phenol red (Bio Whittaker/Cambrex, East Rutherford, NJ), supplemented with 2 mM l-Glutamine in 0.85% (w/v) sodium chloride solution (Bio Whittaker/Cambrex). Medium and supplements are stored at 4°C.

2. Human Dynabeads CD3/CD28 T cell expander (Invitrogen, Carlsbad, CA) and Magnet Particle Concentrator MPC-1 (5–50 mL) (Invitrogen).

3. Sterile FACS-tube with cap (BD Falcon, San Jose, CA).

4. Interleukin-2 (IL-2) (final concentration 100–200 U/mL) (Proleukin, Chiron GmbH, München, Germany) is added to the T-cell culture from a $10^4$ U/mL IL-2 stock solution (stored at –20°C). The desired amount of IL-2 stock solution is defrosted and can be stored at 4°C for at least 3–4 weeks.

5. Human serum (pooled human AB serum (Bio Whittaker/Cambrex)), heat-inactivated at 56°C for 30 min.

6. Rotator for incubation 348 "Assistent" RM5 (Assistent-Präzision-Hecht, Sondheim/Rhön, Germany).

**2.3. Isolation of Murine Mononuclear Cells (MNCs)**

1. Scissor and two tweezers.

2. Dulbecco's Phosphate-Buffered Saline (1×) (PBS) without calcium and magnesium (PAA Laboratories), supplemented with 1% (v/v) of Penicillin/Streptomycin (100×), 10,000 units/10 mg/mL in normal saline (PAA Laboratories).

3. A screen cup (85 mL) with a fixed screen (100 mesh) (Sigma-Aldrich, St. Louis, MO); for later cleaning a special key for the screen cup is needed (Sigma-Aldrich).

4. A Nylon 100-µm cell strainer (BD Falcon).

5. Histopaque for gradient centrifugation (density 1.083) (Sigma-Aldrich).

**2.4. Coupling of Magnetic Beads for Murine T Cell Stimulation and Washing/Blocking**

1. Dynabeads M-450 Epoxy Dynal and magnetic separator Dynal MPC-1 (5–50 mL) (Invitrogen). Beads are stored at 4°C.

2. Sterile FACS-tube with cap (BD Falcon).

3. mAb anti-CD3 (clone: 145-2C11) and mAb anti-CD28 (clone: 37.51) (BD) for coupling.

4. Rotator for incubation 348 "Assistent" RM5 (Assistent-Präzision-Hecht).

5. 0.1% (w/v) solid mouse serum albumin (MSA) in PBS solution must be prepared. 25 mg of MSA (Calbiochem/Merck, Darmstadt, Germany) are dissolved in 25 mL PBS. The suspension is filtered for sterility through a 0.22-µm filter (Millipore, Billerica, MA). 0.1% (w/v) MSA/PBS can be stored at 4°C for several months.

**2.5. Medium Composition/Supplements for Murine T Cell Cultivation/Stimulation and Transduction**

1. *RPMI-mouse medium.* RPMI 1640 with phenol red (Bio Whittaker/Cambrex), supplemented with 10% (v/v) fetal calf serum (FCS) (PAN Biotech GmbH, Aidenbach, Germany), 2% l-glutamine, 2 mM in 0.85% sodium chloride solution (Bio Whittaker/Cambrex), 1% of Penicillin/Streptomycin (Pe/St) (100×),

10,000 units/10 mg/mL in normal saline (PAA Laboratories), 1% Sodium pyruvate (100×) (Gibco/BRL Invitrogen), 1% Nonessential Amino Acids (NEAA) (100×) (Gibco/BRL Invitrogen), 0.1% β-mercaptoethanol (1,000×) (Gibco/BRL Invitrogen). Medium and supplements are stored at 4°C.

2. Interleukin-2 (IL-2) $10^4$ U/mL (Roche Diagnostics, Mannheim), 100 U/mL or 50 U/mL for stimulation. A $10^4$ U/mL stock solution is prepared and stored at –20°C. An appropriate amount of IL-2 stock is thawed and can be stored at 4°C for at least 3–4 weeks.

**2.6. Preparation of Retronectin-Coated NonTissue Culture-Treated Plates**

1. 6- and 24- well nontissue culture-treated plates (BD).

2. By gentle swirling, retronectin (Takara Bio Inc., Tokyo, Japan) lyophilized powder must be thoroughly dissolved in water for tissue culture or water for injection (sterile, free of endotoxin) to a concentration of 1 mg/mL. The solution is filtered through a 0.22-μm filter (Millipore) and adjusted to a concentration of 50 μg/mL by adding an appropriate volume of sterile PBS (PAA Laboratories). For dilution some of the PBS is passed through the same filter as previously used one for complete recovery of retronectin (2.5 mg of retronectin will give a 50-mL suspension). The retronectin solution can be stored at 4°C for at least one month.

3. 2% (w/v) bovine serum albumin (BSA) (20 mg/mL) solution (Sigma-Aldrich) in sterile PBS (PAA Laboratories) must be prepared and should be filtered through a 0.22-μm filter (Millipore). The solution can be stored at 4°C.

4. HBSS/Hepes solution for washing of retronectin-coated plates: Hank's buffered saline solution (HBSS) (Sigma-Aldrich) must be supplemented with 2.5% (v/v) 1 M Hepes (Gibco/BRL Invitrogen). The solution can be stored at room temperature.

# 3. Methods

Extensive in vitro proliferation of T lymphocytes is dependent on appropriate stimulation. In classical protocols mitogenic agents like phytohemagglutinin (PHA) and concanavalin A (Con A) are used for polyclonal T cell stimulation. The use of anti-CD3- and anti-CD28-coated beads for stimulation of T lymphocytes is a more physiological approach, which mimics partially the stimulation by an antigen-presenting cell (APC). It offers a simple alternative for stimulation of T cells and does not require autologous or MHC-matched APC nor any specific antigen. Most importantly expanded clones remain antigen specific after stimulation.

**3.1. Isolation of Human Peripheral Blood Mononuclear Cells (PBMCs) from Fresh Blood or Buffy Coats by Ficoll Density Gradient Centrifugation**

1. Dilute blood sample with equal volume PBS.

2. Carefully load diluted blood onto Ficoll in a 50-mL tube.

3. Centrifuge 30 min at $500 \times g$ at room temperature in a tissue culture centrifuge without break.

4. Carefully collect the interphase containing human PBMCs.

5. Dilute the interphase with 30 mL PBS in a new 50-mL tube.

6. Centrifuge cells 8 min at $500 \times g$ at room temperature.

7. Remove supernatant and resuspend cell pellet with 50 mL of PBS.

8. Count cells before further use.

9. For quality control, if desired, stain PBMCs with a mixture of anti-CD3, anti-CD4 and anti-CD8 antibodies and perform flow cytometric analysis to characterize T helper and T killer cell subpopulations.

**3.2. Isolation of Murine Mononuclear Cells**

1. Remove spleens and lymph nodes from required number of animals (one animal will yield up to $1 \times 10^8$ MNCs) (*see* **Note 1**) and transfer them into a 50-mL tube (spleens and lymph nodes from maximal 3 animals per tube) containing 10–20 mL ice-cold PBS with 1% Pe/St for transport from the animal facility to the lab (using an ice box).

2. Put a screen cup with a fixed screen into a Petri or tissue culture dish and transfer the spleens and lymph nodes with the whole volume of PBS into the screen cup.

3. Homogenize spleens and lymph nodes with a syringe stamp in the screen cup to get a single cell suspension and filtrate the whole volume of the single cell suspension through a cell strainer into a new 50-mL tube.

4. Wash the dish twice with 10 mL PBS and filtrate the washing solutions through the same cell strainer into the same 50-mL tube.

5. Centrifuge cells for 10 min at $500 \times g$ and room temperature.

6. Remove supernatant; cell pellet is resuspended in 10 mL PBS.

7. Centrifuge the removed supernatant again. This time supernatant is discarded.

8. Resuspend the second cell pellet in 10 mL PBS and combine both cell suspensions.

9. Carefully underlay the cell suspension (20 mL) with 10 mL Histopaque using a 10-mL syringe with a long needle or a 10-mL pipette.

10. Centrifuge 30 min at $600 \times g$, room temperature and without break (!) because of gradient stability.

11. Carefully collect the interphase with a 5-mL pipette.

12. Dilute the interphase containing murine MNCs with 10 mL PBS in a new 50-mL tube.

13. Centrifuge cells 10 min at $500 \times g$ and room temperature.

14. Remove supernatant, resuspend cell pellet in 10 mL PBS, and repeat the washing step.

15. Resuspend cell pellet in 10 mL PBS and count cells. Leave cells on ice before further use (cells can be stored for 1–2 h on ice, but should not be stored too extensively before stimulation). For quality control *see* Sec19_6 **Subheading 3.1 step 9**.

### 3.3. Coupling of Anti-CD3 and Anti-CD28 Antibodies to Magnetic Dynal-Epoxy Beads for Murine T Cell Stimulation

1. Resuspend Dynabeads M-450 Epoxy ("uncoated") by vortexing container before opening. After opening the container, continue resuspension by pipetting.

2. Remove required amount of beads (1 mL beads per coupling is equivalent to $4 \times 10^8$ beads) and place in a sterile FACS tube.

3. Place tube in a magnetic separator.

4. Remove and discard all liquid with a 1,000-µL pipette, while tube is still in the magnetic separator.

5. Remove the tube from the separator; add 1 mL PBS and vortex until beads are resuspended.

6. Repeat **steps 3–5** twice. Beads are now suspended in 1 mL PBS.

7. Mix mAb: anti-CD3 (Clone: 145-2C11) and anti-CD28 (Clone: 37.51). 60 µg of each mAb corresponds to a total of 120 µL (60 µL per antibody).

8. Add mAb solution to bead suspension of **step 6** and vortex (final concentration for each mAb should be 60 µg/mL).

9. Close lid tightly and wrap parafilm around the top of tube.

10. Place tube in a 50-mL Falcon and on an appropriate rotator in an incubator at 37°C.

11. Coupling should proceed for 16–24 h, while rotating (*see* **Note 2**).

12. Remove tube with beads from rotator and place tube in a magnetic separator.

13. Remove and discard all liquid with a 1,000-µL pipette.

### 3.4. Washing/Blocking and Storage of Beads

1. Add PBS/0.1% MSA to the original volume of beads removed for coupling (*see* **Subheading 3.3**).

2. Place in appropriately sized rotator.

3. Washing/blocking should proceed for 5 min at room temperature, while rotating.

4. Remove tube with beads from rotator and place in an appropriately sized magnet and discard washing liquid.

5. Repeat washing step twice and resuspend beads in original volume for overnight washing/blocking.

6. Washing/blocking should proceed for 16–24 h at 2–8°C, while rotating.

7. After washing/blocking discard washing liquid.

8. Add PBS/0.1% MSA to the original volume of beads removed for coupling (*see* **Subheading 3.3**) and vortex.

9. Transfer resuspended beads into a new, appropriately sized tube.

10. Label container with contents, date of preparation, bead concentration, and store at 2–8°C (*see* **Note 3**). The vial containing Dynabeads should be stored upright to keep the Dynabeads in liquid suspension. Precautions should be taken to prevent bacterial contamination of opened vials.

11. Before further use, mAb-coupled beads must be washed twice with PBS: After resuspension by vortexing remove required amount of beads ($1.8 \times 10^6$ beads are needed for $6 \times 10^5$ cells (3:1 ratio); if the storage dilution of the beads is $4 \times 10^8$, 4.5 µL of beads are removed and added per mL of cell suspension corresponding to $6 \times 10^5$ cells).

12. Place beads into a sterile FACS tube and in magnetic separator.

13. Remove and discard all liquid with a 1,000-µL pipette.

14. Add 2 mL PBS while the tube is still in the magnetic separator.

15. Remove the tube from the magnet and mix by vortexing.

16. Place beads in magnetic separator and repeat washing step.

17. 1 mL of prepared cell suspension (*see* **Subheading 3.5**) is added and washed beads are resuspended by vortexing.

***3.5. Stimulation of Isolated Murine MNCs with mAb-Coupled Dynal-Epoxy Beads***

1. *Day 0.* Freshly isolated MNCs (*see* **Subheading 3.2**) are prepared in RPMI-mouse medium at a density of $6–7 \times 10^5$ cells per mL.

2. Coupled and washed anti-CD3/anti-CD28 beads (*see* **Subheading 3.4**) are added in a 3:1 beads to cell ratio after resuspension in 1 mL of prepared cell suspension (*see* **Subheading 3.4, step 17**).

3. After mixing of cell suspension and beads by pipetting carefully up and down, the cell/bead suspension is seeded in 6-well culture plates with a density of $4.2–4.9 \times 10^6$ cells/7 mL per well.

4. Cells are kept under standard cell culture conditions for 4 days.

5. Alternatively 100 U/mL IL-2 can be added on day 1, for better stimulation.

6. Cells are observed for proliferation by light microscopy – cluster/rosette initiation is a good indicator for optimal cell growth (**Fig. 1**).

**3.6. Stimulation of Isolated PBMCs with Human Dynabeads CD3/CD28 T Cell Expander**

1. Day 0, CD3/CD28 beads are washed: Add 3 mL of PBS supplemented with 2% (v/v) human serum into a 5-mL tube, add appropriate number of CD3/CD28 beads to the tube depending on the processed cell number (for standard process use a ratio of 3 CD3/CD28 Dynabeads to 1 CD3 + T cell), place the tube on the magnetic separator for 1 min and remove supernatant. Repeat washing step and resuspend to $1.5 \times 10^7$ CD3/CD28 beads per mL in PBS supplemented with 2% (v/v) human serum.

2. Centrifuge the desired number of T cells for 8 min at 500 × $g$, remove supernatant, resuspend cell pellet with PBS supplemented with 2% (v/v) human serum ($1 \times 10^7$ cells/mL), add washed CD3/CD28 beads at a 3:1 ratio (beads: CD3 + cells) for magnetic concentration of CD3 + cells.

3. Incubate for 30 min at 4°C on a rotator, place the tube on the magnetic separator for 1 min, and remove supernatant.

4. Resuspend the bead rosetted cells in 2 mL of PBS supplemented with 2% (v/v) human serum.

5. Count cells, centrifuge cells 8 min at 500 × $g$ and remove supernatant.

6. Adjust volume of cell suspension to give $0.5 \times 10^6$ CD3 + cells/mL with complete medium and add IL-2 (100 U/mL).

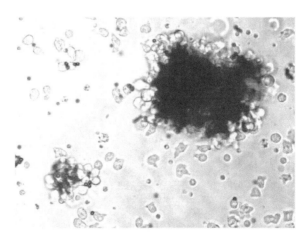

Fig. 1 Light microscopy of bead stimulated T cells. Day 4 after isolation. Stimulated murine T cells grow in large clusters/rosettes with anti-CD3- and anti-CD28-coated beads.

7. Incubate cells at 37°C, 5% $CO_2$ for 4 days.

8. Cells are observed for proliferation by light microscopy – cluster initiation is a good indicator for optimal cell growth (**Fig. 1**).

*3.7. Preparation of Retronectin-Coated Nontissue Culture Plates*

1. Dispense an appropriate volume (dispense 0.4 mL into each well of a 24-well plate or 1.5 mL into each well of a 6-well plate, respectively) of sterile retronectin solution into each well of an appropriate nontissue culture-treated plate. Allow the covered plate to stand for 2 h at room temperature in a clean bench or overnight at 4°C.

2. Remove retronectin solution and put solution into each well of an appropriate second nontissue culture-treated plate, which should be stored at 4°C for second transduction (retronectin solution can be used up to four times for precoating). Then add an appropriate volume (0.4 mL into each well of a 24-well plate or 1.5 mL into each well of a 6-well plate, respectively) of sterile PBS containing 2% (w/v) bovine serum albumin (BSA) into each well of the previously retronectin-precoated plate for blocking. Blocking should proceed for 30 min at room temperature. Alternatively plate can be stored with BSA solution overnight at 4°C.

3. Remove BSA solution, and wash wells once with 0.6 mL for each well of a 24-well plate or 3 mL for each well of a 6-well plate with HBSS/Hepes solution. The plate is now coated with retronectin and ready for use (plate can be stored at least for 1 week at 4°C).

*3.8. Preloading of Retronectin-Coated Plates with Retroviral Vector Supernatants*

1. The following steps "Preparation of gammaretroviral preloaded plates," "Debeading of stimulated cells," and "Retroviral transduction" are identical for human and murine T cells.

2. Thaw an appropriate amount of retroviral vector supernatant in a water bath at 37°C. Thawing should be observed, because vector particles should not be too extensively warmed (*see* **Note 4**).

3. Add 1 mL (or 3 mL) of retroviral vector supernatant per well of a 24-well (or 6-well) retronectin-coated plate.

4. Centrifuge the plates at 1,000 × *g*, 4°C for 30 min (*see* **Note 5**).

5. After centrifugation retroviral vector supernatant is removed completely from the plate and discarded.

6. Preloading with retroviral vector supernatant is repeated two more times.

7. Plates are now precoated with retroviral supernatant and ready for transduction of stimulated T cells.

8. Preloading is repeated on the second day of transduction (*see* **Subheading 3.9**), as described in **steps 2–6**.

### 3.9. Debeading of CD3/CD28-Coated Magnetic Bead Stimulated T Cells and Transduction on Retroviral Vector Preloaded Culture Plates

1. On day 4 of stimulation, cells are collected in an appropriately sized tube (per prestimulated 6-well; cells are pooled in a 50-mL Falcon tube). Pooled prestimulated cells are centrifuged at $500 \times g$ and room temperature for 10 min (**Fig. 1.** shows a light microscopy of bead-stimulated T cells on day 4, growing in big clusters with paramagnetic beads).

2. After centrifugation cell/beads-pellet is resuspended in an appropriate amount of PBS (10 mL per 50 mL Falcon tube). Resuspend thoroughly by rigorous pipetting 4–5 times up and down with a 10-mL pipette.

3. Up to 10 mL of brown shimmering cell/beads suspension are transferred to a 15-mL Falcontube. A second resuspension of 4–5 times up and down pipetting is performed and suspension is placed into an appropriately sized magnetic separator (*see* **Note 6**).

4. Cell/beads solution needs to stand in the magnetic separator for 30 s until most of the magnetic particles accumulate at the edge of the Falcon tube (a brown shimmering slime is visible at the edge of the tube). Debeaded cells remain in the supernatant. Supernatant containing debeaded cells is removed and transferred into a new 15-mL Falcon tube and immediately placed into the magnetic separator for another 90 s to remove remaining magnetic particles.

5. Supernatant with cells is removed again and placed into an appropriately sized Falcon tube. Debeaded cells are counted, centrifuged at $500 \times g$ and room temperature for 10 min.

6. After centrifugation cells are resuspended in RPMI mouse medium (or complete medium X-Vivo 15 for human cells) at acquired density. 100 U/mL IL-2 are added to the cell suspension. For transduction $4–5 \times 10^6$ cells/3 mL per well of a 6-well plate or $1 \times 10^6$ cells/mL per well of a 24-well plate are seeded on retroviral vector preloaded culture plates (*see* **Subheading 3.8**) (*see* **Note 7**).

7. Cells are now kept under standard culture conditions for 16–24 h.

8. On day 5, the second transduction takes place. Retroviral preloaded plates are prepared again, as described in **Subheading 6**. Cells are collected from the plate of first transduction and centrifuged at $500 \times g$ and room temperature for 10 min. Cell pellet is resuspended in fresh medium (in the same volume of medium as before centrifugation) and seeded on new prepared retroviral vector preloaded plates.

9. On day 6, only fresh medium and 50 U/mL IL-2 are added to the cell culture (2 mL medium to each well of a

6-well plate and 500 µL medium to each well of a 24-well plate). If the cells were transduced with fluorescent proteins (for example GFP) the operator can check for transgene expression via fluorescence microscopy or FACS analysis.

10. On day 7 cells can be pooled and adjusted to $5 \times 10^5$ cells/mL for further cultivation or murine T cells can be transplanted into recipient mice (for example Rag-1 deficient animals). Analysis for full transgene expression should be performed on day 7 or day 8 and observed over cultivation time.

## 4. Notes

1. The most favorable age of donor animals is 6–8 weeks.

2. Coupling of beads should be performed on a rotator, not on a shaker.

3. Precoated beads can be stored for at least two months at 4°C, without any failure of stimulation activity.

4. For high transduction efficacies with retroviral vectors it is recommended to use fresh particles. Particles can be stored directly after production at 4°C for 5–7 days without any loss of activity (normally freezing of supernatants diminishes the particle activity).

5. Preloading of vector supernatants can be prolonged for up to 60 min per centrifugation step. Supernatant coating on the plate is enhanced.

6. Alternatively debeading can be skipped and beads are kept in culture. Keeping beads in culture can lead to better stimulation/transduction and is not disadvantageous for cultivation. Nevertheless, beads are removed prior to transplantation.

7. After seeding of debeaded cells on retroviral vector preloaded culture plates another centrifugation step can be included. Cells can be centrifuged at $600 \times g$ for 60 min at 31°C to ensure a better contact of precoated viral particles and seeded T cells.

## References

1. Reddy, M.M., K.O. Goh and C. Poulter (1975) Mitogenic stimulation of lymphocytes in cancer patients. *Oncology.* 32, 47–51

2. Stobo, J.D. (1972) Phytohemagglutin and concanavalin A: probes for murine 'T' cell activation and differentiation. *Transplant Rev.* 11, 60–86

3. Andersson, J., O. Sjoberg, and G. Moller (1972) Mitogens as probes for immunocyte activation and cellular cooperation. *Transplant Rev.* 11, 131–77

4. Damle, N.K., et al. (1992) Differential costimulatory effects of adhesion molecules B7, ICAM-1, LFA-3, and VCAM-1 on resting

and antigen-primed CD4 + T lymphocytes. *J Immunol.* 148(7), 1985–92

5. Wingren, A.G., et-al. (1995) T cell activation pathways: B7, LFA-3, and ICAM-1 shape unique T cell profiles. *Crit Rev Immunol.* 15(3–4), 235–53

6. June, C.H., et al. (1987) T-cell proliferation involving the CD28 pathway is associated with cyclosporine-resistant interleukin 2 gene expression. *Mol Cell Biol.* 7(12), 4472–81

7. Bordignon, C., et al. (1995) Gene therapy in peripheral blood lymphocytes and bone marrow for ADA- immunodeficient patients. *Science.* 270(5235), 470–5

8. Morgan, R.A., et al. (2006) Cancer regression in patients after transfer of genetically engineered lymphocytes. *Science.* 314(5796), 126–9

9. Lunzen, J.V., et al. (2007) Transfer of Autologous Gene-modified T Cells in HIV-infected Patients with Advanced Immunodeficiency and Drug-resistant Virus. *Mol Ther.* 15(5), 1024–33

10. Buchschacher, G.L., Jr. and F. Wong-Staal (2000) Development of lentiviral vectors for gene therapy for human diseases. *Blood.* 95(8), 2499–504

11. Fehse, B., et al. (1998) Highly-efficient gene transfer with retroviral vectors into human T lymphocytes on fibronectin. *Br J Haematol.* 102(2), 566–74

12. Coito, S., et al. (2004) Retrovirus-mediated gene transfer in human primary T lymphocytes induces an activation- and transduction/selection-dependent TCR-B variable chain repertoire skewing of gene-modified cells. *Stem Cells Dev.* 13(1), 71–81

13. Kalamasz, D., et al. (2004) Optimization of human T-cell expansion ex vivo using magnetic beads conjugated with anti-CD3 and Anti-CD28 antibodies. *J Immunother.* 27(5), 405–18

14. Zhou, J., et al. (2005) Telomere length of transferred lymphocytes correlates with in vivo persistence and tumor regression in melanoma patients receiving cell transfer therapy. *J Immunol.* 175(10), 7046–52

# Chapter 7

# Retroviral Transduction of Murine Primary T Lymphocytes

## James Lee, Michel Sadelain, and Renier Brentjens

## Summary

In comparison to human T cells, efficient retroviral gene transfer and subsequent expansion of murine primary T cells is more difficult to achieve. Herein, we describe an optimized gene transfer protocol utilizing an ecotropic viral vector to transduce primary murine T cells activated with magnetic beads coated with agonistic anti-CD3 and CD28 antibodies. Activated T cells are subsequently centrifuged (spinoculated) on RetroNectin-coated tissue culture plates in the context of retroviral supernatant. Variables found to be critical to high gene transfer and subsequent efficient T cell expansion included CD3/CD28 magnetic bead to cell ratio, time from T cell activation to initial spinoculation, frequency of T cell spinoculation, interleukin-2 concentration in the medium, and the initial purity of the T cell preparation.

**Key words:** Retroviral gene transfer, Murine primary T cells, Mouse T lymphocytes, Phoenix-eco packaging cells, Spinoculation, RetroNectin, Interleukin-2 (IL-2), Nylon wool column, CD3/CD28-activating magnetic beads.

## 1. Introduction

Retroviral gene modification of human T cells provides a novel and potentially powerful approach to the treatment of a wide array of diseases including inherited or acquired immunodeficiencies, infection, autoimmune disorders, and cancer (1–5). Similarly, retroviral-mediated gene transfer in murine T cells may enable investigators to better understand in vivo T cell biology as well as study the potential of genetically modified T cells in murine models of disease. However, in contrast to human T cells, which are readily transduced and expandable in the ex vivo setting, retroviral-mediated gene transfer into murine T cells is compromised

Christopher Baum (ed.), *Methods in Molecular Biology, Methods and Protocols, vol. 506*
© Humana Press, a part of Springer Science + Business Media, LLC 2009
DOI: 10.1007/978-1-59745-409-4_7

both by poor gene transfer and inadequate subsequent T cell expansion and survival. Herein, we describe an optimized retroviral gene transfer protocol that allows for the generation of efficiently transduced and highly viable murine T cell populations.

In order to enhance gene transfer as well as T cell viability, we addressed several critical parameters in optimizing our protocol. First, we utilized the Pheonix-eco *(6)* ecotropic packaging cell line, due to the ubiquitous expression of ecotropic receptors on murine cells, including T cells *(7, 8)*. Second, we use the high-titer, Moloney murine leukemia virus-derived vector SFG, a variant of the MFG gamma-retroviral vector *(9)*. Third, we further explored T cell activation conditions and timing of retroviral exposure, which are critical for efficient retroviral integration and T cell survival *(10)*. Fourth, we found that the purity of isolated murine T cells significantly impacted on transduction efficiency. Fifth, we found that repeated transduction through the centrifugation of retroviral supernatant and T cells on RetroNectin-coated nontissue culture plates (spinoculation), which enhances the colocalization of target cells and viral particles, *(11)*, markedly improved gene transfer. Finally, we demonstrate the role of T cell culture conditions, specifically the concentration of IL-2, in the viability and expansion of transduced murine T cells. By optimizing these parameters we were able to achieve consistent and efficient gene transfer with expression of the target gene by highly viable murine T cells. The 19z1 SFG retroviral vector, encoding a chimeric antigen receptor (CAR) that targets the CD19 antigen on normal and malignant B cells *(12)*, was utilized to illustrate the murine T cell transduction protocol described here.

## 2. Materials

1. 1× sterile PBS.
2. Sterile PBS with 0.5% heat-inactivated fetal calf serum (FCS).
3. Complete RPMI-1640 medium with 10% heat-inactivated fetal calf serum (FCS), 20 mM HEPES (Gibco-BRL, Grand Island, NY), 1 mM sodium pyruvate (Gibco-BRL), 0.05 mM 2-ME (Gibco-BRL), 2 mM L-Glutamine (Gibco-BRL), 100 U/mL penicillin (Gibco-BRL), and 100 µg/mL streptomycin (Gibco-BRL).
4. Dulbeccos modified Eagles Medium (DMEM) with 10% FCS, 2 mM l-Glutamine (Gibco-BRL), 100 U/mL penicillin (Gibco-BRL), and 100 µg/mL streptomycin (Gibco-BRL).
5. Poly-l-lysine solution – 4 parts 0.1% gelatin (Fisher Scientific Cat. No. G8-500), 1 part 0.01% poly-l-lysine buffer (Sigma Cat. No. P4832).

6. Six-well flat bottom tissue culture plates with lids (BD Falcon).

7. Six-well flat bottom nontissue culture plates with lids (BD Falcon).

8. 100-mm tissue culture plates (Corning).

9. 12 × 75 mm Polystyrene Round-Bottom Test Tube, 5 mL, snap cap (BD Falcon, Cat. No. 352054).

10. Sterile Nylon Wool Fiber Column (Polysciences Cat. No. 21759).

11. 10 mL Syringe (BD).

12. 0.45 μm filter for syringes (Whatman Cat. No. 6780–2504).

13. ACK Lysing Buffer (BioWhittaker Cat. No. 10-548E).

14. Recombinant Mouse Interleukin-2 (rmIL-2) (Roche Cat. No. 11271164001).

15. Dynabeads Mouse CD3/CD28 T Cell Expander (Invitrogen Cat. No. 114-11D).

16. Dynal MPC-50 Magnet (Invitrogen Cat. No. 120–24).

17. RetroNectin (Takara Biomedicals, Otsu, Japan Cat. No. T100A).

18. Blue Max 50 mL Polypropylene Conical Tube (BD Falcon).

19. 40 μm Nylon Cell Strainer (BD Falcon).

20. 5 mL Syringe (BD).

21. PE-conjugated anti-19z1 antibody (Monoclonal Antibody Facility, MSKCC).

22. FITC-conjugated anti-mouse CD3 antibody (eBioscience).

23. FITC-conjugated anti-mouse CD4 antibody (eBioscience).

24. C57BL/6 mice for T cells.

25. Phoenix Eco Cell Line (ATCC product# SD 3444).

26. HEPES (Sigma Cat. No. H-7006).

27. $CaCl_2$ (Mallinkrodt Cat. No. 4160).

28. $Na_2HPO_4$ (Sigma).

29. dd $H_2O$.

30. 19z1 SFG retroviral plasmid (*see* **ref. 12**).

## 3. Methods

The protocol described later outline *(1)* the transfection of Phoenix-eco retroviral producer cells *(6)*, *(2)* the preparation of murine T cells for retroviral transduction, and *(3)* the transduction of murine T cells with Phoenix-eco retroviral supernatant.

**3.1. Transfection of Phoenix-Eco Cells**

The protocol describes the transfection of Phoenix-eco cells with the 19z1 SFG retroviral plasmid. It includes (a) preparation of the Phoenix-eco producer cells for transfection, (b) transfection of the Phoenix-eco cells, and (c) collection of the retroviral supernatant for gene transfer of murine T cells.

*3.1.1. Preparation of Phoenix-Eco Producer Cells for Transfection (Day –1)*

1. Before transfection, determine the amount of viral supernatant required. Generally 2 mL of viral supernatant (1 mL for each spinoculation) is sufficient for the transduction of $3 \times 10^6$ isolated mouse T cells (*see* **Subheading 3.3**). Each 100-mm plate yields approximately 7 mL of viral supernatant.

2. One day prior to transfection, plate Phoenix-eco cells at 50% confluence per 100-mm plate in 10 mL DMEM + 10% FCS. Make sure to evenly distribute cells by rocking plates forward and backward, then side to side, 3–4 times. Leave the plate in a humidified 37°C, 5% $CO_2$ incubator and do not disturb the cells as they attach to the plate.

3. On the day of transfection (day 0), the Phoenix-eco plate should be 90–100% confluent.

*3.1.2. Calcium Phosphate Transfection (Day 0)*

1. Coat 100-mm tissue culture plates with 10 mL Poly-L-lysine solution (*see* **Note 1**). Immediately remove the solution which may be reused to coat further plates.

2. Resuspend confluent Phoenix-eco cells from day 1 plates by pipetting (these cells adhere very lightly to the plate and therefore do not require trypsin to generate a cell suspension). Add one-third of the resulting cell suspension to each Poly-l-lysine-coated plate in a final volume of 10 mL DMEM + 10% FCS. Ensure even distribution of cells by gently tilting plates forward and backward, then side to side, 3–4 times.

3. Store plates undisturbed at 37°C with 5% $CO_2$ for 8 h.

4. When the Phoenix-eco cells are attached and evenly distributed at 30–50% confluence, they are ready for transfection. Prior to transfection, aspirate medium and very carefully replace with 10 mL of fresh DMEM + 10% FCS being sure not to dislodge adherent cells.

5. Allow 20 min to thaw all transfection reagents. Make 2x HEPES-buffered saline (HBS) by mixing 8.0 g of NaCl, 6.5 g of HEPES, in 10 mL of $Na_2HPO_4$ stock solution (5.25 g $Na_2HPO_4$ in 500 mL of water). Adjust pH to exactly 7.0 using NaOH or HCl and bring final volume of 2× HBS solution to 500 mL.

6. Prepare the transfection cocktail in 12 × 75-mm polystyrene round-bottom test tubes. Each cocktail will provide enough reagent to transfect 3,100 mm Phoenix-eco Poly-l-lysine-coated plates. To each tube, add 30 μg of 19z1-SFG DNA

plasmid, dd $H_2O$ followed by 186 µL of calcium chloride for a final volume of 1,500 µL (*see* **Note 2**).

7. Next, while gently but continuously vortexing the cocktail prepared, slowly add 1,500 µL of 2× HBS dropwise for DNA precipitation (*see* **Note 3**).

Each tube now contains:

| DNA | 30 µg |
|---|---|
| 2 M $CaCl_2$ | 186 µL |
| dd $H_2O$ | 1,314 µL – volume of DNA |
| 2× HBS | 1,500 µL |
| Total volume: | 3,000 µL |

8. Immediately add 1 mL of the resulting transfection mix dropwise onto each of the 100-mm Phoenix-eco Poly-l-lysine-coated plates distributing the cocktail evenly over the cells. Rock plates side to side/back and forth a few times to evenly distribute DNA/$CaPO_4$ particles.

9. Incubate plates at 37°C with 5% $CO_2$ for 16 h.

*3.1.3. Generation of Retroviral Supernatant of Murine T Cell Transduction (Days 1–2)*

1. Aspirate DNA/$CaPO_4$ particles and medium, and carefully replace with 10 mL of fresh DMEM + 10% FCS along side of the plates so as not to dislodge adherent Phoenix-eco cells.

2. Incubate plates at 37°C with 5% $CO_2$ overnight.

3. Aspirate and carefully replace the 10 mL of DMEM + 10% FCS with 7 mL complete RPMI (*see* **Note 4**).

4. Incubate plates at 37°C with 5% $CO_2$ over night.

*3.1.4. Harvesting Retroviral Supernatant for Murine T Cell Transduction (Day 3)*

1. Using a 10-mL syringe, draw up the day 3 retroviral supernatant from transfected Phoenix cells (*see* **Note 5**). Filter supernatant through a 0.45-um filter attached to the end of the syringe to remove cellular debris.

2. Day 3 supernatants can be used either directly on activated T cells (*see* **Subheading 3.2**) or stored at –80°C for later experiments (*see* **Note 6**).

**3.2. Preparation of Murine T Cells for Transduction**

The protocol describes the preparation of murine T cells for retroviral transduction. Steps for murine T cell isolation include (a) harvesting of mouse splenocytes, (b) isolation of the CD3+ T cell fraction, and (c) activation of the resulting purified CD3+ T cell population.

*3.2.1. Harvesting Mouse Splenocytes*

1. Euthanize C57BL/6 mice by $CO_2$ asphyxiation. Soak mouse briefly in 70% EtOH for sterility and surgically remove spleen.

2. To make a single cell suspension of splenocytes from mouse spleen, first place 40-μm nylon cell strainer on 50-mL polypropylene conical tube and wet the cell strainer surface with complete RPMI. Macerate fresh spleen on the cell strainer with a 5-mL syringe plunger.

3. When spleen is completely macerated on cell strainer, rinse cell strainer with 10 mL of complete RPMI to flush splenocytes into 50-mL conical tube.

4. Wash splenocytes with sterile PBS, pellet cells by centrifugation at 400 g for 5 min, repeat wash.

5. To lyse erythrocytes, resuspend washed cell pellet in 1 mL of ACK Lysing Buffer and incubate at room temperature for 7 min. Add 10 mL PBS and pellet cells by centrifugation.

6. Wash twice with PBS.

7. Resuspend splenocytes in complete RPMI to a concentration of $0.5–1 \times 10^8$ viable cells/mL. Each spleen will yield approximately $1 \times 10^8$ splenocytes of which 35–45% are CD3+ T cells *(13)*.

*3.2.2. Enrichment of CD3+ T Cells from Splenocyte Pool*

The purity of the T cell population enhances retroviral transduction efficiency (*see* **Note 7**). To this end, we utilize nylon wool columns to further enrich for the CD3+ T cell population as described (modified from the manufacturers protocol) *(14)*.

1. Load nylon wool column with 5 mL complete RPMI. Tap gently to remove air bubbles.

2. Incubate prepared column for 1 h at 37°C.

3. Drain medium from the column.

4. Add $1–2 \times 10^8$ viable cells in a volume of 2 mL of complete RPMI. Drain to allow entire cell volume to enter into the nylon wool.

5. Add another 2 mL of complete RPMI to the top of the column. Drain column to allow the additional 2 mL to enter into the nylon wool.

6. Add another 4 mL of medium to the column wool to ensure top of the wool is covered.

7. Incubate splenocyte-loaded column for 1 h at 37°C in 5% $CO_2$. Cover the top of the column with parafilm to maintain sterility.

8. Drain the nylon wool column by gravity into a 50-mL polypropylene conical tube to collect the nonadherent CD3+ T cell fraction, washing the column once with 5 mL of complete RPMI to collect residual nonadherent cells. The resulting cell population is routinely 80–90% CD3+ (**Fig. 1**).

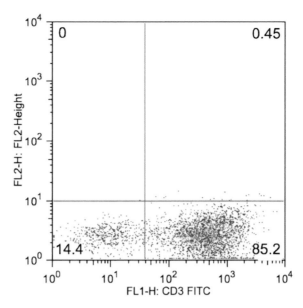

Fig. 1. FACS analysis of splenocytes eluted from a nylon wool column and stained with mouse anti-CD3 FITC-conjugated antibody. As shown here, the CD3+ T cell fraction of nonadherent cells accounts for 85% of the total recovered cell population.

9. Pellet recovered cells by centrifugation and resuspend in complete RPMI to a concentration of $1 \times 10^6$ cells/mL.

*3.2.3. Activation of CD3+ Murine T Cells. (Day 0)*

CD3+ T cells are activated using Dynabeads Mouse CD3/CD28 T Cell Expander magnetic beads coated with agonistic mouse CD3 and CD28 antibodies.

1. Calculate the amount of CD3/CD28 beads required to obtain a 1:1 CD3/CD28 bead to cell ratio.

2. To wash beads prior to activation, dilute the required number of beads in 3 mL of PBS with 0.5% FCS in a FACS tube, and place suspended beads into the MPC-50 magnet and allow for beads to collect on the side of the tube (1 min).

3. Aspirate PBS/FCS solution, remove tube from magnet, resuspend beads in fresh PBS/FCS, and repeat magnetic isolation and aspiration of PBS/FCS for a total of three wash cycles.

4. Resuspend beads in 1 mL complete RPMI.

5. Add washed, resuspended beads to mouse T cells at a final concentration $1 \times 10^6$ cells/mL at a bead to cell ratio of 1:1 in complete RPMI supplemented with 20 IU/mL IL-2 (*see* **Note 8**).

6. Plate the cells in 6-well tissue culture plates at a volume of 3 mL per well at a concentration of $1 \times 10^6$ cells/mL. Incubate cells at 37°C and 5% $CO_2$ overnight.

**3.3. Transduction of Murine T Cells with Viral Supernatant by Spinoculation (Day 1)**

Optimal gene transfer into murine T cells is achieved when the first cycle of spinoculation is performed at 24 h following initial T cell activation (*see* **Note 9**).

1. Coat 6-well nontissue culture plates with RetroNectin at 15 $\mu$g/mL in PBS. Incubate at room temperature for 3 h.

2. Aspirate RetroNectin/PBS solution and coat plates with PBS with 0.5% FCS. Incubate plates at room temperature for 30 min.

3. Just prior to spinoculation, aspirate PBS/FCS solution from RetroNectin-coated plates. The plates are now ready for transduction.

4. Harvest activated CD3+ T cells from day 0, pellet cells and resuspend cells at a concentration of $3 \times 10^6$ cells/mL in complete RPMI supplemented with 80 units of IL-2 (*see* **Note 10**).

5. Add 1 mL of previously prepared fresh or thawed viral supernatant per well.

6. Immediately after adding the viral supernatant, add 1 mL of CD3+ T cells per well. Each well now has a total volume of 2 mL containing $3 \times 10^6$ T cells mixed with viral supernatant at a 1:1 vol/vol ratio.

7. Immediately centrifuge the plate(s) at $2,000 \times g$ and 30°C, for 1 h (first spinoculation)

8. Incubate at 37°C with 5% $CO_2$ overnight.

9. (Day 2) Being careful not to disturb T cells at the bottom of the wells, carefully aspirate 1 mL of medium from each well (*see* **Note 11**).

10. Replenish with 1 mL of fresh or thawed viral supernatant.

11. Centrifuge plate(s) at $2,000\ g$ and 30°C, for 1 h (second spinoculation)

12. Incubate at 37°C with 5% $CO_2$ overnight.

13. (Day 3) Add 1 mL of fresh complete RPMI supplemented with 20 units of IL-2 to each well in 6-well plate. Subsequently, obtain T cell counts daily and maintain a T cell concentration of $1–2 \times 10^6$ cells/mL for optimal expansion and viability (*see* **Note 12**).

14. (Day 5) Analyze transduction efficiency of T cells by FACS (*see* **Fig. 2**).

15. (Days 7) The transduced mouse T cells may be further expanded by restimulation with CD3/CD28 beads at 1:5 bead to cell ratio (**Fig. 3**) making sure to maintain a T cell concentration of $1–2 \times 10^6$ cells/mL for optimal expansion and viability (see **Note 12**).

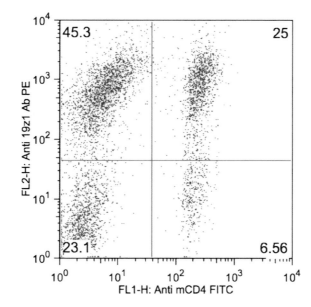

Fig. 2. FACS analysis of transduced mouse T cells stained with PE-conjugated anti-19z1 antibody and FITC-conjugated anti-mouse CD4 antibody demonstrating 70% 19z1 + T cells at day 5 following T cell isolation.

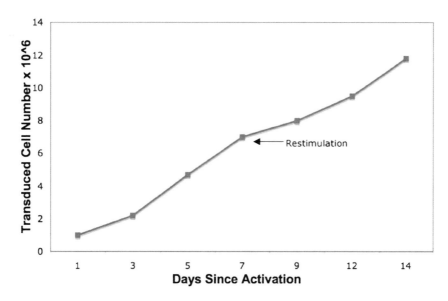

Fig. 3. Murine T cells may be further expanded by restimulation with CD3/CD28 magnetic beads following initial activation. T cells were activated with 1:1 beads to cell ratio on day 0, transduced, and restimulated (*arrow*) on day 7 with beads at a 1:5 bead to T cell ratio. Viable cells were counted via trypan blue exclusion with a hemocytometer and assessed by flow cytometry for 19z1 transduction. The viability of cells was consistently greater than 90% during data collection. Cell number represents the expansion of the transduced fraction over 2 weeks.

## 4. Notes

1. Poly-l-lysine is used to enhance adherence of Phoenix-eco cells to the tissue culture plates. Phoenix cells are otherwise loosely attached to plates and are often easily dislodged.

2. The mechanism of calcium phosphate coprecipitation based transfection is as follows: A controlled mixing of DNA, calcium chloride, and HBS generates a DNA precipitate that is subsequently dispersed onto the cultured cells. The precipitate is then taken up by the cells via endocytosis or phagocytosis. The positively charged calcium ions neutralize negatively charged phosphate backbones of DNA, so they enter into cells through a negatively charged membrane *(15)*.

3. The vortex speed needs to be fast enough to thoroughly agitate the mixture for calcium phosphate coprecipitation, but not spill the contents within the tube. A setting at 60% of full power of the vortexer is usually adequate.

4. Retroviral supernatant in complete RPMI significantly enhances the transduction efficiency and the survival of transduced mouse T cells when compared to retroviral supernatant generated in DMEM with 10% FCS (data not shown).

5. Optimal retroviral titer is obtained at day 3 following Phoenix-eco cell transfection, resulting in the highest gene transfer of murine T cells (**Fig. 4**).

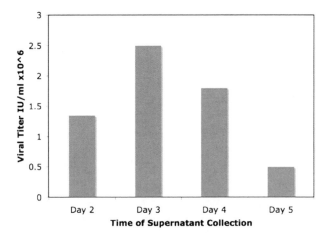

Fig. 4. Optimal viral titer, as assessed by infection of NIH 3T3 cells with serial dilution of viral supernatant, (**a**) obtained from day 3 retroviral supernatant resulted in optimal murine T cell transduction (**b**) as assessed by FACS analysis of T cells at day 7 following T cell transduction.

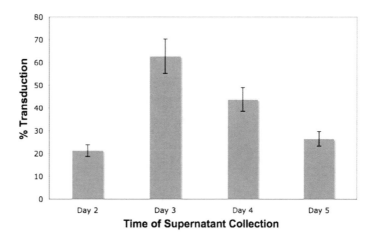

Fig. 4. (continued)

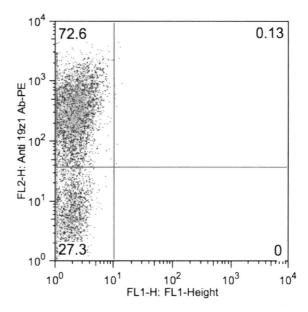

Fig. 5. FACS analysis of transduction efficiencies of (**a**) purified CD3+ T cells and (**b**) whole mouse splenocytes.

6. The retroviral supernatant can be frozen at –80°C for later transduction and stored for up to 3 months, although its titer drops by one-half with each freeze-thaw cycle. The difference in the transduction efficiency of T cells between fresh and frozen supernatants, thawed once, however, is not significant when using this protocol.

7. CD3/CD28 bead activation of enriched CD3+ T cells allows for better transduction efficiency when compared to bead stimulation of unsorted splenocytes (**Fig. 5**).

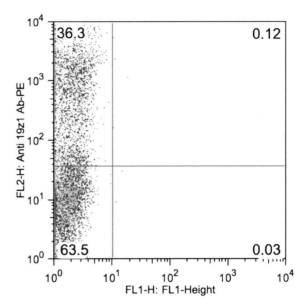

Fig. 5. (continued)

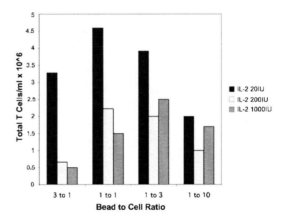

Fig. 6. Representative result of five independent experiments demonstrating comparative T cell count with various CD3/CD28 bead to cell ratios and IL-2 concentrations. The starting concentration of T cells was $1 \times 10^6$ cells/ml and cells were counted at day 5 postactivation. Optimal expansion was obtained using a 1:1 bead to T cell ratio in the context of low dose (20 IU/ml) IL-2.

8. Optimal activation of mouse T cells, as assessed by total T cell number at day 5 following activation, is attained using a CD3/CD28 bead to cell ratio of 1:1 in medium supplemented with low dose IL-2 (20 IU/mL) (**Fig. 6**).

9. Optimal gene transfer is obtained when initial spinoculation occurs at 24 h following CD3/CD28 bead activation (**Fig. 7**).

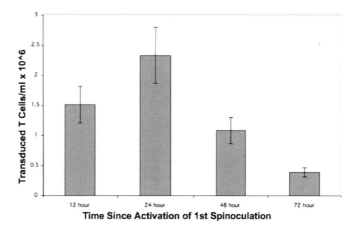

Fig. 7. The number of transduced T cells at day 5 postactivation when the two spinoculation cycles were initiated at different time points following CD3/CD28 bead activation. These data demonstrate optimal gene transfer when mouse T cells were spinoculated 24 h following initial activation.

10. A higher concentration of IL-2 is added here in anticipation of IL-2 dilution due to the addition of IL-2 free viral supernatant. Starting with 80 units/mL, the concentration of IL-2 returns to 20 units/mL following two cycles of spinoculation.

11. A quick spin of the 6-well plates at 100 g for 3 min could be used to settle the T cells to the bottom of the well if the plates are disrupted prior to medium aspiration.

12. Since mouse T cells exhibit density-dependent proliferation and survival preference, be careful to maintain a T cell concentration of $1–2 \times 10^6$ T cells/mL by either supplementing the T cell culture with complete RPMI with 20 IU IL-2/mL or splitting the culture as needed to maintain this T cell concentration.

# Acknowledgments

We thank Dr. Isabelle Riviere for reviewing the manuscript. Our work is supported by the US National Cancer Institute (grants CA95152, CA59350), The Annual Terry Fox Run for Cancer Research (New York, NY) organized by the Canada Club of New York, William H. Goodwin and Alice Goodwin and the Commonwealth Cancer Foundation for Research and the Experimental Therapeutics Center of MSKCC, and the Bocina Cancer Research Fund.

## References

1. Onodera M, Ariga T, Kawamura N, Kobayashi I, Ohtsu M, Yamada M, Tame A, Furuta H, Okano M, Matsumoto S, Kotani H, McGarrity GJ, Blaese RM, Sakiyama Y. (1998) Successful peripheral T-lymphocyte-directed gene transfer for a patient with severe combined immune deficiency caused by adenosine deaminase deficiency. *Blood.* 91, 30Ð6

2. Palù G, Pira GLi, Gennari F, Fenoglio D, Parolin C and Manca F. (2001) Genetically modified immunocompetent cells in HIV infection. *Gene Therapy.* 21, 1593Ð1600

3. Nakajima A. (2006) Application of cellular gene therapy for rheumatoid arthritis. *Mod Rheumatol.* 5, 269Ð275

4. Morgan RA, Dudley ME, Wunderlich JR, Hughes MS, Yang JC, Sherry RM, Royal RE, Topalian SL, Kammula US, Restifo NP, Zheng Z, Nahvi A, de Vries CR, Rogers-Freezer LJ, Mavroukakis SA, Rosenberg SA. (2006) Cancer regression in patients after transfer of genetically engineered lymphocytes. *Science.* 314, 126Ð129

5. Hwu P, Yannelli J, Kriegler M, Anderson WF, Perez C, Chiang Y, Schwarz S, Cowherd R, Delgado C, Mule J, Rosenberg SA. (1993) Functional and molecular characterization of tumor-infiltrating lymphocytes transduced with tumor necrosis factor-alpha cDNA for the gene therapy of cancer in humans. *J Immunol.* 150, 4104Ð4115

6. Pear WS, Scott ML, Nolan GP. (1996) Generation of high titre, helper-free retroviruses by transient transfection, in Robbins PD (ed). Methods in Molecular Medicine: Gene Therapy Protocols. Humana, Totowa, NJ, pp 41Ð57

7. Wang H, Kavanaugh MP, North RA, Kabat D. (1991) Cell-surface receptor for ecotropic murine retroviruses is a basic amino-acid transporter. *Nature.* 352, 729Ð731

8. Sadelain M. (1996) Methods for Retrovirus-Mediated Gene Transfer into Primary T-Lymphocytes, in Robbins PD (ed). Methods in Molecular Medicine: Gene Therapy Protocols. Humana, Totowa, NJ, pp 241Ð248

9. Riviere I, Brose K, Mulligan RC. (1995) Effects of retroviral design on expression of human adenosine deaminase in murine bone marrow transplant recipients engrafted with genetically modified cells. *Proc Natl Acad Sci USA.* 92, 6733Ð6737

10. Hagani AB, Riviere I, Tan C, Krause A, Sadelain M. (1999) Activation conditions determine susceptibility of murine primary T-lymphocytes to retroviral infection. *J Gene Med.* 1, 341Ð351

11. Pollok KE, Hanenberg H, Noblitt TW, Schroeder WL, Kato I, Emanuel D, Williams DA. (1998) High-efficiency gene transfer into normal and adenosine deaminase-deficient T lymphocytes is mediated by transduction on recombinant fibronectin fragments. *J Virol.* 72, 4882Ð4892

12. Brentjens RJ, Latouche JB, Santos E, Marti F, Gong MC, Lyddane C, King PD, Larson S, Weiss M, Riviere I, Sadelain M. (2003) Eradication of systemic B-cell tumors by genetically targeted human T lymphocytes co-stimulated by CD80 and interleukin-15. *Nat Med.* 9, 279Ð286

13. Deane JA, Trifilo MJ, Yballe CM, Choi S, Lane TE, Fruman DA. (2004) Enhanced T cell proliferation in mice lacking the p85beta subunit of phosphoinositide 3-kinase. *J Immunol.* 172, 6615Ð6625

14. Coligan JE, Kruisbeer AM, Marguilies DH, Shevach EM, and Strober W. (1993) T cell enrichment by nonadherence to nylon, in *Current Protocols in Immunology.* Wiley, NY, pp 3.2.1

15. Loyter A, Scangos GA, Ruddle FH. (1982) Mechanisms of DNA uptake by mammalian cells: fate of exogenously added DNA monitored by the use of fluorescent dyes. *Proc Natl Acad Sci USA.* 79, 422Ð426

# Chapter 8

# Lentiviral Vector Gene Transfer into Human T Cells

## Els Verhoeyen, Caroline Costa, and Francois-Loic Cosset

## Summary

Efficient gene transfer into T lymphocytes may allow the treatment of several genetic dysfunctions of the hematopoietic system, such as severe combined immunodeficiency, and the development of novel therapeutic strategies for diseases such as cancers and acquired diseases such as AIDS. Lentiviral vectors can transduce many types of nonproliferating cells, with the exception of some particular quiescent cell types such as resting T cells. Completion of reverse transcription, nuclear import, and subsequent integration of the lentivirus genome do not occur in these cells unless they are activated *via* the T-cell receptor (TCR) and/or by cytokines inducing resting T cells to enter in $G_{1b}$ phase of the cell cycle. In T-cell-based gene therapy trials performed to date, cells have been preactivated via their cognate antigen receptor (TCR). However, TCR stimulation shifts the T cells from naïve to memory phenotype and leads to skewing of the T-cell population. Since, especially the naïve T cells will provide a long-lasting immune reconstitution to patients these are the cells that need to be transduced for effective gene therapy. Now it is clear that use of the survival cytokines, IL-2 or IL-7, allows an efficient lentiviral vector gene transfer and could preserve a functional T-cell repertoire while maintaining an appropriate proportion of naïve and memory T cells. In this protocol we give details on lentiviral transduction of T cells using TCR-stimulation or rIL-7 prestimulation. In addition, we describe the use of a new generation of lentiviral vectors displaying T-cell-activating ligands at their surface for targeted T-cell gene transfer.

**Key words:** IL-7, Lentiviral vector, Gene therapy, TCR, Human T-cell.

## 1. Introduction

### 1.1. T-Cell Gene Therapy

#### 1.1.1. Human T Cells as Targets for Gene Therapy

One of the major advantages of using peripheral blood T cells is that they are easily accessible for genetic modification than other targets such as HSCs. Moreover, they can be isolated in high amounts. T cells most likely have a lower risk of transformation, as up to now a leukemia was not observed in T-cell-based gene therapies *(1–4)*. Recently, it was also shown that retroviral vector

Christopher Baum (ed.), *Methods in Molecular Biology, Methods and Protocols, vol. 506*
© Humana Press, a part of Springer Science + Business Media, LLC 2009
DOI: 10.1007/978-1-59745-409-4_8

integration deregulates gene expression in T cells to some extent. But this had no consequences on the function and biology of the transplanted T cells *(5)*. Of importance, the naïve T-cell subset, which responds to a novel antigen, has a long-term life span and persists over years in the patients, so at least a long-term correction can be envisaged by T-cell gene therapy. Moreover, gene transfer of T cells improved enormously in the last years thanks to the engineering of new gene transfer vehicles, the lentiviral vectors (LVs). These new vehicles enable to obtain a highly efficient transduction of T cells without changing their phenotype or their functional characteristics (*see* **Subheading 1.2.3**).

*1.1.2. T Cell Gene Therapy Trials*

Efficient gene transfer into T lymphocytes may allow the treatment of several genetic dysfunctions of the hematopoietic system, such as severe combined immunodeficiency *(1, 2)*, and the development of novel therapeutic strategies for diseases such as cancers and for acquired diseases such as acquired immunodeficiency syndrome (AIDS) *(6)*.

Indeed, gene therapy has proven over the past years to be a solution for several inherited diseases such as severe combined immunodeficiency (SCID) *(7, 8)*, adenosine deaminase deficiency (ADA), *(9)* and hemophilia *(10)*. All these have been evaluated in clinical trials with success. Very recently a gene therapy trial of chronic granulomatous disease (CGD) resulted in successful treatment of the myeloid compartment *(11)*. These trials, however, were all stem cell-based gene therapy trials. Here we give some examples of T-cell gene therapy applications.

ADA-deficient SCID was the first inherited disease investigated for T-cell gene therapy because of a postulated survival advantage for gene corrected T lymphocytes. Indeed, in an allogeneic BM transplantation normal ADA-expressing T lymphocytes have a selective advantage in SCID patients and develop a protective immune system of donor-derived T lymphocytes *(12–14)*. Aiuti et al. showed immune reconstitution in ADA-SCID patients after T-cell gene therapy *(15)*. Patients received multiple infusions of autologous retroviral vector transduced peripheral blood lymphocytes (PBLs). Discontinuation of ADA replacement therapy led then to a selective growth of the infused ADA expressing lymphocytes, which eventually replaced the non-transduced T-cell population for nearly a 100%. These ADA-corrected T cells were capable of responding to novel antigens and represented a new polyclonal T-cell repertoire. Thus PBL-ADA gene therapy leads to sustained T-cell functions in the absence of enzyme therapy. Recently, also T cells of Wiskott–Aldrich Syndrome (WAS) patients were functionally corrected by transduction with lentiviral vector encoding WAS protein *(16)*.

In the treatment of several blood cancers T-cell gene therapy has now proven to correct severe side effects of bone marrow transplantation (allo-BMT). Allo-BMT is widely used as a curative approach to many hematologic malignancies *(17, 18)*. Treatment with allogeneic T cells, either as a part of an allo-BMT or as an infusion of isolated allogeneic lymphocytes, offers the possibility of cure for patients with chronic myelogenous leukemia (CML) *(19)*. The specificity of this effect, called graft vs. leukemia effect (GVL), is not fully understood. Frequently, but not always, GVL is associated with graft vs. host disease (GVHD). The latter is a very severe side effect of allo-BMT, mediated by allospecific T cells within the graft. A strategy for the prevention of GVHD, now starting to be implemented, is the depletion of the donor T cells in vivo following infusion into the recipient in cases where GVDH becomes severe. This can be achieved by gene transfer of a suicide gene such as herpes simplex virus thymidine kinase (HSV-tk) into the T cell before infusion. Should the need for eradication of these cells arise, administration of the drug ganciclovir will induce apoptotic death of HSV-tk-transduced T cells. This has already proven to be effective in the treatment of GVHD in clinical trials *(20–22)*.

Also for acquired diseases such as AIDS that are in demand of novel therapies, T-cell gene therapy might be an important option. While the treatment of human immunodeficiency virus (HIV)-infected people has been greatly improved by the development of highly active antiretroviral therapy (HAART) several problems remain. Limitations of HAART such as appearance of drug-resistant HIV variants and toxicity show that patients need novel additional immune therapies *(23, 24)*. Here, anti-HIV T-cell gene therapy is an option since it would allow protecting the major reservoir, the CD4+ T cells, against HIV infection and at the same time could protect HIV-specific memory T cells, which are preferentially attacked by HIV.

The first clinical HIV gene therapy trials used genes that inhibit HIV RNA and protein production (e.g., Transdominant rev and tat) and lead to low antiviral activity in the patients. Indeed, genes that inhibit production of viral RNA and protein, but allow the provirus to integrate are expected to mediate selection of cells containing a suppressed HIV provirus *(25, 26)*.

An HIV-1-based lentiviral vector was engineered expressing an HIV envelope antisense that highly protected T cells from healthy and infected patients against HIV infection *(27, 28)*. A clinical phase I trial including patients with chronic HIV infection was performed using this strategy, which improved immune function in four out of five patients *(29)*. Recently, an antiviral gene was developed that encodes for a membrane-anchored peptide, which inhibits HIV entry on the level of virus-cell fusion with great efficacy *(30)*. Efficacy was also shown against several primary isolates in primary lymphocytes from different donors *(31)*. These kinds of genes that inhibit prior to integration of the provirus are expected to lead to an accumulation of noninfected, gene-protected T cells. These kinds of therapies offer a solution for patients who do not respond any more to antiretroviral therapy.

In summary, it becomes clear that T-cell based gene therapy offers a valuable alternative for treating patients.

### 1.2. Lentiviral Vectors for T Cell Gene Transfer

#### 1.2.1. MLV Retroviral Vectors Vs. Lentiviral Vectors

Vectors derived from retroviruses are probably among the most suitable tools to achieve a long-term gene transfer since they allow stable integration of a transgene and its propagation in daughter cells. To date, vectors derived from gamma-retroviruses such as murine leukemia viruses (MLVs) have been widely used for gene transfer into human T cells (32). Perhaps one of the most important drawbacks associated with the use of such vectors is their inability to transduce nonproliferating target cells. Indeed, following internalization of the vector into the target cells' cytoplasm and reverse transcription, transport of the preintegration complex to the nucleus requires the breakdown of the nuclear membranes during mitosis (33, 34). This provides a barrier to the use of MLV-based vectors in the many gene therapy protocols for which target cells such as T cells are quiescent or for which induction of cell proliferation is to be avoided. Lentiviral vectors, derived from HIV-1, have shown promise in the transduction of several resting cell types such as retinal cells, pancreatic islets, cells of the central nervous system, or progenitor and differentiated hematopoietic cells (35). For these reasons, lentiviral vectors should be preferred gene delivery vehicles over vectors derived from gamma-retroviruses such as MLVs that cannot transduce nonproliferating target cells (34). Thus, lentiviral vectors may provide a valuable alternative to overcome this problem owing to the lentivirus mechanism that allows mitosis-independent nuclear import of the preintegration complex and infection of nonproliferating cells (36–38).

#### 1.2.2. Restrictions of Lentiviral-Mediated Gene Transfer in Human T Cells

Several studies have now established the capacity of these HIV-1-derived vectors to transduce various types of nonproliferating cells both in vitro and in vivo (35). However, some cell types that are important gene therapy targets are refractory to gene transfer with lentiviral vectors. This includes, in particular, early progenitor hematopoietic stem cells in $G_0$ (39), monocytes (40, 41) and resting T lymphocytes (42). That the parental virus, HIV-1, can enter into resting T lymphocytes but does not replicate (43–47), has been attributed to multiple post-entry blocks. This includes in particular, (a) defects in initiation and completion of the reverse-transcription process (43, 46–48), (b) lack of ATP-dependent nuclear import (49), and (c) lack of integration of the proviral genome. Low levels of nucleotides in the resting cells do not entirely explain the restricted HIV-1 replication since artificially raising intracellular nucleotide pools increased reverse-transcription products but not the level of productive infection *(50)*.

Inclusion of the HIV-1 central polypurine track (cPPT) in lentiviral vectors has resulted in enhanced transduction of human progenitor stem cells and T cells. However, the improved lentiviral vectors that include the cPPT sequence still fail to transduce nonactivated T lymphocytes *(42)* most likely because the primary block in initiation and/or completion of reverse transcription could not be alleviated with the novel vectors. Thus still activation of these cells, causing $G_0$-to-$G_{1b}$ transition of the cell cycle, is required to relieve the blocks in gene delivery *(42–48)*. It is now reported that inducing the resting T cells to enter into the $G_{1b}$ phase of the cell cycle by stimulation through the T-cell receptor and CD28 costimulation receptor, using anti-CD3 plus anti-CD28 antibodies, was sufficient to render the cells susceptible to HIV-1 infection and replication *(48)*. Alternatively, exposing T cells to cytokines that do not trigger cell division could render them permissive to transduction with HIV-1-vectors *(51)*. These findings suggest that partial activation of resting T cells is sufficient for gene transfer by HIV-1-derived vectors and that DNA synthesis or mitosis of these cells is not necessary.

*1.2.3. Conservation of T Cell Phenotype After Lentiviral Transduction*

The population of mature adult T cells can be divided into two different subsets, namely, memory and naïve T cells. Naïve T cells are especially important as gene therapy target cells since they maintain the capacity to respond to novel antigens. It is also of utmost importance that the responses of T cells to antigens are not dramatically altered by the gene transfer protocol. We and others have reported that inducing cell-cycle entry into $G_{1b}$ via stimulation through the T-cell receptor allows efficient transduction of adult naïve T cells by HIV-1-based vectors and wt HIV-1 *(48, 52)*. However, TCR stimulation of T cells alters their half life and immune competence and often results in an inversion of physiologic CD4/CD8 ratio, enrichment in activated memory cells associated with loss of naïve T-cell subsets and a skewed TCR repertoire *(53–56;* **Fig. 1**). Up to now T-cell gene therapy trials are based on TCR-mediated stimulation of T cells.

However, transduction of naïve T cells is a pre-requisite for any T-cell-mediated gene therapy trial aimed at providing long-lasting immune reconstitution to patients. Therefore protocols were developed that allowed efficient LV transduction of T cells in the absence of TCR triggering. It was shown that IL-7, a master regulator of T-cell survival and homeostatic proliferation *(57–59)*, and also IL-2 and IL-15 promoted long-term survival in vitro of memory and naïve T lymphocytes. Thus it is of utmost interest to note that exposure of adult T cells to cytokines such as IL-2, IL-15, and IL-7 renders them permissive to lentiviral

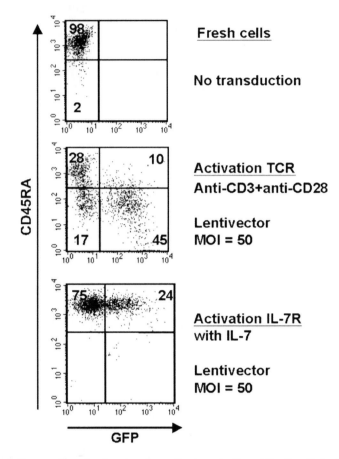

Fig. 1. Conservation of naïve phenotype after transduction of T cells with lentiviral vectors upon IL-7 prestimulation. Freshly isolated naïve adult CD4+ T cells are shown in (**a**). These naïve cells were transduced with VSV-G pseudotyped lentiviral vectors using MOIs of 10–20 after 24 h prestimulation with anti-CD3 and anti-CD28 antibody (1 μg/mL) (**b**) or after 6 days of rIL-7 (10 ng/mL) stimulation (**c**). The percentage of GFP+ naïve T cells (CD45RA+) was determined 6 days after transduction by FACS analysis.

transduction in the absence of TCR activation *(51, 56, 60–62)*. These cytokine-treated T cells move out of $G_0$ into the $G_{1b}$ phase of the cell cycle, the phase in which T cells are susceptible to LV transduction, but did not start to proliferate *(56, 60–62)*.

Nevertheless, it is clear that only use of IL-2 and IL-7, but not IL-15, for lentiviral T-cell transduction could preserve a functional T-cell repertoire and maintained an appropriate proportion of naïve and memory CD4+ and CD8+ T cells with low expression of activation markers. Moreover, functional analysis of immune response to cytomegalovirus showed that IL-2- and IL-7-mediated transduced T cells were highly immunocompetent *(60)*.

Recently it was shown that (recombinant interleukin-7) rIL-7 promotes the extended survival of both naïve and memory CD4+ T cells, whereas cell-cycle progression of these two subsets is distinct and limited. In the continued presence of biologically active cytokine, IL-7-stimulated memory cells enter and exit from the cell cycle much earlier than naïve T cells *(63)*. This is in agreement with the fact that naïve adult T cells need at least 6 days of IL-7 stimulation vs. only 3 days for memory cells to allow efficient LV transduction (**Fig. 2**).

In conclusion, LV transduction of IL-2- or IL-7-stimulated cells overcomes the limitation of TCR-mediated LV gene transfer and may improve the efficacy of T-cell based gene therapy.

*1.2.4. Targeting Gene Transfer to T Cells by Specific Vector-mediated Target Cell Activation*

This targeting strategy consists of an interaction of a ligand displayed on the surface of the vector, which upon binding to its specific receptor will induce signaling and stimulation of the target cells. As a consequence of the specific stimulation, gene transfer into the target cell, in this case the T-cell, is significantly

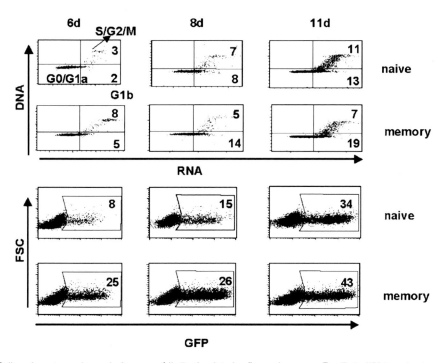

Fig. 2. Cell-cycle entry and permissiveness of IL-7 stimulated naïve and memory T cells to HIV-1 vector transduction. Naïve and memory CD4+ T cells were cultured in 10 ng/mL IL-7. Following 6, 8, 11 days of culture, cells were analyzed for cell-cycle progression by PY/7AAD staining. The percentages of cells in the $G_{1b}$ (*lower right quadrant*) and $S/G_2/M$ (*upper right quadrant*) are indicated. Naïve and memory T cells from the same donor were infected with an HIV-1-based vector expressing enhanced green fluorescence protein (EGFP) and pseudotyped with the pantropic VSV-G envelope following 6, 8, 11 days of culture in rIL-7. Infections were performed at an MOI of 20, and infected cells were detected by monitoring EGFP expression. The percentages of EGFP+ infected cells are indicated.

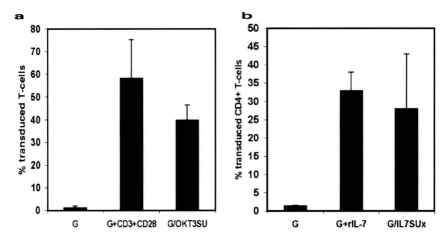

Fig. 3. Transduction of adult T cells by anti-CD3scFV- or IL-7-displaying lentiviral vectors. (**a**) PBLs were transduced with lentiviral vectors encoding for EGFP immediately upon isolation. Cells were transduced with VSV-G-pseudotyped lentiviral vectors (G) in the absence or in the presence of anti-CD3 and anti-CD28 antibodies or with anti-CD3scFv-displaying vectors (G/OK3SU). (**b**) CD4+ freshly isolated T cells were transduced with VSV-G-pseudotyped lentiviral vectors (G) in the absence or in the presence of rIL-7. Alternatively, transduction was performed with IL-7-displaying lentiviral vectors (G/IL7Sux). Multiplicities of infection of 20 were used. The % of GFP+ cells was determined on day 3 post-transduction by FACS.

enhanced. T cells are refractory to gene transfer with LVs as discussed earlier. Upgraded lentiviral vectors have therefore been engineered in order to overcome their inability to transduce nonactivated T cells. A T-cell-activating polypeptide was displayed on HIV-1 vector particles in order to target and stimulate the T cells at time of transduction. A single-chain antibody variable fragment (scFv) derived from the anti-CD3 OKT3 monoclonal antibody, which recognizes and activates the T-cell receptor, was fused to the amino-terminus of the SU subunit of the MLV envelope glycoprotein. This chimeric CD3-targeted MLV glycoprotein demonstrated reduced infectivity; thus, coexpression of an "escorting" wild-type VSV-G envelope protein was necessary to render the LV particles fully infectious. Stimulation by this surface-modified lentiviral vector was sufficient to allow efficient gene transfer in T lymphocytes, i.e., 100-fold more than the performance of unmodified lentiviral vectors in nonactivated T cells (*52*; **Fig. 3a**). However, it was demonstrated that the phenotype of the transduced naïve T cells was modified to memory cells as is the case for anti-CD3 + anti-CD28 antibody stimulation. Therefore, to transduce resting T cells while conserving their phenotype (*64, 65*), human IL-7 gene was fused to the amino-terminus of the MLV envelope glycoprotein. IL-7 displaying LVs allowed efficient transduction of naïve neonatal CD4+ T cells as well as memory CD4+ T cells allowing to maintain the functional characteristics of the naïve T cells *(56;* **Fig. 3b**). Importantly, a recent breakthrough has been made by engineering measles virus gp pseudotyped LVs that allow transduction of quiescent naive and memory T-cells *(66).*

## 2. Materials

### 2.1. Buffers and Solutions

1. *2× Hepes-buffered saline (HBS) and 2 M CaCl₂.* Calphos Mammalian Transfection Kit (Clontech, BD Biosciences, Location San Diego, USA).

2. Phosphate-buffered saline (PBS) without calcium and magnesium, without sodium bicarbonate, sterile.

3. Trypsin-ethylenediminetetraacetric acid (EDTA) 1× Hank's balanced salt solution without calcium and magnesium, sterile.

4. Ficol-Paque Plus, sterile.

5. *Nucleic acid-staining solution (NASS).* 0.15 M NaCl in 0.1 M phosphate-citrate buffer containing 5 mM sodium EDTA and 0.5% bovine serum albumin.

6. *Nucleic acid staining solution (NASS).* 0.15 M NaCl in 0.1 M phosphate-citrate buffer containing 5 mM sodium EDTA and 0.5 bovine serum albumin (BSA, fraction 5, Sigma, Saint Quentin Fallavier, France pH 6.0).

7. *7AAD- (7-amino-actinomycin D-) staining buffer.* 0.03% saponin in NASS buffer containing 20 μM 7AAD.

8. Pyronine-Y (PY) stock solution 100 μM in water.

### 2.2. Media

1. Fetal calf serum (FCS), sterile.

2. DMEM (Dulbecco's modified Eagle medium) with 0.11 g/L sodium pyridoxine and pyridoxine. DMEM is supplemented with 10% FCS, 100 μg/L streptomycin, 100 U/mL penicillin (stored at 4°C).

3. RPMI medium is supplemented with 10% FCS, 100 μg/L streptomycin, 100 U/mL penicillin (stored at 4°C).

### 2.3. Nucleic Acids

1. Lentiviral vector DNA encoding for an HIV-1-derived self-inactivating vector with the internal SFFV (spleen focus foamy virus) promoter driving the reporter gene GFP.

2. Envelope glycoprotein expressing plasmids:

   (a) Fusion glycoprotein: stomatitis virus G glycoprotein (VSV-G).

   (b) *Activating and targeting* glycoproteins for T cells *(1)* OKT3-SU Env (anti-CD3scFv fused to the murine leukemia virus envelope) and *(2)* IL-7SU Env (human interleukin-7 fused to murine leukemia virus envelope glycoprotein).

3. Virus structural protein (gagpol) expressing plasmid (pCMV8.91).

**2.4 Cells and Tissue**

1. 293T cells.

2. *Source of T cells.* Fresh adult or cord blood (*see* **Note 1**).

**2.5. Special Equipment**

Magnetic separation device (Dynal Location Biotech ASA, Oslo, Norway).

**2.6. Additional Reagents**

1. Rosette Sep cocktail from stem cell Technologies for separation of total T cells or CD4+ T cells (*see* **Note 2**).

2. Pan Mouse IgG Kit (Dynal Biotech ASA, Oslo, Norway) (*see* **Note 3**).

3. 24-well cell culture-coated tissue culture plates.

4. 0.45-μm filter.

5. Mouse monoclonal antibodies:

   (a) *For identification of phenotype.* Anti-hCD45RA-phycoerythrin (PE), anti-hCD45RO-PE, anti-hCD3-PE, anti-hCD4-PE, anti-hCD69-PE, anti-HLADR-PE, anti-hCD25-PE, and corresponding PE-conjugated mouse IgG controls.

   (b) *For isolation of nonactivated naïve and memory CD4+ T-cell population.* Anti-hCD69, anti-hHLADR, and anti-hCD45RA or anti-hCD45RO (BD Pharmingen, San Diego, USA).

   (c) *For stimulation via the T-cell receptor.* Anti-CD3 and anti-CD28 (BD Pharmingen, San Diego, USA).

6. *Cytokines.* Human rIL-7 (Preprotech, LocationRocky Hill, USA).

# 3. Methods

**3.1. Production of VSV-G Pseudotyped Lentiviral Vectors and Lentivectors Displaying T-Cell-Activating polypeptides**

1. *Day 0.* $2.5 \times 10^6$ 293T cells are seeded the day before transfection in 10-cm plates in a final volume of 10 mL DMEM.

2. *Day 1.* Cotransfection of HIV packaging construct (8.6 μg) with the lentiviral gene transfer vector (8.6 μg) and (a) the glycoprotein VSV-G (3 μg) for VSV-G pseudotyped lentiviral vectors or (b) the glycoproteins VSV-G (1.5 μg) and OKT3SU env or (c) the glycoproteins VSV-G (1.5 μg) and IL7SUx env (1.5 μg) is performed using the Clontech calcium-phosphate transfection system.

3. *Day 2.* 15 h after transfection, the medium is replaced with 6 mL of fresh DMEM medium (*see* **Note 4**).

4. *Day 3.* 36 h after transfection, vectors are harvested, filtrated through 0.45-μm pore-sized membrane, and stored at –80°C for 2–3 months.

**3.2. Immunoselection of Human T Cells**

1. Add to the adult or cord blood Rossette Sep isolation cocktail for T-cell or CD4+ T-cell (25 μL/mL blood) and incubate for 20 min at room temperature while rocking.

2. Dilute blood + Rosette Sep cocktail 1: 1 with PBS and gently layer 35 mL of this diluted product on 15 mL Ficoll in a 50-mL tube.

3. Centrifuge the cells at 850 g for 30 min, 20°C without brake and collect the layer containing mononuclear cells (*see* **Note 2**).

4. Wash the collected mononuclear cell interface in PBS/2% FCS at 850 g, 20°C for 10 min.

5. If further isolation of memory and naïve T-cell subsets is wanted the T cells are resuspended at $10^8$/mL PBS and incubated for 30 min at 4°C with anti-CD45RA antibody (1 μg/$10^6$ target cells) for isolation of memory T-cell or with anti-CD45RO antibody (1 μg/$10^6$ target cells) for isolation of naïve T cells (*see* **Note 5**).

6. Wash cells to remove the unbound antibody and resuspend in PBS/2% FCS.

7. Add the Pan Mouse beads to the T cells according to the manufacturer's indication (DynalBiotech ASA, Oslo, Norway) and incubate for 30 min while rocking at 4°C.

8. Place the tube into the Dynal magnetic device for 2 min and collect the unbound cells.

9. Remove the tube from the magnet and wash the beads once with PBS/2% FCS.

10. Repeat **step 8**.

11. The purity of the T-cell subsets is routinely 90–95%.

**3.3. Titer Determination**

1. *Day-1.* 293T cells are seeded in DMEM at a density of 2 × $10^5$ cells per well in 6-well plates in a final volume of 2 mL.

2. *Day 0.* Serial dilutions of vector preparations were added to 293T cells and incubated O/N.

3. *Day 1.* Medium on the cells is replaced with 2 mL fresh DMEM and cells are incubated for 72 h.

4. *Day 3.* Cells are trypsinized and transferred in FACS tubes. The percentage of green fluorescent protein (GFP)-positive cells is determined by fluorescence-activated cell sorter (FACS) analysis.

**3.4. Analysis of Transduction and Titer**

1. Transduction efficiency is usually determined as the percentage of GFP-positive cells after transduction of $3 \times 10^5$ target cells with 1 mL of viral supernatant.

2. *Infectious titers.* Are provided as transducing units (TU)/mL and can be calculated by using the formula: Titer = %inf × $(3 \times 10^5/100) \times d_i$; where "$d$" is the dilution factor of the viral supernatant and "%inf" is the percentage of GFP-positive cells as determined by FACS analysis using dilutions of the viral supernatant that transduce between 5 and 10% of GFP-positive cells.

3. *Multiplicities of infection (MOI).* Ratio between infectious particles and target cells that are required to optimally transduce target cells of interest, which are generally much less permissive to transduction than the cells used for titrations

**3.5. Cell-Cycle Fractionation by 7AAD/ Pyronin Y Staining (Fig. 2a)**

1. *First DNA staining with 7AAD is performed.* T cells are resuspended at concentration of $10^6$ cells/mL in 7AAD-staining buffer and cells are incubated for 30 min at room temperature (*see* **Note 6**).

2. Cells are put on ice for at least 10 min.

3. Subsequently, RNA staining is performed by adding to the T cells, resuspended in the 7AAD-staining buffer (no washing is required), PY at a final concentration of 5 µM; cells are kept on ice for 10 min and are immediately analyzed on a FACS Calibur (BD Biosciences, San Diego, USA).

4. The living T cells are gated and in this gate cells in $G_0$ are identified by their minimal RNA (PY) and DNA content (7AAD), whereas cells in $G_{1b}$ are identified by low DNA content but increased RNA content, S + $G_2$ + M phases were defined as those with high or maximal PY staining and increased DNA staining (**Fig. 2a**).

**3.6. Transduction of Human T Cells**

1. $1 \times 10^6$ T cells are seeded in RPMI medium 10% FCS in 24-well plates and are then prestimulated:

   (a) For TCR-mediated activation stimulation with anti-CD3 antibody (1 µg/mL) and anti-CD28 antibody (1 µg/ mL) is performed during 24 h to obtain efficient lentiviral transduction (*see* **Note 7**).

   (b) In the absence of TCR activation stimulation with rhIL-7 (10 ng/mL) was performed during 3–4 days for total or memory adult T cells; naïve T cells need to be prestimulated for 6–12 days to obtain efficient lentiviral transduction (**Fig. 2b**).

   (c) No prestimulation is needed for lentiviral vector displaying T-cell-activating ligands.

2. Prestimulated T cells are transduced at an MOI of 10 overnight and transduced cells are washed and resuspended in RPMI/10% FCS supplemented with IL-2 (1 ng/mL) for the TCR-activated cells or with rhIL-7 (10 ng/mL) for the IL-7 prestimulated cells during a further 72 h before transduction efficiency is determined by flow cytometry. Alternatively, the OKT3SU/VSV-G and IL-7SUx/VSV-G codisplaying vectors are added to the freshly isolated T cells at an MOI of 10. The vector is not removed until analysis by flow cytometry at 72 h post-transduction (**Fig. 3**).

***3.7. Evaluation of Transduced T Cell Phenotype by Cell Surface Marker Staining (Fig. 3)***

1. The transduced T cells were divided into aliquots in PBS/2% FCS to stain for different phenotypic markers: anti-hCD45RA-PE for naïve T-cell identification, anti-hCD45RO-PE, antibodies for memory T-cell phenotype and anti-CD25-PE, anti-CD69PE, and anti-HLADR-PE to verify T-cell activation. In all cases, corresponding PE-conjugated mouse IgG controls need to be used to evaluate specific labeling. Incubation is performed at 4°C for 20 min at concentrations indicated by the manufacturer.

2. Cells are washed once with PBS/2% FCS.

3. GFP+ cells are detected for the different T-cell subsets by two-color flow cytometry analysis.

# 4. Notes

1. T cells from cord blood contain over 90% naïve T cells that reside in a more immature stage than adult naïve T cells which make up only one-third of the adult T-cell population. The former naïve cells are considered as recent thymic emigrants and enter much easier into cell cycle after cytokine stimulation with IL-7 as compared to adult naïve T cells.

2. The Rossette sep cocktail works through negative selection in order not to activate the T cells. This cocktail contains tetrameric antibody complexes which crosslink unwanted cells to the red blood cells. This increases the weight of the unwanted cells that will be pelleted together with erythrocytes after a Ficoll gradient. The layer of mononuclear cells that appears at the top of the Ficoll after centrifugation contains only T cells (purity of T cells is over 95%).

3. The Pan Mouse magnetic beads will allow to retrieve all unwanted cells that are bound by a cell type-specific mouse monoclonal antibody. This methodology allows to remove different cell types from the T-cell population isolated by

Rosette Sep isolation, e.g., all activated cells and memory cell can be removed in a single step through negative selection, without activating the target cells.

4. For the lentiviral vector displaying T-cell-activating polypeptides, extra Hepes is added to the DMEM medium to guarantee stability of the OKTSU and IL-7SU env gps displayed on the lentiviral vectors after freezing at –80°C.

5. If removal of activated T cells is wanted, add anti-CD69 and anti-HLADR mouse monoclonal antibody in addition to the anti-CD45RA or anti-CD45R0 antibody at 1 (1 μg/$10^6$ target cells). Freshly isolated T cells contain approximately 50% memory cells, 50% naïve cells of which maximum 10% express the T-cell activation markers CD69 and HLA-DR.

6. For 7AAD/PY cell-cycle staining it is very important to respect the order of staining: first the DNA is stained with 7AAD followed by staining of the RNA by PY. PY can stain as well DNA as RNA and will only stain the RNA when the DNA was previously stained with 7AAD. This allows to identify the different cell-cycle phases: $G_0$, $G_{1b}$, $S/G_2/M$ as depicted in **Fig. 2a**.

7. Alternatively, TCR stimulation can be performed by (1) precoating the culture plates with anti-CD3 antibody (1 μg/mL PBS) for 2 h at RT. The anti-CD3 antibody solution is then removed and replaced with 1 mL of T cells in RPMI/10% FCS to which anti-CD28 is added; (2) incubating T cells with beads that are coated with optimized amounts of anti-CD3 and anti-CD28 according to the manufacturer's protocol.

## Acknowledgments

This work was supported by the Agence Nationale pour la Recherche contre le SIDA (ANRS), the European community (contract LSHB-CT-2004-005242, "Consert"), and INSERM. We acknowledge the contributions of Naomi Taylor, Louise Swainson, and Valerie Dardahlon to these studies.

## References

1. Blaese, R.M., Culver, K.W., Miller, A.D., Carster, C.S., Fleisher, T., Clerci, M., Shearer, G., Chang, L., Chiang, Y., Tolstoshev, P., Greenblatt, J.J., Rosenberg, S.A., Klein, H., Berger, M., Mullen, C.A., Ramsey, W.J., Muul, L., Morgan, R.A., Anderson, W.F. (1995). T-lymphocyte directed gene therapy for ADA-SCID: initial trial results after 4 years. *Science* 270, 275–280

2. Bordignon, C., Notaranglo, L.D., Nobili, N., Ferrari, G., Casorati, G., Panina, P., Mazzolana, E., Maggioni, D., Rossi, C., Servida, P.,

Ugazio, A.G., Mavillo, F. (1995). Gene therapy in peripheral blood lymphocytes and bone marrow for ADA-immunodeficient patients. *Science* 270, 470–475

3. Mitsuyasu, R.T., Anton, P.A., Deeks, S.G., Scadden, D.T., Connick, E., Downs, M.T., Bakker, A., Roberts, M.R., June, C.H., Jalali, S., Lin, A.A., Pennathur-Das, R., Hege, K.M. (2000). Prolonged survival and tissue trafficking following adoptive transfer of CD4zeta gene-modified autologous CD4(+) and CD8(+) T cells in human immunodeficiency virus-infected subjects. *Blood* 96(3), 785–93

4. Deeks, S.G., Wagner, B., Anton, P.A., Mitsuyasu, R.T., Scadden, D.T., Huang, C., Macken, C., Richman, D.D., Christopherson, C., June, C.H., Lazar, R., Broad, D.F., Jalali, S., Hege, K.M. (2002). A phase II randomized study of HIV-specific T-cell gene therapy in subjects with undetectable plasma viremia on combination antiretroviral therapy. *Mol Ther* 5(6), 788–797

5. Recchia, A., Bonini, C., Magnani, Z., Urbinati, F., Sartori, D., Muraro, S., Tagliafico, E., Bondanza, A., Stanghellini, M.T., Bernardi, M., Pescarollo, A., Ciceri, F., Bordignon, C., Mavilio, F. (2006). Retroviral vector integration deregulates gene expression but has no consequence on the biology and function of transplanted T cells. *Proc Natl Acad Sci U S A* 103(5), 1457–1462

6. Buchschacher, G.L., Wong-Staal, F. (2002). Approaches to gene therapy for human immunodeficiency virus infection. *Hum Gene Ther* 12, 1013–1019

7. Hacein-Bey-Abina, S., Le Deist, F., Carlier, F., Bouneaud, C., Hue, C., De Villartay, J.P., Thrasher, A.J., Wulffraat, N., Sorensen, R., Dupuis-Girod, S., Fischer, A., Davies, E.G., Kuis, W., Leiva, L., Cavazzana-Calvo, M. (2002). Sustained correction of X-linked severe combined immunodeficiency by ex vivo gene therapy. *N. Engl. J. Med* 346, 1185–1193

8. Cavazzana-Calvo, M., Fischer, A. (2004). Efficacy of gene therapy for SCID is being confirmed. *Lancet* 364(9452), 2155–2156

9. Aiuti, A., Slavin, S., Aker, M., Ficara, F., Deola, S., Mortellaro, A., Morecki, S., Andolfi, G., Tabucchi, A., Carlucci, F., Marinello, E., Cattaneo, F., Vai, S., Servida, P., Miniero, R., Roncarolo, M.G., Bordignon, C. (2002) Correction of ADA-SCID by stem cell gene therapy combined with nonmyeloablative conditioning. *Science*, 296(5577), 2410–2413

10. Powell, J.S., Ragni, M.V., White, G.C. 2nd, Lusher, J.M., Hillman-Wiseman, C., Moon, T.E., Cole, V., Ramanathan-Girish, S., Roehl,

H., Sajjadi, N., Jolly, D.J., Hurst, D. (2003). Phase 1 trial of FVIII gene transfer for severe hemophilia A using a retroviral construct administered by peripheral intravenous infusion. *Blood* 102(6), 2038–2045

11. Ott, M.G., Schmidt, M., Schwarzwaelder, K., Stein, S., Siler, U., Koehl, U., Glimm, H., Kuhlcke, K., Schilz, A., Kunkel, H., Naundorf, S., Brinkmann, A., Deichmann, A., Fischer, M., Ball, C., Pilz, I., Dunbar, C., Du, Y., Jenkins, N.A., Copeland, N.G., Luthi, U., Hassan, M., Thrasher, A.J., Hoelzer, D., von Kalle, C., Seger, R., Grez, M. (2006). Correction of X-linked chronic granulomatous disease by gene therapy, augmented by insertional activation of MDS1-EVI1, PRDM16 or SETBP1. *Nat Med* 12(4), 401–409

12. Tjonnfjord, G.E., Steen, R., Veiby, O.P., Friedrich, W., Egeland, T. (1994). Evidence for engraftment of donor-type multipotent CD34+ cells in a patient with selective T-lymphocyte reconstitution after bone marrow transplantation for B-SCID. *Blood* 84(10), 3584–3589

13. Hirschhorn, R., Yang, D.R., Puck, J.M., Huie, M.L., Jiang, C.K., Kurlandsky, L.E. (1996). Spontaneous in vivo reversion to normal of an inherited mutation in a patient with adenosine deaminase deficiency. *Nat Genet* 13(3), 290–295

14. Stephan, V., Wahn, V., Le Deist, F., Dirksen, U., Broker, B., Muller-Fleckenstein, I., Horneff, G., Schroten, H., Fischer, A., de Saint Basile, G. (1996). A typical X-linked severe combined immunodeficiency due to possible spontaneous reversion of the genetic defect in T cells. *N Engl J Med* 335(21), 1563–1567

15. Aiuti, A., Vai, S., Mortellaro, A., Casorati, G., Ficara, F., Andolfi, G., Ferrari, G., Tabucchi, A., Carlucci, F., Ochs, H.D., Notarangelo, L.D., Roncarolo, M.G., Bordignon, C. (2002) Immune reconstitution in ADA-SCID after PBL gene therapy and discontinuation of enzyme replacement. *Nat Med* 8(5), 423–425

16. Dupre, L., Trifari, S., Follenzi, A., Marangoni, F., Lain de Lera, T., Bernad, A., Martino, S., Tsuchiya, S., Bordignon, C., Naldini, L., Aiuti, A., Roncarolo, M.G. (2004). Lentiviral vector-mediated gene transfer in T cells from Wiskott–Aldrich syndrome patients leads to functional correction. *Mol Ther* 10(5), 903–915

17. Szydlo, R., Goldman, J.M., Klein, J.P., Gale, R.P., Ash, R.C., Bach, F.H., Bradley, B.A., Casper, J.T., Flomenberg, N., Gajewski, J.L., Gluckman, E., Henslee-Downey, P.J., Hows,

J.M., Jacobsen, N., Kolb, H.J., Lowenberg, B., Masaoka, T., Rowlings, P.A., Sondel, P.M., van Bekkum, D.W., van Rood, J.J., Vowels, M.R., Zhang, M.J., Horowitz, M.M. (1997). Results of allogeneic bone marrow transplants for leukemia using donors other than HLA-identical siblings. *J Clin Oncol* 15(5), 1767–1777

18. Vigorito, A.C., Azevedo, W.M., Marques, J.F., Azevedo, A.M., Eid, K.A., Aranha, F.J., Lorand-Metze, I., Oliveira, G.B., Correa, M.E., Reis, A.R., Miranda, E.C., de Souza, C.A. (1998). A randomised, prospective comparison of allogeneic bone marrow and peripheral blood progenitor cell transplantation in the treatment of haematological malignancies. *Bone Marrow Transplant* 22(12), 1145–1151

19. Horowitz, M.M., Gale, R.P., Sondel, P.M., Goldman, J.M., Kersey, J., Kolb, H.J., Rimm, A.A., Ringden, O., Rozman, C., Speck, B. (1990). Graft-versus-leukemia reactions after bone marrow transplantation. *Blood*, 75(3), 555–562

20. Tiberghien, P., Ferrand, C., Lioure, B., Milpied, N., Angonin, R., Deconinck, E., Certoux, J.M., Robinet, E., Saas, P., Petracca, B., Juttner, C., Reynolds, C.W., Longo, D.L., Herve, P., Cahn, J.Y. (2001) Administration of herpes simplex-thymidine kinase-expressing donor T cells with a T-cell-depleted allogeneic marrow graft. *Blood* 97(1), 63–72

21. Bonini, C., Ferrari, G., Verzeletti, S., Servida, P., Zappone, E., Ruggieri, L., Ponzoni, M., Rossini, S., Mavilio, F., Traversari, C., Bordignon, C. (1997). HSV-TK gene transfer into donor lymphocytes for control of allogeneic graft-versus-leukemia. *Science* 276(5319), 1719–1724

22. Bondanza, A., Valtolina, V., Magnani, Z., Ponzoni, M., Fleischhauer, K., Bonyhadi, M., Traversari, C., Sanvito, F., Toma, S., Radrizzani, M., La Seta-Catamancio, S., Ciceri, F., Bordignon, C., Bonini, C. (2006). Suicide gene therapy of graft-versus-host disease induced by central memory human T lymphocytes. *Blood* 107(5), 1828–1836

23. Chun, T.W., Stuyver, L., Mizell, S.B., Ehler, L.A., Mican, J.A., Baseler, M., Lloyd, A.L., Nowak, M.A., Fauci, A.S. (1997). Presence of an inducible HIV-1 latent reservoir during highly active antiretroviral therapy. *Proc Natl Acad Sci U S A* 94, 13193–13197

24. Finzi, D., Hermankova, M., Pierson, T., Carruth, L.M., Buck, C., Chaisson, R.E., Quinn, T.C., Chadwick, K., Margolick, J., Brookmeyer, R., Gallant, J., Markowitz, M., Ho, D.D., Richman, D.D., Siliciano, R.F. (1997). Identification of a reservoir for HIV-1 in patients on highly active antiretroviral therapy. *Science* 278, 1295–1300

25. Woffendin, C., Ranga, U., Yang, Z., Xu, L., Nabel, G.J. (1996). Expression of a protective gene-prolongs survival of T cells in human immunodeficiency virus-infected patients. *Proc Natl Acad Sci U S A* 93, 2889–2894

26. Ranga, U., Woffendin, C., Verma, S., Xu, L., June, C.H., Bishop, D.K., Nabel, G.J. (1998). Enhanced T cell engraftment after retroviral delivery of an antiviral gene in HIV-infected individuals. *Proc Natl Acad Sci U S A* 95, 1201–1206

27. Braun, S.E., Wong, F.E., Connole, M., Qiu, G., Lee, L., Gillis, J., Lu, X., Humeau, L., Slepushkin, V., Binder, G.K., Dropulic, B., Johnson, R.P. (2005). Inhibition of simian/human immunodeficiency virus replication in CD4+ T cells derived from lentiviral-transduced CD34+ hematopoietic cells. *Mol Ther* 12(6), 1157–1167

28. Humeau, L.M., Binder, G.K., Lu, X., Slepushkin, V., Merling, R., Echeagaray, P., Pereira, M., Slepushkina, T., Barnett, S., Dropulic, L.K., Carroll, R., Levine, B.L., June, C.H., Dropulic, B. (2004). Efficient lentiviral vector-mediated control of HIV-1 replication in CD4 lymphocytes from diverse HIV+ infected patients grouped according to CD4 count and viral load. *Mol Ther* 9(6), 902–913

29. Levine, B.L., Humeau, L.M., Boyer, J., MacGregor, R.R., Rebello, T., Lu, X., Binder, G.K., Slepushkin, V., Lemiale, F., Mascola, J.R., Bushman, F.D., Dropulic, B., June, C.H. (2006). Gene transfer in humans using a conditionally replicating lentiviral vector. *Proc Natl Acad Sci U S A* 103(46), 17372–17377

30. Hildinger, M., Dittmar, M.T., Schult-Dietrich, P., Fehse, B., Schnierle, B.S., Thaler, S., Stiegler, G., Welker, R., von Laer, D. (2001). Membrane-anchored peptide inhibits human immunodeficiency virus entry. *J. Virol* 75, 3038–3042

31. Egelhofer, M., Brandenburg, G., Martinius, H., Schult-Dietrich, P., Melikyan, G., Kunert, R., Baum, C., Choi, I., Alexandrov, A., von Laer, D. (2004). Inhibition of human immunodeficiency virus type 1 entry in cells expressing gp41-derived peptides. *J Virol* 78(2), 568–575

32. Anderson, W.F. (1998). Human gene therapy. *Nature* 392, 25–30

33. Miller, D.G., Adam, M.A., Miller, A.D. (1990). Gene transfer by retrovirus vectors occurs only in cells that are actively replicating at the time of infection. *Mol Cell Biol* 10, 4239–4242

34. Roe, T., Reynolds, T.C., Yu, G., Brown, P.O. (1993). Integration of murine leukemia virus DNA depends on mitosis. *EMBO J* 12, 2099–2108

35. Vigna, E., Naldini, L. (2000). Lentiviral vectors: excellent tools for experimental gene transfer and promising candidates for gene therapy. *J Gene Med* 2, 308–316

36. Bukrinsky, M.I., Sharova, N., Dempsey, M.P., Stanwick, T.L., Haggerty, S., Stevenson, M. (1992). Active nuclear import of human immunodeficiency virus type 1 preintegration complexes. *Proc Natl Acad Sci U S A* 89, 6580–6584

37. Lewis, P., Hensel, M., Emerman, M. (1992). Human immunodeficiency virus infection of cells arrested in the cell cycle. *EMBO J* 11, 3053–3058

38. Weinberg, J.B., Matthews, T.J., Cullen, B.R., Malim, M.H. (1991). Productive human immunodeficiency virus type 1 (HIV-1) infection of non proliferative human cells. *J Exp Med* 174, 1477–1482

39. Sutton, R.E., Reitsma, M.J., Uchida, N., Brown, P.O. (1999). Transduction of human progenitor hematopoietic stem cells by human immunodeficiency virus type 1-based vectors is cell cycle dependent. *J Virol* 73, 3649–3660

40. Kootstra, N.A., Zwart, B.M., Schuitemaker, H. (2000). Diminished human immunodeficiency virus type 1 reverse transcription and nuclear transport in primary macrophages arrested in early G(1) phase of the cell cycle. *J Virol* 74, 1712–1717

41. Neil, S., Martin, F., Ikeda, Y., Collins, M. (2001). Postentry restriction to human immunodeficiency virus-based vector transduction in human monocytes. *J Virol* 75, 5448–5456

42. Dardalhon, V., Herpers, B., Noraz, N., Pflumio, F., Guetard, D., Leveau, C., Dubart-kupperschmitt, A., Charneau, P., Taylor, N. (2001). Lentivirus-mediated gene transfer in primary T cells is enhanced by a central DNA flap. *Gene ther* 8, 190–198

43. Stevenson, M., Stanwick, T.L., Dempsey, M.P., Lamonica, C.A. (1990). HIV-1 replication is controlled at the level of T cell activation and proviral integration. *EMBO J* 9, 1551–1560

44. Sun, Y., Pinchuk, L.M., Agy, M.B., Clark, E.A. (1997). Nuclear import of HIV-1 DNA in resting CD4+ T cells requires a cyclosporin A-sensitive pathway. *J Immunol* 158, 512–517

46. Zack, J.A., Haislip, A.M., Krogstad, P., Chen, I.S.Y. (1992). Incompletely reverse-transcribed human immunodeficiency virus type 1 genomes in quiescent cells can function as intermediates in the retroviral life cycle. *J Virol* 68, 1717–1725

47. Zack, J.A. (1995). The role of the cell cycle in HIV-1 infection. *Adv.Exp.Med.Biol* 374, 27–31

48. Korin, Y.D., Zack, J.A. (1993). Progression to the G1b phase of the cell cycle is required for completion of human immunodeficiency virus type 1 reverse transcription in T cells. *J Virol* 72, 3161–3168

49. Bukrinsky, M.I., Haggerty, S., Dempsey, M.P., Sharova, N., Adzhubel, A., Spitz, L., Lewis, P., Goldfarb, D., Emerman, M., Stevenson, M. (1993). A nuclear localization signal within HIV-1 matrix protein that governs infection of non-dividing cells. *Nature* 365, 666–669

50. Korin, Y.D., Zack, J.A. (1999). Nonproductive human immunodeficiency virus type 1 infection in nucleoside-treated Go lymphocytes. *J Virol* 73, 6526–6532

51. Unutmaz, D., KewalRamani, V.N., Marmon, S., Littman, D.R. (1999). Cytokine signals are sufficient for HIV-1 infection of resting human T lymphocytes. *J Exp Med* 189, 1735–1746

52. Maurice, M., Verhoeyen, E., Salmon, P., Trono, D., Russell, S.J., Cosset, F.-L. (2002). Efficient gene transfer into human primary blood lymphocytes by surface-engineered lentiviral vectors that display a T cell-activating polypeptide. *Blood* 99, 2342–2350.

53. Roth, M.D. (1994). Interleukin 2 induces the expression of CD45RO and the memory phenotype by CD45RA+ peripheral blood lymphocytes. *J Exp Med* 179(3), 857–864

54. Marktel, S., Magnani, Z., Ciceri, F., Cazzaniga, S., Riddell, S.R., Traversari, C., Bordignon, C., Bonini, C. (2003). Immunologic potential of donor lymphocytes expressing a suicide gene for early immune reconstitution after hematopoietic T-cell-depleted stem cell transplantation. *Blood* 101(4), 1290–1298

55. Ferrand, C., Robinet, E., Contassot, E., Certoux, J.M., Lim, A., Herve, P., Tiberghien, P. (2000). Retrovirus-mediated gene transfer in primary T lymphocytes: influence of the transduction/selection process and of ex vivo expansion on the T cell receptor beta chain hypervariable region repertoire. *Hum Gene Ther* 11(8), 1151–1164

56. Verhoeyen, E., Dardalhon, V., Ducrey-Rundquist, O., Trono, D., Taylor, N., Cosset, F.-L. (2003). IL-7 surface-engineered lentiviral vectors promote survival and efficient gene transfer in resting primary T-lymphocytes. *Blood* 101, 2167–2174

57. Fry, T.J., Mackall, C.L. (2001) Interleukin-7: master regulator of peripheral T-cell homeostasis? *Trends Immunol* 22(10), 564–571

58. Geiselhart, L.A., Humphries, C.A., Gregorio, T.A., Mou, S., Subleski, J., Komschlies, K.L. (2001). IL-7 administration alters the CD4:CD8 ratio, increases T cell numbers, and increases T cell function in the absence of activation. *J Immunol* 166(5), 3019–3027

59. Rathmell, J.C., Farkash, E.A., Gao, W., Thompson, C.B. (2001). IL-7 enhances the survival and maintains the size of naive T cells. *J Immunol* 167(12), 6869–6876

60. Cavalieri, S., Cazzaniga, S., Geuna, M., Magnani, Z., Bordignon, C., Naldini, L., Bonini, C. (2003). Human T lymphocytes transduced by lentiviral vectors in the absence of TCR activation maintain an intact immune competence. *Blood* 102(2), 497–505

61. Dardalhon, V., Jaleco, S., Kinet, S., Herpers, B., Steinberg, M., Ferrand, C., Froger, D., Leveau, C., Tiberghien, P., Charneau, P., Noraz, N., Taylor, N. (2001). IL-7 differentially regulates cell cycle progression and HIV-1-based vector infection in neonatal and adult CD4+ T cells. *Proc Natl Acad Sci U S A* 98(16), 9277–82

62. Ducrey-Rundquist, O., Guyader, M., Trono, D. (2002). Modalities of interleukin-7-induced human immunodeficiency virus permissiveness in quiescent T lymphocytes. *J Virol* 76(18), 9103–9111

63. Swainson, L., Verhoeyen, E., Cosset, F.L., Taylor, N. (2006). IL-7R alpha gene expression is inversely correlated with cell cycle progression in IL-7-stimulated T lymphocytes. *J Immunol* 176(11), 6702–6708

64. Soares, M.V., Borthwick, N.J., Maini, M.K., Janossy, G., Salmon, M., Akbar, A.N. (1998). IL-7-dependent extrathymic expansion of CD45RA+ T cells enables preservation of a naive repertoire. *J Immunol* 161(11), 5909–5917

65. Webb, L.M., Foxwell, B.M., Feldmann, M. (1999) Putative role for interleukin-7 in the maintenance of the recirculating naive CD4+ T-cell pool. *Immunology* 98(3), 400–405

66. Frecha, C., Costa, C., Negre, D., Gauthier, E., Russell, S.J., Cosset, F.L., Verhoeyen, E. (2008) Stable transduction of quiescent T-cells without induction of cycle progression by a novel lentiviral vector pseudotyped with measles virus glycoproteins. *Blood*, Sep 23 [Epub ahead of print]

# Chapter 9

# DNA Transposons for Modification of Human Primary T Lymphocytes[1]

## Xin Huang, Andrew Wilber, R. Scott McIvor, and Xianzheng Zhou

## Summary

Genetic modification of peripheral blood T lymphocytes (PBL) or hematopoietic stem cells (HSC) has been shown to be promising in the treatment of cancer (Nat Rev Cancer 3:35–45, 2003), transplant complications (Curr Opin Hematol 5:478–482, 1998), viral infections (Science 285:546–551, 1999), and immunodeficiencies (Nat Rev Immunol 2:615–621, 2002). There are also significant implications for the study of T cell biology (J Exp Med 191:2031–2037, 2000). Currently, there are three types of vectors that are commonly used for introducing genes into human primary T cells: oncoretroviral vectors, lentiviral vectors, and naked DNA. Oncoretroviral vectors transduce and integrate only in dividing cells. However, it has been shown that extended ex vivo culture, required by oncoretroviral-mediated gene transfer, may alter the biologic properties of T cells (Nat Med 4:775–780, 1998; Int Immunol 9:1073–1083, 1997; Hum Gene Ther 11:1151–1164, 2001; Blood 15:1165–1173, 2002; Proc Natl Acad Sci U S A, 1994). HIV-1-derived lentiviral vectors have been shown to transduce a variety of slowly dividing or nondividing cells, including unstimulated T lymphocytes (Blood 96:1309–1316, 2000; Gene Ther 7:596–604, 2000; Blood 101:2167–2174, 2002; Hum Gene Ther 14:1089–1105, 2003). However, achieving effective gene transfer and expression using lentivirus vectors can be complex, and there is at least a perceived risk associated with clinical application of a vector based on a human pathogen (i.e., HIV-1). Recently it has been found that oncoretroviral and lentiviral vectors show a preference for integration into regulatory sequences and active genes, respectively (Cell 110:521–529, 2002 ; Science 300:1749–1751, 2003). Additionally, insertional mutagenesis has become a serious concern, after several patients treated with an oncoretroviral vector for X-linked SCID developed a leukemia-like syndrome associated with activation of the LMO2 oncogene (Science 302:415–419, 2003). Naked DNA-based genetic engineering of human T lymphocytes also requires T cells to be activated prior to gene transfer (Mol Ther 1:49–55, 2000; Blood 101:1637–1644, 2003; Blood 107:2643–2652, 2006). In addition, random integration by electroporation is of low efficiency. We have recently reported that the *Sleeping Beauty* transposon system can efficiently mediate stable transgene expression in human primary T cells without prior T cell activation (Blood 107:483–491, 2006). This chapter describes methodology for the introduction of SB transposons into human T cell cultures with subsequent integration and stable long-term expression at noticeably high efficiency for a nonviral gene transfer system.

**Key words:** Transposons, *Sleeping Beauty* transposon, Human T lymphocytes, Nonviral DNA, Gene transfer.

Christopher Baum (ed.), *Methods in Molecular Biology, Methods and Protocols, vol. 506*
© Humana Press, a part of Springer Science + Business Media, LLC 2009
DOI: 10.1007/978-1-59745-409-4_9

# 1. Introduction

The *Sleeping Beauty* (SB) transposon system has recently emerged as an effective genetic tool to achieve high-level, persistent transgene expression from a nonviral plasmid vector *(1, 2)*. SB is a "cut-and-paste" DNA transposon of the *Tc1/mariner* superfamily that was reconstructed from sequences of teleost fish *(1)*. The SB transposase mediates transposition by recognition of short inverted/directe repeat (IR/DR) sequences that make up the termini of a constructed SB transposon. SB transposase then excises the transposon from the plasmid substrate and inserts the transposon into a TA dinucleotide sequence in a host chromosome (**Fig. 1**). SB transposase has an N-terminal DNA-binding domain, a nuclear localization signal, and a DDE catalytic domain *(2)*, and is usually provided by codelivery of an SB transposase-encoding DNA. The transposase-encoding sequence can be located on the same DNA molecule as the transposon, termed a *cis* vector, or supplied on separate DNA molecule as a *trans* vector (**Fig. 2**). SB transposons have exhibited efficient transposition in cells of a wide range of vertebrates, including cultured mammalian cells *(1, 3, 4)*, mouse liver *(5)* and lung tissues *(6)*, mouse embryonic stem cells *(7)*, and mouse embryos with applications for germline transgenesis and insertional mutagenesis *(8–12)*. Recently, we

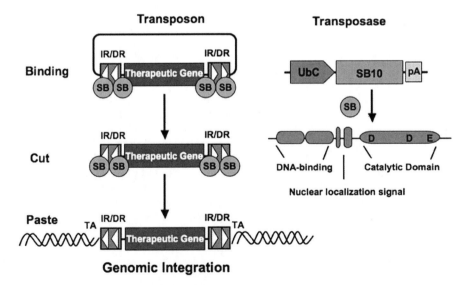

Fig. 1. The *Sleeping Beauty* (SB) transposon system. There are two components in this system: the transposon and the transposase. A schematic drawing of the transposon vector and the transposase-expressing vector is shown. In the transposon vector, a therapeutic gene is flanked by terminal inverted repeat/direct repeats (IR/DR), each containing two binding sites for the transposase. In the transposase expression vector, SB10 transposase is transcriptionally regulated by the human ubiquitin C promoter (UbC). The transposition is a cut-and-paste process in which the transposase binds within the IR/DR repeats; the therapeutic gene is then excised and inserted into a target TA dinucleotide sequence in the genomic DNA.

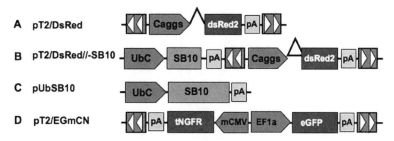

Fig. 2. SB transposon vectors used in this study. Each SB transposon consists of SB inverted repeat/direct repeats (IR/DR, marked as triangles) flanking the gene of interest. An expression cassette encoding SB transposase is also required. Caggs is a chimeric promoter derived from chicken β-actin and cytomegalovirus immediate-early promoter sequences. UbC, human ubiquitin C promoter. SB10, transposase. pT2/EGmCN is an SB vector in which a synthetic bidirectional promoter is used to regulate eGFP and tNGFR (truncated human nerve growth factor receptor) expression. mCMV, minimal CMV promoter; EF1α, human elongation factor 1α promoter. DsRed2, red fluorescent protein. pA, polyadenylation signal. (Reproduced from **ref. 13** with permission from *Blood.*).

demonstrated that the SB transposon system can be used for stable expression of reporter genes in human primary T cells without prior activation or drug selection *(13)*. Molecular analyses showed that SB-mediated stable gene expression in human T cells was through transposition into human chromosomal sequences.

The SB transposon offers many advantages for T cell gene transfer and therapy over widely used virus-based or conventional mammalian DNA vectors *(2)*. First, preparation and storage of SB is straightforward since it is composed of naked plasmid DNA. Second, transgene expression is permanent due to transposon integration. Third, SB transposons integrate into TA sites of the human genome with no preference for active genes or chromosomal destination *(4)*. Thus, the random chance for activation of a proto-oncogene or signaling locus is reduced in comparison with retrovirus and lentivirus vectors. Fourth, SB is less immunogenic than viral vectors, and thus T cells genetically engineered with SB may persist longer after infusion. Fifth, there is no need for prior T cell activation to achieve stable SB-mediated gene transfer. Thus, the duration of ex vivo culture is reduced, and alterations in T cell phenotypes and functions are minimal.

## 2. Materials

### 2.1. SB Transposons and Transposase-encoding Plasmids

1. pT2/DsRed (**Fig. 1a**) is an SB transposon *trans* vector in which the DsRed reporter gene (BD Biosciences, Mountain View, CA) is transcriptionally regulated by a Caggs promoter *(14)*.

2. pT2/DsRed//SB10 (**Fig. 1b**) is an SB *cis* vector in which both the DsRed reporter gene and the SB10 transposase are contained on the same plasmid.

3. pUbSB10 (**Fig. 1c**) contains the human ubiquitin (Ub) promoter (Invitrogen) regulating SB transposase expression.

4. pT2/EGmCN (**Fig. 1d**) contains a synthetic bidirectional promoter in which EF1α and minimal CMV (mCMV) promoters regulate eGFP and tNGFR (truncated nerve growth factor) expression, respectively. The mCMV promoter was obtained from an HIV-1-derived lentiviral vector *(15)* and generously provided by Dr. Luigi Naldini (San Raffaele Telethon Institute for Gene Therapy, Milan, Italy).

5. All SB transposons and the transposase-expressing plasmids are amplified in DH5α E. coli (Cat. No. 18265-D17, Invitrogen) and QIAGEN Endo-free plasmid purification kits (Giga kit: Cat. No. 12391, Maxi kit: Cat. No. 12362) are used to prepare the plasmids.

6. All the plasmids are dissolved in endotoxin-free $H_2O$ (Qiagen) and their concentrations are adjusted to 1–4 µg/µL, and stored at –20°C.

### 2.2. Human T Cell Nucleofection

1. The Nucleofector™ II Device (Amaxa GmbH, Cologne, Germany).

2. Human T cell nucleofector kit (Cat. No. VPA-1002, Amaxa) including 25 cuvettes, 25 plastic pipettes, 10 µg of control pmaxGFP plasmid, 2.25 mL of human T cell nucleofector solution, and 0.5 mL of supplement. Each kit is used for 25 transfections.

3. pmaxGFP is stored at –20°C. Human T cell nucleofector solution and the supplement are stored at –4°C.

4. 10× Dulbecco's Phosphate-Buffered Saline (DPBS, Cat. No. 17-515F, $Ca^{2+}$ and $Mg^{2+}$ free, BioWhittaker, Walkersville, MD). Dilute 1× PBS with double distilled $H_2O$.

5. Bovine serum albumin (BSA, Sigma).

### 2.3. Human T Cell Culture and Expansion

1. *Human T cell medium.* RPMI-1640 (Invitrogen), 10% heat (56°C, 30 min)-inactivated fetal bovine serum (FBS, Hyclone) or human fresh serum, 10 mM HEPES, 2 mM l-glutamine, 50 µM β-mercaptoethanol, 50 U/mL penicillin, and 50 µg/mL streptomycin.

2. Human interleukin-2 (hIL-2) (Proleukin, Chiron, Emeryville, CA) is dissolved in RPMI-1640 at $5 \times 10^6$ IU/mL and stored at –80°C in 50 or 100 µL aliquots. Prior to use, a working stock of $5 \times 10^4$ IU/mL is prepared and stored at 4°C.

3. Human interleukin-7 (hIL-7) is generously provided by the Biological Resources Branch, NCI-Frederick Cancer Research and Development Center, Maryland, or purchased from R&D Systems (Cat. No. 207-IL, Minneapolis, MN). HIL-7 is dissolved at 10 μg/mL in 0.5% BSA/RPMI-1640 and stored at –80°C.

4. OKT3 (Ortho Biotech, Raritan, NJ) 1 mg/mL is stored at –80°C in 100 μL aliquots. Upon thawing, 3.2 mL RPMI-1640 is added to the tube at a working solution of 30 μg/mL and stored at 4°C.

5. Dynabeads® CD3/CD28 (Cat. No. 111-41D) are purchased from Invitrogen or provided by Dr. Bruce Levine (University of Pennsylvania).

6. *Magnet.* Dynal MPC-L (Cat. No. 120.21) (Invitrogen).

7. Buffy coat is purchased from Memorial Blood Center in Minneapolis. Umbilical cord blood (UCB) is provided by the laboratory of Dr. John Wagner (University of Minnesota). Peripheral blood mononuclear cells (PBMC) and UCB mononuclear cells (UCBMC) are purified through Ficoll-Hypaque (Mediatech Cellgro, Herndon, VA) and stored in liquid nitrogen at $4 \times 10^7$ cells per vial.

8. Epstein–Barr-virus-transformed lymphoblastoid cells (EBV-LCL) are grown in Iscove's modified Dulbecco's medium (IMDM, Cat. No. 12440, Invitrogen) supplemented with 10% FBS (Hyclone), 116 μg/mL l-arginine, 36 μg/mL l-asparagine, 216 μg/mL l-glutamine, 10 mM HEPES, 50 U/mL penicillin, and 50 μg/mL streptomycin.

### 2.4. Flow Cytometric Analysis of Transgene Expression

1. *Fluorescence-Activated Cell Sorter (FACS) buffer.* PBS, 2% FBS and 0.05% sodium azide.

2. FACS tubes (Cat. No. 352052, 12 × 75 mm, BD Falcon).

3. FACSCalibur, BD Biosciences.

## 3. Methods

### 3.1. Preparation of PBMC

1. Buffy coat, fresh blood, and umbilical cord blood (UCB) are diluted with 2–4 volumes of PBS.

2. Pipet 15 mL Ficoll-Hypaque (Mediatech Cellgro, Herndon, VA) in a 50-mL conical tube. Slowly pipet 35 mL blood onto the top layer of Ficoll-Hypaque and centrifuge at 913 × *g* (2,000 rpm, Beckman Coulter, Allegra 6R) for 20 min at

room temperature (RT, 20–25°C) in a swinging-bucket rotor without brake.

3. Remove the upper layer and carefully transfer the interphase cells to 2–3 new 50-mL conical tubes.

4. Add PBS-0.5% BSA up to 50 mL, close the cap, and invert the tube to mix cells.

5. Centrifuge at 300 × $g$ for 10 min at RT. Remove the supernatant carefully.

6. Add 10 mL ammonium chloride solution (0.8% $NH_4Cl$ with 0.1 mM EDTA) (Cat. No. 07850, StemCell Technologies, Vancouver, Canada), pipet to mix the cells, and place the tube on ice for 10 min (*see* **Note 1**).

7. Add PBS-0.5% BSA up to 50 mL, centrifuge at 300 × $g$ for 10 min at RT.

8. Remove the supernatant. Wash the cells twice at 300 × $g$ for 10 min.

9. Resuspend cell pellet in 20 mL PBS/0.5 BSA and count the cells with a hemacytometer.

*3.2. Human T Cell Gene Transfer*

1. Add 0.5 mL Supplement to 2.25 mL Human T Cell Nucleofector Solution and mix gently. Prewarm the solution at RT for 30 min.

2. Add 1.5 mL per well human T cell medium to 24-well plates and also dispense additional medium in a 15-mL or 50-mL tube. The extra medium is used to transfer transfected cells from the nucleofection cuvette to 24-well plates. Preincubate plates in a humidified 37°C/5% $CO_2$ chamber for 30 min to warm medium.

3. Dispense the required number of PBMC or UCBMC ($5 \times 10^6$ cells per nucleofection) into a 15-mL conical tube.

4. Centrifuge at 220 × $g$ (1,000 rpm, Beckman Coulter, Allegra 6R) for 5 min. Remove the supernatant completely.

5. Gently resuspend the pellet in prewarmed and supplemented human T cell nucleofector solution to a final concentration of $5 \times 10^6$ cells/100 μL.

6. Mix 100 μL of cell suspension with 5 μg transposon and 5–10 μg transposase-encoding DNA in a sterile eppendorf tube (*see* **Note 2**). In some experiments, 5 μg pmaxGFP is used as a nucleofection control.

7. Transfer cells/DNA mixture into a cuvette. Avoid air bubbles while pipetting. Close cuvette with a cap.

8. Select program U-014. Insert the cuvette into the cuvette holder and press the "X" button to start the program.

9. When the display shows "OK," take the cuvette out of the holder and immediately add 0.5 mL prewarmed human T cell medium to the cuvette (*see* **Note 3**).

10. Use the plastic pipettes provided in the kit to transfer the cell suspension into the preincubated 24-well plates (*see* **Note 4**).

11. Incubate the cells at 37°C overnight.

*3.3. Analysis of Transgene Expression by FACS*

1. 18–24 h after nucleofection, collect 0.5–0.8 mL of cell suspension per well and dispense into a FACS tube.

2. Centrifuge at $300 \times g$ for 10 min and remove the supernatant.

3. Resuspend the pellet in 0.3–0.5 mL FACS buffer and analyze the cells for transgene expression on FACSCalibur (*see* **Note 5**).

*3.4. Human T Cell Activation by CD3/CD28 Beads*

1. For the remaining PBMC in 24-well plates, add Dynabeads CD3/CD28 (3:1 bead/cell ratio) to activate T cells.

2. Dispense the required number of Dynabeads CD3/CD28 into a 5-mL tube.

3. Add PBS to 4 mL. Place the tube in a magnet for 1 min until the Dynabeads are separated.

4. Discard the supernatant. Remove the tube from the magnet.

5. Wash two more times with PBS by following **steps 3** and **4**.

6. Resuspend the beads in human T cell medium without cytokines at 0.5–0.8 mL per well.

7. Transfer 0.5–0.8 mL bead suspensions into individual wells of a 24-well plate. Incubate the cells at 37°C for 3–5 days (*see* **Note 6**).

8. Remove the beads by placing a 5-mL tube in a magnet and transferring cell suspension into the tube.

9. After the beads are separated, transfer the cells into new 24-well plates and incubate the cells at 37°C.

10. Subculture the cells into a new well when they become confluent and add human T cell medium supplemented with hIL-2 (50 IU/mL) and IL-7 (10 ng/mL).

11. Periodically check transgene expression by flow cytometry. Without transposase, transgene expression in human T cells declines and is nearly undetectable at around day 21 posttransfection. However, transgene expression in T cells transfected with both the transposon and the transposase remains significantly high (5–20%) (**Figs. 3** and **4**).

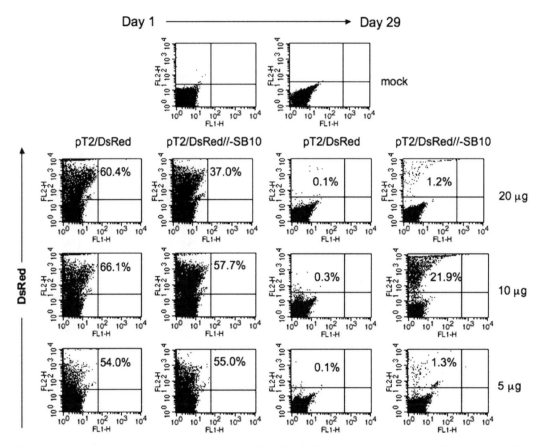

Fig. 3. Stable transgene expression in human primary PBL using an SB transposon *cis* construct. Freshly isolated PBL ($5\times10^6$) were nucleofected at concentrations of 20, 10, or 5 μg *trans* vector pT2/DsRed or *cis* vector pT2/DsRed//-SB10 using the T-cell Nucleofector kit (Amaxa). On day 1 postnucleofection, a fraction of the cells was analyzed for DsRed expression by flow cytometry, and the remaining cells were stimulated with anti-CD3/CD28 beads for 3–5 days. After removal of the beads, the cells were cultured in human T-cell medium containing IL-2 (50 IU/mL) and IL-7 (10 ng/mL). DsRed transgene expression was analyzed on day 29 by flow cytometry. The cells were restimulated once every 10–14 days. The data from one out of at least six individual PBLs are shown.

12. A bidirectional promoter containing EF1α and mCMV in the SB transposon can stably express two genes (**Fig. 5**).

13. T cells are restimulated once every 2 weeks with CD3/CD28 beads *(16)* or OKT3 *(17)*.

### 3.5. T-cell Expansion by OKT3

1. Thaw PBMC from three donors in warm human T cell medium.

2. Centrifuge at 300 × *g* for 8 min. Discard the supernatant.

3. Wash 1× with human T cell medium at 300 × *g* for 8 min.

4. Resuspend PBMC in human T cell medium and count the cells with a hemacytometer.

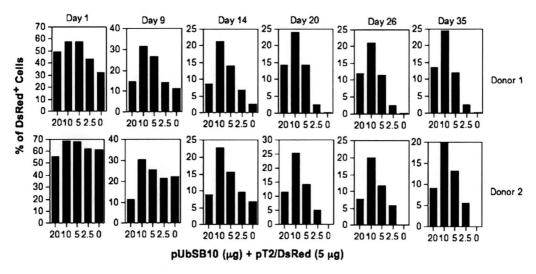

Fig. 4. Stable transgene expression in human primary PBL using an SB transposon *trans* construct. Freshly isolated PBL (5 × 10⁶) were conucleofected with 5 μg of pT2/DsRed and different amounts of pUbSB10 (20, 10, 5, 2.5, and 0 μg). DsRed expression was analyzed on days 1, 9, 14, 20, 26, and 35. The data from two individual PBL are shown.

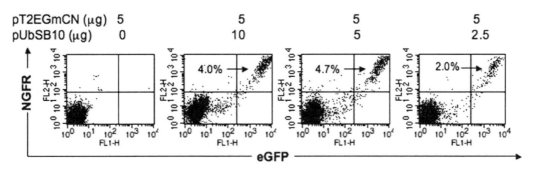

Fig. 5. Dual gene expression in human T cells. PBL from two donors were nucleofected with the SB dual gene plasmid pT2/EGmCN (5 μg) +/− pUbSB10 (10, 5, 2.5, and 0 μg). The cells were activated by anti-CD3/CD28 beads on day 2 and assayed for eGFP and NGFR expression on day 21. Similar data were also obtained from a second donor. (Reproduced from **ref. 13** with permission from *Blood*.).

5. Harvest EBV-LCL cells and spin at 300 × *g* for 8 min.

6. Remove the supernatant. Resuspend the EBV-LCL in human T cell medium and count the cells.

7. Irradiate PBMC (40 Gy) and EBV-LCL (80 Gy).

8. Mix T cells with mixture of three allogeneic PBMC and EBV-LCL (**Table 1**).

9. Add hIL-2 (50 IU/mL) on day 1 after culture at 37°C.

10. Remove OKT3 on day 5. Transfer the cells into centrifuge tubes, spin at 300 × *g* for 8 min, and discard the supernatant.

**Table 1**
**Human T cell expansion by OKT3**

|  | 24-well plate | T-25 flask | T-75 flask |
|---|---|---|---|
| T cells | Few | $0.5–2 \times 10^5$ | $0.2–1 \times 10^6$ |
| PBMC | $3 \times 10^6$ | $25 \times 10^6$ | $125 \times 10^6$ |
| EBV-LCL | $3 \times 10^5$ | $5 \times 10^6$ | $25 \times 10^6$ |
| OKT3 (30 µg/mL) | 2 µL | 25 µL | 100 µL |
| Final volume | 2 mL | 25 mL | 100 mL |

11. Resuspend the cells in human T cell medium with hIL-2 (50 IU/mL) and return to original plates or flasks at the original volume.

12. On days 7–14, subculture the cells into another well when they are confluent in 24-well plates. If the cells are $1–2 \times 10^6$/mL, subculture into two flasks each at the original total volume. For cell densities $> 2 \times 10^6$/mL, subculture into three flasks. Supplement human T cell medium with hIL-2 (50 IU/mL) to the original volume per flask.

13. On days 10–14, the T cells are ready to use for functional assays or infusions. Average expansion is generally 100–500 fold.

*3.6. Molecular Analysis of Transposition in T Cells*

Southern hybridization, splinkerette polymerase chain reaction (PCR), and Hirt (low molecular weight DNA) DNA analyses can be used to characterize transposon integrants in human T cells. These methods have been described in detail (13, 18, 19), and the protocols are available upon request.

# 4. Notes

1. UBCMC contain a substantial amount of red blood (RBC) cells after Ficoll-Hypaque separation. Thus, lysing RBC is needed. For PBMC, this step is not often required.

2. The optimum ratio of transposon to transposase DNA used for T cell nucleofection should be titrated, since it has been shown that the SB transposase exhibits overexpression inhibition (3).

3. Alternatively, add 2 mL human T cell medium per well in 24-well plates instead of the 1.5 mL per well described earlier.

After nucleofection, use the plastic pipettes to add prewarmed medium into the cuvette and transfer the cells back into the well.

4. While transferring cells back to the well, pipetting cells up and down is not recommended. After transferring, gently rinse the cuvette with medium.

5. While transferring the nucleofected T cells into FACS tubes, cell debris is often observed. To avoid clogging FACS tubing, removing cell debris with pipette tips is recommended.

6. Due to nucleofection with high amounts of DNA, transfected PBMC show a 1–2 day delay in activation by CD3/CD28 beads compared with the mock control. We observed that nucleofected UCBMC require 1–2 days longer than PBMC for full activation by CD3/CD28 beads.

## Acknowledgments

This work was supported by a Grant-in-Aid of Research, Artistry and Scholarship from the Graduate School, University of Minnesota, the Minnesota Medical Foundation, the Children's Cancer Research Fund, the Alliance for Cancer Gene Therapy Young Investigator Award, the G&P Foundation for Cancer Research, the Sidney Kimmel Foundation for Cancer Research Kimmel Scholar Award, and the National Blood Foundation. X.Z. is the recipient of an American Society of Hematology Junior Faculty Scholar Award.

## References

1. Ivics, Z., Hackett, P. B., Plasterk, R. H., Izsvak, Z. (1997) Molecular reconstruction of *Sleeping Beauty*, a Tc1-like transposon from fish, and its transposition in human cells. *Cell* 91, 501–510

2. Izsvak, Z., Ivics, Z. (2004) *Sleeping Beauty* transposon: biology and applications for molecular therapy. *Mol Ther* 9, 147–156

3. Geurts, A. M., Yang, Y., Clark, K. J., et al. (2003) Gene transfer into genomes of human cells by the *Sleeping Beauty* transposon system. *Mol Ther* 8, 108–117

4. Yant, S. R., Wu, X., Huang, Y., et al. (2005) High-resolution genome-wide mapping of transposon integration in mammals. *Mol Cell Biol* 25, 2085–2094

5. Yant, S. R., Meuse, L., Chiu, W., Ivics, Z., Izsvak, Z., Kay, M. A. (2000) Somatic integration and long-term transgene expression in normal and haemophilic mice using a DNA transposon system. *Nat Genet* 25, 35–41

6. Belur, L. B., Frandsen, J. L., Dupuy, A. J., et al. (2003) Gene insertion and long-term expression in lung mediated by the *Sleeping Beauty* transposon system. *Mol Ther* 8, 501–507

7. Luo, G., Ivics, Z., Izsvak, Z., Bradley, A. (1998) Chromosomal transposition of a Tc1/mariner-like element in mouse embryonic stem cells. *Proc Natl Acad Sci USA* 95, 10769–10773

8. Dupuy, A. J., Clark, K., Carlson, C. M., et al. (2002) Mammalian germ-line transgenesis by

transposition. *Proc Natl Acad Sci USA* 99, 4495–4499

9. Collier, L. S., Carlson, C. M., Ravimohan, S., Dupuy, A. J., Largaespada, D. A. (2005) Cancer gene discovery in solid tumours using transposon-based somatic mutagenesis in the mouse. *Nature* 436, 272–276

10. Dupuy, A. J., Akagi, K., Largaespada, D. A., Copeland, N. G., Jenkins, N. A. (2005) Mammalian mutagenesis using a highly mobile somatic Sleeping Beauty transposon system. *Nature* 436, 221–226

11. Keng, V. W., Yae, K., Hayakawa, T., Mizuno, S., Uno, Y., Yusa, K., Kokubu, C., Kinoshita, T., Akagi, K., Jenkins, N. A., Copeland, N. G., Horie, K., Takeda, J. (2005) Region-specific saturation germline mutagenesis in mice using the Sleeping Beauty transposon system. *Nat Methods* 2, 763–769

12. Kitada, K., Ishishita, S., Tosaka, K., Takahashi, R., Ueda, M., Keng, V. W., Horie, K., Takeda, J. (2007) Transposon-tagged mutagenesis in the rat. *Nat Methods* 4, 131–133

13. Huang, X., Wilber, A. C., Bao, L., Tuong, D., Tolar, J., Orchard, P., et al., (2006) Stable gene transfer and expression in human primary T cells by the *Sleeping Beauty* transposon system. *Blood* 107, 483–491

14. Akagi, Y., Isaka, Y., Akagi, A., et al. (1999). Transcriptional activation of a hybrid promoter composed of cytomegalovirus enhancer and beta-actin/beta-globin gene in glomerular epithelial cells *in vivo*. *Kidney Int* 1999 51, 1265–1269

15. Amendola, M., Venneri, M. A., Biffi, A., Vigna, E., Naldini, L. (2004) Coordinate dual-gene transgenesis by lentiviral vectors carrying synthetic bidirectional promoters. *Nat Biotechnol* 23, 108–116

16. Levine, B. L., Bernstein, W. B., Aronson, N. E., et al. (2002) Adoptive transfer of costimulated CD4+ T cells induces expansion of peripheral T cells and decreased CCR5 expression in HIV infection. *Nat Med* 8, 47–53

17. Riddell, S. R., Greenberg, P. D. (1990) The use of anti-CD3 and anti-CD28 monoclonal antibodies to clone and expand human antigen-specific T cells. *J Immunol Methods* 128, 189–201

18. Sambrook, J., Russell, D. W. (2001) Molecular cloning: a laboratory manual. 3rd Edition. New York, NY: Cold Spring Harbor Laboratory Press

19. Hirt, B. (1967) Selective extraction of polyoma DNA from infected mouse cell cultures. *J Mol Biol* 26, 365–369

# Chapter 10

## Retroviral Gene Transfer into Primary Human Natural Killer Cells

### Evren Alici, Tolga Sutlu, and M. Sirac Dilber

## Summary

Modulation of intracellular signaling pathways or receptor expression in natural killer (NK) cells by genetic manipulation is an attractive possibility in studies of NK cell specificity and function. Moreover, feasible applications of these genetic manipulations in the context of gene and NK cell therapy regimens may be considered. However, efficient gene modification of primary NK cells has been largely hampered by the absence of an efficient gene-transfer protocol.

A retrovirus-based easy-to-use transduction protocol that can insert the gene of interest permanently into primary NK cells would be an important tool to advance our studies in NK cell biology and NK cell-mediated therapies. We have recently described a protocol for efficient expansion of NK cells under good manufacturing practice (GMP) conditions from the healthy donors and from patients with hematological malignancies. As the active division of cells is a prerequisite for efficient retroviral insertion, the high rate of expansion in this protocol provides more efficient transduction by retroviral vectors. We hereby present this simple and efficient retroviral vector-based gene-transfer protocol for such ex vivo cultured primary human NK cells.

**Key words:** Natural killer cells, Gene modification, Transduction, Retroviral vectors.

## 1. Introduction

Natural killer (NK) cells are lymphocytes of the innate immune system which play an important role in the early response against many types of microorganisms and tumors *(1, 2)*. They develop from precursor cells in the bone marrow and migrate via blood to peripheral lymphoid and nonlymphoid tissues *(3, 4)*. NK cells express multiple cell surface receptors that enable them to scan

Christopher Baum (ed.), *Methods in Molecular Biology, Methods and Protocols, vol. 506*
© Humana Press, a part of Springer Science+Business Media, LLC 2009
DOI: 10.1007/978-1-59745-409-4_10

the environment for virus-infected or transformed cells *(5)*, many of which express stress associated molecules and/or have altered or downregulated expression of MHC class I *(6–8)*. This phenomenon, which was referred to as the "missing-self" hypothesis, has led to the identification of inhibitory killer immunoglobulin-like receptors (KIRs) which, upon recognition of appropriately presented self MHC class I molecules on the target cells, inhibit NK cell cytotoxicity *(9, 10)*. Stimulation by target cell ligands that bind specific activating receptors on the NK cell surface is another prerequisite for triggering degranulation of NK cells and subsequent target cell lysis *(11)*. They also express defined sets of chemokine receptors that regulate their migration in tissues *(12–14)*. Thus, the recognition of target cells by NK cells is a complex event which is controlled through a yet unclear balance of various signals from inhibitory and activating receptors.

Gene transfer into primary NK cells may open new possibilities in immunotherapy of tumors, including leukemias in both autologous and allogeneic transplantation settings. However, primary NK cells have been reported to be difficult to transduce by retroviral vectors, as well as by other viral vectors *(15)*. On the other hand, NK cell lines appear to be more permissive to gene transfer by means of electroporation *(16)*, particle-mediated gene transfer *(17)*, retroviral transduction *(18, 19)*, and adenoviral transduction *(20)*, although rather few reports have demonstrated stable gene expression. More recently, transient transfection of NK cells based on direct transfer of DNA into the nucleus using a modified electroporation technique has also been demonstrated *(21)*.

In order to transduce primary cells or cell lines with retroviral vectors, the cells of interest need to be actively dividing. Thus, cell expansion protocols that are not giving selective expansion advantage to the cells of interest will not be successful in terms of retroviral gene modification. The previous lack of an efficient NK cell expansion protocol may explain some of the past failures or only limited successes in gene transduction of NK cells *(14–17, 20)*.

We here describe a protocol for retrovirus-mediated gene transfer into primary human NK cells. The use of a mixed lymphocyte culture protocol that enables selective expansion of NK cells provides the efficient transduction and overcomes previous difficulties in DNA uptake by primary NK cells using common transduction techniques. The strategy opens up new ways to introduce a variety of genes including genes encoding different cytokines, chemokines, antiapoptotic genes, intracellular signaling intermediates, suicide genes, targeting receptors, as well as genes encoding chemokine, activating and/or inhibitory receptors.

With respect to the possible implications of the present protocol, it has been shown that cytokines like IL-2 *(22–24)*, IL-15 *(24–26)*, and IL-21 *(27, 28)* play important roles in NK cell differentiation, maturation, and expansion. IL-12 has also been shown to be important for induction of NK lytic capacity *(29, 30)*. By gene transfer of such cytokines, it is then possible to create NK or NKT cells that are independent of such cytokines for growth and cytotoxicity. Apart from cytokines, chemokine receptors such as CXCR3 and CXCR4 have been shown to be important for NK cell homing to different organs in mice in vivo *(12)*. Therefore, introduction of different chemokines into NK cells may enable NK cells to migrate *in vivo* to more specific target areas of interest for efficient tumor combat. Moreover, retargeting NK cells to some specific molecules on tumor cells such as CD19 *(31)* or Erb/neu *(32)* may make NK cells more powerful tools in immunotherapy regimens against tumors. Furthermore, due to the dependence of NK cell activity on the balance between inhibitory and activatory receptors on their surface, overexpression of activation receptors, e.g., stress- or transformation-associated target cell molecules, or the silencing of inhibitory receptors on NK cells for target cell MHC class I molecules may serve as attractive possibilities for gene therapy approaches using NK cells.

## 2. Materials

### 2.1. Separation of Mononuclear Cells

1. Lymphoprep (AXIS-SHIELD PoC AS, Dundee, Scotland) Store at +4°C and protect from light. It should be at room temperature during use.
2. PBS (GIBCO, Invitrogen, Carlsbad, CA, USA).
3. Türk's Dye (GIBCO).

### 2.2. NK Cell Expansion

1. Human AB Serum (BioWhittaker, Cambrex Bioscience, Walkersville, MD, USA) Heat inactive and keep in frozen aliquots at –20°C. Thaw as needed prior to use.
2. IL-2 stock solution ($1 \times 10^6$ U/mL), sterile (Proleukin, Chiron, Emeryville, CA, USA) Dissolve in sterile water and keep in frozen aliquots at –20°C.
3. OKT-3 stock solution (1 mg/mL), sterile (Ortho Biotech, Raritan, NJ, USA) Dissolve in sterile water and keep in frozen aliquots at –20°C (*see* **Note 1**).
4. Cellgro SCGM culture medium (Cellgenix, Freiburg, Germany).
5. *NK cell stimulation medium.* Cellgro SCGM with 5% human AB serum, 500 U/mL IL-2 and 10 ng/mL OKT3. Prepare fresh for each use.
6. PBS (GIBCO).
7. Türk's dye (GIBCO).

**2.3. Transduction**

1. Retroviral supernatant (*see* **Note 2**).
2. Human AB Serum (BioWhittaker).
3. IL-2-solution ($1 \times 10^6$ U/mL), sterile (Proleukin, Chiron).
4. Polybrene (0.8 mg/mL) (Sigma-Aldrich, St. Louis, MO, USA) (*see* **Note 3**) Dissolve in PBS, filter, sterilize, and keep in frozen aliquots at –20°C.
5. CellGro SCGM culture medium (Cellgenix).

# 3. Methods

**3.1. Preparation**

1. Heat-inactivate human AB serum (*see* **Note 4**).
   (a) Thaw the serum at +4°C and mix the contents of the bottle thoroughly.
   (b) Place the thawed bottle of serum into a 56°C water bath containing enough water to immerse the bottle to just above the level of the serum.
   (c) Begin timing for 30 min. Swirl the serum every 5–10 min to ensure uniform heating and to prevent protein coagulation at the bottom of the bottle.
   (d) After 30 min at 56°C cool the serum immediately, preferably in an ice bath.
2. Thaw IL-2 solution. Store at +4°C until use.
3. Warm up CellGro SCGM culture medium at +37°C in the incubator.

**3.2. Separation of Mononuclear cells**

1. The lymphoprep solution should be warmed up to room temperature before starting.
2. Transfer all blood into a 50-mL tube.
3. Dilute the blood 1:2 in PBS (*see* **Note 5**).
4. Add 15 mL of lymphoprep into 50-mL tubes.
5. Carefully add 30 mL of the diluted blood suspension on the top of lymphoprep in the 50-mL tube.
6. Centrifuge at $800 \times g$ for 30 min, room temperature (*see* **Note 6**).
7. Go through the serum with a pipette and collect the lymphocyte band (avoid the lymphoprep layer). Transfer to a new 50-mL tube.
8. Fill the tube with PBS.
9. Centrifuge at $800 \times g$ for 10 min, room temperature.
10. Discard the supernatant.

11. Repeat the washing with PBS.

12. Resuspend and pool the pellets in 1–5 mL PBS.

13. Count the cells in Türk's dye (*see* **Note 7**).

14. Calculate the total amount of cells in the tube. Take necessary amount for NK cell expansion and transduction ($10 \times 10^6$ cells).

15. Take a sample for counting, flow cytometry, and cytotoxicity analysis. Freeze the rest of the cells.

### 3.3. NK Cell Expansion Day 0: Initiation of Culture

1. Use the $10 \times 10^6$ cells from lymphoprep separation for NK cell expansion and transduction.

2. Prepare 20 mL NK cell stimulation medium.

3. Resuspend cells in 20 mL medium (final concentration: $0.5 \times 10^6$ cells/mL).

4. Pipette the cells into a T-75 flask.

5. Close the flask and incubate cells at 37°C, 95% humidity, 5% $CO_2$.

### 3.4. NK Cell Expansion Day 2

1. Take the flask out of the incubator. Inspect the cells under a microscope.

2. Add fresh IL-2 to a final concentration of 500 U/µL (Add 10 µL of the stock solution for 20 mL culture volume)

### 3.5. NK Cell Expansion Day 5: First Infection (See Note 8)

1. Take the flask out of the incubator. Inspect the cells under a microscope.

2. Scrape the cells and mix well.

3. Count the cells, centrifuge at $800 \times g$ for 5 min, room temperature and resuspend in PBS.

4. Prepare $4 \times 50$ mL centrifuge tubes with $2.5 \times 10^6$ cells in each. Use the rest of the cells for flowcytometry and cytotoxicity assays.

5. Centrifuge the tubes at $800 \times g$ for 5 min, room temperature.

6. Discard the supernatant and resuspend the pellets in 15 mL (for MOI 3) of appropriate retroviral supernatant ($0.5 \times 10^6$ infectious particles/mL):

   (a) Retroviral vector with the gene of interest

   (b) Mock vector (Same vector without the gene of interest)

   (c) Positive control (Preferentially the same vector with a marker gene such as GFP, YFP, etc.)

   (d) Negative control (Cellgro SCGM medium only)

7. Supplement all samples with 500 U/mL IL-2 and 4–8 µg/mL Polybrene.

8. Centrifuge at $2,000 \times g$ for 2 h, room temperature.

9. Discard supernatants and resuspend the cells in 15 mL PBS and then centrifuge again in RT with $800 \times g$ for 5 min.

10. Discard supernatants and resuspend cells 5 mL NK cell stimulation medium without OKT3 into a T-25 flask, label each flask accordingly (final concentration: $0.5 \times 10^6$ cells/mL).

11. Close the flasks and incubate cells at $+37°C$, 95% humidity, 5% $CO_2$.

*3.6. NK Cell Expansion Day 6: Second Infection (See Note 9)*

1. Take flasks out of the incubator. Inspect the cells under a microscope.

2. Scrape the cells and mix well with a pipette. Take a sample for counting.

3. Count the cells and transfer $2.5 \times 10^6$ cells to a 50-mL tube for the second round of transduction. Use the rest of the cells for flowcytometry and cytotoxicity assays.

4. Centrifuge the 50-mL tube at $800 \times g$ for 5 min, room temperature.

5. Discard the supernatant and resuspend the pellets in 15 mL (for MOI 3) of appropriate retroviral supernatant ($0.5 \times 10^6$ viral particles/mL):
   (a) Retroviral vector with the gene of interest
   (b) Mock vector (Same vector without the gene of interest)
   (c) Positive control (Preferentially the same vector with a marker gene such as GFP, YFP, etc.)
   (d) Negative control (Cellgro SCGM medium only)

6. Supplement all samples with 500 U/mL IL-2 and 4–8 µg/mL Polybrene

7. Carry out **steps 8–11** of Day 5 infection protocol.

*3.7. NK Cell Expansion Day 7: End of Transduction*

1. Take flasks out of the incubator. Inspect the cells under a microscope.

2. Scrape the cells and mix well with a pipette. Take a sample for counting, flow cytometry, and cytotoxicity analysis.

3. Centrifuge the rest of the cells at $800 \times g$ for 5 min, room temperature.

4. Resuspend in fresh NK cell stimulation medium without OKT3 at a concentration of $0.5 \times 10^6$ cells/mL and pipette into T-25 or T-75 flasks according to the final volume.

5. Close the flasks and incubate cells at $+37°C$, 95% humidity, 5% $CO_2$.

6. Continue NK cell expansion until day 20.

*3.8. NK Cell Expansion Day 9–19: Expansion and Quality Control (See Note 10)*

The following steps should be carried out every 2–3 days until the end of culture period:

1. Take the flasks out of the incubator. Inspect the cells under a microscope.

2. Scrape the cells and mix well with a pipette, take samples for counting.

3. Count the cells and take samples for flowcytometry and cytotoxicity analysis, transfer the rest to 50-mL tubes and centrifuge.

4. Discard supernatants and resuspend in fresh NK cell stimulation medium without OKT3 at a final concentration of $1 \times 10^6$ cells/mL.

5. Pipette in appropriate flasks according to the final volume.

6. Close the flasks and incubate cells at +37°C, 95% humidity, 5% $CO_2$.

*3.9. NK Cell Expansion Day20: End of culture*

1. Take the flasks out of the incubator. Inspect the cells under a microscope.

2. Scrape the cells and mix well with a pipette, take samples for counting, flow cytometry, and cytotoxicity assays.

3. Use necessary amount of cells for any further investigation regarding specific effects of the gene of interest (*see* **Note 11**).

4. Freeze the rest of the cells.

    **Figure 1** illustrates a representative example of NK cell expansion and transduction throughout the culture period.

# 4. Notes

1. The usage of anti-CD3 antibody, OKT-3, is indispensable for NK cell expansion, due to the reason that for efficient NK cell expansion, the support of OKT-3 activated-T cells by possibly secreting some cytokines or growth factors was necessary and essential *(33)*.

2. In this protocol, we adjusted the supernatant titer to $0.5 \times 10^6$ infectious particles/mL to standardize the calculations; however, the reader can readjust this simply by modifying the volumes where multiplicity of infections is mentioned. Also, in addition to the cell expansion protocol used, the type of viral envelope may have impact on transduction efficiency. We generally use retroviral vectors with GALV envelope since such envelope has been shown to be efficient for transducing

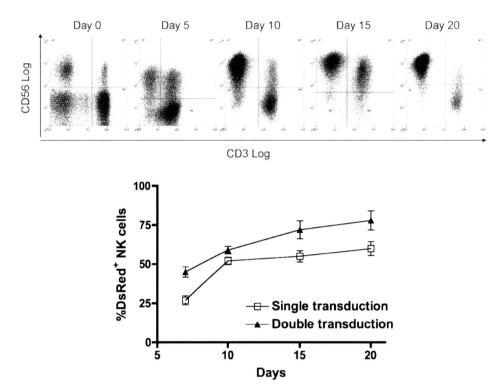

Fig. 1. A representative example of NK cell expansion and transduction. (**a**) Flowcytometry results showing the selective expansion of NK cells which dominate the culture throughout the expansion period. The cells were acquired on a Cyan ADP 8-color flow cytometer. Data were analyzed by gating on lymphocytes in FSC/SSC plots and using CD19⁻ CD14⁻ cells. (**b**) Transduction of primary NK cells with a retroviral vector coding for DsRed protein. As seen, the percentage of transduced cells increases by time due to the higher expansion of NK cells, and double transduction positively affects the overall transduction rate. It should also be noted that the final product also contains low levels of transduced T- and NK-like T cells.

primary lymphocytes *(34–36)*, and in some studies to be even more efficient than vectors with envelopes such as the VSV-G or MLV amphotropic envelope *(37–39)*.

3. Polybrene (1,5-dimethyl-1,5-diazaundecamethylene polymethobromide, hexadimethrine bromide) is a cationic polymer used to increase the efficiency of infection. The efficiency of retroviral infection is enhanced significantly, 100–1,000-fold in some cells, by including polybrene during the infection.

4. Heating the serum for longer than 30 min or higher than 56°C will have an adverse effect on the efficacy of the serum and will most likely cause an increase in the amount of cryoprecipitate that might form. For volumes less than 100 mL of serum to be heat-inactivated, reduce the time of incubation in the 56°C water bath. One should also take into account that when one heat inactivates serum, not only is complement protein degraded, but also all other components such as amino

acids, vitamins, growth factors, etc. are subjected to temperatures that could cause degradation. In many cases, certain important components that are of borderline concentration may be reduced to concentrations less than that required by certain types of cells for growth. Our own experience suggests that heat inactivation in this setting is necessary when collecting supernatant, during culture and transduction.

5. 1:2 dilution with PBS is optimized according to normal blood counts. However, if blood samples with high leukocyte counts (e.g., patients with tumor load) are used, we recommend a 1:3 dilution.

6. This centrifugation step should be carried out with minimum acceleration and no brakes in order to keep the lymphoprep-blood interface intact.

7. Leukocyte counting is a routine method. The basis of all counting methods is the dilution and preparation of a blood sample of known volume. The erythrocytes are hemolyzed by the acetic acid of Türk's solution and the leukocytes are stained by the dye contained. The required cell type in a defined volume is counted and the number of cells per microliter of blood is then calculated.

8. Using the present NK cell expansion protocol, NK cells show the highest expansion rates between days 5 and 10. Thus, we chose to transduce the cells at the beginning of logarithmic cell expansion at day 5. After day 15 of culture, a decline in expansion rate of the NK cells is observed. Accordingly, we found that primary NK cell cultures were efficiently transduced on days 5 and 6, while the transduction rate declined significantly on days 15–21.

9. Although the transduction efficiency can still be increased by multiple transductions, there are indications that even a single round of transduction yields a high percentage of transduced NK cells using the current protocol. Therefore, efficient transduction with single transduction and lower MOI may also prevent the possible problems with gene transfer such as insertional mutagenesis seen by other groups *(40)*.

10. After retroviral transduction, we observed selective NK cell expansion among both nontransduced and transduced cells until the end of the culture. Due to this unique expansion characterization of NK cells in our expansion cocktail, we end up with having transduced cells as the main cell type in the culture. Importantly, we found neither any significant inhibition of NK cell proliferation and/or NK cell death among transduced cells, nor any loss of cytotoxic activity after gene transfer into NK cells, when compared to parental nontransduced cells.

11. This protocol can be scaled up according to the necessary amount of transduced cells at the end of the culture period and the average transduction rate with your specific retroviral vectors.

## References

1. Carayannopoulos, L. N. and Yokoyama, W. M. (2004) Recognition of infected cells by natural killer cells. *Curr Opin Immunol* 16, 26–33

2. Wu, J. and Lanier, L. L. (2003) Natural killer cells and cancer. *Adv Cancer Res* 90, 127–56

3. Yokoyama, W. M., Kim, S., and French, A. R. (2004) The dynamic life of natural killer cells. *Annu Rev Immunol* 22, 405–429

4. Colucci, F., Caligiuri, M. A., and Di Santo, J. P. (2003) What does it take to make a natural killer? *Nat Rev Immunol* 3, 413–425

5. French, A. R. and Yokoyama, W. M. (2003) Natural killer cells and viral infections. *Curr Opin Immunol* 15, 45–51

6. Ljunggren, H. G. and Karre, K. (1990) In search of the 'missing self': MHC molecules and NK cell recognition. *Immunol Today* 11, 237–244

7. Yokoyama, W. M. (2000) Now you see it, now you don't! *Nat Immunol* 1, 95–97

8. Backstrom, E., Kristensson, K., and Ljunggren, H. G. (2004) Activation of natural killer cells: underlying molecular mechanisms revealed. *Scand J Immunol* 60, 14–22

9. Moretta, A., Bottino, C., Pende, D., Tripodi, G., Tambussi, G., Viale, O., Orengo, A., Barbaresi, M., Merli, A., Ciccone, E., and et-al. (1990) Identification of four subsets of human CD3-CD16+ natural killer (NK) cells by the expression of clonally distributed functional surface molecules: correlation between subset assignment of NK clones and ability to mediate specific alloantigen recognition. *J Exp Med* 172, 1589–1598

10. Moretta, A., Tambussi, G., Bottino, C., Tripodi, G., Merli, A., Ciccone, E., Pantaleo, G., and Moretta, L. (1990) A novel surface antigen expressed by a subset of human CD3- CD16+ natural killer cells. Role in cell activation and regulation of cytolytic function. *J Exp Med* 171, 695–714

11. Moretta, A., Bottino, C., Vitale, M., Pende, D., Cantoni, C., Mingari, M. C., Biassoni, R., and Moretta, L. (2001) Activating receptors and coreceptors involved in human natural killer cell-mediated cytolysis. *Annu Rev Immunol* 19, 197–223

12. Beider, K., Nagler, A., Wald, O., Franitza, S., Dagan-Berger, M., Wald, H., Giladi, H., Brocke, S., Hanna, J., Mandelboim, O., Darash-Yahana, M., Galun, E., and Peled, A. (2003) Involvement of CXCR4 and IL-2 in the homing and retention of human NK and NK T cells to the bone marrow and spleen of NOD/SCID mice. *Blood* 102, 1951–1958

13. Robertson, M. J. (2002) Role of chemokines in the biology of natural killer cells. *J Leukoc Biol* 71, 173–183

14. Robertson, M. J., Williams, B. T., Christopherson, K., II, Brahmi, Z., and Hromas, R. (2000) Regulation of human natural killer cell migration and proliferation by the exodus subfamily of CC chemokines. *Cell Immunol* 199, 8–14

15. Nagashima, S., Mailliard, R., Kashii, Y., Reichert, T. E., Herberman, R. B., Robbins, P., and Whiteside, T. L. (1998) Stable transduction of the interleukin-2 gene into human natural killer cell lines and their phenotypic and functional characterization in vitro and in vivo. *Blood* 91, 3850–3861

16. Liu, J. H., Wei, S., Blanchard, D. K., and Djeu, J. Y. (1994) Restoration of lytic function in a human natural killer cell line by gene transfection. *Cell Immunol* 156, 24–35

17. Tam, Y. K., Maki, G., Miyagawa, B., Hennemann, B., Tonn, T., and Klingemann, H. G. (1999) Characterization of genetically altered, interleukin 2-independent natural killer cell lines suitable for adoptive cellular immunotherapy. *Hum Gene Ther* 10, 1359–1373

18. Miller, J. S., Tessmer-Tuck, J., Blake, N., Lund, J., Scott, A., Blazar, B. R., and Orchard, P. J. (1997) Endogenous IL-2 production by natural killer cells maintains cytotoxic and proliferative capacity following retroviral-mediated gene transfer. *Exp Hematol* 25, 1140–1148

19. Tran, A. C., Zhang, D., Byrn, R., and Roberts, M. R. (1995) Chimeric zeta-receptors direct human natural killer (NK) effector function to permit killing of NK-resistant tumor cells and HIV-infected T lymphocytes. *J Immunol* 155, 1000–1009

20. Schroers, R., Hildebrandt, Y., Hasenkamp, J., Glass, B., Lieber, A., Wulf, G., and Piesche, M. (2004) Gene transfer into human T lymphocytes and natural killer cells by Ad5/F35

chimeric adenoviral vectors. *Exp Hematol* 32, 536–546

21. Trompeter, H. I., Weinhold, S., Thiel, C., Wernet, P., and Uhrberg, M. (2003) Rapid and highly efficient gene transfer into natural killer cells by nucleofection. *J Immunol Methods* 274, 245–256

22. Fehniger, T. A., and Caligiuri, M. A. (2001) Ontogeny and expansion of human natural killer cells: clinical implications. *Int Rev Immunol* 20, 503–534

23. Luhm, J., Brand, J. M., Koritke, P., Hoppner, M., Kirchner, H., and Frohn, C. (2002) Large-scale generation of natural killer lymphocytes for clinical application. *J Hematother Stem Cell Res* 11, 651–657

24. Dunne, J., Lynch, S., O'Farrelly, C., Todryk, S., Hegarty, J. E., Feighery, C., and Doherty, D. G. (2001) Selective expansion and partial activation of human NK cells and NK receptor-positive T cells by IL-2 and IL-15. *J Immunol* 167, 3129–3138

25. Mrozek, E., Anderson, P., and Caligiuri, M. A. (1996) Role of interleukin-15 in the development of human CD56+ natural killer cells from CD34+ hematopoietic progenitor cells. *Blood* 87, 2632–2640

26. Liu, C. C., Perussia, B., and Young, J. D. (2000) The emerging role of IL-15 in NK-cell development. *Immunol Today* 21, 113–116

27. Habib, T., Nelson, A., and Kaushansky, K. (2003) IL-21: a novel IL-2-family lymphokine that modulates B, T, and natural killer cell responses. *J Allergy Clin Immunol* 112, 1033–1045

28. Parrish-Novak, J., Foster, D. C., Holly, R. D., and Clegg, C. H. (2002) Interleukin-21 and the IL-21 receptor: novel effectors of NK and T cell responses. *J Leukoc Biol* 72, 856–863

29. Lee, S. M., Suen, Y., Qian, J., Knoppel, E., and Cairo, M. S. (1998) The regulation and biological activity of interleukin 12. *Leuk Lymphoma* 29, 427–438

30. Alli, R. S., and Khar, A. (2004) Interleukin-12 secreted by mature dendritic cells mediates activation of NK cell function. *FEBS Lett* 559, 71–76

31. Roessig, C., Scherer, S. P., Baer, A., Vormoor, J., Rooney, C. M., Brenner, M. K., and Juergens, H. (2002) Targeting CD19 with genetically modified EBV-specific human T lymphocytes. *Ann Hematol* 81 Suppl 2, S42–S43

32. Uherek, C., Tonn, T., Uherek, B., Becker, S., Schnierle, B., Klingemann, H. G., and Wels, W. (2002) Retargeting of natural killer-cell cytolytic activity to ErbB2-expressing cancer cells results in efficient and selective tumor cell destruction. *Blood* 100, 1265–1273

33. Carlens, S., Gilljam, M., Chambers, B. J., Aschan, J., Guven, H., Ljunggren, H. G., Christensson, B., and Dilber, M. S. (2001) A new method for in vitro expansion of cytotoxic human CD3-CD56+ natural killer cells. *Hum Immunol* 62, 1092–1098

34. Movassagh, M., Boyer, O., Burland, M. C., Leclercq, V., Klatzmann, D., and Lemoine, F. M. (2000) Retrovirus-mediated gene transfer into T cells: 95% transduction efficiency without further in vitro selection. *Hum Gene Ther* 11, 1189–1200

35. Bunnell, B. A., Muul, L. M., Donahue, R. E., Blaese, R. M., and Morgan, R. A. (1995) High-efficiency retroviral-mediated gene transfer into human and nonhuman primate peripheral blood lymphocytes. *Proc Natl Acad Sci U S A* 92, 7739–7743

36. Rudoll, T., Phillips, K., Lee, S. W., Hull, S., Gaspar, O., Sucgang, N., Gilboa, E., and Smith, C. (1996) High-efficiency retroviral vector mediated gene transfer into human peripheral blood CD4+ T lymphocytes. *Gene Ther* 3, 695–705

37. Gallardo, H. F., Tan, C., Ory, D., and Sadelain, M. (1997) Recombinant retroviruses pseudotyped with the vesicular stomatis virus G glycoprotein mediate both stable gene transfer and pseudotransduction in human peripheral blood lymphocytes. *Blood* 90, 952–957

38. Bauer, T. R., Jr., Miller, A. D., and Hickstein, D. D. (1995) Improved transfer of the leukocyte integrin CD18 subunit into hematopoietic cell lines by using retroviral vectors having a gibbon ape leukemia virus envelope. *Blood* 86, 2379–2387

39. Lam, J. S., Reeves, M. E., Cowherd, R., Rosenberg, S. A., and Hwu, P. (1996) Improved gene transfer into human lymphocytes using retroviruses with the gibbon ape leukemia virus envelope. *Hum Gene Ther* 7, 1415–1422

40. Hacein-Bey-Abina, S., Von Kalle, C., Schmidt, M., McCormack, M. P., Wulffraat, N., Leboulch, P., Lim, A., Osborne, C. S., Pawliuk, R., Morillon, E., Sorensen, R., Forster, A., Fraser, P., Cohen, J. I., de Saint Basile, G., Alexander, I., Wintergerst, U., Frebourg, T., Aurias, A., Stoppa-Lyonnet, D., Romana, S., Radford-Weiss, I., Gross, F., Valensi, F., Delabesse, E., Macintyre, E., Sigaux, F., Soulier, J., Leiva, L. E., Wissler, M., Prinz, C., Rabbitts, T. H., Le Deist, F., Fischer, A., and Cavazzana-Calvo, M. (2003) LMO2-associated clonal T cell proliferation in two patients after gene therapy for SCID-X1. *Science* 302, 415–419

# Chapter 11

## Lentiviral Vector-Mediated Genetic Programming of Mouse and Human Dendritic Cells

### Renata Stripecke

### Summary

Dendritic cells (DCs) play a key role in the orchestration of immune reactions. Manipulation of DC function through genetic manipulation for vaccine development provides a multitude of applications for active immunotherapy of cancer and chronic infections. Several laboratories have shown that lentiviral vectors (LVs) are efficient and consistent tools for ex vivo gene manipulation of DCs and their precursors. LVs integrate in the genome of target cells resulting in persistent and stable transgene expression, and gene delivery does not result in cytostatic or nonspecific adverse immunomodulatory reactions. Mouse, macaque, and human DCs are efficiently transduced with LVs, allowing preclinical vaccination studies to be gradually implemented into clinical trials. This chapter describes HIV-1-derived LV transduction used for ex vivo gene delivery of marking genes, antigens, and immunomodulatory molecules into mouse and human hematopoietic precursors and DCs. With the perspective of bioengineering DCs from the inside-out, we also describe a one-hit LV transduction method for constitutive expression of GM-CSF and IL-4 genes, which allows self-differentiation of mouse and human hematopoietic precursor cells into highly viable and potent DCs.

**Key words:** Dendritic cells, Lentiviral vectors, Cell vaccines, Immunotherapy, Bioengineered DCs, Self-differentiated DCs.

## 1. Introduction

Dendritic cells (DCs) coordinate the initiation of immune responses by naïve T cells and B cells *(1)*. During the past decade, exciting preclinical studies in mice for ex vivo generation of DC vaccines were translated into phase 1 and 2 clinical trials for active immunotherapy of advanced cancer patients *(2)*. Antigen-specific immune responses were consistently observed, but the

Christopher Baum (ed.), *Methods in Molecular Biology, Methods and Protocols, vol. 506*
© Humana Press, a part of Springer Science+Business Media, LLC 2009
DOI: 10.1007/978-1-59745-409-4_11

clinical responses were quite modest so far (3). Simplification and optimization of the operating procedures and standardization of clinical and immunologic criteria are major ongoing efforts for large-scale trials. Despite these obstacles, the learning curve for DC manipulation has not yet reached plateau and the field is dynamic and persevering. Currently, generation of DC vaccines has expanded toward treatment of chronic infectious diseases (4), which is likely to provide a new boost of basic and clinical information. In due course, the deeper understanding of the biology of DCs and dependable rational programming to enhance their immune stimulation and minimize tolerogenic effects will culminate in highly effective cellular vaccines.

Different subtypes of DCs have been identified, but most clinical studies to date use monocyte-derived DCs, also known as myeloid DCs. Myeloid DCs can be differentiated with growth factors in vitro from cell precursors present in peripheral blood, bone marrow, or cord blood. These ex vivo grown DCs show high endocytic activity and high levels of intracellular major histocompatibility complex (MHC) class I and II molecules, resulting in effective cellular adjuvant for capture and presentation of antigens (5) (**Fig. 1**). The methodology described in this chapter will cover the use of GM-CSF and IL-4 for ex vivo DC generation, since these growth factors are the most commonly used and are largely commercially available. GM-CSF regulates cellular proliferation, differentiation and upregulation of MHC and costimulatory molecules in myeloid cells through binding to its receptor and is required to generate cells with morphologic characteristics of DCs from mouse bone marrow cultures (6). Interleukin-4 downregulates CD14, which is a lipopolysaccharide receptor, and blocks the development of macrophages (7). The combination of GM-CSF and IL-4 promotes effective and consistent differentiation of human peripheral monocytes and mouse bone marrow cells into immature DCs, which became the most traditional method to culture DCs ex vivo (5, 8) (**Fig. 1**). After one week of ex vivo culture, DCs can be loaded with antigenic peptides, proteins, or cell extracts (9, 10). To perform effective antigen presentation, DCs need to undergo a "maturation" process: MHC-peptide complexes are transported to the cell surface, high levels of MHC-I and costimulatory ligands (CD80, CD86) are expressed, and inflammatory cytokines (IL-12, IL-10) are secreted (11) (**Fig. 1**). CD40L, also known as CD154, is a transmembrane protein expressed on activated T-helper cells, which promotes activation of DCs upon its engagement with its CD40 receptor (12). Currently, several protocols for ex vivo production of DCs include a 24–48 h treatment of DCs with soluble trimeric CD40L after they have been loaded with immunologically relevant antigenic molecules to boost DC activity in immunizations (13, 14).

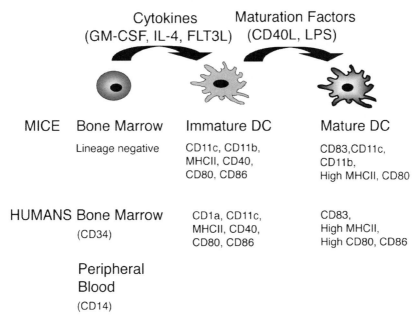

Fig. 1. Diagrammatic representation of ex vivo generation of mouse and human immature and mature dendritic cells and corresponding immunophenotypic markers.

Genetic reprogramming of DCs and their precursors has emerged as a strategy by which, independent of the inflammatory milieu, DCs can autonomously reach the necessary steps of development: differentiation, antigen capture, maturation, and antigen presentation (**Fig. 2**). Several gene delivery systems have been tested for DC bioengineering. In contrast to nonviral or lytic viral vector modalities, lentiviral vectors may offer consistent, efficient, persistent, nontoxic, and nonhazardous gene delivery into hematopoietic cells (for a review *see (15)*. Lentiviral infection occurs naturally in cells of the immune system such as DCs, macrophages, and T cells. The molecular machinery that promotes nuclear uptake of the lentiviral preintegration complex allows efficient transduction of nondividing cells (primary quiescent, growth-arrested, and terminally differentiated). In most cases, lentiviral vectors are pseudotyped, i.e., encoated with a heterologous envelope protein. The vesicular stomatitis virus glycoprotein (VSV-G) envelope is commonly used and binds to cell surface phospholipids, thereby achieving a wide host range of infection. Notably, VSV-G pseudotyped LVs injected intravenously in mice showed high infectivity of splenic APCs, such as DCs and B cells *(16, 17)*. VSV-G pseudotyped LVs also provide a very robust system for efficient ex vivo gene delivery into hematopoietic precursors and cultured DCs *(18–25)*. DC transduction with LVs

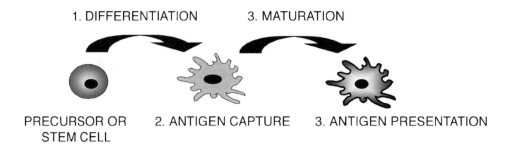

1. DIFFERENTIATION          3. MATURATION

PRECURSOR OR          2. ANTIGEN CAPTURE          3. ANTIGEN PRESENTATION
STEM CELL

LV-TRANSDUCTION:
- MARKER GENES (GFP, LUC)
- ANTIGENS (OVA, TRP2)
- CYTOKINES FOR
  DIFFERENTIATION
  (GM-CSF, IL-4)

LV-TRANSDUCTION:
- MARKER GENES
- ANTIGENS
- MATURATION FACTORS
  (CD40L)

Fig. 2. Steps of DC development susceptible to lentiviral gene delivery and different possible types of genetic manipulation.

at multiplicity of infection (M.O.I) of 10–50 leads consistently to high levels (70–100%) of gene transfer without altering their viability, immunophenotype, or maturation status *(19, 26, 27)*.

Examples of tumor-associated antigens (TAAs) expressed and processed in DCs after LV transduction include several melanoma-associated antigens such as TRP2 *(28, 29)*, tyrosinase *(30)*, MAGE-A3 *(31)*, MART-1 *(29)*, and antigenic polyepitopes *(32)*. Viral antigens such as a Flu peptide *(22)*, an immunodominant epitope of LCMV *(33)*, HIV antigenic epitopes *(34)*, and SIV Gag *(35)* have also been expressed via lentiviral vectors in DCs, leading to potent antigen-specific responses.

In addition to antigen expression, LVs can deliver growth factors and immunomodulatory molecules involved in DC differentiation and maturation. We have established the proof of concept that lentiviral vector-mediated expression of GM-CSF and IL-4 in CD14+ purified human monocytes cells was sufficient to promote their self-differentiation in vitro into cells with typical DC morphology and immunophenotype, which were named "DC/LVs" *(36)*. Mixed lymphocyte reactions showed that the T-cell stimulating activity of DC/LVs was superior to that of DCs grown by conventional methods. DC/LVs displayed efficient antigen-specific, MHC Class-I restricted stimulation of autologous CD8+ T cells, as shown by IFN-gamma production and CTL assays. Importantly, DC/LVs exhibited a longer lifespan in culture and could be maintained metabolically active and viable in culture for 2–3 weeks in the absence of exogenously added growth factors, compared to DCs cultured by conventional methods *(36)*.

We subsequently evaluated self-differentiation of mouse bone marrow (BM) cells lentivirally transduced for coexpressing GM-CSF/IL-4 in vivo after transplantation into C57BL/6 mice: the genetically programmed cells self-differentiated autonomously into DCs in vivo, showed robust viability, migrated efficiently to adjacent lymph nodes, and stimulated a dramatic influx of host DCs *(29)*. The therapeutic efficacy of the DC/LV vaccines was evaluated by lentiviral vector codelivery of melanoma antigens (MART1, TRP2). C57BL/6 mice vaccinated with DC/LV produced higher antigen-specific CD8$^+$ T-cell responses than conventional DCs generated ex vivo, which correlated with protective and therapeutic effects against B16 melanoma development *(29)*. Thus, here we also cover novel methods by which DC precursors genetically engineered after a single ex vivo lentiviral gene transfer result into self-differentiated DCs with enhanced immune potency. We also describe DC transduction with a vector expressing the CD40L, which correlated with potent maturation effects as shown by immunophenotype, production of IL-12, and enhancement of allogeneic and antigen-specific immune responses *(26)*.

The methods presented here are very feasible and straightforward for genetic programming DC vaccines for preclinical studies. Since CD34$^+$ DC precursors are more difficult to be obtained, they were not included in the methods.

## 2. Materials

### 2.1. Solutions, Buffers, and Cell Culture

1. *2× HBS (pH 7.12)*. 280 mM NaCl, 1.5 mM Na$_2$HPO$_4$, 100 mM HEPES. For 500 mL: Mix 28 mL 5 M NaCl, 1.5 mL 0.5 M Na$_2$HPO$_4$, 50 mL 1 M HEPES, precisely set pH to 7.12, filter sterilize through 0.2-μm filter, dispense in 10 mL aliquots, freeze solution at –20°C (stable for several months).

2. 2.5 M CaCl$_2$ (cell culture grade, Sigma) filter sterilize through 0.2-μm filter, dispense in 10 mL aliquots, freeze solution at –20°C (stable for several months).

3. *n*-Butyric acid (Sodium salt; Sigma). Prepare 0.5 M solution in water: Dissolve 1 g of sodium butyrate in 18.17 mL of water, filter sterilize through 0.2 μm filter, keep 1 mL aliquot frozen at –20°C (stable for several months).

4. Poly-l-lysine (0.01% Solution; Sigma).

5. *D10 culture medium*. DMEM (BioWhittaker, Walkersville, MA), 10% heat-inactivated fetal bovine serum (FBS), 2 mM

L-glutamine, penicillin/streptomycin (50 U/mL) (*see* **Note 1**).

6. *Medium for culture of mouse BM.* RPMI (BioWhittaker), 10% fetal bovine serum (FBS), l-glutamine (2 mmol/L), penicillin/streptomycin (50 U/mL).

7. *Materials for PBMC cryopreservation.* Ficoll-Hypaque, 10% dimethyl sulfoxide (DMSO) and 90% human AB serum (Omega, Tulare, CA).

8. Serum-free clinical grade X-VIVO 15 medium (Cambrex, Walkersville, MD).

9. Dulbecco's phosphate-buffered saline (DPBS).

10. *Recombinant cytokines.* Human/mouse GM-CSF; human/mouse IL-4; human/mouse soluble trimeric CD40L (all from R&D Systems, Minneapolis, MN).

11. LPS (Sigma).

12. *Protamine sulfate (Fujisawa, Deerfield, IL).* 1 mg/mL solution stored at 4°C (stable for several months).

13. *Blocking solution for immune staining.* Mouse γ-immunoglobulins, at 0.1 mg/mL in PBS (maintained in aliquots at 4°C).

14. *Fixative solution for immune staining.* 1% paraformaldehyde in PBS, pH 7.3 (stable for 1 month at 4°C).

### 2.2. Kits

1. Endotoxin-free Maxi Kit (Qiagen, Valencia, CA).

2. HIV-1 p24 ELISA Kit (Cell BioLabs Inc.).

3. CD14⁺ positive immunomagnetic cell selection (Miltenyi Biotec, Auburn, CA).

### 2.3. Vector Constructs

We recommend the production of HIV-1-derived third-generation self-inactivating (SIN) lentiviral vectors, which are approved by most institutions for use in Biosafety Level 2 (BL-2) (*see* **Note 2**). The plasmids required for the packaging of third-generation self-inactivating lentiviral vectors consist of the plasmid pMD.G (for the production of the VSV-G viral envelope), pMDLg/pRRE, and RSV-REV (for expression of the structural proteins, enzymes, and regulatory proteins of HIV) *(37)*, and were provided by Dr. Luigi Naldini (San Raffaele Telethon Institute for Gene Therapy, Milano, Italy). The reference SIN vector pRRL-sin.cPPT-hCMV-GFP-pre containing the cytomegalovirus (CMV) promoter, the central polypurine tract and termination sequences (CPPT/CTS), and the woodchuck hepatitis B post-transcriptional regulatory element (WPRE) were developed by Dr. Luigi Naldini and colleagues (*see* **Note 3**). LVs for expression of marking genes (green fluorescent protein, firefly luciferase), immunomodulatory molecules (GM-CSF, IL-4, CD40L) and TAAs (TRP-2, MART-1) have been described *(26, 29, 36, 38)*.

Plasmids are amplified in XL-10 blue *Escherichia coli* (Stratagene, La Jolla, CA) and purified using Endotoxin-free Kits. After the isopropanol precipitation, the DNA is reprecipitated in 70% ice-cold Ethanol and resuspended in sterile TE buffer pH 8.0 using aseptic techniques.

### 2.4. Cells

1. 293T (CRL-11268) cells are obtained from the American Type Culture Collection (Rockville, MD) (*see* **Note 4**).

2. *Mouse bone marrow precursor cells.* Bone marrow is aseptically flushed from femurs and tibias of 6–10 week-old C57BL/6 mice. The dissociated cells are washed with RPMI and incubated in 10-cm diameter plastic dishes with RPMI + 10% FCS overnight. For optional immunodepletion of granulocytes, T, B, and NK cells, nonadherent bone marrow cells can be treated with antibodies (1 μg/million of cells) against GR-1, B220, Thy 1.2, NK1.1, and cell lysis is induced with 10% rabbit complement (Sigma) in RPMI for 30 min (*see* **Note 5**).

3. PBMCs are obtained from donors in accordance with informed consent protocols approved by the Institutional Review Board. PBMCs are separated by density gradient centrifugation, and cryopreserved in 10% DMSO and 90% human AB serum. Immediately before use, frozen cells are thawed at 37°C and washed with PBS.

4. CD14$^+$ monocytes are obtained by positive immunomagnetic cell selection (Miltenyi Biotec, Auburn, CA).

5. For selection of adherent PBMCs, thawed cells are placed in 3 mL of X-VIVO 15 medium in a 6-well plate at a concentration of $5 \times 10^6$ cell/well, and allowed to adhere to the plate at 37°C with 5% $CO_2$ for 1 h. Nonadherent cells are then removed by gently washing the surface of the plate with PBS.

### 2.5. Other Supplies

1. 14-mL polystyrene tube (Becton Dickinson, San Jose, CA).

2. 0.2-mm filter (Nalgene, Rochester, NY).

3. Polyalomer ultracentrifugation tubes (Beckman, Fullerton, CA).

### 2.6. Equipment

1. L7 Ultracentrifuge (Beckman, Fullerton, CA) or equivalent.

2. SW 28 or SW30.1 rotors and buckets (Beckman, Fullerton, CA) or equivalent.

## 3. Methods

### 3.1. Production of Lentiviral Vectors

#### 3.1.1. Day 0

1. Coat 8× T175 tissue culture flasks with poly-l-lysine. This is done by pipetting 5 mL of a 0.01% solution into a flask, removing, adding to next flask, etc. Remove excess liquid thoroughly from the flasks, let air-dry for 15 min.

2. Plate $18 \times 10^6$ 293T cells per flask and add D10 medium to complete 40 mL. Incubate the cells in a 5% $CO_2$ incubator at 37°C overnight.

#### 3.1.2. Day 1

1. In the morning, remove medium from the 293T flasks and add 40 mL of fresh D10.

2. In the late afternoon, calculate the volume of each plasmid necessary for each flask of 293T cells for the transient triple transfection: RRL-cPPT-XYZ vector plasmid: 60 µg; pMDLGg/p packaging plasmid: 39 µg; RSV-REV plasmid: 15 µg; PMD.G envelope plasmid: 21 µg

3. Mix the necessary volumes of the three plasmids plus $ddH_2O$ to a total volume of 2.7 mL. Add 300 µl of 2.5 M $CaCl_2$ per transfection sample of DNA and mix gently.

4. Transfer 2.8 mL of 2× HBS (pH = 7.12) into a 14-mL polystyrene tube. Add approximately 2.8 mL DNA/$CaCl_2$ mix dropwise, under constant vortexing. Incubate at room temperature for 3–4 min (see **Note 6**).

5. Take the 293T-cell flasks out of the incubator and slowly add the DNA/$CaPO_4$ suspension to each flask. Incubate the flasks in a 5% $CO_2$ incubator at 37°C overnight (12–15 h).

#### 3.1.3. Day 2

1. In the morning, carefully remove the medium from the cells and discard. Carefully wash the cells once with DMEM. Add 40 mL D10 medium containing 20 mM HEPES and 10 mM Sodium Butyrate.

2. Incubate the cells in a 5% $CO_2$ incubator at 37°C for 8–12 h. After that, wash the cells once with DMEM and add 40 mL fresh D10 with 20 mM HEPES onto the 293T cells. Incubate in a 5% $CO_2$ incubator at 37°C for overnight.

#### 3.1.4. DAY 3: Virus Concentration by Ultracentrifugation

1. Collect medium supernatant in 50-mL conical tubes.

2. Centrifuge the harvested supernatant $1{,}000 \times g$ for 5 min to remove cellular debris.

3. Filter the supernatants through a 0.45-µm filter.

4. Aliquot the filtered supernatant in autoclaved polyalomer tubes. Add sterile DPBS to balance if necessary.

5. Ultracentrifuge for 2 h and 20 min at room temperature (around 15–22°C) at 19,000 rpm (50,000 × $g$). The brake should be set to slow.

6. Carefully remove the supernatant without disturbing the loose white/brown pellet in the center of the tube. Leave a small amount of supernatant (1 mL) behind so as not to disturb the pellet.

7. Suspend pellets with the remaining supernatant, combine, and place in a 50-mL tube. Vortex this 50 mL tube at low speed for 2–3 h at room temperature to make sure the virus particles are resuspended. Avoid foaming.

8. Transfer the virus suspension to a microcentrifuge tube and spin at 700 × $g$ for 10 min to remove any macroscopic debris formed during concentration.

9. Aliquot the cleared concentrated supernatant into cryovials at a convenient volume (0.5 mL) and freeze at –80°C.

*3.1.5. Day 4: p24 Titer Determinationa*

1. Thaw and vortex sample (without foaming) for 1 h at room temperature.

2. Perform $10^{-3}$, $10^{-4}$ dilutions. Since the virus is very sticky, change pipette tips every time you perform a dilution.

3. Load the virus samples onto the p24 ELISA wells and follow the manufacturer's instructions (*see* **Note 7**).

### 3.2. Generation of Mouse-Transduced DCs

*3.2.1. Transduction of Mouse Precursor Cells and Subsequent DC Differentiation*

1. Nonadheren t BM cells flushed from C57BL/6 mice femur and tibia are plated at 5 × 10⁶ cells/well of 6-well plates in RPMI with 10% FCS in the presence of recombinant GM-CSF and IL-4 (50 ng/mL each). This preconditioning step is optional and increases the LV transduction efficiency (Fig. 11.3) (*see* **Note 8**).

2. Lentiviral vectors (1–5 µg p24 equivalent/mL) and Protamine Sulfate (5 µg/mL) are added in a final volume of 1 mL and transduction is carried out for 16 h at 37°C. Since variations in the p24 measurement can occur, we recommend that a dose curve be performed. In our hands, lentiviral vectors corresponding to concentration of 1 µg/mL p24, yielded approximately 50% GFP positive cells, which increased to > 90% GFP⁺ at a concentration of 5 µg/mL p24 (Fig. 11.3) (*see* **Note 9**).

3. Cells are resuspended, washed with PBS, reseeded in 6-well plates with 2 mL RPMI/FBS/GM-CSF/IL-4, and fresh medium is replenished every 3 days.

4. If DC maturation is required, soluble trimeric CD40L (500 ng/mL) or LPS (1 µg/mL) can be added 24–48 h prior to analyses or vaccinations.

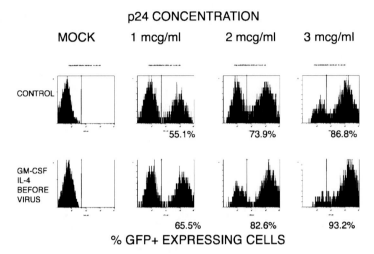

Fig. 3. Transduction of mouse nonadherent bone marrow cells with the vector pRRL-sin. cPPT-hCMV-GFP-pre, differentiation into DCs and evaluation of transduction efficiency by flow cytometry analyses of GFP expression.

5. On day 7 day of culture, cells are harvested and flow cytometry analyses can be performed to confirm the DC immunophenotype CD11b$^{low}$, CD11c$^{high}$, MHCI$^{high}$, MHCII$^{high}$ (**Fig. 4**). A simplified protocol is provided in the subsequent steps.

6. Transfer supernatant with nonadherent cells into a 15-mL tube.

7. Deattach cells on the wells by incubating them in PBS at 37°C for 30–60 min. Join adherent and nonadherent cells.

8. Centrifuge 800 × $g$ 5 min, resuspend cells in PBS at 5 × 10$^6$ cells/mL.

9. Transfer 100 µl of cell suspension for each staining arm (always include isotype controls) to a 5-mL flow cytometry tube.

10. Add blocking Ig at the recommended concentration, incubate 15 min on ice.

11. Add conjugated antibodies at the recommended concentration, incubate 30 min in the dark.

12. Wash cells with 4 mL PBS, centrifuge 800 × $g$ 5 min.

13. Add 100 µl of 1% paraformaldehyde to cell pellet and keep cells in dark at 4°C until FACS analyses.

14. Expression of transgenes can be evaluated by flow cytometry (e.g., GFP, tNGFR), optical imaging analyses (luciferase), western blot (intracellular proteins), or ELISA (for secreted proteins). Alternatively, transduced DCs can be used for functional characterization such as T cell priming in vitro, vaccinations, or biodistribution analyses.

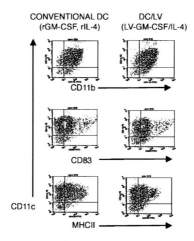

Fig. 4. Immunophenotypic characterization by flow cytometry of relevant immunologic markers of day 7 mouse bone marrow-derived conventional DCs and DCs generated through lentiviral transduction (DC/LVs).

*3.2.2. Transduction of Differentiated Mouse DCs*

1. For some purposes, such as expression of maturation or signal transduction factors that can deregulate DC differentiation, gene delivery into differentiated DCs may be preferential. Although less susceptible for transduction than their precursors, differentiated DCs can also be efficiently transduced with LVs. Nonadherent BM cells are manipulated essentially like described in **Subheading 3.2.1**, except that transduction is carried out on later days of culture (days 3–5).

2. Transduction is optimally performed with LVs at > 5 μg p24 equivalent/mL and Protamine Sulfate (5 μg/mL). Since at later time points some of the differentiating DCs are attached to the wells, we recommend to collect the cells in the supernatant (leaving some medium behind such that the cells attached to the plate do not dry out), centrifuge them, resuspend them in the indicated volume/concentration of virus, and then add them to the cells attached to the well (*see* **Note 10**).

3. After transduction, cells are incubated for 16 h at 37°C, after which the cells were washed twice with PBS and maintained in RPMI/FBS/GM-CSF, IL-4 medium for 1–2 days prior to analyses.

4. If DC maturation is required, soluble trimeric CD40L (500 ng/mL) or LPS (1 μg/mL) can be added 24–48 h prior to analyses or vaccinations.

*3.2.3. Generation of LV-Transduced Mouse DCs Expressing GM-CSF and IL-4 that Self-Differentiate In Vitro and In Vivo*

1. Combinatorial LV cotransduction can be achieved using a mixture of high-titer LVs. This approach can be used, for example, for gene codelivery of GM-CSF, IL-4 and marking or antigenic genes into mouse BM precursors. Transduction of mouse BM cells with a bicistronic vector

RRL-cPPT-hCMV-mGMCSF-IRES-mIL4 plus other vectors expressing marking genes or antigens is performed following the optimized protocol described earlier (including the pre-conditioning of mouse BM cells with recombinant GM-CSF and IL-4 prior to transduction) (29) (see **Note 11**).

2. BM cells are cotransduced with 5 μg/mL of each lentiviral vector plus Protamine Sulfate (5 μg/mL) in 1 mL final volume on a well of a 6-well plate. Transduction is carried out for 16 h at 37°C. Subsequently, cells are resuspended, washed twice with PBS, and can be used directly for vaccinations since they have the capability to self-differentiate in vivo.

3. Alternatively, the transduced cells can be returned to culture in the absence of recombinant cytokine (endogenously produced GM-CSF is approximately 30 ng/$10^6$ cells/mL/24 h and IL-4 is 3 ng/$10^6$ cells/mL/24 h). The cells can be maintained in culture with weekly replenishment of RPMI/FBS. Viability is very high in the initial 3 weeks after transduction and it typically decreases after 4 weeks (29).

4. LV-generated DCs (DC/LVs) show similar immunophenotype as conventional DCs (**Fig. 4**).

### 3.3. Generation of Human-Transduced DCs

*3.3.1. Transduction of Human Monocytes or Adherent PBMC and Subsequent DC Differentiation*

1. Fresh or thawed monocytes are efficiently transduced with LVs and therefore can be genetically manipulated prior to DC differentiation. Monocytes obtained through positive selection of CD14+ cells produce more pure and homogeneous populations of DCs than adherent PBMCs. One group has reported that positive selection of monocytes using anti-CD14 antibody-coated magnetic beads triggers cellular activation, enhancing lentiviral transduction (39).

2. 3–5 × $10^6$ cells are placed in wells of a sterile 6-well plate containing 2 mL of serum-free, clinical-grade X-VIVO 15 medium plus 80 ng/mL human IL-4 and 80 ng/mL GM-CSF and incubated at 37°C with 5% $CO_2$ for 8 h prior to transduction (see **Note 8**).

3. Lentiviral vectors (1–5 μg p24 equivalent/mL) and Protamine Sulfate (5 μg/mL) are added in a final volume of 1 mL and transduction is carried out for 16 h at 37°C at 5% $CO_2$ (see **Note 9**).

4. Cells are resuspended, washed with PBS, reseeded in 6-well plates with 2 mL RPMI/FBS/GM-CSF/IL-4 and fresh medium is replenished every 3 days.

5. If DC maturation is required, soluble trimeric CD40L (500 ng/mL) or LPS (1 μg/mL) can be added 24–48 h prior to analyses or vaccinations.

6. On day 7 day of culture, cells are harvested and flow cytometry analyses can be performed to confirm the DC immunophenotype CD1a⁺, CD11c$^{high}$, CD80$^{high}$, MHCII$^{high}$. A protocol for staining was described in **Subheading 3.2.1**.

7. Expression of transgenes can be evaluated by flow cytometry (e.g., GFP, tNGFR), optical imaging analyses (luciferase), western blot (intracellular proteins), or ELISA (for secreted proteins). The persistence of transgene expression after transduction of monocytes was assessed by GFP expression 7, 14, and 21 days after transduction. Kinetic analysis of GFP expression demonstrated consistently high percentages of GFP⁺ cells throughout the 21-day period (**Fig. 6**).

8. The transduced DCs can also be used to stimulate T cells in vitro (e.g., mixed lymphocyte reactions, ELISPOT assays, or Chromium release assays).

*3.3.2. Transduction of Differentiated Human DCs*

1. Like in the mouse system, human differentiated DCs can also be efficiently transduced with LVs. CD14⁺-selected monocytes or adherent PBMCs are manipulated essentially like described

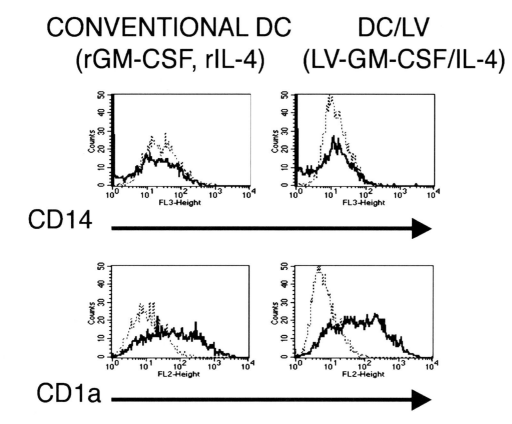

## CONVENTIONAL DC (rGM-CSF, rIL-4)        DC/LV (LV-GM-CSF/IL-4)

Fig. 5. Immunophenotypic characterization by flow cytometry of relevant immunologic markers of day 7 human conventional monocyte-derived DCs and DCs generated through lentiviral transduction (DC/LVs).

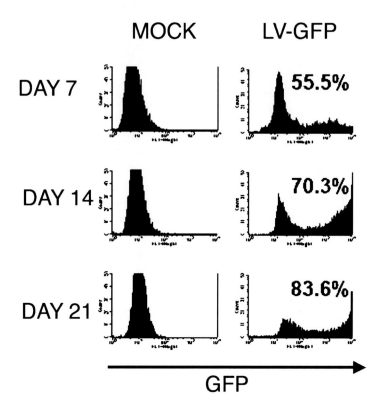

Fig. 6. Transduction of human CD14+ monocytes with the vector pRRL-sin.cPPT-hCMV-GFP-pre and evaluation of transduction efficiency by flow cytometry analyses of GFP expression.

in **Subheading 3.3.1**, except that transduction is carried out on later days of culture (days 3–5).

2. Transduction should be performed with LVs at > 5 μg p24 equivalent/mL and Protamine Sulfate (5 μg/mL). Since some differentiating DCs are attached to the wells and some are loose, cells in the supernatant are collected (leaving some medium behind such that the cells attached to the plate do not dry out), centrifuged, resuspended in the indicated concentration/volume of virus, and then added to the cells attached to the well (*see* **Note 10**).

3. After transduction, cells are incubated for 16 h at 37°C, after which the cells were washed twice with PBS and maintained in RPMI/FBS/GM-CSF, IL-4 medium for 1–2 days prior to analyses of immunophenotype or transgene expression (**Fig. 7**).

4. If DC maturation is required, soluble trimeric CD40L (500 ng/mL) or LPS (1 μg/mL) can be added 24–48 h prior to analyses or vaccinations. Alternatively, lentiviral transduction for CD40L is very effective for maturation, which can

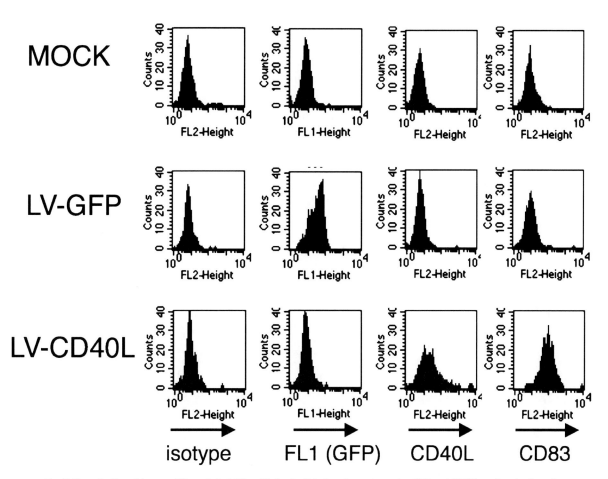

Fig. 7. Transduction of human differentiated DCs with the lentiviral vectors expressing GFP and CD40L and evaluation of transduction efficiency and maturation by flow cytometry analyses of GFP and CD83 expression, respectively.

be followed by expression of the CD83 maturation marker *(26)*(Fig. 11.7).

5. Transduced DCs are ready for T-cell stimulation assays.

*3.3.3. Generation of LV-Transduced Human DCs Expressing GM-CSF and IL-4 that Self-Differentiate In Vitro*

1. Positively selected CD14+ cells (> 90% pure) are maintained in culture in serum-free X-VIVO 15 medium in the presence of recombinant human GM-CSF and recombinant human IL-4 for 8 h prior to transduction.

2. Monocytes are transduced with 5 µg/mL p24 equivalent of each lentiviral vector (RRL-hCMV-GM-CSF, RRL-hCMV-IL4 and vectors expressing marking genes or antigenic proteins) plus Protamine Sulfate (5 µg/mL) added in a final volume of 1 mL in a well of a 6 well. Transduction is carried out for 16 h at 37°C at 5% $CO_2$.

3. Cells are resuspended, washed twice with PBS, and are returned to culture in X-VIVO 15 medium (no cytokines required, since endogenously produced GM-CSF and IL-4 accumulate in the cell supernatants at concentrations of approximately $10 \text{ ng}/10^6$ cells/mL/24 h).

4. The LV-generated DCs (DC/LVs) cells can be maintained in culture with weekly replenishment of X-VIVO 15.

5. Similar to mouse DC/LVs, viability is very high in the initial 3 weeks after transduction and typically decreases after 4 weeks *(36)*.

6. Human DC/LVs show similar immunophenotype as conventional DCs (Fig. 11.5) and stimulate potent antigen-specific T-cell responses in vitro *(36)*.

## 4. Notes

1. High-quality serum with low levels of endotoxin is essential for consistent virus production. Test 2–3 different lots of heat-inactivated characterized FBS (100 mL is usually provided by the vendors at no cost) in a small-scale virus production and purchase a good reserve of bottles from the best lot.

2. The first-generation packaging vectors express *gag* and *pol* and all the HIV-1 pathogenicity genes (*vif, vpr, vpu, nef*) and regulatory genes (*tat, rev*); the second-generation packaging vectors express *gag* and *pol* and the HIV-1 regulatory genes (*tat, rev*); the third-generation packaging vectors are split into two plasmids for *gag/pol* expression and *rev* expression. Minimizing the number of HIV-1 genes in the packaging and splitting them in different vectors decreases the risk of generation of replication-competent lentivirus. In parallel with improvements in the packaging system has been the development of self-inactivating (SIN) lentivirus vector designs which generally contain a 400-nucleotide deletion in the 3′ long terminal repeat (LTR) *(37)*. Through the process of reverse transcription, this deletion is copied to the 5′ LTR, thereby abolishing the 5′ LTR promoter activity and hampering recombination with wild-type HIV in an infected host *(37)*. Therefore, the risk of vector mobilization with the wild-type virus, and subsequent production of replication competent lentiviral vectors is drastically reduced for the SIN vectors. Third-generation SIN-LVs are approved in most institutions for use in BL-2 conditions, whereas restrictions may apply for earlier LV versions.

3. The CMV promoter provides high expression of transgenes during the different steps of DC differentiation (precursors, immature, mature). Some groups have tested weaker, housekeeping constitutive promoters such as elongation factor 1 alpha (EF1-α) or phosphoglycerate kinase (PGK). For selective LV expression in DCs, promoters such as HLA-DR-a, B7-DC and CCL17 have been described *(40, 41)*. An important contribution for maximizing lentiviral vector infectivity is the presence of the central polypurine tract and termination sequences (cPPT/CTS, a 118-bp element) within *pol*, which is required in cis to promote high integration of lentiviral vectors in primary cells *(38, 42, 43)*. The inclusion of the WPRE in the SIN LVs is supposed to increase the viral RNA transport, but some groups have implicated detrimental effects of WPRE in DC transduction *(31, 44)*.

4. *Maintenance conditions of 293T stocks.* Earlier passage 293T cells (< 3 months in culture) give better transfection results. 293T cells should never be allowed to overgrow confluency, because this leads to selection of subpopulations with altered transfection efficiencies. Furthermore, because 293T cells are not very adherent, care should be taken not to disturb the cells when medium is added onto the dishes. The addition of poly-L-lysine to coat the dishes, as described earlier, will allow the cells to adhere to the plate and avoid cell loss during transfection and virus harvesting.

5. Positive selection systems for mouse monocytes are not commercially available. We do not recommend the use of CD11b selection procedures for enrichment of DC precursors, as granulocytes are highly represented in the resulting selected population.

6. The 2× HBS buffer should be carefully set at pH 7.12. Different batches of 2× HBS can give high variability of transfection. Before using a new batch of 2× HBS it is advisable to compare it to an older, successful batch.

7. For our reference vector RRL-sin.cPPT-hCMV-GFP-pre, 1 μg/mL of p24 measured in the preparation corresponds to approximately $1-5 \times 10^7$ GFP transduction units/mL, as assessed by titration in 293T cells.

8. Both for the mouse and the human systems, we included this preconditioning step after performing pilot experiments showing that *(1)* GM-CSF improves lentiviral transduction; *(2)* IL-4 prevents the outgrowth of contaminating macrophages in the culture and favors DC growth.

9. Protamine Sulfate is a cationic adjuvant that neutralizes the electrostatic repulsion between negatively charged cells and virus. Protamine Sulfate is approved for clinical use and in

our hands was more effective than polybrene, an alternative cationic adjuvant. Additional procedures such as use of fibronectin-coated plates or centrifugation during transduction (spinoculation) did not improve significantly transduction with highly concentrated LVs.

10. Resuspending attached DCs can be detrimental to their viability and function and should be performed only as a last step, prior to analyses.

11. Combinatorial cotransduction leads to superinfection of target cells and integration of several copies of lentiviral vectors in the genome. Alternatively, multicistronic LVs expressing various genes simultaneously can also be designed.

## Acknowledgments

The author would like to thank the previous and current members of her group and of the UCLA Vector Core for the hard work and support. This work was supported by The Margareth E. Early Research Trust, Stop Cancer, and NIH grants (UCLA Center for in vivo Imaging in Cancer Biology/2P50-CA086306-06, Rebirth/DFG and SFB 738/DFG.

## References

1. Banchereau, J., Steinman, R.M. (1998) Dendritic cells and the control of immunity. *Nature* 392: 245–252

2. O'Neill, D.W., Adams, S., Bhardwaj, N. (2004) Manipulating dendritic cell biology for the active immunotherapy of cancer. *Blood* 104: 2235–2246

3. Figdor, C.G., de Vries, I.J., Lesterhuis, W.J., Melief, C.J. (2004) Dendritic cell immunotherapy: mapping the way. *Nat Med* 10: 475–480

4. Van den Bosch, G.A., Ponsaerts, P., Vanham, G., Van Bockstaele, D.R., Berneman, Z.N., Van Tendeloo, V.F. (2006) Cellular immunotherapy for cytomegalovirus and HIV-1 infection. *J Immunother* 29: 107–121

5. Sallusto, F., Lanzavecchia, A. (1994) Efficient presentation of soluble antigen by cultured human dendritic cells is maintained by granulocyte/macrophage colony-stimulating factor plus interleukin 4 and downregulated by tumor necrosis factor alpha. *J Exp Med* 179: 1109–1118

6. Inaba, K., Steinman, R.M., Pack, M.W., Aya, H., Inaba, M., Sudo, T., Wolpe, S., Schuler, G. (1992) Identification of proliferating dendritic cell precursors in mouse blood. *J Exp Med* 175: 1157–1167

7. Heidenreich, S. (1999) Monocyte CD14: a multifunctional receptor engaged in apoptosis from both sides. *J Leukoc Biol* 65: 737–743

8. Lutz, M.B., Suri, R.M., Niimi, M., Ogilvie, A.L., Kukutsch, N.A., Rossner, S., Schuler, G., Austyn, J.M. (2000) Immature dendritic cells generated with low doses of GM-CSF in the absence of IL-4 are maturation resistant and prolong allograft survival in vivo. *Eur J Immunol* 30: 1813–22

9. Schuler, G., Steinman, R.M. (1997) Dendritic cells as adjuvants for immune-mediated resistance to tumors. *J Exp Med* 186: 1183–1187

10. Ribas, A., Butterfield, L.H., Glaspy, J.A., Economou, J.S. (2003) Current developments in cancer vaccines and cellular immunotherapy. *J Clin Oncol* 21: 2415–2432

11. Cella, M., Scheidegger, D., Palmer-Lehmann, K., Lane, P., Lanzavecchia, A., Alber, G. (1996) Ligation of CD40 on dendritic cells triggers production of high levels of interleukin-12 and enhances T cell stimulatory capacity: T-T help via APC activation. *J Exp Med* 184: 747–752

12. Banchereau, J., Bazan, F., Blanchard, D., Briere, F., Galizzi, J.P., van Kooten, C., Liu, Y.J., Rousset, F., Saeland, S. (1994) The CD40 antigen and its ligand. *Annu Rev Immunol* 12: 881–922

13. Ribas, A., Butterfield, L.H., Amarnani, S.N., Dissette, V.B., Kim, D., Meng, W.S., Miranda, G.A., Wang, H.J., McBride, W.H., Glaspy, J.A., Economou, J.S. (2001) CD40 cross-linking bypasses the absolute requirement for CD4 T cells during immunization with melanoma antigen gene-modified dendritic cells. *Cancer Res* 61: 8787–93

14. Lau, R., Wang, F., Jeffery, G., Marty, V., Kuniyoshi, J., Bade, E., Ryback, M.E., Weber, J. (2001) Phase I trial of intravenous peptide-pulsed dendritic cells in patients with metastatic melanoma. *J Immunother* 24: 66–78

15. Dullaers, M., Thielemans, K. (2006) From pathogen to medicine: HIV-1-derived lentiviral vectors as vehicles for dendritic cell based cancer immunotherapy. *J Gene Med* 8: 3–17

16. VandenDriessche, T., Thorrez, L., Naldini, L., Follenzi, A., Moons, L., Berneman, Z., Collen, D., Chuah, M.K. (2002) Lentiviral vectors containing the human immunodeficiency virus type-1 central polypurine tract can efficiently transduce nondividing hepatocytes and antigen-presenting cells in vivo. *Blood* 100: 813–822

17. Follenzi, A., Battaglia, M., Lombardo, A., Annoni, A., Roncarolo, M.G., Naldini, L. (2004) Targeting lentiviral vector expression to hepatocytes limits transgene-specific immune response and establishes long-term expression of human antihemophilic factor IX in mice. *Blood* 103: 3700–3709

18. Schroers, R., Sinha, I., Segall, H., Schmidt-Wolf, I.G., Rooney, C.M., Brenner, M.K., Sutton, R.E., Chen, S.Y. (2000) Transduction of human PBMC-derived dendritic cells and macrophages by an HIV-1-based lentiviral vector system. *Mol Ther* 1: 171–179

19. Gruber, A., Kan-Mitchell, J., Kuhen, K.L., Mukai, T., Wong-Staal, F. (2000) Dendritic cells transduced by multiply deleted HIV-1 vectors exhibit normal phenotypes and functions and elicit an HIV-specific cytotoxic T- lymphocyte response in vitro. *Blood* 96: 1327–1333

20. Chinnasamy, N., Chinnasamy, D., Toso, J.F., Lapointe, R., Candotti, F., Morgan, R.A., Hwu, P. (2000) Efficient gene transfer to human peripheral blood monocyte-derived dendritic cells using human immunodeficiency virus type 1-based lentiviral vectors. *Hum Gene Ther* 11: 1901–1909

21. Granelli-Piperno, A., Zhong, L., Haslett, P., Jacobson, J., Steinman, R.M. (2000) Dendritic cells, infected with vesicular stomatitis virus-pseudotyped HIV-1, present viral antigens to CD4+ and CD8+ T cells from HIV-1-infected individuals. *J Immunol* 165: 6620–6626

22. Dyall, J., Latouche, J.B., Schnell, S., Sadelain, M. (2001) Lentivirus-transduced human monocyte-derived dendritic cells efficiently stimulate antigen-specific cytotoxic T lymphocytes. *Blood* 97: 114–121

23. Salmon, P., Arrighi, J.F., Piguet, V., Chapuis, B., Zubler, R.H., Trono, D., Kindler, V. (2001) Transduction of CD34+ cells with lentiviral vectors enables the production of large quantities of transgene-expressing immature and mature dendritic cells. *J Gene Med* 3: 311–320

24. Esslinger, C., Romero, P., MacDonald, H.R. (2002) Efficient transduction of dendritic cells and induction of a T-cell response by third-generation lentivectors. *Hum Gene Ther* 13: 1091–1100

25. Rouas, R., Uch, R., Cleuter, Y., Jordier, F., Bagnis, C., Mannoni, P., Lewalle, P., Martiat, P., Van den Broeke, A. (2002) Lentiviral-mediated gene delivery in human monocyte-derived dendritic cells: optimized design and procedures for highly efficient transduction compatible with clinical constraints. *Cancer Gene Ther* 9: 715–724

26. Koya, R.C., Kasahara, N., Favaro, P.M., Lau, R., Ta, H.Q., Weber, J.S., Stripecke, R. (2003) Potent maturation of monocyte-derived dendritic cells after CD40L lentiviral gene delivery. *J Immunother* 26: 451–460

27. Veron, P., Boutin, S., Bernard, J., Danos, O., Davoust, J., Masurier, C. (2006) Efficient transduction of monocyte- and CD34+-derived Langerhans cells with lentiviral vectors in the absence of phenotypic and functional maturation. *J Gene Med* 8: 951–961

28. Metharom, P., Ellem, K.A., Schmidt, C., Wei, M.Q. (2001) Lentiviral vector-mediated tyrosinase-related protein 2 gene transfer to dendritic cells for the therapy of melanoma. *Hum Gene Ther* 12: 2203–2213

29. Koya, R.C., Kimura, T., Ribas, A., Rozengurt, N., Lawson, G.W., Faure-Kumar, E., Wang, H.J., Herschman, H., Kasahara, N., Stripecke, R.

(2007) Lentiviral vector-mediated autonomous differentiation of mouse bone marrow cells into immunologically potent dendritic cell vaccines. *Mol Ther* 15: 971–980

30. Lizee, G., Gonzales, M.I., Topalian, S.L. (2004) Lentivirus vector-mediated expression of tumor-associated epitopes by human antigen presenting cells. *Hum Gene Ther* 15: 393–404

31. Breckpot, K., Dullaers, M., Bonehill, A., van Meirvenne, S., Heirman, C., de Greef, C., van der Bruggen, P., Thielemans, K. (2003) Lentivirally transduced dendritic cells as a tool for cancer immunotherapy. *J Gene Med* 5: 654–667

32. Firat, H., Zennou, V., Garcia-Pons, F., Ginhoux, F., Cochet, M., Danos, O., Lemonnier, F.A., Langlade-Demoyen, P., Charneau, P. (2002) Use of a lentiviral flap vector for induction of CTL immunity against melanoma. Perspectives for immunotherapy. *J Gene Med* 4: 38–45

33. Zarei, S., Abraham, S., Arrighi, J.F., Haller, O., Calzascia, T., Walker, P.R., Kundig, T.M., Hauser, C., Piguet, V. (2004) Lentiviral transduction of dendritic cells confers protective antiviral immunity in vivo. *J Virol* 78: 7843–7845

34. Chen, X., Wang, B., Chang, L.J. (2006) Induction of primary anti-HIV CD4 and CD8 T cell responses by dendritic cells transduced with self-inactivating lentiviral vectors. *Cell Immunol* 243: 10–18

35. Buffa, V., Negri, D.R., Leone, P., Borghi, M., Bona, R., Michelini, Z., Compagnoni, D., Sgadari, C., Ensoli, B., Cara, A. (2006) Evaluation of a self-inactivating lentiviral vector expressing simian immunodeficiency virus gag for induction of specific immune responses in vitro and in vivo. *Viral Immunol* 19: 690–701

36. Koya, R.C., Weber, J.S., Kasahara, N., Lau, R., Villacres, M.C., Levine, A.M., Stripecke, R. (2004) Making dendritic cells from the inside out: lentiviral vector-mediated gene delivery of granulocyte-macrophage colony-stimulating factor and interleukin 4 into CD14+ monocytes generates dendritic cells in vitro. *Hum Gene Ther* 15: 733–748

37. Dull, T., Zufferey, R., Kelly, M., Mandel, R.J., Nguyen, M., Trono, D., Naldini, L. (1998) A third-generation lentivirus vector with a conditional packaging system. *J Virol* 72: 8463–8471

38. Stripecke, R., Koya, R.C., Ta, H.Q., Kasahara, N., Levine, A.M. (2003) The use of lentiviral vectors in gene therapy of leukemia: combinatorial gene delivery of immunomodulators into leukemia cells by state-of-the-art vectors. *Blood Cells Mol Dis* 31: 28–37

39. Breckpot, K., Corthals, J., Heirman, C., Bonehill, A., Michiels, A., Tuyaerts, S., De Greef, C., Thielemans, K. (2004) Activation of monocytes via the CD14 receptor leads to the enhanced lentiviral transduction of immature dendritic cells. *Hum Gene Ther* 15: 562–573

40. Gorski, K.S., Shin, T., Crafton, E., Otsuji, M., Rattis, F.M., Huang, X., Kelleher, E., Francisco, L., Pardoll, D., Tsuchiya, H. (2003) A set of genes selectively expressed in murine dendritic cells: utility of related cis-acting sequences for lentiviral gene transfer. *Mol Immunol* 40: 35–47

41. Cui, Y., Golob, J., Kelleher, E., Ye, Z., Pardoll, D., Cheng, L. (2002) Targeting transgene expression to antigen-presenting cells derived from lentivirus-transduced engrafting human hematopoietic stem/progenitor cells. *Blood* 99: 399–408

42. Follenzi, A., Ailles, L.E., Bakovic, S., Geuna, M., Naldini, L. (2000) Gene transfer by lentiviral vectors is limited by nuclear translocation and rescued by HIV-1 pol sequences. *Nat Genet* 25: 217–222

43. Sirven, A., Pflumio, F., Zennou, V., Titeux, M., Vainchenker, W., Coulombel, L., Dubart-Kupperschmitt, A., Charneau, P. (2000) The human immunodeficiency virus type-1 central DNA flap is a crucial determinant for lentiviral vector nuclear import and gene transduction of human hematopoietic stem cells. *Blood* 96: 4103–4110

44. Mangeot, P.E., Duperrier, K., Negre, D., Boson, B., Rigal, D., Cosset, F.L., Darlix, J.L. (2002) High levels of transduction of human dendritic cells with optimized SIV vectors. *Mol Ther* 5: 283–290

# Chapter 12

## In Situ (In Vivo) Gene Transfer into Murine Bone Marrow Stem Cells

## Dao Pan

### Summary

Adult bone marrow stem cell is an ideal target for gene therapy of genetic diseases, selected malignant diseases, and AIDS. The in vivo approach of lentivirus vector (LV)-mediated stem cell gene transfer by intrafemoral (IF) injection can take full advantage of any source of stem cells residing in the bone cavity. Such an approach may avoid several difficulties encountered by ex vivo hematopoietic stem cell (HSC) gene transfer. We have shown that both HSC and mesenchymal stem/progenitor cells (MSC) can be genetically modified successfully by a single "in situ" IF injection in their natural "niche" in mice without any preconditioning. This approach may provide a novel application for treatment of human diseases, and represent an interesting new tool to study adult stem cell plasticity and the nature of unperturbed hematopoiesis.

**Key words:** Intrafemoral injection, In vivo gene transfer, Lentiviral vectors, Hematopoietic stem cells, Mesenchymal stem/progenitor cells, Anesthesia, Survival surgery.

---

## 1. Introduction

Bone marrow stem cells from adults have been viewed as the ideal target for gene- and cell-based therapy of genetic diseases, selected malignant diseases, and AIDS. Under steady-state conditions, i.e., normal hematopoietic turnover and an intact bone marrow (BM) "niche", the majority of hematopoietic stem cells (HSC) in humans cycle slowly, yet continuously *(1)*. In vivo vector delivery can take advantage of both the natural HSC cycling for efficient transduction, and the supportive microenvironment in bone cavity for maintaining stem cell viability and capacities.

Christopher Baum (ed.), *Methods in Molecular Biology, Methods and Protocols, vol. 506*
© Humana Press, a part of Springer Science+Business Media, LLC 2009
DOI: 10.1007/978-1-59745-409-4_12

We and others have demonstrated the feasibility of in vivo gene transfer into murine adult BM HSC and/or mesenchymal stem/progenitor cells (MSC) by a single intravenous injection of a lentiviral vector (LV) (*2*, *3* ), or by a relatively localized "in situ" intrafemoral (IF) injection of a retroviral vector into mice pretreated with sublethal dose of 5-fluorouracil (5-FU) *(4)* or an LV into unconditioned mice *(5)*.

The in vivo approach of LV-mediated stem cell gene transfer may have a significant impact on disease treatment and stem cell research. First, direct gene delivery would take full advantage of any source of stem cells present in the bone cavity, including those highly "plastic" cells that can differentiate into mature, nonhematopoietic cells of multiple tissues including epithelial cells of the liver, kidney, lung, skin, and GI tract *(6)*(see review *(7)*). Second, this approach would avoid the difficulties encountered by ex vivo HSC gene transfer, including maintaining stem cell properties and the loss of engraftment potential. Third, it would also avoid cytokine stimulation, which may activate unwanted signaling pathways that could potentially increase the risk of nonrandom mutagenic events during provirus insertion. Therefore, this approach may provide a novel application for treatment of human diseases. It may also represent an interesting new tool to study adult stem cell plasticity and the nature of unperturbed hematopoiesis without the complications by in vitro manipulation or in vivo preconditioning.

Concentrated lentiviral vector stocks with high infectivity are essential for in vivo gene transfer. Survival surgery with intrafemoral injection is performed in mice under continuous inhalation anesthesia using aseptic techniques. Postprocedural care of animals is needed to monitor and minimize any postinjection pain. Examples of in situ gene transfer into murine HSC and MSC by IF injection are shown in **Figs. 1** and **2**.

## 2. Materials

### 2.1. Concentration of VSVG-Pseudotyped LV

1. Ethanol-sterilizing solution (70%) is made freshly from Absolute Ethyl Alcohol (AAPER Alcohol and Chemical Co., Shelbyville, KY) and autoclaved distilled $H_2O$.

2. Ultracentrifuge polymer tubes for SW28 rotor (36 mL) and SW55 rotor (3.2 mL) (Beckman Instrument Inc., Palo Alto, CA).

3. X-VIVO 10 serum-free medium (Lonza Baltimore Inc., Baltimore, MA).

4. *Tris-buffered saline (TBS) made in distilled $H_2O$:* 50 mM Tris (2-Carboxyethyl) phosphine.HCl, 0.9% NaCl, and 10 mM

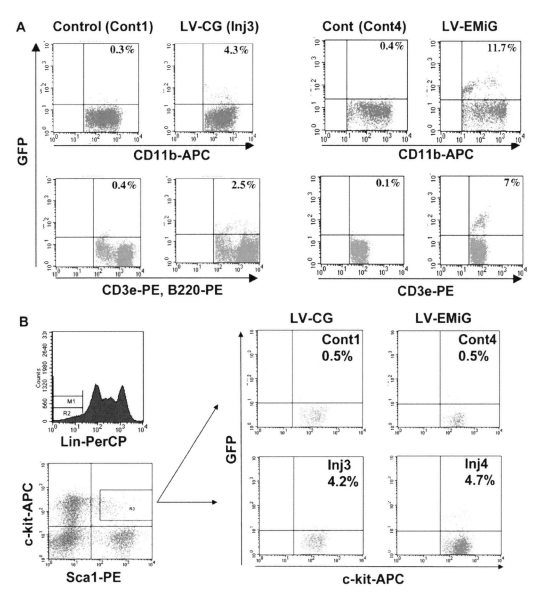

Fig. 1. GFP⁺ PBL and HSC were detected up to 4 months in primary iBM-injected mice. Data shown are representative results. (**a**) PBL from mice 3-month post injection were stained with CD11b-APC for myeloid, and CD3e-PE (and B220-PE) for lymphoid lineages. We used 7-AAD staining to gate out dead cells. (**b**) Total BM cells were harvested from femur and tibia of both (for LV-CG) or injected (for LV-EMiG) hind legs 4-months post injection, and stained with Sca1-PE, c-kit-APC and lineage-markers-PerCP for HSC-enriched (Lin⁻c-kit⁻Sca1⁺) cells (i.e., R3, 0.2–0.6% of total BM). (Reproduced from ref. *5* with permission from Elsevier Science.).

MgCl$_2$. The pH is adjusted with HCl to pH 7.3. The TBS is sterilized by filtration through a 0.2-µM Corning syringe filter (Sigma-Aldrich, St. Louis, MO), and could be stored at 4°C for several months.

5. Cryogenic vials (Fisher Scientific).

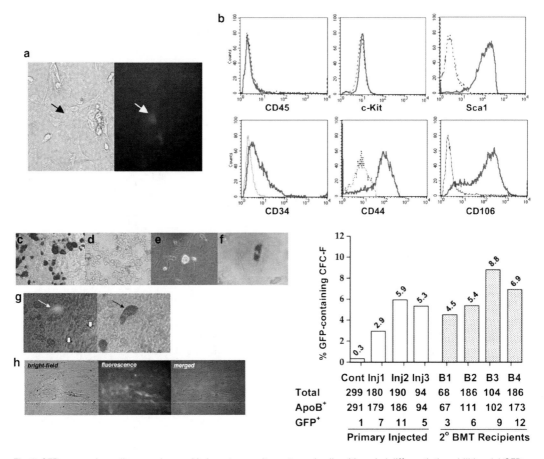

Fig. 2. GFP-expressing cells were observed in long-term culture stromal cells with varied differentiation abilities. (**a**) GFP[+] adherent cells were observed in long-term stromal culture derived from iBM-injected mice. Cells shown (from Inj2) have been in culture for 2 months at passage 5. (**b**) FACS analysis of MSC cell-surface antigens was performed after in vitro stromal culture for 7–8 passages. Dashed lines represent IgG controls. (**c–f**) In vitro inductive culture was conducted using MSC at 8–9 passages and stained with Alizarin Red S for osteogenic (**c**), with Alcian-blue for chondrogenic (**d**) and phase contrast microscopy or Oil Red-O for adipocytic (**e** and **f**) differentiation. (**g**) GFP[+] MSCs were observed with (*arrow*) or without (*flat arrow*) osteogenic potential. (**h**) Representative GFP-expressing CFU-fibroblastoid unit derived from CFC-F assay using MSC-Inj2 at passage 8. Colonies with diameter more than 2 mm were scored. (**i**) CFU-F colonies were stained with Crystal Violet and randomly collected into PCR-direct lysis buffer. Duplex real-time PCR was performed on 96-well plates, each including 2–3 no template controls and one background control derived from apparently empty space. (Reproduced from ref. *5* with permission from Elsevier Science.).

## 2.2. Vector Preparation

1. Insulin syringe with 27-gauge needle (1/2 length) (Becton Dickinson and Company, Franklin Lakes, New Jersey).
2. Sterile Alcohol Prep Pads (Fisher HealthCare).

## 2.3. General Anesthesia and Continuous Inhalation Anesthesia

1. Survival IF injection is performed using sterile instruments, sterile surgical gloves, mask, cap, gown, and aseptic procedures.

2. *Nose-cone assembly.* 20-mL syringe cover (Becton Dickinson and Company), rubber lid from Vacuum collection blood tubes, catheter, and surgical tapes (Fisher Scientific).

3. Isoflurane liquid (IsoFlo, halogenated agent) is from the Abbott Laboratories (North Chicago, IL), and Buprenorphine Hydrochloride (Buprenex[R]) is a Schedule III narcotic obtained through Veterinary Service.

**2.4. Intrafemoral Injection and Injection-Related Pain Management**

1. Twenty-five-gauge needle with 3-mL syringe (Becton Dickinson and Company).
2. Betadine surgical scrubs (Abbott Laboratories).
3. Electric clippers (Golden A-5 Clipper, Oster, McMinnville, TN).

# 3. Methods

Wild-type vesicular stomatitis virus (VSV) has a broad host range extending from insects to nearly all mammals (8). The VSV glycoprotein (VSVG) envelope also has the unique ability to withstand the shearing forces encountered during ultracentrifugation. Therefore, VSVG has been utilized to "pseudotype" gene therapy vectors such as retroviral (9), HIV-1-based (10), and FIV-based (11) lentiviral vectors to broaden vector host range (tropism) and increase the potency in vector stocks by concentration. In addition to the ultracentrifugation method described here, other interventions have also been reported including low-speed overnight centrifugation and filtration. The high infectivity in concentrated vector stocks provides the basis for efficient in vivo HSC gene transfer by IF injection.

The intrafemoral injection can be accomplished without open-wound surgery in adult mice (10 weeks and older). It can also be applied for intrafemoral cell transplantation procedure (12) with minor modifications. To reduce the potential postoperational complications, a biobubble or a biosafety cabinet is used during the operation, and the injected mice are accommodated in a specific pathogen-free facility.

The IF injection approach is limited by the applicable volume per injection and relatively low transduction frequency for clinical application. It is likely that LV-mediated stem cell gene transfer by IF injection could be further improved by pretreatment with 5-FU or sublethal irradiation to reduce the numbers of HSC staying at $G_0$ phase and total numbers of BM cells, by temporarily stopping blood flow (by tourniquet) or by successful repeated LV administration.

**3.1. Concentration of VSVG-Pseudotyped Lentiviral Vector**

1. Vector should be kept on ice whenever it is possible during all process to reduce the loss of infectivity.

2. Sanitize rotor buckets and caps with 70% ethanol and let them air dry in a Biosafety cabinet.

3. Load the presterilized ultracentrifuge tubes with vector supernatant that has been filtered through a 0.45-μM filter (*see* **Note 1**).

4. Insert tubes into rotor buckets for SW28 rotor, and put the corresponding caps on. Each pair of tube-bucket sets (number 1 with 4, 2 with 5, and 3 with 6) should be balanced, followed by tightening the cap with the alignment of the two identical numbers on cap and bucket.

5. Hook the buckets in place onto the rotor by their assigned numbers (*see* **Note 2**).

6. Position the rotor inside the Beckman L-90 Ultracentrifuge. Spin rotor counterclockwise gently by hand to ensure a secure and correct connection.

7. Ultracentrifuge at 20,000 rpm and 4°C for 2 h with brake set at "low" (*see* **Note 3**).

8. Remove the tubes from buckets promptly in the hood without shaking. Discard the supernatant into bleach by aspirating carefully with pipettes to avoid disturbing viral pellets.

9. Resuspend each pellet in 0.5% of starting medium volume (~180 μL) of X-VIVO 10 medium by gently shaking the tubes (inserted in a sterile 50-mL Conical tube) in a shaker for 15 min. Pipette up and down to fully resuspend each pellet and pool several pellets together.

10. Rinse each tube with 300–400 μL X-VIVO 10 medium and collect them to the pooled vector tube.

11. A second ultracentrifugation is performed using a SW55 rotor at 21,000 rpm and 4°C for 1 h.

12. Remove the supernatant as much as possible and discard into the bleach.

13. Resuspend the vector pellets in TBS at the volume of 1:2,000 of initial vector amount by gently shaking at 4°C for 30 min to loosen the pellet, followed by pipetting to obtain complete suspension.

14. Aliquot the concentrated LV stocks into cryogenic vials with 2–3 diluted aliquots (1:100) for titration assay and replication competent lentivirus (RCL) test (*see* **Note 4**).

15. Vector stocks should be stored at –80°C for further usage.

***3.2. Vector Preparation for Injection***

1. At the day of IF injection, the concentrated LV stocks should be quickly thawed with cap-side up at 30°C in a water bath.

2. As soon as they are thawed, mix the vector by pipetting up and down 2–3 times, and pool vector as needed, with care to avoid extensive bubbles.

3. The vector supernatant should be transferred into a 27-gauge needle insulin syringe (with 1/2 length) by pipetting into the unplugged syringe (*see* **Note 5**).

4. Plug the insert back carefully and eliminate all the air bubbles. Point the needle to an alcohol wipe during this step to collect any potential discharge.

5. The vector-containing syringe should be stored on ice during the rest of the experiment.

*3.3 General Anesthesia and Continuous Inhalation Anesthesia*

1. Connect the presterilized sedating chamber with the outlet tubing of an anesthesia machine.

2. Complete the assembly of the nose-cone with the rubber lid on and connect it to the outlet tubing of a North American Drager NarkoMed Anesthesia System (*see* **Note 6**). Multiple sets can be connected for simultaneous operation.

3. Secure the tubing(s) on the surgery plane with heat pad to keep the animal body temperature during operation.

4. Turn on the NarkoMed Anesthesia System and set the oxygen rate at 0.6% (or 1.2% for double unites) and isoflurane at 2.5.

5. The initial general anesthesia is performed in the sedating chamber prefilled with isoflurane at the rate of 3.

6. To control the significant postoperational pain, an analgesic agent Buprenorphine (Buprenex[R]) is administered at 0.5–1 mg/kg by subcutaneous injection in the back of the neck right after the subject showed no response to touching (*see* **Note 7**).

7. The animal is put back to the sedating chamber until the smooth/slow breathing stage is observed for deep anesthesia phase.

8. Shave the hair around the knee joints of both hind limbs with an electric clipper for a better view of the kneecaps. The left knee is prepared in case of the failure with attempts in the right knee (*see* **Note 8**).

9. The animal is transferred to surgery plane and laid on its back with nose securely inside the nose-cone. Secure the nose-cone in position on the heat pad with surgical tapes.

10. Use surgical tapes to attach both forelegs by the side on the heat pad.

11. Watch the breath of the mouse and gently pinch a leg with a twister to make sure the continuous anesthesia of the subject.

*3.4. Intrafemoral Injection*

1. These injection procedures should be practiced with a coloring solution (such as Trypan blue) on freshly exterminated mice.

2. Use alcohol wipes to clean the knee areas of both hind limbs.

3. Prepare a 25-gauge needle attached to a 3-mL empty syringe. Make sure to use a new needle for each knee.

4. Bend the right knee with the femur at about 75° angle with the heat pad.

5. Gently feel the kneecap with the needle to find the highest point of the femur and lodge between the condyles (*see* **Note 9**).

6. Apply gentle twisting and pressure to gain access to the intrafemoral space. Always check the insertion by turning the needle around and slightly back and forth to ensure the needle is inside the marrow cavity (*see* **Note 10**).

7. Aspirate the marrow to reduce the number of marrow cells to be transduced for higher multiplicity of infection (*see* **Note 11**).

8. Hold the femur and tibia tightly in place while retrieving the penetrating needle.

9. Reinsert the vector-containing insulin syringe needle into the marrow cavity (*see* **Note 12**). Recheck the insertion as in **step 6**.

10. Inject 40–50 µL of vector supernatant slowly in 2–3 min with simultaneously retrieving the needle. Wait for another 2 min before removing the needle completely.

11. Immediately straighten the injected limb and use Betadine surgical scrubs to clean knee area (*see* **Note 13**).

12. Clip ear(s) to appoint identification for the injected mouse right after removing it from the nose-cone.

***3.5. Postinjection Pain Management***

1. The injected mice are allowed to emerge from general anesthesia in their cage with their mouth facing sideway (*see* **Note 14**).

2. All mice are observed for ambulation before transferring back to husbandry.

3. Twenty-four hours later, a thorough observation for behavioral signs of acute pain is performed for all injected mice, including attitude, porphyrin staining, gait and posture, and appetite (**Table 1**). No additional intervention is needed if the score is under 0.8.

4. For mice with score of 0.8–1.2, Ibuprofen is added in drinking water at 7.5 mg/kg per oral ad libitum (*see* **Note 15**). Mice with score > 1.2 and obvious limping are removed from the study.

**Table 1**
**Postoperational evaluation criteria for pain management**

| Observation | Scoring benchmark | Score |
| --- | --- | --- |
| Attitude | Bright, active, and alert (BAR) | 0 |
| | Burrowing/hiding, quiet, but rouses when touched | 0.1 |
| | No cage exploration when lid off, burrowing/hiding, may vocalize when head pressed, or be unusually aggressive when touched | 0.4 |
| Porphyrin staining | None | 0 |
| | Mild around eyes and/or nostrils | 0.1 |
| | Obvious on face and/or paws | 0.4 |
| Gaiting and posture | Normal | 0 |
| | Mild incoordination when stimulated, hunched posture, mild piloerection | 0.1 |
| | Obvious ataxia or head tilt, dragging one or both limbs, severe piloerection | 0.4 |
| Appetite | Normal, eats dry food, evidence of urine and feces | 0 |
| | No evidence of eating food, drinks but appears hydrated (by skin tent test)[a] | 0.1 |
| | No interest in food or treats, and/or appears dehydrated | 0.4 |

[a] The loose skin on the side, back, or shoulders is gently pulled outward to create the "tent" in the skin. If skin tent does not immediately return to position, then the mouse is dehydrated

5. A repeated evaluation is conducted on Day 2 post injection.

6. A weekly weight measurement should be maintained to monitor potential chronic pain or any sign of abnormal development.

## 4. Notes

1. The tubes should be filled in full with 35–36 mL supernatant (add fresh media if needed); otherwise they may collapse during centrifugation.

2. All six buckets with full-filled tubes should be in place to make sure a stable ultracentrifugation even if tube(s) with water is needed as "balancer."

3. The initiation of centrifugation should be observed to make sure that no imbalance occurs, and that the speed and vacuum reach the desired settings.

4. Prior to injection, 0.5% of the total LV production should be tested to be "free" of RCL.

5. The use of insulin syringe can minimize the loss of vector stock in the needle's dead space.

6. To cut cost, the nose-cone unite can be made using a 20-mL syringe cover with both cover pieces secured by surgical tape to avoid gas leakage. A hole is generated on the longitudinal side with the size that can fit the lid of 14-mL blood collecting tubes. An open catheter needle is inserted into the syringe cover with the entrance taped to prevent gas leakage. The catheter is connected to the outlet tubing of the Anesthesia System.

7. Surgery on the femur is painful because of large muscle mass trauma and direct bone manipulation. An average weight can be obtained from 2 to 3 randomly selected (male or female, separately) mice and used as a guideline for the dose of Buprenorphine injection.

8. Hair shaving has a risk of skin rupture that increases the chance of infection. For the well being of the subject and the reduction of anesthesia time, this step can be skipped if injection on mice with hair-on can be achieved during practice.

9. A small incision can be made over the kneecap for easier observation of the condyles, especially when long-term survival of the mice is not required.

10. It is important to insert the needle with appropriate angle along with the femur. A maximum of two insertion attempts can be applied to one kneecap without serious damage to the limb.

11. The marrow aspiration step is not always successful due to the blockage of the needle during knee penetration. If the marrow aspiration is required, a second attempt can be performed with a new needle-syringe set.

12. It is essential to keep the femur and the tibia in place during needle change. Make sure the vector-containing syringe is within easy reach or have a coworker to assist on this step.

13. The execution of injection can vary from mouse to mouse. For better reference of follow-up assays, it is important to document the actual injection process, such as side of injection, number of attempts, success of marrow aspiration, and any visible leakage after removing the injecting needle.

14. Mice should regain consciousness within 5 min. They are generally quiet but responsive when touched.

15. Inclusion of Ibuprofen in water can lead to decreased water consumption and should be stopped as soon as unnecessary. A close observation is needed for signs of dehydration during treatment period.

## Acknowledgments

The author would like to thank the Veterinary Service at Cincinnati Children's Hospital Medical Center for their support and consultation. This work was supported in part by Translational Research Initiative grant from Cincinnati Children's Foundation, and the National Institutes of Health (AI061703).

## References

1. Cheshier, S.H., S.J. Morrison, X. Liao, and I.L. Weissman, *In vivo proliferation and cell cycle kinetics of long-term self-renewing hematopoietic stem cells. Proc Natl Acad Sci U S A*, 1999. 96(6): pp. 3120–3125

2. Pan, D., R. Gunther, W. Duan, S. Wendell, W. Kaemmerer, T. Kafri, I.M. Verma, and C.B. Whitley, *Biodistribution and toxicity studies of VSVG-pseudotyped lentiviral vector after intravenous administration in mice with the observation of in vivo transduction of bone marrow. Mol Ther*, 2002. 6(1): pp. 19–29

3. Follenzi, A., G. Sabatino, A. Lombardo, C. Boccaccio, and L. Naldini, *Efficient gene delivery and targeted expression to hepatocytes in vivo by improved lentiviral vectors. Hum Gene Ther*, 2002. 13(2): pp. 243–260

4. McCauslin, C.S., J. Wine, L. Cheng, K.D. Klarmann, F. Candotti, P.A. Clausen, S.E. Spence, and J.R. Keller, *In vivo retroviral gene transfer by direct intrafemoral injection results in correction of the SCID phenotype in Jak3 knock-out animals. Blood*, 2003. 102(3): pp. 843–848

5. Worsham, D.N., T. Schuesler, C. von Kalle, and D. Pan, *In vivo gene transfer into adult stem cells in unconditioned mice by in situ delivery of a lentiviral vector. Mol Ther*, 2006. 14(4): pp. 514–524

6. Jiang, Y., B.N. Jahagirdar, R.L. Reinhardt, R.E. Schwartz, C.D. Keene, X.R. Ortiz-Gonzalez, M. Reyes, T. Lenvik, T. Lund, M. Blackstad, J. Du, S. Aldrich, A. Lisberg, W.C. Low, D.A. Largaespada, and C.M. Verfaillie, *Pluripotency of mesenchymal stem cells derived from adult marrow. Nature*, 2002. 418(6893): pp. 41–49

7. Herzog, E.L., L. Chai, and D.S. Krause, *Plasticity of marrow derived stem cells. Blood*, 2003. 102(10): pp. 3483–3493

8. Schnitzlein, W.M. and M.E. Reichmann, *Characterization of New Jersey vesicular stomatitis virus isolates from horses and black flies during the 1982 outbreak in Colorado. Virology*, 1985. 142(2): pp. 426–431

9. Burns, J.C., T. Friedmann, W. Driever, M. Burrascano, and J.K. Yee, *Vesicular stomatitis virus G glycoprotein pseudotyped retroviral vectors: concentration to very high titer and efficient gene transfer into mammalian and nonmammalian cells. Proc Natl Acad Sci U S A*, 1993. 90(17): pp. 8033–8037

10. Naldini, L., U. Blomer, P. Gallay, D. Ory, R. Mulligan, F.H. Gage, I.M. Verma, and D. Trono, *In vivo gene delivery and stable transduction of nondividing cells by a lentiviral vector. Science*, 1996. 272(5259): pp. 263–267

11. Poeschla, E.M., F. Wong-Staal, and D.J. Looney, *Efficient transduction of nondividing human cells by feline immunodeficiency virus lentiviral vectors. Nat Med*, 1998. 4(3): pp. 354–357

12. Mazurier, F., M. Doedens, O.I. Gan, and J.E. Dick, *Rapid myeloerythroid repopulation after intrafemoral transplantation of NOD-SCID mice reveals a new class of human stem cells. Nat Med*, 2003. 9(7): pp. 959–963

# Chapter 13

# In Vivo and Ex Vivo Gene Transfer in Thymocytes and Thymocyte Precursors

Oumeya Adjali, Amélie Montel-Hagen, Louise Swainson, Sophie Marty, Rita Vicente, Cedric Mongellaz, Chantal Jacquet, Valérie Zimmermann, and Naomi Taylor

## Summary

The thymus provides a specialized environment allowing the differentiation of T lymphocytes from bone marrow-derived progenitor cells. We and others have demonstrated that gene transfer into distinct thymocyte populations can be obtained, both in vivo and ex vivo, using lentiviral vectors. Here, we describe techniques for intrathymic lentiviral transduction in mice, using a surgical approach wherein the thoracic cavity is exposed as well as a significantly less invasive strategy wherein virions are directly injected through the skin. Moreover, thymocyte differentiation from murine and human progenitors is now feasible in vitro, under conditions wherein the Notch and IL-7 signaling pathways are activated. We describe methods allowing transduction of murine and human progenitors and their subsequent differentiation into more mature thymocytes. Conditions for lentiviral gene transfer into more differentiated human thymocyte subsets are also presented. Optimization of technologies for HIV-based gene transfer into murine and human thymocyte progenitors will advance strategies aimed at modulating T-cell differentiation and function in-vivo; approaches potentially targeting patients with genetic and acquired immunodeficiencies as well as immune-sensitive tumors. Furthermore, this technology will foster the progression of basic research aimed at elucidating molecular aspects of T-cell differentiation in mice and humans.

**Key words:** Lentiviral vectors, Thymus, HIV, Intrathymic gene transfer, Transduction, Progenitors, OP9-DL1.

## 1. Introduction

The major site of T-cell maturation is the thymus, and as such, differentiation of the T lymphocyte lineage can potentially be modulated by directly targeting this organ. A few years ago, we,

Christopher Baum (ed.), *Methods in Molecular Biology, Methods and Protocols, vol. 506*
© Humana Press, a part of Springer Science + Business Media, LLC 2009
DOI: 10.1007/978-1-59745-409-4_13

together with the group of D. Klatzmann, hypothesized that under the extreme selective conditions of severe combined immunodeficiency (SCID), an *in situ* gene correction of, even a few, T lymphoid progenitors by lentiviral-mediated gene transfer could be sufficient to alleviate the symptoms associated with SCID. The potential for in vivo lentiviral-mediated gene transfer is supported by the following observations (a) Lentiviral vectors have resulted in efficient in vivo gene transfer in hepatocytes, antigen-presenting cells, muscle cells, as well as cells in the central nervous system *(1–6)*; (b) Klatzmann and colleagues found that intrathymic (IT) injection of a ubiquitous lentiviral vector results in the transduction of thymic epithelial cells as well as a low number of immature thymocytes *(7)*; and (c) Early T lymphoid progenitors in the thymus appear to be derived from early bone marrow progenitor cells, more immature than the common lymphoid progenitors found in the BM *(8)*. Thus, we reasoned that IT injection of a T-cell-specific lentiviral vector encoding a gene that provides a selective advantage for transduced prothymocytes might result in the generation of functional T lymphocyte progeny allowing long-term immune reconstitution.

Our data demonstrated that a single intrathymic injection of a T-cell-specific ZAP-70 lentiviral vector results in a low efficiency transduction of T progenitor cells in ZAP-70-deficient SCID mice, which subsequently expand in the periphery *(9)*. This approach, using adult mice, was performed following surgical exposure of the thoracic cage, but only 4 of 18 mice were reconstituted. We hypothesized that intrathymic gene transfer in infant mice might result in a higher success rate since lentiviral vector-mediated gene transfer in T lineage cells requires cell cycle entry *(10, 11)* and a higher percentage of prothymocytes in infant thymi are in cycle as compared to their adult counterparts. Nevertheless, surgical manipulation of infant mice is significantly more complex than in adults, and as such we developed a direct blind approach for intrathymic injections, not requiring exposure of the thoracic cavity. This simplified technique was validated by injection of a dye as shown in **Fig. 1**. Indeed, using this strategy, we obtained a significantly higher T-cell recovery in ZAP-70-deficient mice, with reconstitution observed in 15 of 20 intrathymically treated animals *(9)*.

In vivo intrathymic lentiviral transduction, under conditions where there is no selective advantage for transduced thymocytes, results in a very low efficiency gene transfer in the thymus (**Fig. 2**). This is not extremely surprising given the relatively low multiplicity of infection dictated by the maximum volume that can be injected into the thymus. A murine thymus contains approximately $200 \times 10^6$ thymocytes and can be injected with a maximum volume of 10–20 μL. With a high titer lentiviral stock corresponding to $> 10^{10}$ transducing units (TU)/mL, this still

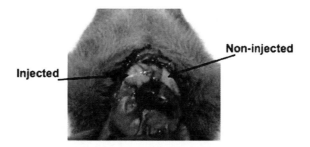

Fig. 1. In vivo intrathymic gene transfer. Intrathymic injections were performed on either infant mice, aged 10–21 days, or adult mice, ranging from 6 to 9 weeks of age. In the latter case, mice were anesthetized and injections were performed following incision of the sternum. In infant mice, virions, in a total volume of 10 µL, were injected directly through the skin into the thoracic cavity immediately above the sternum at an approximately 30° dorsal angle, using a 0.3-mL 28-gauge 8-mm insulin syringe. To assess the efficacy of this latter "blind" approach, infant mice were directly injected with 10 µL of trypan blue dye and killed immediately to visualize the localization of the dye. Trypan blue-injected and -uninjected thymic lobes are indicated in this representative thymus. (*See Color Plates*)

translates to $10^8$ TU and a multiplicity of infection of < 1. Under these conditions, transduction is detected in the most immature double negative thymocyte population and, at slightly lower levels, in the immature CD4+ CD8+ double positive population (**Fig. 2**).

As lentiviral transduction requires cell cycle entry *(12)*, these data are consistent with the finding that a subset of immature double negative cells cycle whereas more mature single positive thymocytes are quiescent. Notably though, under conditions wherein there is a significant selective advantage for transduced thymocytes, as in the case of ZAP-70-deficient mice, ZAP-70--transduced thymocytes are detected at levels of up to 3% for more than one year following intrathymic lentiviral gene transfer (**Fig. 3**). Thus, this in vivo approach can result in a long-term stable gene transfer in the thymus, at least under conditions where there is a selective advantage for transduced cells.

Until recently, it has been very difficult to achieve thymocyte differentiation in vitro, in large part due to an incomplete understanding of the factors necessary for induction of T-lineage commitment. Indeed, it was only possible to obtain reliable thymocyte differentiation in the context of thymus organ cultures. The finding that the Notch pathway is critical in promoting the emergence of T lineage cells *(13, 14)*, allowed the development of a stromal system wherein a critical Notch receptor ligand, Delta-like-1, is expressed. The development of this system has greatly simplified in vitro studies of T-cell differentiation (reviewed in *15)*, and moreover Notch signals promote the metabolism and

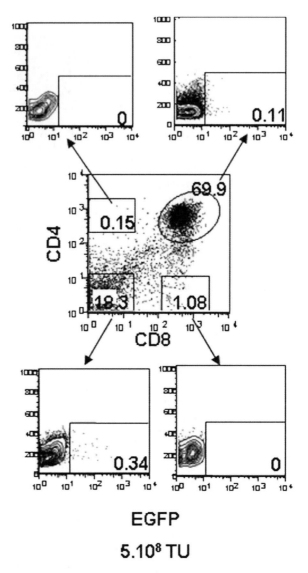

Fig. 2. Thymocyte gene transfer immediately following in situ injection of an EGFP-expressing lentiviral vector. Virions harboring a control EGFP-expressing lentiviral vector were injected intrathymically into ZAP-70$^{-/-}$ mice and thymocytes were harvested from euthanized animals 3 days later. A representative thymus from one animal injected with $5 \times 10^8$ transducing units of virus, in a total volume of 20 μL, is shown 3 days following injection. The percentages of double negative, double positive, and CD4+ and CD8+ single positive thymocytes are indicated, and the percentages of EGFP+ cells within each of these subsets are presented as SSC/EGFP dot plots.

survival of murine and human thymocytes ex vivo. Thus, this system has been instrumental for ex vivo gene transfer studies in thymocytes and thymocyte progenitors (**Fig. 4**) and furthermore, has allowed the fate of transduced cells to be evaluated.

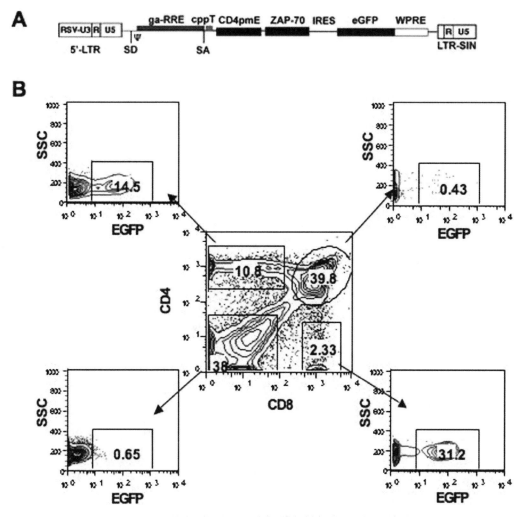

**55 weeks post lentiviral gene transfer**

Fig. 3. Thymic expression of a ZAP-70 lentiviral vector one year following in vivo intrathymic gene transfer. (**a**) Schematic representation of the ZAP-70 lentiviral vector (pT-ZAP) initially described by Adjali et al. *(9)*. The relative positions of the elements contained within the pCD4 lentiviral vector are indicated. *5′ LTR* long terminal repeat (LTR), *SD* splice donor, *SA* splice acceptor, Ψ, packaging signal, *GA-RRE* truncated *gag* sequence with the *rev* responsive element, *cppT* central polypurine tract of HIV, *CD4pmE* human CD4 minimal promoter/murine enhancer cassette (590 bp), ZAP-70 cDNA, *IRES* internal ribosome entry site, *eGFP* enhanced green fluorescent protein cDNA, *WPRE* post-transcriptional cis-acting regulatory element of the woodchuck hepatitis virus (587 bp), and *LTR-SIN* self-inactivating 3′ LTR (deleted of 400 bp in the U3 region). (**b**) The CD4/CD8 profile of a thymus from a ZAP-70-deficient mouse intrathymically injected with the pT-ZAP lentiviral vector 55 weeks earlier. The percentage of EGFP+ cells in each thymocyte population is shown with maturation proceeding from double negative CD4-CD8-, to double positive CD4+ CD8+, and finally to mature single positive CD4+ and CD8+ thymocytes.

HIV-based lentiviral vectors have been used for the last 10 years to achieve gene transfer in immune cells *(16)* (and reviewed in *(17, 18)*. Notably, the efficiency of gene transfer has been significantly augmented by the introduction of the HIV-1 central

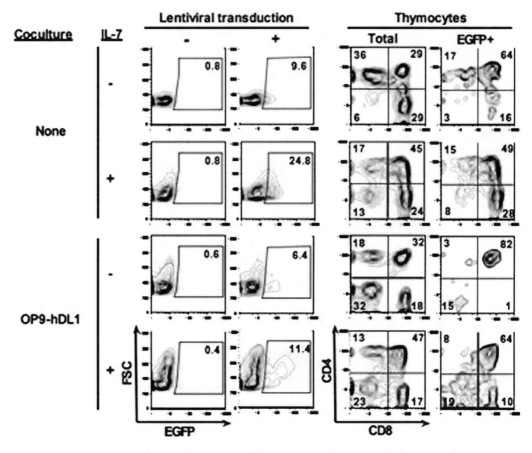

Fig. 4. Gene transfer in human thymocytes using an EGFP-expressing lentiviral vector. Human thymocytes were teased from thymi obtained from pediatric cardiac surgery patients and seeded in RPMI containing 10% FCS, in the absence or presence of rIL-7 (10 ng/mL). Alternatively, thymocytes were seeded on confluent OP9 stromal cells stabling expressing the human Notch ligand, Delta-like-1 (hDL1) *(29)*. Thymocytes were immediately transduced with a VSV-G pseudotyped HIV-based lentiviral vector harboring an EGFP transgene driven by the EF1α transgene at an MOI of 30. Thymocytes were harvested 72 h later and stained for EGFP expression in combination with surface expression of CD4 and CD8. Representative *dot plots* showing EGFP expression in mock and virion-transduced cultures are shown on the left with the percentages of EGFP+ cells indicated. The CD4/CD8 profiles of the total recovered thymocyte populations and the EGFP+ subset are shown on the right. The percentages of double negative, double positive, and CD4+ and CD8+ single positive thymocytes present in each condition are indicated.

polypurine tract and central termination sequence *(11, 19–22)*. This sequence creates a plus strand overlap, the central DNA flap, which is a cis-determinant of nuclear import *(19)*. Two major packaging plasmids have been used to generate virions harboring HIV-based lentiviral virions. One of the first packaging plasmids, pCMVΔR8.2, expresses all the HIV proteins with the exception of *env (16)*, while newer generation plasmids such as pCMVΔR8.91 have removed the virulence genes *vif, vpu, vpr,* and *nef (23)* and are favored for the techniques described here.

In the studies presented here, the HIV-based virions are pseudotyped with the vesicular stomatitis G protein (VSV-G) which provide the advantage of allowing efficient concentration of viral particles. Nevertheless, it is important to note that pseudotyping virions with envelopes that allow activation of specific cell types, including thymocytes and progenitor cells, can specifically target gene transfer to those cells that are activated *(24, 25)*.

The level of transgene expression will be influenced by the site of vector integration. In addition, the average expression can be significantly modulated by the use of different promoters. While the use of tissue-specific promoters in MuLV-based retroviral vectors has met with multiple difficulties, this approach has been successfully used in lentiviral vectors *(26, 27)*, including a T-cell promoter used in our ZAP-70 gene transfer studies (Fig. 13.3a) *(9)*. Notably though, the level of transgene expression is generally significantly higher under conditions wherein the internal lentiviral promoter is derived from a murine leukemia virus (MLV) as compared to a ubiquitous or tissue-specific cellular promoter and as such, the choice of promoter should be based on the expression level that is desired.

This methods section will focus on in vivo and ex vivo transduction of murine and human thymocytes subsets and progenitors using HIV-based lentiviral vectors.

## 2. Materials

### 2.1. Plasmid Preparation

1. HIV-derived lentiviral vectors can be obtained from various commercial sources including Invitrogen or, alternatively, are generously provided by Didier Trono's laboratory following completion of a material transfer agreement which can be downloaded from the following website <http://tronolab. epfl.ch/page58114.htmL>. Other academic sources include the laboratories of Pierre Charneau and Luigi Naldini. SIV-based lentiviral vectors can be obtained from the laboratory of FL Cosset. Other core facilities that can provide aid include, in the USA, the Penn vector core <http://www.uphs.upenn. edu/penngen/gtp/vcore_lv.html and National gene vector laboratory <http://www.ngvl.org>, in France, Genethon <http://www.genethon.fr>, in Sweden, Lund University <http://www.rvec.lu.se/services.html>, and the National gene vector laboratory of Ireland <http://www.nuigalway. ie/remedi/index.php>.

2. Bacterial strains used for transformation should lack the recombinase gene (rec) to prevent recombination of plasmid

and genomic DNA. Electrocompetent *E. coli* DH5α or Top10 are commonly used (purchased from ATCC or Eurogentec).

3. Bacteria are grown in 2× LB media (20 g Bacto Tryptone, 10 g Bacto Yeast Extract, 10 g NaCl, $H_2O$ to 1 l, pH from 7.2 to 7.5).

4. Ampicillin is dissolved in milliQ water at a concentration of 100 mg/mL, stored at –20°C, and used at a final concentration of 50 μg/mL.

5. Plasmid amplification kits are available from commercial sources, e.g., Nucleobond PC500, Macherey-Nagel, Easton, PA.

## 2.2. Cell Culture and Transfections

1. 293T cells can be obtained from the American Tissue Type Culture Collection, ref CRL-11268.

2. Dulbecco's Modified Eagle's Medium (DMEM) supplemented with 10% fetal calf serum and 1% antibiotics (penicillin/streptomycin).

3. Dulbecco's Phosphate-Buffered Saline (D-PBS) without calcium and magnesium.

4. 2.5% Trypsin and 0.5 M ethylenediamine tetraacetic acid (EDTA).

## 2.3. Virus Production and Titering

1. 1× HBS solution for transfection can be made in 500-mL batches by adding 4.1 g NaCl, 0.186 g $Na_2HPO_4 \cdot 2H_2O$, 0.6 g dextrose, 3 g Hepes, qsp deionized $H_2O$. Buffer the solution to four different pHs ranging from 7.01 to 7.10 and pass through a 0.2-μm pore filter (*see* **Note 1**).

2. $CaCl_2$ is dissolved in deionized water to a final concentration of 2 M and also passed through a 0.2-μm pore filter.

3. Serum-free media for virion production can include CellGro or Optimem (Gibco/BRL).

4. 0.45-μm pore filters (Sartorius, Goettingen, Germany).

5. 20- or 50-mL syringes.

6. Ultracentrifugation is performed on an SW28 rotor with swinging bucket with adapted ultraclear centrifuge tubes (Beckman Coulter, ref. 344058, Fullerton, CA).

7. Ultrafiltration can be performed using columns such as Centricon Plus-70 100,000 MWCO (Millipore, ref UFC710008, Billenca, MA).

8. Sucrose is dissolved in TNE buffer (10 mM Tris–HCl, pH 7.5; 100 mM NaCl; 1 mM EDTA).

9. p24 ELISA kits are commercialized by multiple companies including Beckman Coulter.

10. Primers can be obtained from multiple companies including Sigma-Genosys.

11. Genomic DNA is extracted by a commercial DNA extraction kit.

12. Real-time amplifications are performed using the QuantiTect SybrGreen kit (Qiagen) on a LightCycler system (Roche).

**2.4. In Vivo Murine Thymocyte Transductions**

1. Thymocyte injections performed in mice wherein the thymus is exposed prior to injection were treated with 40 mg/kg pentobarbital (Sanofi-Synthelabo) prior to surgery.

2. Direct thymic injections through the skin are performed after inhalation of isoflurane.

3. Injections are performed using 0.3-mL 28-gauge 8-mm insulin needles.

**2.5. Ex Vivo Human and Murine Thymocyte Transductions**

1. Human thymi can be obtained from fetus or neonatal/pediatric patients undergoing cardiac surgery after obtaining approval of the appropriate hospital review and ethics boards.

2. Total thymocytes can be used for transductions or cell types of interest can be purified by commercial antibody selection kits (Miltenyi, Dynal, StemCell) using either negative or positive depletion methods with anti-CD3, anti-CD4, and/or anti-D8 antibodies or anti-CD34 antibodies for more immature progenitor cells.

3. Lineage-negative murine progenitor cells are obtained from mouse bone marrow. To increase the relative percentage of progenitor cells, mice can be treated with 5-FU (6.25 mg per mouse via IP administration) 4 days prior to bone marrow harvest.

4. Cytokines are purchased from Peprotech (Paris, France).

5. Recombinant fibronectin is available from Takara (Otsu, Shiga, Japan).

6. For the culture of progenitor cells, cells are resuspended in either StemSpan (StemCell Technologies) or X-Vivo-20 (Cambrex BioSciences). Culture of murine cells requires the addition of beta-mercaptoethanol (50 $\mu$M final concentration).

7. OP9 stromal cells harboring the Notch ligand DL1 are cultured in $\alpha$MEM supplemented with 20% FCS. These cells can be obtained from the group of Zuniga-Pflucker, who stably introduced murine DL1 into OP9 cells and showed that they could support murine and human T-cell differentiation (28). OP9 cells stably expressing human DL1 can be obtained from the group of H. Spits (29).

8. OP9-DL1/precursor cocultures may be performed either in supplemented αMEM media, or a mixture consisting of 50% supplemented RPMI and 50% supplemented αMEM. For separating thymocytes/precursors from OP9 stromal cells after coculture, the cell suspension should be passed through a 40-μm filter to remove stromal cells.

## 3. Methods

### 3.1. Preparation of Plasmids

1. Add 200 ng of plasmid DNA to 50 μL of electrocompetent *E. coli* (DH5-α) bacteria on ice under sterile conditions.

2. Mix and transfer to a chilled electroporation cuvette and electroporate at 2.5 kV, 200 Ω, 25 μF. The time constant reading should be > 4.5.

3. Immediately add 500 μL of ice-cold 2× LB.

4. Spread dilutions of transformed bacteria on LB agar-ampicillin plates and incubate at 37°C overnight.

5. Pick three colonies and use each to inoculate 2 mL of 2× LB-ampicillin medium (containing 50 μg/mL ampicillin). Incubate for 6 h at 37°C with agitation (200 rev/min).

6. Use each cultured colony to inoculate 500 mL 2× LB-ampicillin medium in a conical flask of at least 2-L capacity and incubate overnight at 37°C with agitation.

7. Transfer 2 mL of the 500-mL culture volume to an Eppendorf tube for miniprep verification of plasmid.

8. Centrifuge the remaining 500 mL of bacterial culture at $4,500 \times g$ for 10 min at 4°C.

9. Discard the supernatant. Pellets may be stocked at –20°C until maxipreps are made.

10. Perform a miniprep with a commercial kit. Verify the amplified plasmid with restriction enzyme digestion and agarose gel electrophoresis.

11. Large-scale preparation of plasmid DNA from the frozen bacterial pellet can then be carried out (*see* **Note 2**).

12. Verify the amplified plasmid with restriction enzyme digestion and store at –20°C at a concentration of 1 mg/mL.

### 3.2. Culture and Transfection of HEK 293T Cells

1. HEK 293T is the most commonly used packaging cell line for production of VSV-G pseudotyped lentiviral vectors. 293T are adherent cells cultured in DMEM High Glucose (4.5 mg/L) supplemented with 10% FCS and 1% antibiotics (penicillin/streptomycin).

2. Recover cells from plates using a 0.4% trypsin-PBS solution. Incubate for 5 min at 37°C (*see* **Note 3**).

3. Ensure that the cells are visibly detached and homogenize thoroughly by pipetting to dissociate cellular aggregates.

4. Add 5 mL of complete medium. The presence of FCS at this step is necessary to stop the trypsin enzymatic reaction. Resuspend vigorously with a 5-mL pipette to again dissociate cellular aggregates and transfer to a centrifuge tube.

5. Centrifuge cells at $370 \times g$ for 5 min, then discard the supernatant, and resuspend thoroughly in complete medium. Seed $3 \times 10^6$ 293T cells on a 10-cm plate in 7 mL of complete DMEM.

6. Following overnight culture, cells should be 60–70% confluent. This level of confluence is essential for optimal transfection efficiency.

7. Mix the three vectors, lentiviral vector, Gag/Pol- and Env-expressing plasmids, in a 1.5-mL tube at a 3:3:1 ratio. Do not exceed 20 μg of total plasmid DNA per reaction to minimize toxicity. Add 500 μL of 1× HBS solution to the plasmid mix, followed by 33 μL of 2 M $CaCl_2$ and vortex quickly for 10 s (*see* **Note 4**).

8. Gently add the transfection mix to the plate dropwise, taking care not to detach the producer cells. Incubate overnight at 37°C (*see* **Notes 5** and **6**).

### 3.3. Production of Lentiviral Virions

1. Following overnight culture after transfection, replace the medium by gently adding 6 mL of serum-free prewarmed medium (i.e., CellGro or Optimem) to the side of the plate. Incubate for 30–48 h at 37°C.

2. Harvest the supernatant in a 50-mL tube, centrifuge at $400 \times g$ for 5 min to pellet cellular debris, and then pass the viral supernatant through a 0.45-μm pore filters (*see* **Note 7**).

3. VSV-G pseudotyped virions can be concentrated from the supernatant by ultracentrifugation (*see* **Note 8**). At this stage it is also possible to centrifuge over a 20% sucrose cushion or a double 20–50% sucrose cushion to obtain a supernatant of significantly higher purity.

4. Transfer the supernatant to an ultracentrifuge tube (Ultraclear 38.5 mL Beckman Coulter). Use serum-free medium to balance tubes to within $+/- 0.01 \times g$, then load in an SW28 rotor. Centrifuge at $13,000 \times g$ at 4°C for 2 h. The brake must be inactivated to prevent disturbance of the viral pellet during deceleration.

5. Discard the supernatant by inversion and leave the tube inverted for 2 min on absorbent tissue paper.

6. Add the desired volume of resuspension medium (PBS or serum-free medium) and incubate for 1 h on ice to allow the pellet to soften.

7. Resuspend carefully by pipetting, then store at –80°C in small aliquots (*see* **Note 9**).

**3.4. Virus Titering**

1. The viral "titer" may relate to the concentration of either physical particles or transducing units. The concentration of physical particles may be determined from the p24 content using a commercial ELISA kit. For determining the concentration of transducing units, either adherent (e.g., 293T) or suspension (e.g., Jurkat) target cells may be used (*see* **Note 10**).

2. Count and seed target cells. For Jurkats, seed $1 \times 10^6$ cells in 1 mL per well on a 24-well plate (*see* **Note 11**).

3. Thaw an aliquot of virions on ice, then dilute 1 μL in 100 μL of culture medium. Mix serial dilutions of the diluted virions with target cells and incubate for 72 h.

4. Analyze the percentage of cells expressing the transgene marker by FACS. Calculate the titer from a dilution generating no more than 30% of transgene-expressing cells according to the formula indicated later wherein the number of cells refers to the number of cells present in the well at the time point when the virions are added to the culture:

$$\text{Titer (transduction units/mL)} = \frac{(\% \text{ transgene cells}^+ / 100) \times \text{No. cells per well}}{\text{total volume of virions (mL)}}$$

Alternatively, the transducing units' titer may be calculated by q-PCR analysis of the number of viral DNA copies in each dilution. Use a cell line expressing one viral copy per cell or plasmid dilutions for the standard control. Extract DNA from the transduced cells using a commercial DNA extraction kit and perform q-PCR using primers to amplify an exogenous transgene such as EGFP, or the 5′-LTR-Gag product. Commonly used primers for the 5′LTR-Gag are M667 5′-GGCTAACTAGGGAACCCACTG-3′, and M661 5′-CCTGCGTCGAGAGAGCTCCTCTGG-3′ (*see* **Note 12**)

**3.5. In Vivo Transduction of the Murine Thymus Following Surgical Incision**

1. Adult mice of 6–9 weeks of age, from any strain, can be used for these experiments (*see* **Note 13**).

2. Mice are anesthetized with 40 mg/kg of pentobarbital.

3. Access to the trachea is gained by midincision of the lower neck.

4. The sternum is then incised at the level of the first two ribs, and the sternum is pulled aside allowing the thymus to be visualized.

5. Virions are injected in a total volume of 10–20 µL (*see* **Note 14**).

**3.6 In Vivo Transduction of the Murine Thymus by a Direct "Blind" Approach**

1. Infant mice ranging in age from 1–3 weeks are used for these experiments (*see* **Note 15**).

2. Mice are anesthetized by inhalation of isoflurane.

3. Virions are injected in a total volume of 10–20 µL from a 0.3 mL 28 gauge 8 mm insulin syringe.

4. Place mice in a dorsal position in order to situate the needle at a 30–45° angle directly above the sternum.

5. The needle is introduced directly through the skin into the thoracic cavity immediately above the sternum and the contents injected into the thymus.

6. The specific localization of the injection can be assessed by injecting 10 µL of trypan blue dye in control mice. Under conditions where the thymus is correctly targeted, at least one of the lobes will be blue (**Fig. 1**).

**3.7. Lentiviral Transduction of Human Thymocytes on Tissue-Culture Plates or on OP9-DL1 Stromal Cells**

1. Human thymus needs to be disrupted by mechanical means in order to obtain a single cell suspension. To this end, the thymus is teased apart in a 5–10 mL volume of RPMI supplemented with 10% FCS using two 1.5 in. needles. After at least 10–15 min of teasing, cells are passed through a 40 µm filter (Becton Dickinson) in order to eliminate clumps and the single cell suspension is washed and centrifuged twice for 5 min at $370 \times g$.

2. Thymocytes may be seeded directly on a 24 well tissue-culture plate at a concentration of $1 \times 10^5$–$1 \times 10^6$ thymocytes or progenitors per well. Alternatively, they may be seeded on confluent OP9-DL1 stromal cells in a 24-well plate in supplemented αMEM medium (*see* **Note 16**). The stromal layer should be confluent when thymocytes or progenitor cell populations are seeded.

3. Thymocytes should be maintained in supplemented RPMI medium, resulting in complete medium containing 50% supplemented αMEM and 50% supplemented RPMI containing a final volume of 20% FCS. Add cytokines as appropriate (*see* **Note 17**).

4. Add virions directly to wells at the appropriate MOI (*see* **Note 18**) and transduce for a minimum of 24 h. After this time, virions may be either left in the wells, or if virion removal is desired, thymocytes must be separated from the stromal layer and reseeded on fresh confluent OP9 cells

5. OP9-DL1 stromal cells need to be passed every 3–5 days, and this also applies to OP9-DL1/thymocyte cocultures.

Prepare wells containing confluent OP9-DL1 stromal layers in advance so they are at confluency at the necessary time points. Remove thymocytes by pipetting, and then pass through a 40-μm filter (Becton Dickinson) to remove OP9-DL1 cells from the previous well. Wash the filter with 5 mL PBS and recover thymocytes. Centrifuge thymocytes to remove PBS, resuspend in medium, and reseed on the fresh OP9-DL1 stromal layers.

6. Transduction efficiency can be analyzed 72 h after addition of virions by monitoring transgene expression.

### 3.8. Lentiviral Transduction of CD34+ Human Thymocyte Precursors on OP-DL1 Stromal Cells

1. Prepare retronectin-coated plates. Reconstitute lyophilized retronectin in sterilized water at a concentration of 1 mg/mL. This stock solution should be stored at –20°C. Dilute the stock to 32 μg/mL in PBS and add 500 μl of this working solution per well on nontissue-culture treated 24-well plates. Incubate at 37°C for 2 h. Remove the retronectin and store at 4°C; this working solution can be reused up to three times. Rinse wells with PBS, then saturate with PBS containing 5% FCS for 20 min at 37°C.

2. CD34+ progenitor cells from human thymus can be obtained using commercially available kits (see **Note 19**), and cells can be further purified into T-progenitor cells by FACS-sorting the CD1a- population.

3. CD34+ cells are plated on the retronectin-coated plates in X-Vivo or StemSpan media containing 10% FCS and supplemented with SCF (20 ng/mL) and IL-7 (10 ng/mL).

4. Following overnight cytokine stimulation, virions are added directly to the wells and culture is continued for an additional 24 h. A multiplicity of infection (MOI) of at least a 50 is generally used to obtain significant transduction efficiencies.

5. Cells are then washed and resuspended in αMEM media containing 20% FCS (see **Note 20**) supplemented with Flt3L (5 ng/mL) and IL-7 (10 ng/mL).

6. Cells are then seeded on confluent OP9-DL1 cells as indicated earlier.

7. As indicated, progenitor cells must be transferred to fresh confluent OP9-DL1 cells every 3–5 days with addition of fresh media and cytokines.

8. Transduction can be assessed after 48 h (if sufficient cells are recovered), but T-cell differentiation generally requires 21–28 days of culture.

***3.9. Transduction of Murine T Precursor Cells***

1. Seed lineage-negative (lin⁻) progenitor cells at $1 \times 10^6$ cells/well in 500 µl of medium on tissue supplemented with Flt3L (5 ng/mL) and IL-7 (5 ng/mL).

2. After overnight culture, the stromal cells will have adhered to the culture plate. Recover the nonstromal suspension lin⁻ cells by thorough pipetting, and rinse the well twice to collect any remaining cells. Wash cells with PBS containing 5% FCS and resuspend in fresh complete medium at $2 \times 10^6/$mL.

3. Coculture lineage-negative progenitor cells on confluent retroviral-producing cell lines for 24 h (*see* **Note 21**).

4. Cells are removed and resuspended in αMEM media containing 20% FCS (*see* **Note 20**) supplemented with Flt3L (5 ng/mL) and IL-7 (10 ng/mL).

5. Cells are then seeded on confluent OP9-DL1 cells as indicated.

6. As indicated, progenitor cells must be transferred to fresh confluent OP9-DL1 cells every 3–5 days with addition of fresh media and cytokines.

7. Transduction can be assessed after 48 h (if sufficient cells have been recovered) while T-cell differentiation, as monitored by the appearance of TCRαβ + CD4+ CD8+ thymocytes, requires at least 14 days of culture.

# 4. Notes

1. Transfection efficiencies are extremely sensitive to the pH of the HBS-Ca solution. It is thus recommended that four batches of HBS be made with pHs of 7.01, 7.03, 7.07, and 7.10. The efficiency of transfection with the four batches can then be tested and one batch maintained. HBS can then be aliquoted and stored at –20°C for long-term use.

2. Plasmids can easily be purified using the column kits sold by a large number of different biotech companies. Nevertheless, the purity of the plasmid preparations is often higher following purification by cesium chloride gradient.

3. 293T are not overtly adherent cells and may be detached from the plate using less toxic 0.5 mM EDTA instead of trypsin.

4. Safety specifications for culture of 293T cells transfected with plasmids resulting in the production of virions whose enve-

lope allows transduction of human cells vary from country to country. In France, for example, these experiments must be performed in a BL3 facility if the transgene encoded in the lentiviral vector is an oncogene. Virions can be handled in a BL2 facility only following testing to ascertain that there are no replication-competent virus recombinants.

5. Low transfection efficiencies can be due to mycoplasma infection. Importantly, greater than 30% of cells in most cell culture facilities are estimated to be mycoplasma-positive. Routine testing, either by commercially available PCR kits or Hoescht staining to assess whether there is DNA in the cytoplasm, is highly encouraged. In the event of a positive result, the two options are to repurchase the 293T cell line from a cell culture collection or treat the cells with an antimycoplasma agent such as Baytril (Bayer) *(30)*.

6. Even if your lentiviral vector does not encode a marker gene, it is important to assure that your levels of transfection are high (> 80%) before continuing to virion production. This is most easily done by transfecting an EGFP-harboring vector and assessing fluorescence of the tissue-culture plate by microscopy.

7. For pilot experiments wherein titers of > $10^6$ are not necessary, supernatants harboring virions may be used immediately without any further concentration. In this case, it is advantageous to change the media on the 293T cells carefully 24 h post-transfection to the media that is used for culture of the target cells that will be transduced.

8. Due to the secondary structure of the envelope, virions harboring VSV-G-envelopes are easily concentrated via ultracentrifugation. Although titers of virions harboring retroviral envelopes with a SU-TM structure will also be increased by ultracentrifugation, the overall yield may be reduced as some virions will be rendered noninfectious due to shedding of the SU.

9. Due to the stability of the VSV-G envelope, titers of VSV-G-pseudotyped virions decrease by approximately 10% upon freeze-thawing. In contrast, the titers of virions pseudotyped with retroviral envelopes with an SU/TM structure tend to decrease by approximately 1-log upon freeze-thawing. It is therefore optimal to freeze virion preps in small aliquots.

10 The titers of virions can be assessed in various manners. Titers based on assessing expression of a transgene are generally simplest, especially if the transgene can be monitored by flow cytometry. To avoid problems due to multiple integrations, it is best to determine titers at dilutions wherein the percentage of cells expressing the transgene is less than 30%. At these transduction efficiencies, the number of integrated copies is generally 1 *(31)*. Importantly though, virions can

infect cells without resulting in transgene expression and as such it might be more accurate to monitor reverse transcription of the lentiviral vector. Finally, reverse transcription can also be modulated by the cellular environment and quantification of physical particles can be determined by monitoring levels of the p24 protein incorporated in the virions. This last test, however, will not provide any information regarding the ability of the virions to transduce a target cell.

11. Titers, or transducing units, are very difficult to "translate" from one laboratory to another and are biased by the target cells used to test transduction. Thus, the titer that will be obtained on NIH3T3, 293T, HeLa, or Jurkat cells can differ by up to 1-log. Therefore, it is important to standardize within your own laboratory and use the same cell type plated at the same density for titering.

12. The virus titer must be elevated, with at least $1 \times 10^9$ transducing units per mL. This is critical as only 10–20 μL of virions can be injected into a murine thymus. As the number of thymocytes in a murine thymus is approximately $2 \times 10^8$, a titer of $1 \times 10^9$ TU will result in a relative multiplicity of infection of < 0.1. Increasing titers to greater than $1 \times 10^{10}$ will allow an effective MOI approaching 1. As such, it is crucial to begin intrathymic injection studies with high titer virion stocks, which, if need be, can be obtained by further concentration. For in vivo experiments, virion purities are important and as such, it is preferable to concentrate via ultrafiltration or centrifugation over a sucrose cushion as compared to ultracentrifugation alone.

13. In order to achieve lentiviral gene transfer in the thymus, it is optimal to use mice with a large thymus wherein thymopoiesis is ongoing. The size of the murine thymus increases considerably between birth and 4–6 weeks of life, with a corresponding increase in thymocyte numbers. In older mice, the thymus regresses and therefore is not optimal for the described experiments.

14. Virions can be administered in a single injection or as multiple injections. If the goal is to compare an uninjected and injected lobe, it is preferable to perform only a single injection. However, if the goal is to obtain optimal gene transfer, frequencies can be improved by performing 3–4 injections rather than a single injection.

15. It is difficult to use mice of less than 1 week of age because the relative size of the thymus is small. Mice thymi contain approximately $1 \times 10^7$ thymocytes at birth, and this number increases to $1 \times 10^8$ thymocytes by two weeks of age.

16. OP9 stromal cells are generally superior to other stromal cell lines in their capacity to support lymphocyte differentiation. This is most likely due to the absence of secreted M-CSF; this stromal cell line is derived from the M-CSF-deficient op/op mouse *(32)*.

17. A small proportion of human thymocytes can survive for 3 days in the absence of cytokines or Notch-mediated signals, but thymocyte survival is significantly augmented when cells are plated on OP9 cells harboring the Notch ligand DL1 and supplemented with rIL-7 (10 ng/mL). Culture and stimulation will clearly influence the subsets transduced (Fig. *see*13.4). For preferential targeting of DN cells, culture on OP9 stromal cells expressing DL1, together with IL-7 stimulation, is recommended. For transducing mature SP populations, IL-7 cytokine should be used, but contact with OP9 DL1 stromal cells is not necessary. Moreover, the relative survival of immature double positive thymocytes is improved in the presence of IL-7 (Fig. 13.4).

18. To obtain a transduction efficiency of greater than 10%, it is generally advisable to use an MOI of at least 20.

19. Cells can be purified by either positive or negative selection kits, but given the low percentage of CD34+ cells, the purity is significantly higher using kits positively selecting for expression of the CD34 antigen.

20. Different batches of fetal calf sera are NOT equivalent, and batches need to be tested for their ability to support T-cell differentiation. It is known that high steroid levels in sera inhibit T-cell differentiation and, if needed, steroids can be eliminated by charcoal stripping. Fetal calf sera commercialized by Hyclone often have lower steroid levels than other commercial brands and have been successfully used for T-cell differentiation studies.

21. Transduction of lineage-negative murine progenitor cells, for the goal of T-cell differentiation, can be achieved using either VSV-G-pseudoypted lentiviral vectors or ectoropic envelope-pseudotyped MuLV-based vectors. In our hands, transduction efficiency is greater following coculture on cell lines producing MuLV-based vectors, with levels approaching 60%.

## Acknowledgments

We are very grateful to all members of our lab, past and present, for their precious input into improving gene transfer technology. A special thanks to Drs. Els Verhoeyen, F.L Cosset, and

M. Sitbon for their generosity in sharing theoretical and practical guidance on lentiviral gene transfer technology. This work has been supported by the AFM and by the European Community (contract LSHC-CT-2005-018914 "ATTACK").

## References

1. Esslinger, C., L. Chapatte, D. Finke, I. Miconnet, P. Guillaume, F. Levy and H.R. MacDonald. 2003. In vivo administration of a lentiviral vaccine targets DCs and induces efficient CD8(+) T cell responses. *J Clin Invest* 111:1673–1681

2. VandenDriessche, T., L. Thorrez, L. Naldini, A. Follenzi, L. Moons, Z. Berneman, D. Collen and M.K. Chuah. 2002. Lentiviral vectors containing the human immunodeficiency virus type-1 central polypurine tract can efficiently transduce nondividing hepatocytes and antigen-presenting cells in vivo. *Blood* 100:813–822

3. Yanay, O., S.C. Barry, L.J. Katen, M. Brzezinski, L.Y. Flint, J. Christensen, D. Liggitt, D.C. Dale, et al. 2003. Treatment of canine cyclic neutropenia by lentivirus-mediated G-CSF delivery. *Blood* 102:2046–2052

4. Palmowski, M.J., L. Lopes, Y. Ikeda, M. Salio, V. Cerundolo and M.K. Collins. 2004. Intravenous injection of a lentiviral vector encoding NY-ESO-1 induces an effective CTL response. *J Immunol* 172:1582–1587

5. Kobinger, G.P., J.P. Louboutin, E.R. Barton, H.L. Sweeney and J.M. Wilson. 2003. Correction of the dystrophic phenotype by in vivo targeting of muscle progenitor cells. *Hum Gene Ther* 14:1441–1449

6. Baekelandt, V., K. Eggermont, M. Michiels, B. Nuttin and Z. Debyser. 2003. Optimized lentiviral vector production and purification procedure prevents immune response after transduction of mouse brain. *Gene Ther* 10:1933–1940

7. Marodon, G. and D. Klatzmann. 2004. In situ transduction of stromal cells and thymocytes upon intrathymic injection of lentiviral vectors. *BMC Immunol* 5:18

8. Allman, D., A. Sambandam, S. Kim, J.P. Miller, A. Pagan, D. Well, A. Meraz and A. Bhandoola. 2003. Thymopoiesis independent of common lymphoid progenitors. *Nat Immunol* 4: 168–174

9. Adjali, O., G. Marodon, M. Steinberg, C. Mongellaz, V. Thomas-Vaslin, C. Jacquet, N. Taylor and D. Klatzmann. 2005. In vivo correction of ZAP-70 immunodeficiency by intrathymic gene transfer. *J Clin Invest* 115:2287–2295

10. Unutmaz, D., V.N. KewalRamani, S. Marmon and D.R. Littman. 1999. Cytokine signals are sufficient for HIV-1 infection of resting human T lymphocytes. *J Exp Med* 189: 1735–1746

11. Dardalhon, V., B. Herpers, N. Noraz, F. Pflumio, D. Guetard, C. Leveau, A. Dubart-Kupperschmitt, P. Charneau, et al. 2001. Lentivirus-mediated gene transfer in primary T cells is enhanced by a central DNA flap. *Gene Ther* 8:190–198

12. Zack, J.A., S.J. Arrigo, S.R. Weitsman, A.S. Go, A. Haislip and I.S. Chen. 1990. HIV-1 entry into quiescent primary lymphocytes: molecular analysis reveals a labile, latent viral structure. *Cell* 61:213–222

13. Jaleco, A.C., H. Neves, E. Hooijberg, P. Gameiro, N. Clode, M. Haury, D. Henrique and L. Parreira. 2001. Differential effects of Notch ligands Delta-1 and Jagged-1 in human lymphoid differentiation. *J Exp Med* 194:991–1002

14. Schmitt, T.M. and J.C. Zuniga-Pflucker. 2002. Induction of T cell development from hematopoietic progenitor cells by delta-like-1 in vitro. *Immunity* 17:749–756

15. Schmitt, T.M. and J.C. Zuniga-Pflucker. 2006. T-cell development, doing it in a dish. *Immunol Rev* 209:95–102

16. Naldini, L., U. Blomer, P. Gallay, D. Ory, R. Mulligan, F.H. Gage, I.M. Verma and D. Trono. 1996. In vivo gene delivery and stable transduction of nondividing cells by a lentiviral vector. *Science* 272:263–267

17. Delenda, C. 2004. Lentiviral vectors: optimization of packaging, transduction and gene expression. *J Gene Med* 6 Suppl 1:S125–S138

18. Sinn, P.L., S.L. Sauter and P.B. McCray, Jr. 2005. Gene therapy progress and prospects: development of improved lentiviral and retroviral vectors – design, biosafety, and production. *Gene Ther* 12:1089–1098

19. Zennou, V., C. Petit, D. Guetard, U. Nerhbass, L. Montagnier and P. Charneau. 2000. HIV-1 genome nuclear import is mediated by a central DNA flap. *Cell* 101:173–185

20. Follenzi, A., L.E. Ailles, S. Bakovic, M. Geuna and L. Naldini. 2000. Gene transfer by lenti-

viral vectors is limited by nuclear translocation and rescued by HIV-1 pol sequences. *Nat Genet* 25:217–222

21. Sirven, A., F. Pflumio, V. Zennou, M. Titeux, W. Vainchenker, L. Coulombel, A. Dubart-Kupperschmitt and P. Charneau. 2000. The human immunodeficiency virus type-1 central DNA flap is a crucial determinant for lentiviral vector nuclear import and gene transduction of human hematopoietic stem cells. *Blood* 96:4103–4110

22. Manganini, M., M. Serafini, F. Bambacioni, C. Casati, E. Erba, A. Follenzi, L. Naldini, S. Bernasconi, et al. 2002. A human immunodeficiency virus type 1 pol gene-derived sequence (cPPT/CTS) increases the efficiency of transduction of human nondividing monocytes and T lymphocytes by lentiviral vectors. *Hum Gene Ther* 13:1793–1807

23. Zufferey, R., D. Nagy, R.J. Mandel, L. Naldini and D. Trono. 1997. Multiply attenuated lentiviral vector achieves efficient gene delivery in vivo. *Nat Biotechnol* 15:871–875

24. Verhoeyen, E. and F.L. Cosset. 2004. Surface-engineering of lentiviral vectors. *J Gene Med* 6 Suppl 1:S83–94

25. Verhoeyen, E., V. Dardalhon, O. Ducrey-Rundquist, D. Trono, N. Taylor and F.L. Cosset. 2003. IL-7 surface-engineered lentiviral vectors promote survival and efficient gene transfer in resting primary T lymphocytes. *Blood* 101:2167–2174

26. Cui, Y., J. Golob, E. Kelleher, Z. Ye, D. Pardoll and L. Cheng. 2002. Targeting transgene expression to antigen-presenting cells derived from lentivirus-transduced engrafting human hematopoietic stem/progenitor cells. *Blood* 99:399–408

27. May, C., S. Rivella, J. Callegari, G. Heller, K.M. Gaensler, L. Luzzatto and M. Sadelain. 2000. Therapeutic haemoglobin synthesis in beta-thalassaemic mice expressing lentivirus-encoded human beta-globin. *Nature* 406:82–86

28. Zuniga-Pflucker, J.C. 2004. T-cell development made simple. *Nat Rev Immunol* 4:67–72

29. Dontje, W., R. Schotte, T. Cupedo, M. Nagasawa, F. Scheeren, R. Gimeno, H. Spits and B. Blom. 2006. Delta-like1-induced Notch1 signaling regulates the human plasmacytoid dendritic cell versus T-cell lineage decision through control of GATA-3 and Spi-B. *Blood* 107:2446–2452

30. Uphoff, C.C., C. Meyer and H.G. Drexler. 2002. Elimination of mycoplasma from leukemia-lymphoma cell lines using antibiotics. *Leukemia* 16:284–288

31. Kustikova, O.S., A. Wahlers, K. Kuhlcke, B. Stahle, A.R. Zander, C. Baum and B. Fehse. 2003. Dose finding with retroviral vectors: correlation of retroviral vector copy numbers in single cells with gene transfer efficiency in a cell population. *Blood* 102:3934–3937

32. Yoshida, H., S. Hayashi, T. Kunisada, M. Ogawa, S. Nishikawa, H. Okamura, T. Sudo, L.D. Shultz, et al. 1990. The murine mutation osteopetrosis is in the coding region of the macrophage colony stimulating factor gene. *Nature* 345:442–444

# Chapter 14

# Design and Production of Retro- and Lentiviral Vectors for Gene Expression in Hematopoietic Cells

## Axel Schambach, William P. Swaney, and Johannes C.M. van der Loo

## Summary

Successful retroviral gene transfer into hematopoietic cells has been demonstrated in a number of small and large animal models and clinical trials. However, severe adverse events related to insertional mutagenesis in a recent clinical trial for X-linked severe combined immunodeficiency reinforced the need to develop novel retroviral vectors with improved biosafety. Improvements include the use of self-inactivating (SIN) vectors as well as improvements in vector design. This chapter describes the basic design of gamma-retroviral and lentiviral SIN vectors that utilize a split-packaging system and includes a description of the various cloning modules frequently used in the design of such vectors that impact biosafety, titer, and transgene expression. In addition, this chapter describes the methods used for high titer vector production using calcium phosphate transfection both at research scale and at large scale for clinical application using a closed system bioreactor.

**Key words:** Retrovirus, Lentivirus, Vector production, SIN vectors, Gene therapy, Split-packaging system, 293T cells, Calcium phosphate transfection, Wave Bioreactor, Closed System, Large scale.

## 1. Introduction

### 1.1. Modular Control Elements for Vector Design

Adverse events related to insertional mutagenesis and trans-activation of cellular proto-oncogenes in a recent clinical trial for the treatment of X-linked severe combined immunodeficiency (1, 2) have boosted interest in the use of self-inactivating (SIN) gammaretroviral (3, 4) and lentiviral vectors (5, 6). SIN vectors lack enhancer–promoter sequences in the U3 region of the long terminal repeats and have been shown in an in vitro transformation assay to have a reduced transforming capacity as compared to LTR vectors (7). An additional benefit of the SIN design is

Christopher Baum (ed.), *Methods in Molecular Biology, Methods and Protocols, vol. 506*
© Humana Press, a part of Springer Science + Business Media, LLC 2009
DOI: 10.1007/978-1-59745-409-4_14

that it allows incorporation of weaker cellular or lineage-specific promoters to drive the transgene, thereby reducing the likelihood of trans-activating neighboring genes. Several modular control elements have been described that can be incorporated into SIN vectors to modulate safety, titer, and transgene expression. The first part of this chapter describes several of such modular control elements including the 5′ promoter, the leader region, the internal promoter, the 3′ UTR, and the 3′ U3 region, cryptic polyA and splice sites, as well as options for coexpression of two genes.

### 1.2. Transfection vs. Generation of Stable Producer Lines

The increased interest in SIN vectors for clinical use has prompted a shift in vector production methodology from the traditional use of stable packaging cell lines to the use of transient production by transfection. This is primarily due to the fact that transfection-based methodology offers greater flexibility as changes to plasmids can be made without having to generate another stable producer cell line. Secondly, while LTR vectors allow for efficient generation of clones of vector producing cells using transduction, the same approach cannot be used for SIN vectors as the SIN U3 deletion prevents genomic RNA to be packaged after transduction. For transfection, the HEK293 (human embryonic kidney)-derived 293T cell line is frequently used for the production of both retroviral and lentiviral vectors as it has been well characterized, is highly transfectable, and expresses viral restriction factors only at low levels *(8)*. To reduce the chance of generating replication-competent retrovirus (RCR) or lentivirus (RCL), vector and packaging sequences are generally introduced into the producer cell using a split-(genome) packaging system (**Fig. 1**) *(3, 5, 9)*. The second part of this chapter describes a standard transfection protocol for the generation of research grade vector at small scale.

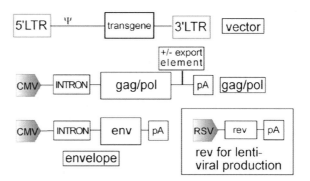

Fig. 1. Split-packaging system. Three/four plasmid split-(genome) packaging system for retro- and lentiviral production consisting of vector, gag/pol, envelope (env), and rev (for lentiviral production) expression constructs. The vector harboring the DNA to be transferred into the target cell is shown with its long terminal repeats (LTRs), packaging signal (ψ), and transgene. Expression plasmids for the viral proteins gag/pol, env, and rev are depicted. These are driven by a CMV or an RSV promoter and contain an intronic sequence and a strong polyA (pA) signal for optimal expression.

**1.3. Generation of Vector for Clinical Application**

Because most gene therapy vectors are not terminally sterilized, products for use in clinical trials need to be produced under aseptic manufacturing conditions. This requires specialized clean rooms with defined levels of airborne particulate contamination in both the supporting areas and the aseptic core, timely and costly media fills to validate the aseptic nature of the manufacturing process, and frequent cleaning of the facility processing equipment. Together, these measures seek to control the environment in which open manipulations occur, thereby limiting the potential exposure of the product to adventitious agents *(10)*. Therefore, a closed manufacturing system for the production of biological products for clinical use is highly desirable *(11)*. The Cincinnati Children's Hospital Vector Production Facility (VPF), a cleanroom facility for the production of cGMP viral vectors for clinical use, has developed such a closed system for the production of viral vectors by transfection. The process utilizes the Wave Bioreactor® manufactured by Wave Biotech. The process includes a single-use disposable BioProcessing bag with FibraCel® disks (New Brunswick), bagged media, and custom-made sterile fluid path collection, filtration, and distribution sets. The Wave Bioreactor provides for controlled temperature and gas and creates a rocking motion within the cell bag that facilitates gas exchange and promotes cell growth. Because the Wave Bioreactor itself does not actually contact the product, no specialized cleaning or sterilization of the bioreactor is required. FibraCel disks are made of a polyester nonwoven fiber and polypropylene and provide a platform for high-density culture of anchorage-dependent cells. Each gram of FibraCel has a surface area of approximately 1,200 cm². Using this method, almost $1 \times 10^{10}$ cells can be cultured within a small space, while almost 150 Roller bottles (850 cm²) would be required to do the same in conventional cell culture *(11, 12)*. An added benefit is the single-use disposable nature of the bioreactor bag. This feature obviates the need to clean and sterilize the inside of conventional bioreactors and eliminates the need to validate cleaning processes to demonstrate that product crosscontamination between successive batches does not exist. Finally, unlike conventional cell culture techniques utilizing roller bottles or multiple stacked culture flasks like Nunc Cell Factories or Corning Cell Stacks, the Wave Bioreactor system is a scaleable for productions ranging from as small as 100 mL up to 1,000 L. The third part of this chapter describes a protocol for the production of a retroviral vector by transfection at a 3-L scale using the Wave Bioreactor.

## 2. Materials

1. *Vector plasmids.* Gammaretroviral (RV) SIN vector *(3)*, lentiviral (LV) SIN vector *(5, 9)*.

2. *Gag/pol plasmids.* RV gag/pol *(9)*, LV gag/pol with Rev-Responsive Element (RRE) *(5)*, LV gag/pol with the Constitutive Transport Element (4× CTE) *(9)*.

3. Rev plasmid for LV production *(5)*.

4. *Envelope plasmids.* Ecotropic murine leukemia virus (MLV) envelope (env) *(13)*, Amphotropic MLV env *(14)*, Gibbon-Ape Leukemia Virus (GALV) env *(15)*, RD114 (feline endogenous virus) env *(16)*, Vesicular Stomatitis Virus glycoprotein (VSVg) *(17)* (*see* **Note 1**).

5. *E. coli strains.* XL-1 blue (Stratagene; La Jolla, CA), DH5alpha (Invitrogen; Carlsbad, CA), Top10 (Invitrogen).

6. *DNA modifiers.* Restriction enzymes, T4 DNA ligase, alkaline phosphatase, Klenow polymerase, T4 DNA polymerase (Fermentas; St. Leon-Rot, Germany).

7. *Miscellaneous cloning materials and equipment.* Agarose (Cambrex; Rockland, ME); DNA purification supplies (Qiagen; Valencia, CA); DNA sequencing equipment (Beckman Coulter; Fullerton, CA).

8. *Producer cells.* HEK 293T/17 (ATCC; Manassas, VA) from a certified Master Cell Bank (MCB).

9. *Complete culture medium (CCM).* High Glucose Dulbecco's Modified Eagles Medium (DMEM) with GlutaMAX™-I and HEPES buffer (Invitrogen 10,564) supplemented with 10% (vol/vol) fetal bovine serum (FBS; Hyclone, Logan, UT), and 1 mM sodium pyruvate (Invitrogen) (*see* **Note 2**).

10. *To passage cells.* Dulbecco's phosphate-buffered saline (D-PBS; Invitrogen or similar) and TrypLESelect (Invitrogen).

11. Cells are cryopreserved in CCM containing 15% (vol/vol) FBS and 10% (vol/vol) dimethyl sulfoxide (DMSO, Sigma).

12. *Transfection reagents.* Calcium phosphate transfection kit (Sigma; St. Louis, MO); chloroquine diphosphate salt (Sigma), diluted to 25 mM in PBS, filter sterilized, and stored frozen at $\leq -20°C$.

13. *Cell culture plastic ware.* T-75, T-175, and T-225 culture flasks (Corning; Lowell, MA, Becton-Dickinson (B-D); Franklin Lakes, NJ, or similar); 850-cm$^2$ Roller bottles (Corning, B-D or similar); Single-use disposable serological pipets at 2, 5, 10, 25, 50, and 100 mL (Corning, B-D or similar); Conical tubes at 6, 15, 50, 250, and 500 mL (Corning, B-D or similar); 1.5 mL-Eppendorf tubes (Sarstedt; Newton, NC, or similar); Storage bottles at 250, 500, and 1,000 mL (Corning or similar); 1.8 mL Cryovials (Nalge Nunc; Rochester, NY, or similar); Aspiration pipettes (Corning, B-D or similar); Pipette tips for P20, P200, and P1000 pipettors (Rainin; Oakland, CA).

14. *For counting cells.* Hemacytometer and 0.4% Trypan Blue Solution (Sigma or similar).

15. *Closed processing supplies.* Cellbag10L/FC (Wave Biotech; Somerset, NJ); Baxter R4R9955 Cryocyte Freezing Bags (Miltenyi Biotec; Auburn, CA); Baxter Cryocyte Manifold

set R4R9960 (Miltenyi Biotec); Transfer Pack with eight Couplers 4R2027 (Baxter; Deerfield, IL); Plasma Transfer Set with two Couplers 4C2243 (Baxter); 2,000 mL Transfer Packs 4R2041 (Baxter); RCQT Leukocyte Reduction Filter (Pall; East Hills, NY); Sebra Model 1,105 Hand-Held Sealing Head and Model 2,600 OMNI™ Sealer Power Source (Sebra; Tucson, AZ); Terumo Sterile Tubing Welder TSCD (SC-201A, Terumo) and Terumo Welding Wafers (Terumo; Summerset, NJ); Cryocyte Freezing Cassettes (Biomedical Marketing Associates; Wexford, PA).

16. Bioreactor Feed and Harvest Set (*see* **Notes 3** and **4**); Filter Sets (*see* **Notes 3** and **5**); Distribution Sets (*see* **Notes 3** and **6**).

17. *Pump and supplies.* Repeater Pump (Baxa; Englewood, CO); H938 21 3 Fluid Transfer Tube Set (Baxa); H938 11 3 Fluid Transfer Tube Set with vented Spike (Baxa); H938 86 3 Fluid Transfer Extension Set (Baxa); Luer Lock Universal Spike (Baxa).

18. *Miscellaneous.* 10-mL Luer Pipets (Ashton Pumpmatic; Dayton, OH); Syringe 60 mL (BD, or similar).

## 3. Methods

The methods below outline the modular control elements available for retroviral and lentiviral SIN vectors (**3.1**), small-scale production of retroviral and lentiviral vectors (**3.2**), and large-scale production of such vectors in a closed system bioreactor (**3.3**).

### 3.1. Modular Control Elements for SIN Vectors

SIN vectors have a deletion of the promoter/enhancer elements in the 3′ U3 region (**Fig. 2**). Expression of the transgene is thus dependent on an internal promoter. Different vector modules that impact vector performance and safety are described as:

1. *5′ promoter.* Heterologous promoters can be fused to the start of the R region (**Fig. 2**) to improve vector titer. For instance, use of the RSV U3 (+/− additional enhancers) increases the amount of genomic RNA that can be packaged in gammaretroviral SIN vectors and thus increase vector titers *(3)*. In lentiviral vectors, the substitution of the 5′ LTR with a CMV or RSV promoter makes virus production independent of tat *(6)*. Note that any modification made to the 5′ promoter is not part of the transcribed RNA and is thus not present in transduced cells.

2. *Leader region.* The leader region contains the packaging signal ψ, which is essential for recognition by the structural

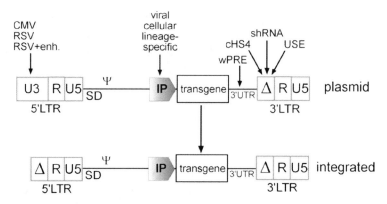

Fig. 2. SIN vector design. The modular composition of the vector (plasmid) is given with the respective LTRs (U3, R, U5), splice donor (SD) in the 5′ UTR, packaging signal ψ, internal promoter (IP) and transgene. The SIN configuration is indicated by a 3′ U3 deletion removing the promoter/enhancer elements. As shown (integrated), this deletion is copied into the 5′ LTR after reverse transcription, thereby making the gene expression in target cells dependent on an internal promoter of choice (viral, cellular or lineage-specific). To increase vector titers, the 5′ U3 region can be exchanged by cytomegalovirus (CMV) or Rous sarcoma virus (RSV, optionally plus additional enhancer) promoters to increase the amount of packageable genomic RNA. To improve titer and gene expression, the woodchuck hepatitis virus post-transcriptional regulatory element (wPRE) and an upstream polyadenylation enhancer (USE, derived from SV40) can be added into 3′ UTR or the 3′ U3 deletion, respectively. The USE also improves efficiency of termination. To shield the genome from transactivation mediated by the enhancers in the vector cassette, chromatin insulators (e.g., chicken HS4) can be introduced. To downregulate a specific gene, short hairpin RNAs (shRNA) driven by a Pol. III promoter can be added into the U3 deletion resulting in two copies in transduced cells.

proteins (i.e., nucleocapsid), several important secondary structure elements, and splice sites (e.g., splice donor). While the gammaretroviral leader can be free of gag sequences, lentiviral vectors require approximately 300 bp of *gag* and additional *env* sequences that contain the Rev-responsive element (RRE) for optimal performance *(5)*. For lentiviral vectors, the addition of the central polypurine tract (cPPT) sequences has been shown to improve transduction efficiency by facilitating nuclear import *(18)*.

3. *Internal promoter.* The level of gene expression needed for a specific application can be controlled by choosing a stronger or a weaker promoter. As previously shown, cellular promoters (e.g., elongation factor 1a, phosphoglycerokinase) are weaker trans-activators as compared to viral promoters and are frequently preferred when medium to lower levels of expression are required. Other promoters, such as inducible promoters (e.g., tetracycline-inducible) *(19)* or lineage specific promoters (e.g., globin promoter) *(20, 21)*, can be incorporated and further reduce the likelihood of trans-activation.

4. *Cryptic polyA sites, cryptic splice sites, transgene orientation, and codon optimization*. It is important to screen the incorporated sequences for cryptic polyA and splice sites. Cryptic polyA sites may lead to inefficient gene expression or to premature termination. If the 3′ R-region (retroviral polyA) is not part of the mRNA, reverse transcription cannot function properly. Cryptic splice sites, which can act alone or together with the splice sites of the leader region, could result in the loss of essential vector sequences. Therefore, to ensure a proper titer, it is advisable to do a splice site prediction on the retroviral vector (e.g., at www.fruitfly.org). To avoid unpredictable splicing, the transgene usually contains only the coding sequence without intron. However, if introns are important for efficient gene expression, as is the case for globin gene expression, the expression cassette (plus a heterologous polyA signal) may be introduced in antisense orientation. In cases, where a gene is difficult to express (e.g., HIV genes, *(22)*), codon optimization may help to achieve higher expression levels *(23)*.

5. *3′ UTR*. Several relatively large sequences can be added into the 3′ UTR to improve titer or modulate transgene expression. Addition of the post-transcriptional regulatory element of the woodchuck hepatitis virus (wPRE) has been shown to enhance both titer and gene expression levels *(24, 25)*. miRNAs (driven by Polymerase II promoter) *(26)* and miRNA target sequences *(27)* have also been shown to modulate gene expression or even direct lineage specific gene expression, respectively. Finally, introduction of a scaffold attachment region (SAR) has been shown to stabilize gene expression *(28, 29)*.

6. *3′ U3 deletion*. The 3′ U3 deletion area is an interesting location as any elements introduced in this area will be present in two copies after reverse transcription. This feature can be utilized when generating functional knock downs using short hairpin RNA (shRNA) cassettes, since the effect of shRNA (normally driven by a Pol. III promoter) is concentration dependent *(26, 30)*. In addition, chromatin insulators (e.g., cHS4) can be inserted into the 3′ U3 deletion area to stabilize gene expression and shield the genome from enhancer effects of the vector cassette *(31)*. Other interesting sequences that can be inserted in this area are UpStream polyadenylation Enhancers (USE elements) which improve gene expression, titer, and termination efficiency *(32)*.

7. *Coexpression*. Finally, coexpression of multiple expression units is often desirable. This can be achieved using IRES (internal ribosomal entry site) sequences, which act as a ribosome-landing platform for reinitiation of translation ensuring coexpression of a second gene. The use of IRES2 (Clontech, Mountain View, CA) and the addition of a 30–70 bp spacer in front

of the IRES sequences, have shown to improve coexpression of two genes *(33)*. As a second principle, 2A esterase moieties mediate cotranslational cleavage of two proteins *(34, 35)*. While IRES and 2A often mediate disproportionate expression of the two proteins, bidirectional promoters, which share one enhancer region driving two core promoters *(36)*, generate a more equal coexpression.

### 3.2. Small-Scale Retroviral and Lentiviral Vector Production

The production of vector by transfection in 10-cm dishes is described. This generates 32 mL of vector supernatant (four harvests of 8 mL each). Depending on the design of the vector, titers generally range from $1 \times 10^6$ to $4 \times 10^7$ Infectious Units (IU)/mL *(3)*.

1. On day 1 seed $5 \times 10^6$ 293T cells per 10-cm dish in 10–15 mL of CCM.

2. On day 2 dilute vector, gag/pol, envelope (and Rev, for lentiviral vectors) plasmids (*see* **Table 1**) in water (included in the Sigma CAPHOS kit) to a total volume of 450 µL in a 1.5-mL microtube and add 50 µL calcium chloride solution (CAPHOS kit). Add 500 µL of 2× HEPES-buffered saline (HeBS; CAPHOS kit) to a 14-mL conical tube. Add the DNA/calcium chloride mixture to the 2× HeBS while bubbling according to the manufacturer's instructions and incubate at ambient temperature for 20 min. Remove the used media from the 293T cells and add 9 mL CCM including 25 µM chloroquine (inactivates lysosomal DNases to increase transfection efficiency). Add the DNA/calcium phosphate mixture dropwise to the cells. Incubate for 6–12 h in an incubator at 37°C and 5% $CO_2$, and replace the used medium with fresh CCM (8 mL).

3. On days 3, 4 and 5 viral supernatants can be harvested every 12 h, beginning at 24 h and up to 72 h after transfection. To

### Table 1
### Small-scale production using cotransfection

| Plasmid | Gamma-retroviral production | Lentiviral production |
|---|---|---|
| SIN vector | 2–8 µg | 5–12 µg |
| Gag/pol | 6–10 µg | 6–15 µg |
| Rev | – | 3–5 µg |
| Envelope | 1–2 µg | 1–3.5 µg |
| *Expected titer (IU/mL)* | $1 \times 10^6$–$4 \times 10^7$ | $1 \times 10^6$–$4 \times 10^7$ |

The amounts of plasmid listed indicate microgram of DNA needed per 10-cm plate. Titers indicated are as expected when titered on murine fibroblast SC-1.

do so, remove supernatants carefully with a syringe and filter through a 0.45-μm filter into a 14-mL Falcon tube (aliquot a small amount in a 1.5-mL Eppendorf tube for titration). After each harvest, replace media with 8 mL CCM.

4. Store supernatants at or below −70°C.

**3.2. Large-Scale Production in a Closed System Bioreactor**

The method below outlines the manufacturing of a retroviral vector by transient transfection at a 3-L culture volume for a total of three vector harvests using the Wave Bioreactor (Wave Biotech). It includes a description of (1) the cells and media, (2) preparation of bagged media, (3) cell expansion, (4) preparation of the Wave Bioreactor, (5) transfection, (6) media change, and (7) vector supernatant harvest.

*3.2.1. Cells and Media*

1. Human Embryonic Kidney (293T) cells are cultured and maintained subconfluent at 37°C and 5% $CO_2$ in Complete Culture Media (CCM).

2. For large-scale productions, Hyclone serum used is of US/Canada origin and was gamma irradiated at a delivered dose of 25.7–32.3 kGy to reduce virus contamination.

3. Cells are passaged in CCM after rinsing with Dulbecco's Phosphate-Buffered Saline and dissociation with animal-origin free Trypsin-like enzyme (TrypLE Select).

4. A master cell bank is prepared by cryopreserving cells in CCM supplemented with 25% FBS, and 10% Dimethyl Sulfoxide (DMSO).

*3.2.2. Preparation of Bagged Media*

In preparation for bioreactor culture, prepare three bags of CCM (5-L volume) as follows.
1. From each bag of High Glucose DMEM, with GlutaMAX™-I and HEPES buffer (Invitrogen 10564-037), remove approximately 550 mL of media and add 500 mL of FBS and 50 mL of Sodium Pyruvate.

2. Materials are transferred in and out of bags using a Baxa Repeater Pump and a Fluid Transfer Tube Set (Baxa H938 21 3) connected to an Ashton Pumpmatic luer pipette.

*3.2.3. Cell Expansion*

1. 293T cells are thawed and seeded at 1–2 × $10^4$ cells/cm² in two 225-cm² tissue culture flasks (T225; Corning).

2. After 4–6 days, cells are harvested and seeded at 1–2 × $10^4$ cells/cm² in eight T225 flasks.

3. After 4–6 days, cells are harvested and seeded at 2–3 × $10^4$ cells/cm² in twenty-four 850-cm² Roller.

4. After 3–4 days, a total of 6 × $10^9$ cells in approximately 1 L of CCM are harvested for transfection the same day.

### 3.2.4. Preparation of the Wave Bioreactor

The Wave Bioreactor® system utilized in this application contains a BASE20/50EH rocker base unit, a KIT20EH rocker platform, and has been equipped with a LOADCELL20 for estimation of weight/volume, and a CO2MIX20 $CO_2$/air controller.

1. Prior to transfection, mount a 10-L size Wave Cellbag® containing 100 g Fibra-Cel® carriers onto the Wave Bioreactor® System 20/50.

2. To allow transfers in and out of the Cellbag for closed processing without reusing connections, connect a custom Bioreactor Feed and Harvest Set to the Wave Cellbag (*see* **Notes 3** and **4**).

3. Inflate the Cellbag with air and 5% $CO_2$ until rigid, use the Baxa Repeater pump to pump in approximately 1 L of CCM.

4. Set the system to operate at 37°C, with a rocking speed of 22 rpm and angle of 6°.

### 3.2.5. Transfection

#### Preparing the DNA

The amount of DNA needed to efficiently transfect cells in suspension is significantly higher as compared to transfection in a standard tissue culture flask. Prepare a mixture of DNA and calcium chloride ($CaCl_2$) in a 250-mL storage bottle as follows:

1. 12.0 mL of plasmid DNA encoding for the retroviral vector (1 mg/mL)

2. 10.8 mL of plasmid DNA encoding for the *gag/pol* genes (1 mg/mL)

3. 4.8 mL of plasmid DNA encoding for the viral envelope (1 mg/mL)

4. 107.4 mL of sterile distilled water

5. 15.0 mL of $CaCl_2$ (2.5 M; Sigma CAPHOS transfection kit)

#### Preparing Transfection Mixture and Cell Transfection

1. Place 150 mL of 2× HEPES-Buffered Saline (HeBS), pH 7.05 (CAPHOS transfection kit) in a 500-mL conical tube.

2. Slowly add the mixture of DNA and calcium chloride dropwise to the solution while bubbling air through the HeBS.

3. Incubate at ambient temperature for 20 min.

4. Add the 300 mL transfection mixture and 3 mL of a 25-mM Chloroquine solution to the cells (for a final concentration of 25 µM in a 3-L volume).

5. Mix by pipetting/swirling.

6. Pump into the Wave Bioreactor through a clean line of the Bioreactor Feed and Harvest Set.

7. Fill the Wave Cellbag with additional CCM to achieve a final volume of 3 L.

8. Incubate overnight.

### 3.2.6. Media Change

The next morning, the media may appear cloudy due to the transfection mixture and cells that failed to seed the Fibra-Cel carriers. Replace the media in the Cellbag with 3 L of ambient temperature or prewarmed CCM and continue incubation.

*3.2.7. Vector Supernatant Harvest*

Vector supernatant is harvested three times at 10–12 h intervals starting 10–12 h after the media change.

1. To harvest, pump vector supernatant into the first two interconnected Baxter Transfer Packs of the Filter Set (*see* **Note 5**).

2. Prior to filtration of the harvested material, pump 3 L of ambient temperature or prewarmed CCM back into the Wave Cellbag for continued culture.

3. Filter the harvested vector by pumping the vector through a Pall PureCell® RCQ High Efficiency Rapid Flow Leukocyte Reduction Filter of the filter set and collect vector in a second set of Transfer Packs 2,000 mL (Baxter 4R2041).

4. Connect a Distribution Set (*see* **Note 6**) to the bags containing the supernatant and aliquot the vector supernatant into individual 500-mL size Baxter Cryocyte bags. Each bag may be filled with up to 120 mL of supernatant.

5. Place each bag into a Freezing Canister and freeze at or below –70°C until use. To allow for functional testing, aliquot and separately freeze a number of small aliquots from each harvest.

*3.2.8. Wave Bioreactor Production Runs*

The protocol described was recently used to generate several batches of retroviral vector. Titers were determined by Flow Cytometry or by bulk PCR on infected HT1080 (for GALV envelope) or 3T3 (for Ecotropic envelope) target cells (*see* **Table 2**).

## Table 2
## Titers of retroviral vectors produced at large scale at the Cincinnati Children's Hospital Vector Production Facility

| Vector | Envelope | Total volume (L) | Titer harvest 1 | Titer harvest 2 | Titer harvest 3 |
|---|---|---|---|---|---|
| 1. SRS11.SF.Ds Red2.pre* | Ecotropic | $4 \times 1$ | $1.4 \times 10^7$ | $1.5 \times 10^7$ | $8.0 \times 10^6$ |
| 2. SRS11.EFS.IL2 Rg.pre* | GALV | $4 \times 1$ | $1.5 \times 10^6$ | $5.0 \times 10^5$ | $1.0 \times 10^5$ |
| 3. RSF91.GFP. pre* | K73-Eco | $3 \times 3$ | $4.0 \times 10^6$ | $3.3 \times 10^6$ | $1.7 \times 10^6$ |
| 4. SERS11.SF.GFP. pre* | K73-Eco | $3 \times 3$ | NA | $3.1 \times 10^6$ | $1.5 \times 10^6$ |
| Mock (packaging plasmids only) | K73-Eco | $3 \times 3$ | NA | NA | NA |

*Gamma-retroviral vector descriptions* (1) SIN vector with internal SFFV (spleen focus-forming virus) and coding sequence for DsRed2 (red fluorescent protein). (2) SIN vector with internal EFS (EF1a short) promoter and interleukin-2 receptor gamma chain (IL2RG). (3) LTR-driven vector with GFP (green fluorescent protein). (4) SIN vector with internal SFFV promoter and GFP. (1–4) All constructs harbor a modified woodchuck hepatitis virus post-transcriptional regulatory element (pre*) to improve titer and gene expression.

## 4. Notes

1. For LV production, modified envelope constructs (GALV/ TR, RD114/TR) are available that yield a higher titer *(37)*.

2. Antibiotics such as penicillin/streptomycin are never utilized in the production of viral vectors for clinical use, but are frequently added in the research laboratory. If DMEM without HEPES is used, add a final concentration of 20 mM HEPES at the time of transfection and at the media change post-transfection as it stabilizes the pH and improves viral viability in freeze/thaw cycles.

3. For closed processing, custom bag and tubing sets are prepared to prevent opening of an already used wet line and reusing connectors. Custom sets are prepared from Baxter components and require a Sebra Model 1,105 Hand-Held Sealing Head, a Sebra Model 2,600 OMNI™ Sealer Power Source and a Terumo Sterile Tubing Welder TSCD (SC-201A, Terumo) and Terumo Welding Wafers (1SC*W017, Terumo).

4. The standard Bioreactor Feed and Harvest Set is prepared using one Cryocyte Bag R4R9955, one Transfer Pack with eight Couplers 4R2027, and one Cryocyte manifold set R4R9960 to construct a set with a female luer connector on one side and four male luer connectors and seven couplers on the other side.

5. A Filter Set is prepared from four 2,000-mL Transfer packs (Baxter 4R2041), one Pall Leukocyte reduction filter (Pall RCQT), one Baxa fluid transfer set with spike and luer (Baxa Ref. #11), and one Universal Spike Luer set (Baxa Ref. #29). Link two transfer packs together by spiking one 2,000 mL Transfer pack with the coupler from another 2,000 mL Transfer pack. Hand seal and remove the coupler from the transfer pack which was spiked. Repeat once for the remaining two transfer packs. Spike the linked transfer pack of one of the two linked transfer packs sets with the Leukocyte reduction filter. Connect the Baxa fluid transfer set to the leukocyte reduction filter. Spike the nonlinked transfer pack of the remaining two bag set with the universal spike luer set (Baxa Ref. #29). Connect the luer end of the leukocyte reduction filter to the luer end of the universal spike luer set.

6. A Distribution Set is prepared from 20 Cryocyte Bags (Baxter R4R9955), three Transfer Packs with 8 Couplers (Baxter 4R2027), and one Plasma Transfer Set with 2 Couplers (Baxter 4C2243). Sterile weld a coupler Plasma Transfer Set with two Couplers to a set of eight couplers. Connect two additional sets with eight couplers to the original set with eight couplers and connect 20 cryocyte bags to each of the couplers

(two remain unused). For harvest into 36 containers, two distribution sets are needed per vector harvest.

## Acknowledgments

The authors thank Dr. Christopher Baum (Hannover Medical School, Germany) for providing the retroviral and Dr. Luigi Naldini (Telethon Institute for Gene Therapy, Milano, Italy) for lentiviral vectors referenced in this chapter. This work was supported by the Else-Kröner Foundation and the REBIRTH Excellence Cluster. The Cincinnati Children's Hospital Medical Center (CCHMC) Vector Production Facility is supported by internal funding provided by the Division of Experimental Hematology and Cancer Biology, and the Cincinnati Children's Research Foundation.

## References

1. Hacein-Bey-Abina, S., Von Kalle, C., Schmidt, M., McCormack, M. P., Wulffraat, N., Leboulch, P., Lim, A., Osborne, C. S., Pawliuk, R., Morillon, E., Sorensen, R., Forster, A., Fraser, P., Cohen, J. I., de Saint Basile, G., Alexander, I., Wintergerst, U., Frebourg, T., Aurias, A., Stoppa-Lyonnet, D., Romana, S., Radford-Weiss, I., Gross, F., Valensi, F., Delabesse, E., Macintyre, E., Sigaux, F., Soulier, J., Leiva, L. E., Wissler, M., Prinz, C., Rabbitts, T. H., Le Deist, F., Fischer, A., and Cavazzana-Calvo, M. (2003) LMO2-associated clonal T cell proliferation in two patients after gene therapy for SCID-X1. *Science* 302, 415–419

2. Williams, D. A., and Baum, C. (2003) Medicine. Gene therapy – new challenges ahead. *Science* 302, 400–401

3. Schambach, A., Mueller, D., Galla, M., Verstegen, M. M., Wagemaker, G., Loew, R., Baum, C., and Bohne, J. (2006) Overcoming promoter competition in packaging cells improves production of self-inactivating retroviral vectors. *Gene Ther.* 13, 1524–1533

4. Yu, S. F., von Ruden, T., Kantoff, P. W., Garber, C., Seiberg, M., Ruther, U., Anderson, W. F., Wagner, E. F., and Gilboa, E. (1986) Self-inactivating retroviral vectors designed for transfer of whole genes into mammalian cells. *Proc. Natl. Acad. Sci. U S A* 83, 3194–3198

5. Dull, T., Zufferey, R., Kelly, M., Mandel, R. J., Nguyen, M., Trono, D., and Naldini, L. (1998) A third-generation lentivirus vector with a conditional packaging system. *J. Virol.* 72, 8463–8471

6. Zufferey, R., Dull, T., Mandel, R. J., Bukovsky, A., Quiroz, D., Naldini, L., and Trono, D. (1998) Self-inactivating lentivirus vector for safe and efficient in vivo gene delivery. *J. Virol.* 72, 9873–9880

7. Modlich, U., Bohne, J., Schmidt, M., von Kalle, C., Knoss, S., Schambach, A., and Baum, C. (2006) Cell-culture assays reveal the importance of retroviral vector design for insertional genotoxicity. *Blood* 108, 2545–2553

8. Pion, M., Granelli-Piperno, A., Mangeat, B., Stalder, R., Correa, R., Steinman, R. M., and Piguet, V. (2006) APOBEC3G/3F mediates intrinsic resistance of monocyte-derived dendritic cells to HIV-1 infection. *J. Exp. Med.* 203, 2887–2893

9. Schambach, A., Bohne, J., Chandra, S., Will, E., Margison, G. P., Williams, D. A., and Baum, C. (2006) Equal potency of gammaretroviral and lentiviral SIN vectors for expression of O6-methylguanine-DNA methyltransferase in hematopoietic cells. *Mol. Ther.* 13, 391–400

11. Przybylowski, M., Hakakha, A., Stefanski, J., Hodges, J., Sadelain, M., and Riviere, I. (2006) Production scale-up and validation of packaging cell clearance of clinical-grade retroviral vector stocks produced in cell factories. *Gene Ther.* 13, 95–100

12. Merten, O. W., Cruz, P. E., Rochette, C., Geny-Fiamma, C., Bouquet, C., Goncalves, D., Danos, O., and Carrondo, M. J. (2001) Comparison of different bioreactor systems for the production of high titer retroviral vectors. *Biotechnol. Prog.* 17, 326–335

13. Morita, S., Kojima, T., and Kitamura, T. (2000) Plat-E: an efficient and stable system for transient packaging of retroviruses. *Gene Ther.* 7, 1063–1070

14. Cone, R. D., and Mulligan, R. C. (1984) High-efficiency gene transfer into mammalian cells: generation of helper-free recombinant retrovirus with broad mammalian host range. *Proc. Natl. Acad. Sci. U S A* 81, 6349–6353

15. Miller, A. D., Garcia, J. V., von Suhr, N., Lynch, C. M., Wilson, C., and Eiden, M. V. (1991) Construction and properties of retrovirus packaging cells based on gibbon ape leukemia virus. *J. Virol.* 65, 2220–2224

16. Takeuchi, Y., Cosset, F. L., Lachmann, P. J., Okada, H., Weiss, R. A., and Collins, M. K. (1994) Type C retrovirus inactivation by human complement is determined by both the viral genome and the producer cell. *J. Virol.* 68, 8001–8007

17. Yang, Y., Vanin, E. F., Whitt, M. A., Fornerod, M., Zwart, R., Schneiderman, R. D., Grosveld, G., and Nienhuis, A. W. (1995) Inducible, high-level production of infectious murine leukemia retroviral vector particles pseudotyped with vesicular stomatitis virus G envelope protein. *Hum. Gene Ther.* 6, 1203–1213

18. Sirven, A., Pflumio, F., Zennou, V., Titeux, M., Vainchenker, W., Coulombel, L., Dubart-Kupperschmitt, A., and Charneau, P. (2000) The human immunodeficiency virus type-1 central DNA flap is a crucial determinant for lentiviral vector nuclear import and gene transduction of human hematopoietic stem cells. *Blood* 96, 4103–4110

19. Goverdhana, S., Puntel, M., Xiong, W., Zirger, J. M., Barcia, C., Curtin, J. F., Soffer, E. B., Mondkar, S., King, G. D., Hu, J., Sciascia, S. A., Candolfi, M., Greengold, D. S., Lowenstein, P. R., and Castro, M. G. (2005) Regulatable gene expression systems for gene therapy applications: progress and future challenges. *Mol. Ther.* 12, 189–211

20. Puthenveetil, G., Scholes, J., Carbonell, D., Qureshi, N., Xia, P., Zeng, L., Li, S., Yu, Y., Hiti, A. L., Yee, J. K., and Malik, P. (2004) Successful correction of the human beta-thalassemia major phenotype using a lentiviral vector. *Blood* 104, 3445–3453

21. Sadelain, M. (2006) Recent advances in globin gene transfer for the treatment of beta-thalassemia and sickle cell anemia. *Curr. Opin. Hematol.* 13, 142–148

22. Schwartz, S., Campbell, M., Nasioulas, G., Harrison, J., Felber, B. K., and Pavlakis, G. N. (1992) Mutational inactivation of an inhibitory sequence in human immunodeficiency

virus type 1 results in Rev-independent gag expression. *J. Virol.* 66, 7176–7182

23. Graf, M., Bojak, A., Deml, L., Bieler, K., Wolf, H., and Wagner, R. (2000) Concerted action of multiple cis-acting sequences is required for Rev dependence of late human immunodeficiency virus type 1 gene expression. *J. Virol.* 74, 10822–10826

24. Zufferey, R., Donello, J. E., Trono, D., and Hope, T. J. (1999) Woodchuck hepatitis virus posttranscriptional regulatory element enhances expression of transgenes delivered by retroviral vectors. *J. Virol.* 73, 2886–2892

25. Schambach, A., Bohne, J., Baum, C., Hermann, F. G., Egerer, L., von Laer, D., and Giroglou, T. (2006) Woodchuck hepatitis virus post-transcriptional regulatory element deleted from X protein and promoter sequences enhances retroviral vector titer and expression. *Gene Ther.* 13, 641–645

26. Scherr, M., and Eder, M. (2007) Gene Silencing by Small Regulatory RNAs in Mammalian Cells. *Cell Cycle* 6(4), 444–449

27. Brown, B. D., Venneri, M. A., Zingale, A., Sergi Sergi, L., and Naldini, L. (2006) Endogenous microRNA regulation suppresses transgene expression in hematopoietic lineages and enables stable gene transfer. *Nat. Med.* 12, 585–591

28. Schubeler, D., Mielke, C., Maass, K., and Bode, J. (1996) Scaffold/matrix-attached regions act upon transcription in a context-dependent manner. *Biochemistry* 35, 11160–11169

29. Ramezani, A., Hawley, T. S., and Hawley, R. G. (2003) Performance- and safety-enhanced lentiviral vectors containing the human interferon-beta scaffold attachment region and the chicken beta-globin insulator. Blood. 2003 June 15;101(12):4717–4724. *Blood* 101, 4717–4724

30. Wiznerowicz, M., Szulc, J., and Trono, D. (2006) Tuning silence: conditional systems for RNA interference. *Nat. Methods* 3, 682–688

31. Gaszner, M., and Felsenfeld, G. (2006) Insulators: exploiting transcriptional and epigenetic mechanisms. *Nat. Rev. Genet.* 7, 703–713

32. Schambach, A., Galla, M., Maetzig, T., Loew, R., and Baum, C. (2007) Improving transcriptional termination of self-inactivating gammaretroviral and lentiviral vectors. *Mol. Ther.* 15(6), 1167–1173

33. Schambach, A., Schiedlmeier, B., Kuhlcke, K., Verstegen, M., Margison, G. P., Li, Z., Kamino, K., Bohne, J., Alexandrov, A., Hermann, F. G., von Laer, D., and Baum, C. (2006) Towards hematopoietic stem

cell-mediated protection against infection with human immunodeficiency virus. *Gene Ther.* 13, 1037–1047

34. Klump, H., Schiedlmeier, B., Vogt, B., Ryan, M., Ostertag, W., and Baum, C. (2001) Retroviral vector-mediated expression of HoxB4 in hematopoietic cells using a novel coexpression strategy. *Gene Ther.* 8, 811–817

35. Szymczak, A. L., Workman, C. J., Wang, Y., Vignali, K. M., Dilioglou, S., Vanin, E. F., and Vignali, D. A. (2004) Correction of multi-gene deficiency in vivo using a single 'self-cleaving' 2A peptide-based retroviral vector. *Nat. Biotechnol.* 22, 589–594

36. Amendola, M., Venneri, M. A., Biffi, A., Vigna, E., and Naldini, L. (2005) Coordinate dual-gene transgenesis by lentiviral vectors carrying synthetic bidirectional promoters. *Nat. Biotechnol.* 23, 108–116

37. Sandrin, V., Boson, B., Salmon, P., Gay, W., Negre, D., Le Grand, R., Trono, D., and Cosset, F. L. (2002) Lentiviral vectors pseudotyped with a modified RD114 envelope glycoprotein show increased stability in sera and augmented transduction of primary lymphocytes and CD34+ cells derived from human and nonhuman primates. *Blood* 100, 823–832

# Chapter 15

## Knock-Down of Gene Expression in Hematopoietic Cells

### Michaela Scherr, Letizia Venturini, and Matthias Eder

## Summary

RNA interference (RNAi) is an evolutionarily conserved sequence-specific post-transcriptional gene silencing mechanism triggered by double-stranded RNA (dsRNA) that results either in degradation of homologues mRNAs or inhibition of mRNA translation. The effector molecules which activate the RNAi pathway are small regulatory RNAs including small interfering RNAs (siRNAs) which are processed from longer dsRNAs by the RNAse III enzyme Dicer, and microRNAs (miRNAs) generated in a regulated multistep process from endogenous primary transcripts (pri-miRNA). Since, in principle, any gene can be silenced, RNAi provides a powerful tool to investigate gene function, and it is therefore a widely used gene silencing method in functional genomics. This chapter provides a collection of protocols for specific gene knock-down in hematopoietic cells by the application of short-hairpin RNAs (shRNAs) transcribed by RNA polymerase III (pol III) promoters or artificial-miRNAs (art-miRNAs) expressed from RNA pol II promoters using lentiviral vectors, respectively.

**Key words:** RNAi, si/shRNA, miRNA, Gene silencing, Gene transfer, Lentivirus, Hematopoietic cells.

## 1. Introduction

RNA interference (RNAi) is a biological process first discovered in the nematode worm *C. elegans* as a response to double-stranded RNA (dsRNA). Fire and Mello initially described that injection of long dsRNA into *C. elegans* led to sequence-specific degradation of the corresponding mRNA *(1)*. Later on it became clear that this silencing response due to RNAi is present in many eukaryotes *(2)*. RNAi is a regulated multistep process triggered by dsRNAs (**Fig. 1**) . Based on their origin and function, small regulatory RNAs are classified into different categories: (a) synthetic siRNAs, (b) pol III transcribed shRNAs consisting of a perfectly double-stranded stem of 19–29 bp, which is identical in

Christopher Baum (ed.), *Methods in Molecular Biology, Methods and Protocols, vol. 506*
© Humana Press, a part of Springer Science + Business Media, LLC 2009
DOI: 10.1007/978-1-59745-409-4_15

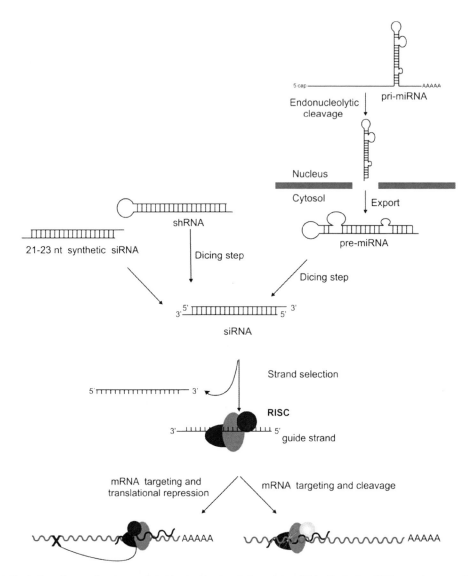

Fig. 1. Illustration of gene silencing mediated by double-stranded regulatory RNAs. Synthetic siRNAs are shown on the *left*, shRNAs which are processed by Dicer in the middle, and miRNAs cleaved by Drosha in the nucleus and exported into the cytoplasm where they are processed by Dicer are shown on the *right*. Mature siRNAs and miRNAs are incorporated into RISC (RNA-induced silencing complex) which finally mediates degradation or translational repression of target mRNAs.

sequence to the targeted mRNA linked to a loop of 6–9 nt, and (c) endogenous (or artificial) miRNAs. miRNAs are encoded in the genome and initially transcribed by RNA pol II as a part of longer primary transcripts (~1 kb or even longer, pri-miRNA). Mature miRNAs species (~22 nt) are generated from pri-miRNAs by sequential processing steps, including endonucleolytic cleavage by RNAses such as Drosha, export from the nucleus, Dicer cleavage, and strand selection *(3)*. Upon processing both siRNAs and miRNAs are loaded into the RNA-induced silencing complex (RISC) and mediate either degradation or translational attenuation of mRNA transcripts. siRNAs usually base-pair to target mRNAs with high complementarity resulting in mRNA degradation, whereas miRNAs usually bind mRNAs with imperfect complementarity leading predominantly to translational attenuation or mRNA destabilization *(4)*.

The ability to replace miRNA by shRNA sequences in miRNA genes has led to the generation of a collection of miR-30-based shRNA retroviral and lentiviral expression systems *(5, 6)*. These ectopically expressed artificial RNAs (miRNA-based shRNAs) resemble the short stem-loop structure of endogenous miRNA precursors allowing the shRNA to enter the miRNA pathway to be processed into functional siRNAs. miRNA-based shRNAs can be transcribed by either pol II or pol III promoters which allows their regulated and conditional as well as tissue-specific expression.

## 2. Materials

### 2.1. Selection of si/shRNA Sequences for RNAi

Websites that offer algorithms for site selection for a given target mRNA are http://www.dharmacon.com and http://www.ambion.com

### 2.2. Cloning of shRNA Expression Cassettes for DNA-Based RNAi

1. Synthetic oligodeoxynucleotides (ODN; sense and antisense strand).
2. T4 Polynucleotidekinase (PNK).
3. 10 mM dATP.
4. Plasmid pSUPER (DNA$_{engine}$ Inc.,Seattle, USA).
5. T4 DNA Ligase.
6. Alkaline phosphatase.
7. HB101 *E. coli* competent bacteria.
8. Plasmid DNA purification kit.
9. Restriction enzymes: BglII, SalI, XhoI, EcoRI, SnaBI, SmaI, HincII.

### 2.3. Functional Analysis of Selected shRNAs

1. BHK-21 cells (Syrian hamster kidney; DSMZ No. ACC61).
2. Dulbecco's modified Eagle's medium (DMEM).
3. Penicillin/streptomycin.
4. Fetal calf serum (FCS).
5. LipofektAMINE™ 2,000 (*see* **Note 1**).
6. Plasmids-DNA's containing (a) expression cassette for shRNA (pSUPER-shRNA or pSUPER-miR-30-shRNA), (b) pDSRed$_{EXPRESS}$ (Clontech, Mountain View, USA) encoding red fluorescent protein (or equivalent gene) as reporter gene, and (c) bi-cistronic plasmid containing the target gene X of interest pX-IRES2-EGFP (Clontech).

**2.4. Virus Generation**

Cell lines: the human embryonal kidney 293 cell line (DSMZ No. ACC 305, 293 T cells can be used as well) is used for virus production; the human chronic myeloid leukemia K562 cell line (DSMZ No. ACC 10) can be used for virus titration and functional studies. The cells are cultured in DMEM (293 cells) and RPMI (K562 cells) supplemented with 10% FCS. For studies with primary cells, CD34+ cells are cultured in X-VIVO-10 medium supplemented with 1% human serum albumin (HSA).

1. Viral vectors . Transgene plasmid carrying the shRNA transcriptional unit, packaging plasmid, and envelope plasmid.

2. Transfection buffer . (a) Make a stock solution of $Na_2HPO_4$ dibasic (5.25 g in 500 mL water). (b) Make 2 × HBS: 8 g NaCl, 6.5 g HEPES (sodium salt), 10 mL $Na_2HPO_4$ from the stock solution. (c) bring to pH 6.95–7.0 using NaOH or HCl and bring volume up to 500 mL.

3. 2 M $CaCl_2$ .

4. 100 × protaminesulfate (400 µg/mL).

5. 0.01% (w/v) polylysine.

6. Phosphate-buffered saline (PBS; pH 7.4, $Ca^{2+}$- and $Mg^{2+}$-free).

**2.5. Northern Blot Analysis of si/shRNA Expression**

1. Trizol reagent.

2. 8 M Urea, polyacrylamide gel and running buffer (1 × TBE): 89 mM Tris-borate, pH 8.3, 2 mM EDTA.

3. Gel-loading buffer: formamide containing 0.1% (w/v) xylene cyanol, 0.1% (w/v) bromophenol blue, and 20 mM EDTA.

# 3. Methods

Designing specific si/shRNAs for specific gene silencing only requires the knowledge of the target sequence (at least 20 nt) which can usually be obtained from suitable databases [http://www.ncbi.nlm.nih.gov]]. Target site selection is still a trial-and-error process. It is best performed by functional analysis of a variety of potential target sequences, although there are some websites available to offer algorithms for site selection in a given target mRNA [e.g., http://www.dharmacon.com;http://www.ambion.com]. The most efficient RNAi triggers today are double-stranded RNAs of 21–23 nt containing overhangs of 2 nt at both 3 '-ends that are selected according to noncomplementarity to off-target sequences and to thermodynamic differences between the two siRNA ends to preferentially drive antisense strand recruitment into RISC (7,8).

siRNA-based RNAi can be triggered from mammalian expression vectors that direct the synthesis of shRNA transcripts. The most commonly used polymerase III promoters are the eukaryotic H1 and U6 snRNA promoters to transcribe shRNAs (*see* **Note 2**). They produce high levels of small, noncoding RNA transcripts lacking the polyA-tail and initiate transcription from position +1 without any restriction by an inhibitory 5′-nucleotide. As shown in **Fig. 2a** , the shRNA contains a perfectly double-stranded stem of 19–21 bp that is identical in sequence to the target mRNA, the two strands of the stems are linked by a usually 9-bp long loop sequence, and the shRNA is processed by Dicer to generate functional active siRNA as described *(9)*.

miRNA-based shRNA expression cassettes can be used to express artificial miRNAs consisting of a stem of 19–29 bp linked to a 5–15 nt loop (**Fig. 2b**). The resulting RNA mimics an

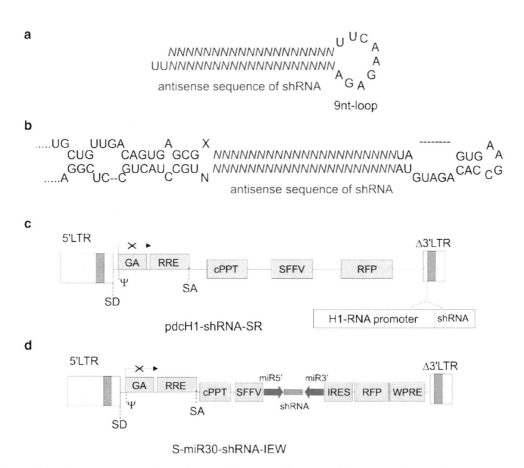

Fig. 2. Schematic representation of heterologous si/shRNA variants and lentiviral transgene plasmids (**a**) illustration of a shRNA transcript, (**b**) illustration of a heterologous si/shRNA sequence embedded in a miRNA backbone, (**c**) illustration of a lentiviral vector encoding a pol III expression cassette in the 3′LTR, (**d**) illustration of a lentiviral vector encoding a pol II miRNA expression cassette (here SFFV: spleen focus forming virus LTR).

intermediate in the miRNA maturation pathway - the precursor miRNA (pre-miRNA) – allowing these artificial miRNA to enter the miRNA pathway and to be processed by Dicer to produce siRNA/miRNAs. We use the miR-30-based shRNA expression system described by several groups (6, 10–12).

Before si/shRNAs are cloned into viral vectors to be used for gene silencing si/shRNAs should be functionally evaluated. For this we developed a rapid and reliable protocol to examine the functional activity of si/shRNAs based on a cotransfection assay (13). This assay is based on FACS analysis of GFP-fluorescence with green fluorescent protein (GFP) encoded along with the target gene on a bi-cistronic transcript. (Similar systems to study shRNA efficacy are commercially available.) These bi-cistronic reporter plasmids and plasmids encoding the red fluorescent protein (RFP, for normalization) are cotransfected into BHK-21 cells with either control- or specific-shRNA or miR-30-shRNA expression cassettes, respectively. After incubation for 24 h cells are monitored by flow cytometry (FACS) or fluorescence microscopy. Based on the functional testing, the two most efficient RNAi triggers should be used to silence gene expression.

Although the si/shRNA expression vector can be delivered with the aid of lipid-based reagents, induction of gene silencing is only transient which may last from 3 to 8 days depending on cell cycle kinetics of the target cells. More promising and efficient technologies employ retro-/lentiviral vectors. Retro/lentiviruses can infect a wide variety of cell types where they integrate into the host genome. In addition, expression of si/shRNAs upon retro/lentiviral gene transfer may be advantageous as it can be used for studies in primary hematopoietic cells which are usually difficult to transfect. Furthermore, as lentiviral vectors are able to integrate into the host genome of nondividing cells, such as stem cells, the range of target cells is expanded as compared to retroviral delivery of RNAi triggers. As shown in **Fig. 2c**, the H1-shRNA expression cassettes can be inserted into lentiviral transgene plasmids dcH1-shRNA-SR or dcH1-shRNA-SEW (dc double-copy, S SFFV promoter, R RFP, E EGFP, W WPRE) within the U3 region of the 3′-LTR (dcH1-shRNA-SEW or dcH1-shRNA-SR, with the GFP or RFP reporter gene). The location in the U3 region of the 3′-LTR leads to a duplication of the H1-cassette during reverse transcription. In contrast, the miR-30-shRNA expression cassette should be inserted into the bi-cistronic lentiviral transgene plasmid SIEW (I internal ribosome entry site) 3′ of the internal SFFV promoter (**Fig. 2d**).

Reporter gene activity allows the rapid and quantitative analysis of transduction rates and, indirectly, si/shRNA expression in living cells. Recently, we demonstrated that reporter gene activity correlates to the number of lentiviral integrations into the host cell genome, to the level of siRNA expression, and to the extent of target gene silencing *(13)*. Furthermore, lentiviral transduction of target cells results in a "random" distribution of viral integrations in the host cell genome and leads to cell populations with heterogenous gene silencing efficacy. This can be monitored by FACS and may allow to track selection processes of gene-modified cells *(14)*. The expression of si/sh RNA can be tested by Northern blotting (*see* **Subheading 3.5**), and the ability of si/shRNA to inhibit target mRNA and protein expression can be tested by quantitative RT-PCR and Western blot or FACS analysis, respectively (*see* **Subheading 3.6.**)

### 3.1. Selection of si/shRNA Sequences for RNAi

1. Sequence requirements for the design of effective si/shRNAs have been described *(15, 16)*. Regions of the mRNA such as introns, 5′ and 3′ untranslated regions (UTRs), and regions around the AUG start codon should not be selected as si/shRNAs.

2. Sequences with a GC content of > 50% should not be chosen as si/shRNAs.

3. 5′-antisense strand of a highly effective siRNAs should be A or U.

4. 5′-sense strands are preferably G or C.

5. At least four out of seven nucleotides at the 5′-end of the antisense strand may be A or U.

6. Avoid any GC stretch of more than nine nucleotides in length.

7. BLAST search [http://www.ncbi.nlm.nih.gov/BLAST] the selected si/shRNA sequences against mRNA sequences or EST libraries carefully to prevent complementarity to other related or unrelated sequences in cellular mRNAs.

8. Synthesize at least two different si/shRNAs against a given target mRNA to control for specificity. Specific and effective si/shRNAs should give identical phenotypes. Furthermore, at least one scrambled or irrelevant si/shRNA is usually needed as negative control (*see* **Note 3**).

### 3.2 Transcription of RNAi Trigger in Mammalian Cells

#### 3.2.1. Cloning of Pol III-shRNA Expression Cassettes for DNA-based RNAi

Single-stranded ODNs (~65–69 nt in length) encoding (a) a 19–21 nt siRNA (sense strand italic), (b) a noncomplementary 9-nt loop-sequence (underlined), (c) a 19–21 nt siRNA (antisense strand italic), (d) a termination signal consisting of six thymidines, and (e) overhang sequences from a 5′ *Bgl*II- and a 3′ *Sal*I- restriction site, are phosphorylated and hybridized with the corresponding complementary single-stranded ODN. The ODN sequences can be designed as follows: sense-shRNA: 5′-GATCCC($N$)$_{19-21}$ TTCAAGAGA ($N$ ′)$_{19-21}$ TTTTTT-GGAAG-3′, antisense-shRNA: 5′-TCGACTTCCAAAAAA ($N′$)$_{19-21}$ TCTCTTGAA ($N$)$_{19-21}$ GGG -3′. The resulting dsDNA is inserted into the *Bgl*II–*Sal*I site of the dephosphorylated plasmid pSUPER (suppression of endogenous RNA, (17)) harboring the H1-RNA promoter to generate pSUPER-shRNA. *E. coli* HB101 competent cells are transformed with the resultant plasmids, followed by plasmid DNA purification, and the correct sequence and insertion is confirmed by DNA sequencing for each plasmid.

*3.2.2. MicroRNA-based Systems*

Single-stranded ODN (~110 nt in length) encoding (a) 28 nt 5′-miR-30 miRNA sequence, (b) a 21 nt siRNA followed by TA (sense strand *italic*), (c) a noncomplementary 15-nt loop-sequence followed by TA (underlined), (d) a 22 nt siRNA (antisense strand italic), (e) 14 nt 3′-miR-30 miRNA sequence, and (f) overhang sequences from a 5′ *Xho*I- and a 3′ *Eco*RI- restriction site are phosphorylated and hybridized with the corresponding complementary single-stranded ODN (**Fig. 2b**). The ODN sequences can be designed as follows: sense-art-miRNA: 5′-TC GAGAAGGTATATTGCTGTTGACAGTGAGCGA($N$)$_{21}$TA$\underline{G}$ $\underline{TGAAGCCACAGATGTA}$($N'$)$_{22}$ TGCCTACTGCCTCGG-3′, antisense-art-mi-RNA: 5′-AATTCCGAGGCAGTAG GCA($N'$)$_2$ $_2\underline{TACATCTGTGGCTTCAC}$TA($N$)$_{21}$TCGCTCACTGTCAACA GCAATATACCTTC-3 ′. The resulting dsDNA is inserted into the *XhoI –EcoRI* site of the dephosphorylated plasmid SUPER-miR-30 containing the H1 promoter and the miR-30 cassette to generate pSUPER-miR-30-shRNA. E. coli HB101 competent cells are transformed with the resultant plasmids, followed by plasmid DNA purification, and the correct sequence and insertion is confirmed by DNA sequencing for each plasmid.

Phosphorylation and Annealing of ODN

1. Dissolve ODN in water to obtain a final concentration of 1 mM.
2. Take 1 µL from each ODN (sense- and antisense-shRNA or –miR-30-shRNA).
3. Add 1 µL 10× T4 PNK buffer.
4. Add 1 µL 10 mM ATP.
5. Add 1 µL T4 PNK (10 U/µL).
6. Add 5 µL H$_2$O.
7. Incubate 30–45 min at 37°C, followed by a denaturing step at 100°C for 3 min and cool slowly the annealed ODN down to room temperature.

Ligation of the Duplex in pSUPER or pSUPER-miR-30 Plasmid

1. Take 3 µL of the duplex (annealing reaction).
2. Add 1 µL 10× T4 ligase buffer.
3. Add 1 µL dephosphorylated plasmid pSUPER (digested with BglII and SalI) or pSUPER-miR30 (digested with XhoI and EcoRI).
4. Add 1 µL T4 DNA Ligase (400 U/µL).
5. Add 4 µL H$_2$O.
6. Incubate 1 h at room temperature.

7. After transformation in E. coli HB101 competent cells and plasmid preparation, the insert can be analyzed by EcoRI/XhoI digestion with positive clones identified by inserts of ~360 bp for the plasmid pSUPER-shRNA and ~110 bp for pSUPER-miR-30-shRNA, respectively.

**3.3. Functional Analysis of Selected shRNAs Based on a Cotransfection Assay in Mammalian Cells**

1. BHK-21 cells are grown in a 5% $CO_2$-humidified incubator at 37°C in DMEM supplemented with 10% FCS, 100 U/mL penicillin, and 100 µg/mL streptomycin. $1 \times 10^5$ cells/wells are seeded in a 24-well plate. 16 h after seeding the cells should have a confluency of ~60–80%.

2. Mix three different plasmids. 1 µg pX-IRES2-EGFP containing the target gene of interest X, 0.1 µg pDSRed$_{EXPRESS}$, and 1 µg shRNA expression vector pSUPER-shRNA or pSUPER-miR-30-shRNA in 50 µL DMEM without serum. In a separate tube add 2 µL LipofectAMINE™2000 (*see* **Note 1**) to 50 µL DMEM and mix the tube gently by inverting followed by incubation for 5 min at room temperature.

3. Combine both suspensions, mix gently and incubate for 20 min at room temperature to allow the formulation of the liposome complex. Add the DNA/liposome complex to the cells without replacing the medium and shake the well-plate gently. Incubate the plate for 24–40 h at 37°C in the incubator.

4. To quantify reporter fluorescence, the cells are removed from the well and GFP/RFP fluorescence can be monitored by FACS or fluorescence microscopy.

**3.4. Stable Knock-down of Target Gene Expression by si/shRNAs**

*3.4.1. Lentivirus-mediated Delivery of shRNA or miR-30-based-shRNA Expression Cassettes*

To construct the double-copy variant dcH1-shRNA-SR or dcH1-shRNA-SEW, the plasmids dcH1-SR or dcH1-SEW should be digested with SnaBI followed by dephosphorylation. H1-shRNA cassettes are released from pSUPER-shRNA by digestion with SmaI and HincII. To construct the lentiviral miR-30-based-shRNA expression cassette, the plasmid SIEW is digested with BamHI followed by a polymeraseI fill-in reaction and dephosphorylation. The miR-30-shRNA cassette is released from pSUPER-miR-30-shRNA by digestion with BglII and EcoO109I followed by a polymerase-fill in reaction.

1. Take 1 µL (~400 ng) H1-shRNA or miR30-shRNA-fragment.

2. Add 1 µL 10× T4 ligase buffer.

3. Add 1 µL (100 ng) dephosphorylated plasmid dcH1-SR/dcH1-SEW or SIEW.

4. Add 1 µL T4 DNA Ligase (400 U/µL).

5. Add 6 µL $H_2O$.

6. Incubate 1 h at room temperature.

7. After transformation in E. coli HB101 competent cells and plasmid preparation, the insert can be analyzed by EcoRI

digestion with positive clones identified by inserts of ~1,300 bp for the plasmid dcH1-shRNA and ~600 bp for S-miR30-shRNA-IEW, respectively

*3.4.2. Generation of Lentiviral Vector Particles*

1. 293 cells are grown in a 5% $CO_2$-humidified incubator at 37°C in DMEM supplemented with 10% FCS, 100 U/mL penicillin, and 100 µg/mL streptomycin.

2. Three days before transfection, plate $6 \times 10^6$ 293 cells onto a poly-l-lysine (0.01% solution)-coated T175 flask in 25 mL DMEM.

3. Transfect the subconfluent 293 cells with 60 µg of the lentiviral transgene plasmid, 45 µg packaging-, and 30 µg envelope-plasmid DNA using the calcium phosphate method *(18)*.

4. After 16 h wash the cells with 1× PBS (pH 7.4, $Ca^{2+}$- and $Mg^{2+}$-free) and incubate for additional 8 h in fresh medium.

5. Collect the culture supernatant containing the lentiviral particles 24–72 h after transfection.

6. Filter through a 0.45-µm pore size filter to remove cell debris, and concentrate the viral particles by low-speed centrifugation at 10.000 × g, 10°C for 16 h.

7. Titrate the viral particles, e.g., on K562 cells and store at –80°C.

8. Use culture medium as indicated in **Subheading 2**. (*see* **Note 5**).

*3.4.3. Transduction and Stable Knock-down of Gene Expression in Hematopoietic Cells*

1. $1 \times 10^6$ K562 cells or CD34+ cells are transduced twice in a 24-well plate by adding 50 µL of lentivirus ($10^7$ viral particles), 5 µL protamine sulfate (final concentration 4 µg/mL) in a total volume of 500 µL (multiplicity of infection; MOI ~10). Transduction of CD34+ cells should be done on retronectin-coated well plates.

2. Spin occulation at 2.000 × g at 32°C for 90 min followed by an overnight incubation at 37°C in an incubator.

3. After transduction, remove the supernatants from the cells, maintain them in RPMI/10%FCS, and monitor the transduction efficacy on day 4.

*3.4.4. FACS Analysis*

As shown in **Fig. 3** (an example according to Scherr et al. *(19)*), FACS analysis of RFP expression in K562 cells transduced with specific anti-bcr-abl (lower panels) and control (upper panels) shRNA lentiviruses reveals a distribution of RFP fluorescence with a range of approximately 1,000 in transduced cells ($10^1$–$10^4$) on day 4 (left panels). This heterogeneity in RFP expression remains unchanged for ~3 weeks in K562 cells transduced with control shRNA encoding lentiviruses (MFI = 735 and 732). In contrast, in cultures of K562 cells transduced with anti-bcr-abl shRNAs the percentage of RFP-negative cells increases, and RFP- mean

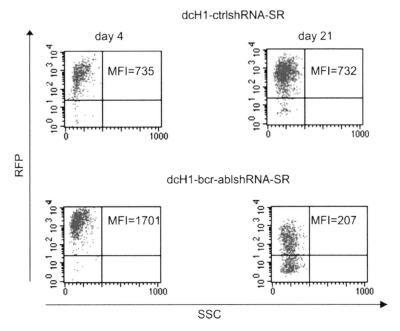

Fig. 3. FACS profile of reporter fluorescence of control (top) and anti-bcr-abl (bottom) shRNA expressing K562 cells at early and late time points. Note the change in RFP expression of K562 cells upon anti bcr-abl-RNAi *(14, 18)*.

fluorescence intensities (MFI) are diminished by ~eightfold during suspension culture (MFI = 1,701 and 207). Almost no RFP-high expressing cells (> $10^3$) are found after 3 weeks (negative selection during cell culture, for more details *see* refs. *14, 19*).

**3.5. Northern Blot Analysis of si/shRNA Expression**

Total cellular RNA is extracted from vector-transduced K562 cells and analyzed for the expression of si/shRNA by Northern blotting.

1. Isolate total cellular RNA from the transduced K562 cells, for example using the Trizol reagent.

2. Subject 10–15 µg of total cellular RNA to electrophoresis in a 16% denaturing polyacrylamide gel containing 8 M urea.

3. Transfer to Hybond-N nylon membrane by electroblotting.

4. Probe the blots with $^{32}$P-labeled ODNs specific for the corresponding siRNA and specific for the endogenous U6 snRNA serving as an internal size standard as well as a loading control, respectively (U6sn probe: 5′-TATGGAACGCTTCACGAATTTGC-3′).

**3.6. Detection of Lentiviral-Mediated Gene Silencing**

For in vivo analysis of gene silencing activity, standard techniques are used such as quantitative RT-PCR and Northern blotting at the levels of RNA and, e.g., Western blotting at the protein level. Phenotypic analysis of proliferation, survival, and differentiation of transduced cells can be easily monitored by several standard techniques such as trypan-blue exclusion, BrdU incorporation or apoptose-assays, or morphological and immunophenotyping analyses, respectively.

## 4. Notes

1. Reagents functionally similar to LipofectAMINE™2000 like DOTAP (Roche), Transfectam RM (Promega), or Effectene (QIAGEN) can also be utilized.

2. It is important to consider the type of pol II or pol III promoter to be used depending on the study design. Both pol II promoters such as CMV, EF-1α and pol III promoters like U6 and H1-RNA have been successfully employed for si/shRNA expression and knock-down experiments.

3. It is recommended to include an si/shRNA directed against an endogenous mRNA which shows no or a different phenotype and a vector-alone control in initial testing of si/shRNA constructs.

4. The use of recombination deficient bacteria (such as HB101) is crucial for successful cloning of lentiviral plasmids.

5. Production of high-titer lentiviral supernatants may require some training and experience (see related protocols). Make sure that the cells used for virus production are free of mycoplasm contamination.

## Acknowledgments

Supported in part by grants of the "Deutsche Forschungsgemeinschaft" (SFB 566), H.W. & J. Hector-Stiftung, Wilhelm Sanders-Stiftung

## References

1. Fire , A., Xu, S., Montgomery, M.K., Kostas, S.A., Driver, S.E., and Mello, C.C. (1998) Potent and specific genetic interference by double-stranded RNA in Caenorhabditis elegans. *Nature* 391, 806–811

2. Hannon, G.J. (2002) RNA interference. *Nature* 418, 244–251

3. Cullen, B.R. (2004) Transcription and processing of human microRNA precursors. *Mol Cell* 16, 861–865

4. Wu, L., Fan, J., and Belasco, J.G. (2006) Micro-RNAs direct rapid deadenylation of mRNA. *Proc Natl Acad Sci U S A* 103, 4034–4039

5. Zeng, Y., Cai, X., and Cullen, B.R. (2005) Use of RNA polymerase II to transcribe artificial microRNAs. *Methods Enzymol* 392, 371–380

6. Stegmeier, F., Hi, G., Rickles, R.J., Hannon, G.J., and Elledge, S.J. (2005) A lenti-viral microRNA-based system for single-copy polymerase II-regulated RNA interference in mammalian cells. *Proc Natl Acad Sci U S A* 102, 13212–13217

7. Schwarz, D.S., Hutvagner, G., Du, T., Xu, Z., Aronin, N., and Zamore, P.D. (2003) Asymmetry in the assembly of the RNAi enzyme complex. *Cell* 115, 199–208

8. Khvorova, A., Reynolds, A., and Jayasena, S.D. (2003) Functional siRNAs and miRNAs exhibit strand bias. *Cell* 115, 209–216

9. Paddison, P.J. and Hannon, G.J. (2002) RNA interference: the new somatic cell genetics? *Cancer Cell* 2, 17–23

10. Zeng, Y., Wagner, E.J., and Cullen, B.R. (2002) Both natural and designed micro RNAs can inhibit the expression of cognate mRNAs when expressed in human cells. *Mol Cell* 9, 1327–333

11. Lee, Y., Kim, M., Han, J., Yeom, K.H., Lee, S., Baek, S.H., and Kim, V.N. (2004) MicroRNA genes are transcribed by RNA polymerase II. *EMBO J* 23, 4051–4060

12. Silva, J.M., Li, M.Z., Chang, K., Ge, W., Golding, M.C., Rickles, M.C., Siloas, D., Hu, G., Paddison, P.J.et al. (2005) Second-generation shRNA libraries covering the mouse and human genomes. *Nat Genet* 37, 12811288

13. Scherr, M., Battmer, K., Ganser, A., and Eder, M. (2003) Modulation of gene expression by lentiviral-mediated delivery of small interfering RNA. *Cell Cycle* 2, 251–257

14. Scherr, M. and Eder, M. (2007) Gene silencing by small regulatory RNAs in mammalian cells. *Cell Cycle* 6, 444–449

15. Elbashir, S.M., Harborth, J., Weber, K., and Tuschl, T. (2002) Analysis of gene function in somatic mammalian cells using small interfering RNAs. *Methods* 26, 199–213

16. Ui-Tei, K., Naito, Y., Takahashi, F., Haraguchi, T., Ohki-Hamazaki, H., Juni, A., Ueda, R., and Saigo, K. (2004) Guidelines for the selection of highly effective siRNA sequences for mammalian and chick RNA interference. *Nucleic Acids Res* 32, 936–948

17. Brummelkamp, T.R., Bernards, R., and Agami, R. (2002) A system for stable expression of short interfering RNAs in mammalian cells. *Science* 296, 550–553

18. Graham, F.L. and van der Eb, A.J. (1973) Transformation of rat cells by DNA of human adenovirus 5. *Virology* 54, 536–539

19. Scherr, M., Battmer, K., Schultheis, B., Ganser, A., and Eder, M. (2005) Stable RNA interference (RNAi) as an option for anti bcr-abl therapy. *Gene Ther* 1, 12–21

# Chapter 16

# The Use of Retroviral Vectors for tet-Regulated Gene Expression in Cell Populations

## Rainer Löw

## Summary

Today the treatment of inherited diseases holds a major field in gene therapy, and γ-retroviral vectors are often the preferred tool for stable introduction of the therapeutic gene(s) into the host cell genome. In many cases, the newly introduced gene has to be constitutively expressed, since enzyme function often is required at all times. However, in some cases gene function might be demanded only transiently, making a strict control of gene expression necessary. For more than a decade, the tet-system has proven to facilitate such strict control by tightly regulating gene expression, thereby assuring high expression levels in almost all organs and tissues. Yet, most of these results were obtained from the analysis of either selected cell clones or transgenic animals. On the contrary, in case of conditional gene expression, as necessary for gene therapy approaches, the use of genetically modified cell populations, where the majority of cells display similar regulatory properties, is required. Therefore, great effort has been undertaken to design viral vectors carrying the response unit that enables homogenous regulation of gene expression in transduced cell populations. This article summarizes critical points that have to be considered for the conditional regulation of gene expression in cell populations mediated by the tet-system. Examples of the required vector elements and tet-system components as well as advice on the handling of the system are given. These tools have been specifically developed to improve population-based gene regulation.

**Key words:** tet-system, Retroviral vector, Uni and bidirectional vectors, pol/env fragment, tet-transactivators, Minimal promoters, Dose response, tet-resistance.

## 1. Introduction

It is reasonable to assume that in future regulated expression of therapeutic genes will gain more importance, at least for some applications in gene therapy, and that the already widely established tet-system *(1)* will be further extended. In the tet-system, control elements of the tetracycline-resistance operon of *E. coli*

Christopher Baum (ed.), *Methods in Molecular Biology, Methods and Protocols, vol. 506*
© Humana Press, a part of Springer Science+Business Media, LLC 2009
DOI: 10.1007/978-1-59745-409-4_16

were converted into eukaryotic transcription activation elements by fusions between the *E. coli* tet-repressor and a the VP16 trans-activation domain. This transactivator binds to a tet-operator sequence fused to a minimal promoter ($P_{tet}$) either in the absence (tet-off) or the presence (tet-on) of the effector tetracycline or its derivative doxycycline (Dox). Fusion of a second minimal promoter to $P_{tet}$ generates bidirectional responsive promoters ($P_{tet}$bi) that allow simultaneous expression of two genes *(2)*. The widespread use of the tet-system demonstrates its unique properties among the conditional expression systems available, in particular its ability to tightly regulate gene expression in almost all cells, while still exhibiting cell-type-specific characteristics. Its function solely depends on the availability of the respective transactivator and effector (tetracycline or derivatives of it, e.g., Dox). The latest design of tet-dependent transactivators (e.g., stTA-2, S2, M2) adds further options on the use of the tet-system, since it is now available as either "tet-on" or "tet-off" system, each showing equal sensitivity and regulatory properties *(3)*.

The application of the tet-system in the investigation of gene function differs significantly from its use in gene therapy approaches. Although the former involves the selection of either cells or transgenic animals, thereby sorting out an individual that bears the desired properties, a successful gene therapy approach does not allow for the selection of an individual but instead results in cell populations where individual cells display similar and stable properties in terms of gene expression and regulation. In case of gene therapy approaches, the level of unregulated background expression as well as the maximum level of induced expression is of importance, whereas tight regulation and high expression levels are not always essential.

Transfer of the tet-system components can be achieved by several techniques, among which retroviral vectors, based on the murine leukemia virus (MLV), or lentiviral vectors, based on HIV-1 or HIV-2, are the most promising since they integrate stably into the target cell genome. Moreover, gene dosage can be controlled by multiplicity of infection (MOI) and rearrangements of the host chromosome upon introduction of the transgene are avoided, since its integration is a highly specific, enzyme-mediated process.

Tight regulation as well as high level gene expression can be achieved by the use of a two-vector system, where one vector carries a constitutively expressed tet-dependent transactivator, while the other bears the response unit for conditional expression of the therapeutic gene. The major hurdle of the two vector system is its requirement for two independently occurring integration events, a fact that might not be sensible for therapeutic approaches, as it can only be achieved through prolonged "in vitro" culture of the primary target cells. Today the only solution to this problem is to apply both vectors at the same time, which will result basically in the generation of four different cell populations: untransduced wild type cells, cells transduced with only one of the two vectors, and cells that carry both vectors. By using a high MOI, the risk of generating varying populations can be minimized; however, multiple integration of the response unit is favored, which will eventually lead to higher background expression thereby reducing the regulation factor *(4)*.

To overcome this problem, one-vector systems simultaneously transferring both tet-system components, transactivator as well as response unit, have been designed. In general, two types can be distinguished (i) the "self-contained" one-vector systems carrying both a $P_{tet}$-controlled response unit as well as an tTA or rtTA encoding unit driven by an independent promoter in the same viral backbone *(5, 6)* and (ii) the "autoregulated" one-vector systems where the transactivator expression is controlled via a positive feedback loop *(7–9)*. Nevertheless, both systems pose problems. The self-contained system might be hampered due to promoter interference between the constitutive and the regulated promoter leading to low regulation and expression levels, whereas the autoregulated system suffers from its dependence on read-through to get the system started, thus inhibiting tight regulation. Furthermore, since the amount of transactivator can not be controlled independent of the transgene product, selected transduced cells may be subject to squelching *(10, 11)*.

In the following, methods that address critical issues in the successful application of tet-regulated gene expression following retroviral vector transfer into cell populations are described. All methods focus on the optimal design of the response unit and rely on cell lines providing constitutively expressed tet-dependent transactivators.

## 2. Materials

### 2.1. Cells and Cell Culture

1. 293T cells (ATCC CRL-11268), large T transformed human embryonic kidney (HEK293) cell line.

2. Hela cells (ATCC CCL-2), cervix epithelial cell line.

3. HtTA-1 cells, Hela cells modified to constitutively express the authentic tTA transactivator, which is inactive in the presence of 100 ng Dox/mL ("tet-off").

4. Hela-M2 cells, Hela cells modified to constitutively express the reverse M2 transactivator, which is fully active in the presence of 300 ng Dox/mL ("tet-on").

5. B16F0.R-S2/M2 cells, mouse melanoma cell line (ATCC CRL-6322) modified to constitutively express either the S2 reverse transactivator or the M2 reverse transactivator (M2 fully active in the presence of 3,000 ng Dox/mL ("tet-on").

6. MDA-MB231.Ro cells, human breast cancer cell line (ATCC HTB-26), modified to constitutively express the S2 reverse transactivator, which is fully active in the presence of 1,000 ng Dox/mL ("tet-on"). Cultivation is performed at 37°C without $CO_2$.

7. Fetal bovine serum (FBS; PAA, Pasching, Austria), tested for absence of tetracyclines in the X-1.5 cell line (12) by measurement of luciferase induction.

8. Dulbecco's Modified Eagle's Medium (DMEM, Invitrogen, Carlsbad, USA) supplemented with 10% fetal bovine serum, for cultivation of 293T cells.

9. Earle's Modified Eagle's Medium (MEM, Invitrogen, 41090–028) supplemented with 10% fetal bovine serum, for cultivation of HtTA-1 and Hela-M2 cells.

10. Dulbecco's Modified Eagle's Medium (DMEM, Invitrogen, 61965) supplemented with 10% fetal bovine serum, 4 mM l-glutamine (Invitrogen, 25030–024), 1.5 g/L sodium bicarbonate and 4.5 g/L glucose, for cultivation of B16-F0 cells.

11. Leibovitz L-15 medium (Invitrogen, 11415–049), supplemented with 10% FCS and 2 mM l-glutamine, for cultivation of MDA-MB231 cells.

12. Phosphate buffered Saline, (1× PBS, Invitrogen, 70013–016).

13. Sodium ethylenediamine tetraacetic acid (Na$_2$-EDTA, SIGMA, St. Louis, USA, E-6511), stock solution 0.5 M in distilled water and sterile filtrated (*see* **Note 1**).

14. PBS/EDTA: EDTA is added to a final concentration of 0.8 mM in 1× PBS. Both the reagents were stored at room temperature.

15. Penicillin/Streptomycin solution (Invitrogen, 15140–122) is exclusively added to culture media (100 U/mL) used for the enrichment of cell populations by fluorescence-activated cell sorting (FACS) and during transfection experiments.

16. Ciprofloxacine solution (Fagron GmbH, Barsbüttel, Germany, 133659) is exclusively added to culture media (10 μg/mL) used for the enrichment of cell populations by FACS.

17. G418 (Bio Whittaker, Walkersville, USA, 15–394N) working solution (100 mg/mL active substance) in distilled water to be used for selection of cell populations.

18. Doxycycline hydrochloride (Merck, Darmstadt, Germany, 324385) is prepared as a 10 mg/mL stock solution in distilled water, filter sterilized, and aliqouted to 1 mL batches that are stored at –20°C. The working solution (1 mg/mL), obtained by diluting the stock solution in sterile, distilled water, can be stored at 4°C for about four weeks.

**2.2. Preparation of Viral Vector Stocks**

1. Dulbecco's Modified Eagle's Medium (DMEM, Invitrogen, 61965) supplemented with 10% fetal bovine serum.

2. Transfection reagent, TransIT-293 (Mirrus, Fisher Scientific, Pittsburgh, USA, MIR-2700).

3. Na-Butyrate (SIGMA, 135887) working solution 0.5 M prepared in distilled water and filter sterilized. The solution can be stored at 4°C for about 6 month. Long-term storage is carried out at –20°C. Final concentration in cell culture medium is 5 mM.

4. Polybrene (Hexadimethrine Bromide, SIGMA, H-9268) working solution 8 mg/mL in distilled $H_2O$, filter sterilized. Solution can be stored at 4°C for about one year. The reagent is used at 5–8 µg/mL final concentration in cell culture medium.

5. 0.45 µm filter (PALL, PN-4614) and 5–20 mL syringe.

**2.3 Fluorescence-Based Purification and Analysis of Cell Populations**

1. Propidium Iodide (SIGMA, P-4864) working solution is 100 µg/mL in PBS.

2. Primary antibody: Anti-erbB2 mAB (Alexis, ALX-804–573). For the working solution the primary antibody is diluted at 2 µg/mL in PBS, supplemented with 2% (w/v) bovine serum albumin (BSA, SIGMA, A-3059).

3. Secondary antibody: donkey anti mouse-IgG FITC conjugated (Affipure, Jackson Immuno Res., Suffolk, UK). For the working solution 10 µL secondary antibody are added to 100 µL PBS, supplemented with 2% (w/v) bovine serum albumin.

4. Analysis of labeled cells is done on FACS-Calibur (Beckton Dickinson Heidelberg, Germany). Data analysis is performed with Cellquest-Pro (Beckton Dickinson) software.

**2.4. Determination of Luciferase Activity and Total Protein Content**

1. Lysis buffer: 25 mM Tris-phosphate, pH 7.8 (Tris-base, pH-adjusted with $NaH_2PO_4$), 2 mM EDTA, 5% (v/v) gycerol, 1% (v/v) Triton X-100. The basic solution can be stored at –20°C for more than 5 years. Before use, the lysis buffer is supplemented with 20 mM DTT (from 1 M stock).

2. Determination of luciferase activity requires the following solutions:

    a. The assay buffer for measurement of luciferase activity (15 mM MgSO$_4$, 25 mM Glycylglycin, 5 mM rATP). (*see* **Notes 2** and **3**)

    b. rATP (SIGMA, A-6419) stock stock solution (250 mM) in distilled H$_2$O, pH 7.5 (*see* **Note 4**).

    c. A stock solution (25 mM) of luciferin (Applichem, A-1029) is prepared by dissolving the reagent in 25 mM NaOH. Aliquots (200 μL) are stored at –20°C. The working solution (125 μM) is obtained by diluting an aliquot in 40 mL of distilled water. It can be stored at –20°C for at least 12 months.

3. Determination of total protein content is carried out by use of Coomassie Plus reagent (Pierce, Thermo Fisher Scientific, Karlsruke, Germany, 1856210) according to the method of Bradford *(13)*. Measurement is performed at 595 nm in a photometer.

4. Berthold Lumat LB9507 (Applied Biosystems) is used for determination of luciferase activity. In detail, 125 μL of substrate are injected into 250 μL assay buffer. Sample volume ranges from 0.1 to 10 μL, depending on luciferase activity. Measure time is set to 10 s.

### 2.5. Preparation of Total RNA for Northern Analysis

1. RNA-extraction: Trizol reagent (Invitrogen 15596–026), Chloroform (SIGMA C-2432), Formamide (SIGMA, F-9037).

2. Denaturing agarose gel: 1% Agarose (Invitrogen, 15510–027) in 1× MOPS (20 mM MOPS (SIGMA, M1254), 5 mM Na-Acetate, 1 mM Na$_2$-EDTA) and 2.2 M formaldehyde (Merck, K-29480703–130).

3. Sample preparation: Following extraction, the total RNA is dissolved in formamide (SIGMA, F-7503) instead of water, and the appropriate amount of RNA is adjusted to 10.25 μL with formamide. After addition of 4.75 μL loading buffer (2.5 μL formaldehyde, 0.75 μL 20× MOPS and 1.5 μL 50% gylcerol/Bromphenol blue), samples are denatured at 65°C for 10 min and snap cooled on ice for about 5 min prior loading of the RNA onto a denaturing agarose gel.

## 3. Methods

Tet-regulated gene expression can be influenced at various levels. For example, the regulatory properties of the system are strongly affected by the design of both the transactivator expression unit and the corresponding tet-regulated promoter unit. Another important aspect is the chromosomal insertion site. Since $P_{tet}$ is highly susceptible to activation from neighboring enhancers, insertion at a particular site might result in unregulated background activity. At the same time, the integration site might as well affect the level to which $P_{tet}$ can be activated by tTA or rtTA.

Unregulated background expression, also known as "leakiness," has often been regarded as an intrinsic property of the tet-regulated minimal promoter in use. However, one has to keep in mind that tightly regulated clones can almost always be selected out of those "leaky" cell populations. Therefore, we assume that rather than being an intrinsic property of the system, "leakiness" reflects the promoter's sensitivity to outside activation (1, 14).

Recent findings revealed that integration of retroviral vectors within the host genome preferably occurs in proximity to promoter sites (15), thereby increasing the risk of unwanted activation of the $P_{tet}$ in the off-state by cellular enhancers. Likewise, the probability to obtain tightly-regulated cell populations might decrease when multiple copies of a given response unit have been integrated, because the number of promoters able to respond to activation from outside is increased. Consequently, transduction of target cells should be performed at low MOI, ensuring that only a small copy number (usually 1–2) per genome is obtained. However, this procedure can not readily be applied to gene therapeutic ex-vivo approaches, where enrichment of the transduced cells often can not be achieved – although this might be overcome by use of a therapeutic transgene that provides a selective advantage. The severity of the problem becomes clearly evident in cases where either large numbers of transduced cells are required or the use of an in vivo strategy is demanded.

Since we started to work on the use of retroviral vectors, we have developed various tools that will improve the overall performance of tet-regulation in cell populations (Loew and Bujard, unpublished). For instance, viral elements, such as the constitutive transport element (CTE) of simian retrovirus 1 (16), have been identified as necessary for antisense orientated transcription (**Fig. 1a**). With respect to the pol/env fragment of MLV, which harbors the native splice acceptor, a function at DNA-level has been revealed. So far the latter function of the fragment has not been described elsewhere, but its relevance is strongly supported by the following observations: (i) it has been shown to protect multiple tet-regulated minimal promoters from outside activation in both uni and bidirectional vectors (**Fig. 1b**); and (ii) it has been shown to direct the transcriptional machinery to the correct transcriptional start site in such promoters (**Fig. 1c**). Unlike the elements described so far, the use of the following elements might be optional. For instance, the posttranscriptional regulatory element of woodchuck hepatitis virus (17) was shown to improve RNA polyadenylation, but did not influence the regulatory

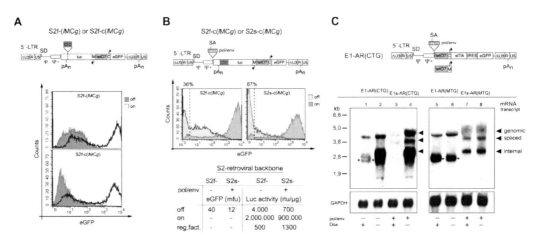

Fig. 1. Vector backbone elements found to improve the regulatory and/or inducibility properties of tet-regulated retroviral vectors in general. Data shown were obtained from HtTA-1 cell populations (**a** and **b**) or Hela cells (**c**), transduced with the indicated vectors and purified via FACS to 95%. For description of common abbreviations used to describe the viral vectors *see* Fig. 2. (**a**) Insertion of the constitutive transport element (cte) of Simian retrovirus 1 (SRV-1) into the S2f-(*MC*g) bidirectional self inactivating (SIN) retroviral vector resulted in the S2f-c(*MC*g) viral backbone. The antisense orientated luciferase expression is strongly enhanced (>100-fold in transient experiments, not shown). Expression from the sense orientated eGFP expression cassette is influenced too, although a positive effect is only evident in the off-state cells, indicated by a highly condensed cell population. (**b**) Insertion of the retroviral pol/env fragment (indicated by "s" in the vector designation) 3' to the packaging region (Ψ and Ψ+) of S2f-c(*MC*g) resulted in the S2s-c(*MC*g) bidirectional retroviral vector. It protects both minimal promoters from outside activation, thereby significantly reducing unregulated background expression of both genes. In case of the CMV driven eGFP this led to an increased accumulation of transduced cells (67% vs. 36%) at the basal fluorescence level of the parental HtTA-1 cells (*dashed line*) combined with a lower mean fluorescence value (mfu) when compared with the cells transduced with the S2f-c(*MC*g) vector (12 vs 40). Similar effects were observed for the antisense orientated expression cassette, where the regulation factor (reg. fact.) based on luciferase activity was increased by about 2.5-fold. It should be noted that the maximum expression level was decreased to a similar extend. (**c**) Analysis of total RNA obtained from Hela cell populations transduced with autoregulated unidirectional viral vectors E1-AR(CTG)/-(MTG) or, after the introduction of the pol/env fragment, E1s-AR(CTG)/-(MTG). These vectors differ with respect to their minimal promoters derived from either CMV or MMTV, and the absence or presence of the pol/env fragment. The results reveal that independent of the induction state (i) a reduced unregulated background (compare lane 1 and 3) and (ii) an unexpected correction of the transcriptional start site occur. In the absence of the pol/env fragment (lanes 1, 2 and 5, 6) a smaller transcript appears (*asterisk*) with similar intensity to the correctly initiated internal transcript (*arrow head*). The shorter transcripts completely disappeared after introduction of the pol/env fragment, which becomes particularly obvious for the MMTV promoter (lane 6 and 8) in the induced state (-Dox), while in the case of the CMV minimal promoter (lane 2 and 4) distinct bands can not be discriminated due to strong signal intensity. For signal detection the blot was hybridized with a biotinylated probe directed against the synthetic (s)tTA transactivator.

properties, so that both inducible activity as well as background expression were enhanced.

In addition to the above-mentioned viral elements, tet-system components themselves, such as minimal promoters, tet-operators as well as transactivators, strongly influence the regulatory properties of transduced cell populations – as well as their subclones. For instance, employing minimal promoters that are less sensitive to outside activation will lead to an increase in the proportion of tightly regulated cells within a transduced cell

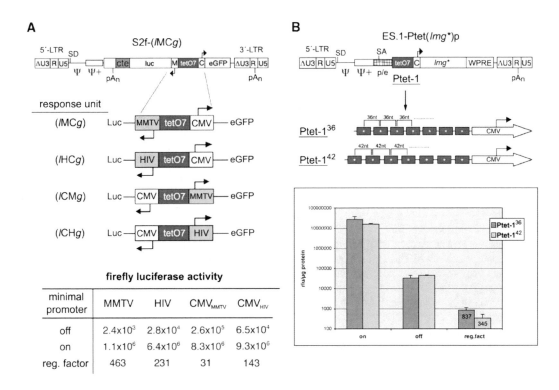

**Fig. 2.** Tet-system components that improve the regulatory properties of the response unit in general when integrated into retroviral vectors. Data shown were obtained from HtTA-1 (**a**) or Hela-M2 (**b**) cell populations, transduced with the indicated vectors and selected via FACS to ≥95%. (**a**) The use of different minimal promoters fused to the various tet-operators allows to adjust the expression level dependent on the experimental needs. The Minimal promoters can be listed by their inducibility: CMV > HIV > MMTV, whereas the reverse order is valid for the unregulated background expression in the off-state. Interestingly, when the two strong promoters CMV/HIV were combined in the bidirectional viral vector, the regulation factor of the CMV promoter is enhanced about fivefold, indicating that promoter interference even between minimal promoters has to be considered. (**b**) Reduction of the tet-operator spacing from 42 nucleotide (Ptet-1$^{42}$) in the original setting of the Ptet-1 regulatory unit *(12)* to 36 nucleotide spacing (Ptet-1$^{36}$) led to an increased induction level and a reduced unregulated background expression resulting in a 2.4-fold improved regulation of the CMV-based tet-response unit. For transduction a unidirectional SIN-vector was used, expressing a dual reportergene (*Img**) providing luciferase activity and eGFP fluorescence which was used for purification of the Hela-M2 population. Abbrevations used for the viral backbones (provirus): 5'- and 3'-LTR, long terminal repeat, consisting of U3, R and U5 region; ΔU3, Deletion of large part of the MLV U3 promoter region (−412/−68) and the TATA-box in the case of SIN vectors. R and U5 region of MLV, responsible for polyadenylation (pA$_n$) of the mRNA transcripts and necessary for viral replication; Ψ- packaging region, Ψ+ enlarged packaging region, comprising the approximately first 400 nucleotides of the gag-orf, with deleted start codon; p/e - pol/env fragment harboring the native splice acceptor of MLV; WPRE, Woodchuck post transcriptional regulatory element, transgene expression; tetO7, tet-operator heptamer consisting of the bacterial wildtype tetO2 operator site, with the central nucleotide labeled (*); M or MMTV, minimal promoter of mouse mammary tumor virus (−89/+120); C or CMV, human Cytomegalus Virus (−53/+75); H or HIV-human immunodeficiency virus type 1 (−81/+76).

population. At this point it has to be noted that promoters displaying low unregulated background in general are not highly inducible and vice versa. Thus, one has to decide whether high expression levels of the therapeutic gene are required to exert its beneficial effects in the transduced cell population – rather than

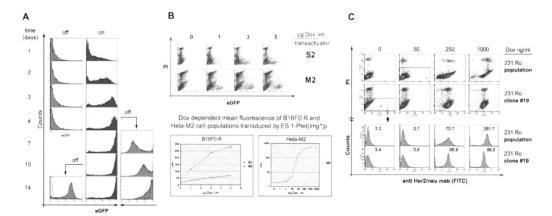

Fig. 3. Regulatory properties of the tet-system in cell populations when transferred by retroviral vectors. (**a**) Induction and shut-off kinetics based on eGFP fluorescence determined in purified S2f-c(*IMCg*)-transduced HtTA-1 cells. With proteins, having very long half lifes (>24 h) a steady state is reached earliest four days after induction, while the cells need very long (>10 days) to come down to the level of the starting cells (*left* and *right row*). It should be mentioned here, that also with short living proteins like luciferase the steady state is reached slower (48 h) when compared with transient experiments. (**b**) An unspecific tetracycline resistance was shown for B16F0.R cells. Cells selected for constitutive expression of the S2-transactivator showed only low sensitivity toward tetracycline upon induction. The S2-transactivator usually is fully activated at around 1–2 μg/mL Dox, but induction of gene expression in cells transduced with a tet-regulated ES.1-Ptet[36](*Img\**)p viral vector required much higher levels of Dox (≥5 μg/mL). The same could be demonstrated when B16F0 cells were generated to constitutively express the more sensitive M2 transactivator, which is usually fully activated at 300 ng/mL Dox in Hela cells, but required ≥3 μg/mL Dox in the analysed cells. (**c**) Dox-dependent adjustment of protein abundance is shown in MDA-MB231.Ro cells. To address the central problem of adjusting protein levels in cell populations, the Her2/neu transgene was introduced by the ES.1-Ptet[36](erb) viral vector and protein levels were adjusted and analyzed, both in the original cell population and a clonal population derived from it. Although the expression of Her2/neu in the clone occurred at well-defined levels at every Dox-concentration analyzed, expression in the "parental" cell population appeared to be more inhomogenous. Especially since no defined expression level could be observed at the medium induction range. This result is certainly due to differences in viral integration sites between individual cells.

demanding a tight off-state, or adversely only minor amounts of the transgene are demanded, however, accompanied by a strong cellular sensitivity of its expression in the off-state. **Figure 2a** shows various minimal promoters that display different properties as described above. Unexpectedly profound was the effect of the tet-operator organization within the response unit (**Fig. 2b**), which might be due to steric hindrance at the time of induction (Loew and Bujard unpublished, *(18)*). Dependent on the experimental demands, one can choose between a variety of different tet-regulated transactivators.

As outlined before, gene therapy approaches will most certainly rely on the induction of gene expression, thus the "tet-on" system will be the preferred one. In this respect, two transactivators are of major importance: on the one hand the S2-version, which displays a relatively low sensitivity to the effector (Dox), and on the other hand the M2-transactivator, which is fully activated in the presence of 100 ng Dox/mL *(3)*. The latter might be the preferred variant in cases where target cells might not be readily accessible by the effector or otherwise might be protected by an unspecific tet resistance, as could be demonstrated in experiments analyzing the conditional expression of Her2/neu (erbB2, *(22)*) receptor in melanoma cells (**Fig. 3**).

Finally, the individual expression kinetics have to be considered, such as induction or deinduction, both strongly dependent on the half life of the protein expressed (**Fig. 3a**). As could be demonstrated for the conditional expression of Her2/neu (**Fig. 3c**), cell populations and their respective clonal derivatives might vary significantly with respect to their adjustment of the protein level.

*3.1. Preparation of Viral Vector Stocks*

1. 293T cells are cultivated at 37°C. Upon reaching 80–90% confluency, cells are transferred to fresh 90 mm dishes. For transient vector production, about $1.5 \times 10^6$ cells are transferred to a fresh 60 mm dish and further diluted with 4 mL cell culture medium, one day prior to transfection (day 0). Before starting the transfection experiment, cells should be allowed to settle for at least 24 h and must have reached about 80–85% confluency.

2. Transfection (day 1) is carried out via a three plasmid system, consisting of 2.5 μg of the gag/pol expression vector (pHIT60, *(19)*), 2.5 μg of the VSV-G envelope expression vector (pcz-VSV-G, *(20)*), and 2.5 μg of the transfer vector (e.g., any of the vectors used in the experiments described). 200 μL of FCS-free culture medium and 15 μL of TransIT-293 reagent (ratio of 2 μL reagent/1 μg DNA) are suspended and carefully mixed by bubbling. After 10 min incubation at RT, the DNA is added to the premix and incubated for 5–10 min at RT (*see* **Note 5**).

3. Cell culture medium is reduced to 2 mL and DNA-solution is added drop wise (day 1). After 4–6 h of incubation 2 mL of fresh medium is added and cells are further incubated over night.

4. After 16–18 h (day 2) 2 mL of the culture medium is replaced by fresh medium containing 10 mM Na-Butyrate (final conc. 5 mM) and cells are incubated for at least additional 6–8 h.

5. On the evening of day 2, the medium is completely replaced by 3 mL of fresh medium and cells are incubated over night (*see* **Note 6**).

6. Following incubation for 16–18 h (day3), cell supernatants are harvested as follows: In general, after removal of the medium from the dish by use of a syringe, a 0.45 μM filter is attached to the syringe and the supernatant is filtrated

directly into a Falcon tube, containing 3 µL of Polybrene (final conc. 8 µg/mL).

7. All supernatants are aliquoted as 0.5 mL batches and frozen at –20°C. Following a freeze/thaw cycle the titer of infectious particles is determined on a suitable cell line (e.g., HtTA-1 for response unit vectors) for one of the obtained aliquots as follows:

8. About $2 \times 10^5$ cells are transferred to each well of a six-well plate the day before titration (day 0) in a total volume of 2 mL culture medium. The following day (day 1) serial dilutions (e.g., 1:5, 1:10, 1:20, 1:40….) of viral supernatants are prepared each in a total volume of 1 mL.

9. The cell medium is replaced by the dilution of the supernatant, and cells are further incubated over night (16–18 h). On day 2, the medium is replaced by 2 mL of fresh culture medium and infection is carried out under inducing conditions, e.g., in the absence of Dox, if HtTA-1 cells are used in the titration experiment (*see* **Note 7**).

10. Analysis is done based by measuring GFP-fluorescence (*see* **Note 8**).

11. Earliest at 48 h after induction of gene expression. Preferably, analysis should be carried out 96 h following induction, since it can be assumed that even the accumulation of long living proteins such as GFP will have reached an almost stable steady-state level by then. The calculation of the titer is done by taking the mean of dilutions that display a linear infection as determined by FACS (*see* **Note 9**).

### 3.2. Fluorescence-Based Purification and Analysis of Cell Populations

#### 3.2.1. Preparation of Transduced Cell Populations

1. On day 0 about $1 \times 10^5$ of the HtTA-1 target cells are transferred to a six-well dish. The tTA transactivator is kept inactive during infection by supplementing the medium with 100 ng Dox/mL (*see* **Note 10**).

2. On day 1, the viral vector containing supernatant (*see* **Subheading 3.1**) is added to the medium (100 ng Dox/mL) at a MOI ≤ 0.1 (i.e. 1 infectious particle/10 cells) and incubation is carried out for at least 16–18 h.

3. On day 2, the culture medium is replaced by fresh medium (100 ng Dox/mL).

4. On day 3, cells are split to two 60 mm dishes as follows. To remove cells from the plate, medium is aspirated and cells are washed with 1× PBS. Upon removal of PBS and addition of 1 mL of PBS-EDTA, cells detach within 2–3 min. Following resuspension of the cells by gentle pipetting ½ of the suspension is transferred to the dish with (uninduced) or without

Dox (induced). Cells are cultivated under these conditions for a further 72 h.

5. On day 7, culture medium is renewed (4 mL). Cells are harvested on day 8 for analysis or otherwise enrichment/purification (*see* **Notes 11** and **12**)

*3.2.2. FACS-Analysis and Purification of Regulatable GFP_{pos} Cell Populations*

1. Transduced HtTA-1 cells (*see* **Subheading 3.2.1, steps 1–5**), cultured either with and without Dox, are harvested on addition of 2 mL of PBS-EDTA (*see* **Subheading3.2.1, steps 1–4**), following 96 h incubation (*see* **Note 13**).

2. For GFP-based FACS-analysis as well as enrichment via cell sorting, the transduced cells are carefully resuspended by pipetting and transferred to appropriate tubes, already containing supplemented with 10 μL propidium iodide per ml cell suspension (final conc. 1 μg/mL) (*see* **Note 14**).

3. GFP-positive cells are detected in the FLH-1 channel, while dead cells, stained by propidium iodide, are detected in the FLH-3 channel of the FACS-Calibur (Beckton Dickinson). **Figure 3b** shows an example for the analysis of MDA-MB231.Ro cells.

4. For each population to be enriched, cells are expanded on a 90 mm dish (*see* **Note 15**). For FACS, a single tube has to be prepared containing 2 mL of the cell culture medium supplemented with antibiotic-mix (Ciprofloxacine at 10 μg/mL, Pen/Strep 100 U/mL).

5. During the sorting procedure, cells are injected directly into the prepared tube. The sorted cells (usually $2–5 \times 10^4$) are transferred to a six-well dish and expanded as described above (*see* **Subheading 3.2.1, steps 1–5**).

6. In general, purified cells are kept in the off-state of the system irrespective of the transactivator employed, since the basal level of gene activity will only be reached after prolonged incubation, especially if proteins with long half-life such as eGFP are to be expressed. In this case, cells have to be kept in the off-state for a minimum of 10 days before a precise on/off situation can be defined. Moreover, using the tet-system, proteins are usually expressed at high concentrations, a condition which is rarely tolerated but rather leads to the loss of the highest expressing cells.

*3.2.3. FACS-Analysis and Purification of Regulatable Cell Populations by Immunostaining*

1. Transduced B16F0-S2 or -M2 cells cultured with and without Dox prepared as described in **Subheading 3.2.1, steps 1–5** are harvested after 96 h with 2 mL of PBS-EDTA (*see* **Subheading 3.2.1, steps 1–4**). The off-state cells (+Dox) are harvested in PBS-EDTA and are kept in the presence of Dox (100 ng/mL) throughout the staining procedure, since in the prolonged

absence of the effector gene expression will be induced, thus increasing the "putative" unregulated background.

2. Cells are transferred to Falcon tubes (15 mL) and centrifuged at 300 g for 3 min at room temperature (RT). Supernatant is discarded and cells are washed once in 500 μL PBS supplemented with 2% BSA, 100 ng Dox/mL. Again, the supernatant is discarded and the cell pellet is resuspended in 100 μL PBS, 2% BSA, and 100 ng Dox/mL.

3. The primary antibody directed against human Her2/neu is diluted in the cell suspension to a final concentration of 2 μg/mL (*see* **Note 16**), mixed by finger flicking and incubated for 10 min at RT.

4. After incubation cells are centrifuged for 3 min with 300 g at RT. Cells are washed once with 1 ml PBS, 2% BSA, 100 ng Dox/mL and centrifugation as described before.

5. The resulting cell pellet is resuspended in 100 μL PBS, 2%BSA 100 ng Dox/mL and secondary antibody is diluted into the cell suspension at 10 μL/1 × 10$^6$ cells (*see* **Note 16**). Suspension is gently mixed by finger flicking and incubated for 10 min at RT.

6. Incubation is followed by a washing step with 1 mL PBS, 2% BSA, and centrifugation is carried out as described above. The supernatant is discarded and cells are carefully resuspended in 1 mL PBS, 2% BSA. PI is diluted into the cell suspension to a final concentration of 1 μg/mL.

7. Her2/neu-positive cells are detected in the FLH-1 channel (the secondary antibody being FITC-labeled), while dead cells are stained by propidium iodide and are therefore detected in the FLH-3 channel. **Figure 3c** shows an example based on the analysis of B16F0 cells.

8. For the enrichment of stained cells by cell sorting, cells are carefully resuspended by pipetting and transferred to appropriate tubes, containing 10 μL propidium iodide (PI) per mL of cell suspension (final conc. 1 μg/mL) (*see* **Note 15**).

9. For each population to be enriched, one tube has to be prepared containing 2 mL of the cell culture medium supplemented with antibiotic-mix (Ciprofloxacine at 10 μg/mL, Pen/Strep 100 U/mL).

10. The sorted cells are transferred to six-well dish and are further expanded as described (*see* **Subheading 3.2.1, steps 1–5**). If further purification is required, the cells are induced for at least 96 h and the above steps are repeated, respectively.

11. The purified cells are always kept in the off-state (*see* **Subheading 3.2.2, step 6**).

# Color Plates

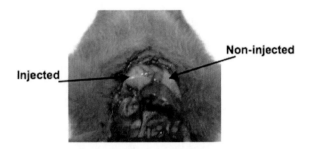

Chapter 13, Fig. 1. In vivo intrathymic gene transfer. Intrathymic injections were performed on either infant mice, aged 10–21 days, or adult mice, ranging from 6 to 9 weeks of age. In the latter case, mice were anesthetized and injections were performed following incision of the sternum. In infant mice, virions, in a total volume of 10 µL, were injected directly through the skin into the thoracic cavity immediately above the sternum at an approximately 30° dorsal angle, using a 0.3-mL 28-gauge 8-mm insulin syringe. To assess the efficacy of this latter "blind" approach, infant mice were directly injected with 10 µL of trypan blue dye and killed immediately to visualize the localization of the dye. Trypan blue-injected and -uninjected thymic lobes are indicated in this representative thymus.

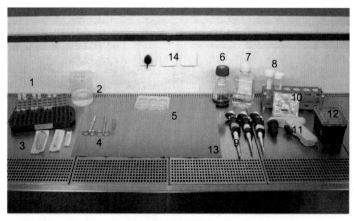

| | | |
|---|---|---|
| 1 cryotubes | 6 DMEM medium | 11 pipette boy |
| 2 ethanol | 7 PBS | 12 sterile tips |
| 3 syringes and needles | 8 sterile falcon tubes | 13 sterile drape sheet |
| 4 scissors and forceps | 9 pipettes | 14 power socket (possible |
| 5 tissue culture plate | 10 cell strainer | to plug in balance) |

Chapter 21, Fig. 1. Typical set up for the endpoint analysis under a flow hood for sterile work.

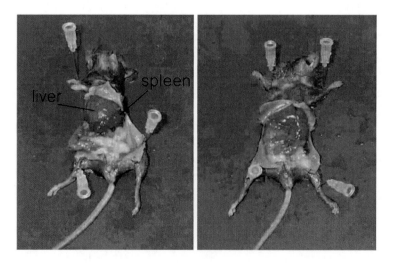

Chapter 21, Fig. 2. Typical sectio of a mouse with enlarged liver and spleen.

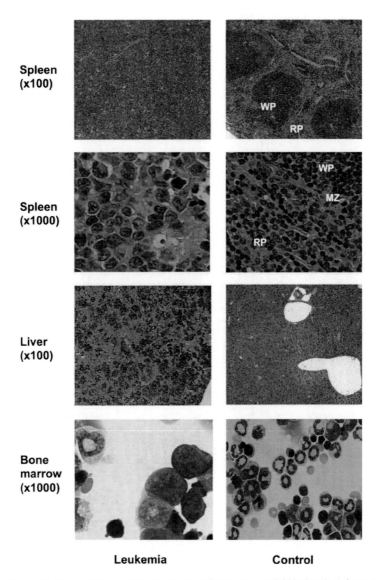

**Spleen (x100)**

**Spleen (x1000)**

**Liver (x100)**

**Bone marrow (x1000)**

**Leukemia**          **Control**

Chapter 21, Fig. 4. Histopathology and cytology of mice with myeloid leukemia (*left panel*) and healthy controls (*right panel*). Histopathology of spleen (low and high power views): normal architecture of spleen in a leukemia mouse is totally abolished and replaced by myeloid blasts with poor differentiation (diagnosis: myeloid leukemia without maturation). In comparison, splenic architecture is normal in the control mouse. *WP* white pulp, *MZ* marginal zone, *RP* red pulp. Histopathology of liver (low power view): Extensive infiltration of leukemic cells in liver of a diseased mouse, while none of hematopoietic cells is present in the liver of the control mouse. Cytospin of bone marrow (high power view): There are >50% myeloblasts in bone marrow of a diseased mouse. Note that myeloblasts have fine chromatin with fairly distinct nucleoli. The control mouse shows a normal myelogram with approximately 60% myeloid cells, 20% erythroid cells, and 15% lymphoid cells. All hematopoitic cells, at all stages of maturation, are present together.

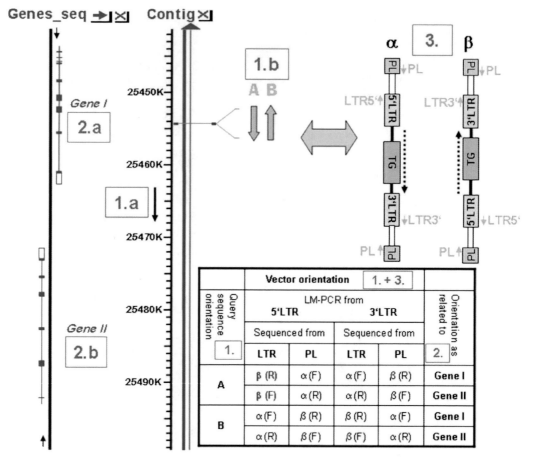

| | Vector orientation | | 1. + 3. | | | |
|---|---|---|---|---|---|---|
| Query sequence orientation | LM-PCR from | | | | | Orientation related to |
| | 5'LTR | | 3'LTR | | | |
| | Sequenced from | | Sequenced from | | | |
| 1. | LTR | PL | LTR | PL | 2. | |
| A | β (R) | α (F) | α (F) | β (R) | | Gene I |
| | β (F) | α (R) | α (R) | β (F) | | Gene II |
| B | α (F) | β (R) | β (R) | α (F) | | Gene I |
| | α (R) | β (F) | β (F) | α (R) | | Gene II |

Chapter 25, Fig. 3. Schematic representation of identification of provirus orientation regarding a gene of interest based on results of an NCBI genomic blast search. The *left* part of the figure corresponds to the schematic representation of results of a genomic BLAST search as provided by the NCBI database. Under 'Contig', a nucleotide scale (in kb) is provided for the chromosome to which your query sequence was aligned. You can use the following algorithm to define orientation of the integrated vector. [1.] To determine the Query sequence orientation: [1a.] The chromosome's orientation/scale in those schemas is always up-to-down (see the respective *arrow*). [1b.] If nucleotide numbers of your blasted Query sequence are increasing together with those of the chromosome, the aligned sequences are directed in the same direction (**A**) as the chromosome; in the opposite case it is directed in an anti-parallel orientation (**B**). [2.] Gene directions are indicated by arrows besides the 'Genes_seq' bar: [2a] Genes located on the right side of the bar (e.g. *Gene I*) are always directed co-linear with the chromosome, [2b] genes outlined on the left side (e.g. *Gene II*) are always directed contra-linear. [3.] To finally determine the orientation of the inserted vector (*dotted arrow*) as related to the chromosome (α = parallel, β = anti-parallel) and the genes of interest (*F* forward or *R* reverse) you have to take into account the location of the adjoining sequences retrieved by LM PCR (5' vs. 3') *and* the direction of your initial sequence reaction (started from the LTR or the polylinker (PL)). All possible variants of sequencing primers are indicated in the *right* part of the figure. The colours of the primers indicate the sequence direction as related to the chromosome (A vs. B). All orientations towards genes and chromosomes are summarised in the Table in the lower part of the picture. For example, after LM PCR you obtained the 5' adjacent region of your integrated vector and sequenced it using an LTR-specific primer. You first determine the query sequence orientation with respect to the chromosome, which turns out to be co-linear (A). Consequently, your vector is oriented in the opposite direction to the chromosome (β). Finally, as related to *Gene I* the vector is oriented in a reverse (R), as related to *Gene II* in a forward (F) orientation (see table, highlighted in *blue* and *bold*).

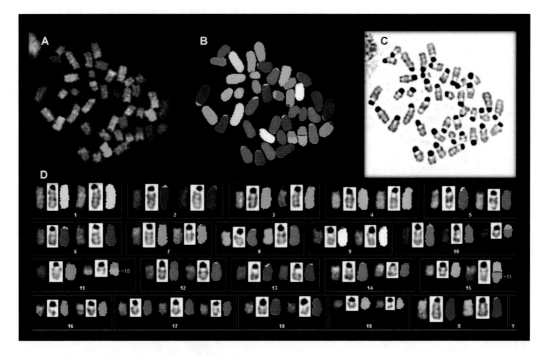

Chapter 30, Fig. 1. SKY analysis of metaphase chromosomes prepared from a murine B-cell lymphoma. Detection of a balanced translocation T(11B1;15D2–D3), an additional chromosome 10 with terminal deletion, and a trisomy 17. ((**a**) RGB (red-green-blue) image; (**b**) RGB image after classification; (**c**) inverted DAPI image; (**d**) karyotype).

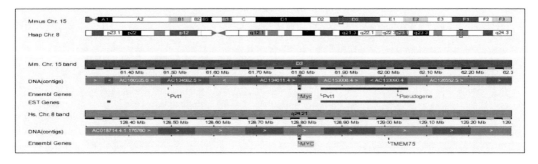

Chapter 30, Fig. 3. The murine chromosome 15 shows synteny to the human chromosomes 5, 8, 12, and 22.

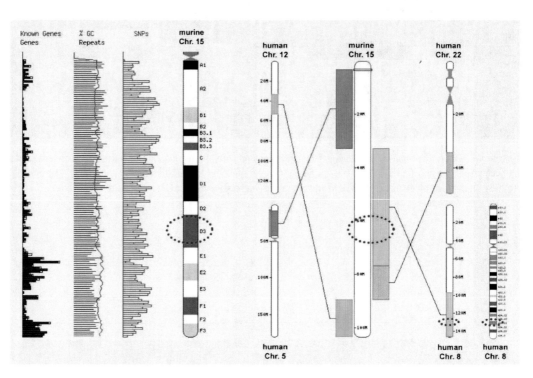

Chapter 30, Fig. 4. Synteny of the c-Myc locus (http://www.ensembl.org(. In mice, the proto-oncogene c-myc is located in chromosomal region 15D3. The syntenic human region is 8q24.21.

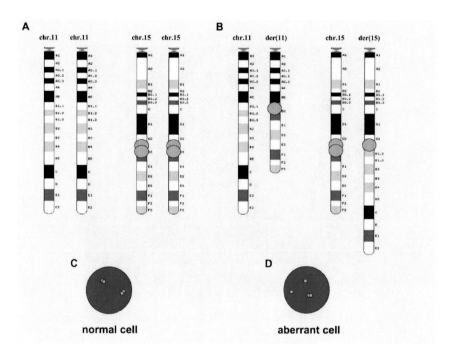

Chapter 30, Fig. 5. Presentation of the ideograms of murine chromosomes 11 and 15 and confirmation of a balanced translocation T(11B1;15D3) by FISH using a single-color break-apart probe. One probe is generated to hybridize centromerically close to the breakpoint in 15D3, the other probe to hybridize telomerically close to the breakpoint in 15D3. Thus, two normal chromosomes 15 show 2 fusion signals (**a**). A rearrangement results in the split of one fusion signal into two single signals, one on Der(11) and the other on Der(15) (**b**). (**c**) and (**d**) show the signal constellation as expected in interphase nuclei in a normal cell and in a cell with T(11B1;15D3).

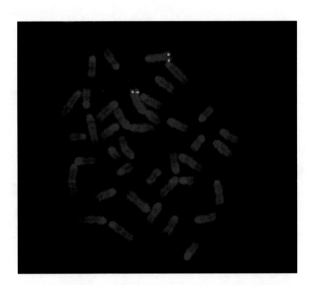

Chapter 30, Fig. 6. FISH on murine metaphase chromosomes. The SpectrumGreen labeled DNA probe has been generated by random priming using the BAC clone RP23-382J5. The probe hybridizes to region XF2 and shows two signals representing the expected signal constellation of a normal cell.

**3.3. Determination of Luciferase Activity and Total Protein Content**

1. Purified HtTA-1 cells (*see* **Subheading 3.2.2, steps 1–6** and **Subheading 3.2.3, steps 1–11**) are cultured with and without Dox. Cell harvesting is done after 96 h incubation with 2 mL of PBS-EDTA (*see* **Subheading 3.2.1, steps 1–4**). Off-state cells (+Dox) are harvested in PBS-EDTA supplemented with Dox (100 ng/mL), since the absence of the effector will result in short term induction of the system, thereby increasing the "putative" unregulated background *(1)*.

2. About 1–5 × 10$^5$ cells are transferred to an eppendorf tube, and centrifuged with 300 g for 3 min at 4°C. After the supernatant has been discarded, 100 µL lysis buffer is added to the cell pellets (*see* **Note 17**). After careful resuspension (cells should dissolve immediately), the cell lysate is placed at –20°C for a minimum of 30 min. The frozen tube is centrifuged at 14,000 g for 10 min at 4°C, to remove cell debris. Throughout the measurement samples are kept on ice.

3. The assay buffer can be prepared, while samples are kept frozen. The total volume of buffer can be calculated (2/sample) based on the following: 500 µL buffer are needed per measurement and each sample is analyzed twice. Thus each 500 µL assay buffer is supplemented with 10 µL ATP stock solution. 250 µL of the "ready to use" assay buffer is decanted into the appropriate plastic vials provided with the Luminometer (*see* **Note 18**). Make sure that the assay buffer has reached room temperature on addition of cell lysate. A minimum of 10 independent measurements is required to determine the instrument background (usually between 50 and 200 rlu).

4. The substrate buffer is prepared also during the freezing time. One 200 µL aliquot is thawed and diluted into 40 mL distilled water. The solution is connected to the injector of the Luminometer. Residual substrate buffer can be stored at –20°C for at least 3 months. The buffer should be kept at room temperature throughout the experiment.

5. 1–20 µL of the cell lysate (or respective dilutions) are used for determination of luciferase activity. The sample is transferred into the prepared assay buffer, mixed by finger flicking, and directly measured in the luminometer. The second sample can be prepared while the first is being analyzed.

6. Protein content of the samples is determined by diluting 1–2 µL of the cell lysate (*see* **Note 17**) in 100 µL of distilled water (*see* **Note 19**). About 900 µL of the Coomassie Plus reagent are mixed with the protein sample and incubated for 10 min at RT. Absorbance is determined in a photometer at 595 nm. For estimation of total protein concentration in

samples, a standard curve is included and standards are prepared by the same method (*see* **Note 20**).

### 3.4. Preparation of Total RNA for Northern Blot Analysis

1. Purified off-state HtTA-1 cells (*see* **Subheading 3.2.2, steps 1–6** and **Subheading 3.2.3, steps 1–11**) are expanded to about 75–80% confluency on 90 mm dishes. The cells are split into medium with and without 100 ng Dox/mL, and cells are allowed to grow for 72 h. On the day before harvest the medium is renewed.

2. After 96 h Induction, cells are detached with 4 mL of PBS-EDTA (*see* **Subheading 16.3.2.1, steps 1–4**). Generally, at this stage, cell quality is determined for a small aliquot of the cells by FACS analysis (at least the percentage of dead cells should be evaluated by propidium iodide staining). The off-state cells were detached in PBS-EDTA supplemented with 100 ng Dox/mL (*see* **Note 10**) to maintain inactivation of the transactivator. Cells sometimes require longer than 2–3 min detach from the dish. In this case, it is helpful to keep the dish on ice during this procedure, since removal is enhanced at low temperatures and massive degradation of RNA is prevented

3. Cells are transferred to a 15 mL Falcon tube and centrifuged at 300 g for 3 min at 4°C. The supernatant is discarded. At this stage, cells can be shock-frozen and stored at –20°C or lysed in Trizol reagent (1 mL/$10^6$ cells).

4. Following lysis, cells are mixed with 0.2 volumes chloroform and vortexed vigorously for a minimum of 1 min. The cell suspension is kept on ice for another 10 min and subsequently is divided into 2 mL Eppendorf tubes.

5. The suspension is centrifuged at 14,000 g for 5 min at RT and the clear supernatant transferred to a fresh eppendorf tube. Following addition of 1 volume of isopropanol and careful mixing, tubes are kept at –20°C for 1 h.

6. Total RNA is precipitated by centrifugation at 14,000 g for 10 min at 4°C and supernatant is discarded. Cells are washed with 70% EtOH, and centrifuged as described before.

7. Following centrifugation, the supernatant is discarded and the RNA pellet air dried. The optimal state is reached, when the rim of the pellet changes from white to translucent.

8. Pure formamide is added to the dried pellet at a ratio of 10 μL formamide/$1 \times 10^6$ cells and the RNA is allowed to dissolve by keeping the tubes for 30 min at 50°C in a thermomixer under vigorous shaking (700 rpm). The resuspended RNA is shock frozen and thawed. RNA is completely dissolved by repeated heating for 30 min at 50°C under constant shaking.

Insoluble particles are pelleted by centrifugation at 14,000 g for 5 min at RT.

9. RNA is quantitated by addition of 2 µL of RNA-formamide to 500 µL distilled water and the RNA concentration is determined at $OD_{260 \ nm}$; optimal ration 260/280 is 1.6–2.0. For the control sample, RNA is replaced by pure formamide to normalize the measurement.

10. At this point, the RNA is ready for Northern analysis (**Fig. 1c**) or other RNA-based methods like RT-PCR. Northern blot preparation and signal detection might be carried out following established protocols. We generally apply a nonradioactive detection method described by Loew and Rausch *(21)* (*see* **Note 21**).

### 3.5. Induction Kinetic

1. Purified off-state HtTA-1 cells (*see* **Subheading 3.2.2, steps 1–6** and **Subheading 3.2.3, steps 1–11**) are cultured with Dox (off-state) on 130 mm dishes to about 75–80% confluency. For each time point, a separate dish of cells has to be prepared, taking into account the cell division during the experimental time scale. The duration of the experiment shown (**Fig. 3a**) was 14 days.

2. The initial split is done at day 0. Cells are counted and about $3 \times 10^6$ cells were transferred to two 90 mm dishes for the harvest at day 1 (one with and one without 100 ng Dox/mL). For the harvest at day 2, 3, and 4 about 1.5, 0.75, and $0.375 \times 10^6$ cells are transferred to two 90 mm dishes. For the later harvests (e.g., day 7, 10, and 14) $1.5 \times 10^6$ cells are splitted at day 4, day 7, and day 11, propagated in medium with and without Dox. The cell culture medium is exchanged every second day, considering that the halflife of Dox in the medium is around 24 h.

3. For each time point cell harvest is done with 4 mL of PBS-EDTA (*see* **Subheading 3.2.1, steps 1–4**). Off-state cells (+Dox) are harvested in PBS-EDTA supplemented with Dox (100 ng/mL), as the absence of the effector will lead to short-term induction of the system and thus increase the "putative" unregulated background. The cells are divided for the parallel analysis of eGFP expression (*see* **Subheading 16.3.2.2**), luciferase (*see* **Subheading 3.3**) and RNA via northern blot (*see* **Subheading 3.4**).

### 3.6. Dose Response Experiments and Induction Level Adjustment

1. Purified off-state MDA-MB231.Ro cell population (*see* **Subheading 3.2.2, steps 1–6** and **Subheading 3.2.3, steps 1–11**) or a clone derived from it are cultured without Dox (off-state for S2 transactivator) on a 90 mm dish to about 75–80% confluency. For each Dox concentration (**Fig. 3c**), a separate dish is prepared with $0.5 \times 10^6$ cells.

2. The cells are propagated for 4 days (*see* **Note 22**), with medium exchange every day (try to keep 24 h rhythm) to counteract fluctuations of induction triggered by destruction of Dox in the culture medium.

3. For each time point cell harvest is done with 4 mL of PBS-EDTA (*see* **Subheading 3.2.1, steps 1–4**). Off-state cells (–Dox) are harvested in PBS-EDTA. The cells are divided for the parallel analysis of eGFP expression (*see* **Subheading 3.2.2**), luciferase (*see* **Subheading 3.3**) and RNA via northern blot (*see* **Subheading 3.4**) (*see* **Note 23**).

## 4. Notes

1. For dissolving EDTA, it is necessary to bring the pH to at least 6.5 with NaOH. When completely dissolved, the stock solution is adjusted to pH 8.0.

2. The working solution is prepared from 10× stock solutions of $MgSO_4$ (SIGMA, M5921) and Glycylglycin (Calbiochem, 3630) and a 50× stock of rATP.

3. ATP is added only to the required amount of buffer for a single experiment. Repeated freeze/thaw cycles of the rATP stock solution should be avoided, since they will severely affect the quality of the solution.

4. The final pH of 7.5 is adjusted by dropwise addition of 5 M NaOH until pH 6.5 is reached, from this point onward the titration is continued with 1 M NaOH, and the volume finally adjusted with distilled water. 1 mL aliquots of the solution are kept at –20°C.

5. Throughout our studies transient vectors have been obtained by the same method, differing only with respect to the transfer vector used. Since introduction of DNA into 293T does not require any particular method, most commercially available transfection reagents can be employed and $CaPO_4$ coprecipitation works perfectly as well.

6. Care should be taken during the handling of the cells, since 293T cells tend to readily detach from the culture dish. In case part of the confluent layer has already been detached, try to remove those cells from the dish, otherwise the adherent cell layer might be completely lost upon movement of the dish and harvesting of infectious particles might no longer be possible.

7. If Hela M2 cells are being used, Dox has to be added to monitor the induction of gene expression. The functionality

of the on/off-state of the response unit, which has been transferred by the viral vector, can be tested directly when performing the transduction in the off-state of the system. On day 2 or likewise day 3 the cells can be split into fresh plates for induction with and without Dox, dependent on the transactivator.

8. Depending on the regulated transgene used any fluorescent label can be detected, (*see* **Subheading 3.2.3**).

9. The titer (infectious particles per mL) is calculated as follows: (number of cells × %GFPpos. cells) × 2* × dilution factor × (1/incubation volume**). The titer is calculated as the mean in the linear range of infection (* = a factor of 2 is included to indicate the fact that cell division is necessary for infection, and in case of a moderate MOI e.g. 1, one infected cell is accompanied by an uninfected sister; **since the usual infection volume is 1 mL in titration assays this factor can be neglected).

10. These conditions are kept constant for the preparation of all other cell populations described herein. Please note: to keep the reverse M2 or S2 transactivators inactive, Dox-free medium has to be used. For full induction either 300 ng Dox/mL for the M2 transactivator or 1,000 ng Dox/mL for the S2 transactivator has to be added. It is also possible of course, to do infection under inducing conditions, e.g., if the functionality of the vector is known. However, to determine the background activity in the off-state will require very long time especially if eGFP or other enzymes with comparable long half lifes were used for detection.

11. At this stage, the induced cells can be expanded by transfer to a 90 mm dish and cells are allowed to grow for an additional day. Thereby a sufficient amount of the transduced population can be provided that allows either purification via cell sorting or simultaneous analysis of eGFP/Luc expression and RNA preparation. When using cells constitutively providing the reverse transactivators, Dox at appropriate concentrations (*see* **Note 10**) has to be added to the culture medium for induction of gene expression.

12. *General note* for handling of tet-regulatable cells: the measurement of regulated gene expression, especially analysis of luciferase activity, requires "perfect cells." Undoubtedly, any sign of overgrowth or medium-starvation of the cells will lead to an increase in background expression and will decrease the level of induced activity. Therefore, exchanging the culture medium on the day before analysis is detrimental. Information on the quality of cells can be obtained by FACS analysis of a small aliquot, whereby cells are incubated in the

presence of propidium iodide to determine the percentage of dead cells in the preparation. We usually do not use results obtained from experiments displaying more than 10–15% of dead cells.

13. Off-state cells (+Dox) are harvested on addition of PBS-EDTA. When luciferase activity is determined, the solution should be supplemented with Dox (100 ng/mL), since absence of the effector will result in short-term induction of the system, thereby increasing the "putative" unregulated background. This effect is not observed when eGFP fluorescence is determined, since in this case short-term induction will not be visible because of the fluorescence background of the cells.

14. Upon addition of propidium iodide, dead cells can be readily excluded from the measurement, since the dye is not taken up by living cells.

15. About $5 \times 10^6$ cells per sample should be available, assuming a proportion of ≤10% positive cells per transduced cell population.

16. The indicated concentrations used for incubation of samples with the primary and secondary antibodies were tested in separate experiments and shown to give best results. However, optimal concentrations may vary if different batches of antibodies or products from different suppliers are being used and therefore have to be determined before starting with the experiment.

17. Since protein determination via the Bradford assay is sensitive to Triton X-100, dilution of cells in lysis buffer is generally carried out at a ratio of 5,000 cells/µL, resulting in a protein concentration of about 1 µg/µL within the final solution.

18. Avoid changing the tubes during one measurement, since the absorbance of plastic might vary between different batches of tubes, making analysis of the results difficult.

19. Care has to be taken not to touch the inner parts of the eppendorf tubes, since the proteins transferred by skin might lead to inconsistent protein concentrations between duplicates. In addition, when duplicates of samples are analyzed the pipette tip should be replaced by a fresh one for every transfer of cell lysate. Thereby inaccuracies resulting from detergent present in the lysis buffer can be minimized.

20. A commercial available BSA-solution (Pierce, 2 mg/mL) is used for preparation of a standard curve. In detail, BSA is diluted in distilled water to obtain working solutions of 0.05,

0.1, 0.2, 0.4, and 0.8 µg BSA/µL each. Thereafter, 10 µL of the respective BSA dilution is mixed with 90 µL of distilled water and 1–2 µL lysis buffer (lysis buffer is added to each standard in order to account for the possible presence of interfering substances in protein samples that are due to lysis buffer compounds.).

21. The presence of formamide is tolerated by almost all enzymes up to 10% final concentration. Formamide is favored over water as a dilution medium for RNA, since RNA can be protected from degradation by RNAse and remains highly stable even when stored at 4°C over prolonged periods.

22. In all cases, were long enzyme half-lifes have to be considered, at least in vitro experiments should be terminated after defined induction times (e.g. 96 h), as the proteins tend to accumulate into nonlinear fashion, thus making the interpretation of the results difficult.

23. The procedure is identical when performing dose response experiments to figure out the properties of the cell-type and transactivator-specific induction of gene expression as exemplified in **Fig. 3b** for Hela M2 and B16F0.R cells.

## Acknowledgments

The author would like to thank Prof Bujard for giving the opportunity to analyze so many different aspects of the tet-system in the retroviral vector context and Angelika Lehr for critical reading of the manuscript.

## References

1. Gossen, M. and Bujard, H. (2002). Studying gene function in eukaryotes by conditional gene inactivation. *Annu. Rev. Genet.* 36, 153–173

2. Baron, U., Freundlieb, S., Gossen, M., Bujard, H. (1995). Co-regulation of two gene activities by tetracycline via a bidirectional promoter. *Nucleic Acids Res* 23, 3605–3606.

3. Urlinger, S., Baron, U., Thellmann, M., Hasan, M.T., Bujard, H., and Hillen, W. (2000). Exploring the sequence space for tetracycline-dependent transcriptional activators: novel mutations yield expanded range and sensitivity. *Proc Natl Acad Sci USA* 97, 7963–7968

4. Vigna, E., Cavalieri, S., Ailles, L., Geuna, M., Loew, R., Bujard, H., and Naldini, L. (2002). Robust and efficient regulation of transgene expression *in vivo* by improved tetracycline-dependent lentiviral vectors. *Mol. Ther.* 5, 252–261

5. Paulus, W., Baur, I., Boyce, M.F., Breakefield, X.O., Reeves, S.A. (1996). Self-contained, tetracycline-regulated retroviral vector system for gene delivery to mammalian cells. *J Virol.* 70, 62–67

6. Hwang, J.-J., Li, L., Anderson, W.F. (1997). A conditional self-inactivating retrovirus vector that uses a tetracycline-responsive expression system. *J Virol.* 71, 7128–7131

7. Hoffmann, A., Nolan, G.P., Blau, H.M. (1996). Rapid retroviral delivery of tetracycline-inducible genes in a single autoregulatory cassette. *Proc Natl Acad Sci USA* 93, 5185–5190

8. Lindemann, D., Patriquin, E., Feng, S., and Mulligan, R.C. (1997). Versatile retrovirus vector systems for regulated gene expression in vitro and in vivo. *Mol. Med.* 3, 466–476

9. Unsinger, J., Kroger, A., Hauser, H., and Wirth, D. (2001). Retroviral vectors for transduction of autoregulated, bidirectional expression cassettes. *Mol. Ther.* 4, 484–489

10. Gill, G. and Ptashne, M. (1988). Negative effect of the transcriptional activator GAL4. *Nature* 334, 721–724.

11. Berger, S.L., Pina, B., Silverman, N., Marcus, G.A., Agapite, J., Regier, J.L., et-al., (1992). Genetic isolation of ADA2: A potential transcriptional adaptor required for function of certain acidic activation domains. *Cell* 70, 251–265

12. Gossen, M. and Bujard, H. (1992). Tight control of gene expression in mammalian cells by tetracycline responsive promoters. *Proc Natl Acad Sci USA* 89, 5547–5551

13. Bradford, M.M. (1976). A rapid and sensitive method for the quantitation of microgram quantities of protein utilizing the principle of protein-dye binding. *Anal. Biochem.* 72, 248–254

14. Loew, R., Vigna, E., Lindemann, D., Naldini, L., and Bujard, H. (2006). Retroviral vectors containing Tet-controlled bidirectional transcription units for simultaneous regulation of two gene activities. *J. Mol.Gen.Med* 2, 107–118

15. Wu, X., Li, Y., Crise, B., and Burgess, S.M. (2003). Transcription start regions in the human genome are favored targets for MLV integration. *Science* 300, 1749–1751

16. Saavedra, C., Felber, B., and Izaurralde, E. (1997). The simian retrovirus-1 constitutive transport element, unlike the HIV-RRE, uses factors required for cellular mRNA export. *Curr. Biol.* 7, 619–28

17. Zufferey, R., Donello, J.E., Trono, D., and Hope, T.J. (1999). Woodchuck hepatitis virus posttranscriptional regulatory element enhances expression of transgenes delivered by retroviral vectors. *J Virol.* 73, 2886–2892

18. Agha-Mohammadi, S., O'Malley, M., Etemad, A., Wang, Z., Xiao, X., and Lotze, M.T. (2004). Second-generation tetracycline-regulatable promoter: repositioned tet-operator elements optimize transactivator synergy while shorter minimal promoter offers tight basal leakiness. *J. Gene Med.* 6, 817–28

19. Soneoka, Y., Cannon, P.M., Ransdale, E.E., Griffiths, J.C., Romano, G., Kingsman, S.M., et-al., (1995). A transient three-plasmid expression system for the production of high titer retroviral vectors. *Nucl. Acids Res.* 23, 628–633

20. Pietschmann, T., Heinkelein, M., Zentgraf, H., Rethwilm, A., and Lindemann, D. (1999). Foamy virus capsids require the cognate envelope protein for particle export. *J. Virol.* 73, 2613–2621

21. Löw, R., and Rausch, T. (1994). Sensitive, non-radioactive northern blots using alkaline transfer of total RNA and PCR-amplified biotinylated probes. *Biotechniques* 17, 1026–1028, 1030

22. Gross, M.E., Shazer, R.L., and Agus, D.B. (2004). Targeting the Her-kinase axis in cancer. *Seminars in Oncology* 31, 9–20

# Chapter 17

# Detection of Replication Competent Retrovirus and Lentivirus

## Lakshmi Sastry and Kenneth Cornetta

## Summary

Retroviral vectors based on murine leukemia viruses (MuLV) have been used in clinical investigations for over a decade. Alternative retroviruses, most notably vectors based on HIV-1 and other lentiviruses, are now entering into clinical trials. Although vectors are designed to be replication defective, recombination events during vector production could lead to the generation of replication competent retroviruses (RCR) or replication competent lentiviruses (RCL). Careful screening of vector prior to human use must insure that patients are not inadvertently exposed to RCR or RCL. We describe methods capable of detecting low levels of virus contamination and discuss the current regulatory guidelines for screening gene therapy products intended for human use.

**Key words:** Retroviral vectors, Lentiviral vectors, Gene therapy, Replication competent retrovirus, Replication competent lentivirus.

---

## 1. Introduction

### 1.1. Risks Associated with RCR and RCL

Gammaretrovirus-based retroviral vectors (subsequently referred to as "retroviral vectors") are membrane bound RNA viruses initially based on murine leukemia viruses (1–3). These were the first viral vectors to enter clinical trials (4) and remain an attractive gene delivery tool when integration of the transgene into the target cells is sought. Unfortunately, integration is also associated with insertional mutagenesis whereby the risk of malignant transformation occurs due to dysregulation of cellular gene expression (reviewed in (5)). Insertional mutagenesis is a rare but known occurrence after insertion of a replication defective vector. Integration has led to activation of the Evi-1 oncogene with

Christopher Baum (ed.), *Methods in Molecular Biology, Methods and Protocols, vol. 506*
© Humana Press, a part of Springer Science + Business Media, LLC 2009
DOI: 10.1007/978-1-59745-409-4_17

subsequent development of myeloid leukemia in a murine bone marrow transplantation model *(6)*. Most concerning, integration of a retroviral vector led to the development of leukemias in subjects treated for X-linked severe combined immunodeficiency (SCID) *(7–10)*. Why these children have developed leukemia is complex, but preliminary evidence suggests the possibility that unregulated expression of the vector transgene (the common gamma chain cytokine receptor) may add to vector-induced dysregulation of the LMO-2 gene (i.e., a second "hit") *(11, 12)*.

While replication defective vector carry risk, the risk of insertional mutagenesis is believed to be greatest if RCR is present, since ongoing viral infection is likely to result in a much greater number of insertional events. RCR contaminating retroviral vector preparations have been shown to cause malignancy in both mice and nonhuman primates *(13, 14)*.

The risks associated with lentiviral vectors are currently unknown. The MLVs are known to cause leukemia related to regulatory regions with the MLV long-terminal repeats (LTR). HIV-1 is not known to cause leukemia, and third-generation lentiviral vectors have deleted the regulatory regions within the LTR (SIN, or self-inactivating vectors). Nevertheless, cellular gene dysregulation mediated by vector promoters remain a theoretical concern. Whether RCL can produce a HIV-1 like syndrome is unknown, although HIV-1 accessory proteins normally required for virulence have been deleted from vector constructs to improve their safety profile.

*1.2. Technical Considerations*

Retroviral and lentiviral vectors are generated by deletion of the viral protein coding sequences (*gag*, *pol*, and *env*) and substitution of this region with an exogenous gene(s) of interest. Vector particles must be generated by coexpressing the vector sequences along with the viral genes. Traditionally, retroviral vectors are generated in packaging cell lines that stably express the *gag*, *pol*, and *env* genes along with the vector plasmid (see Miller *(15)* for review). With packaging cells, RCR arises by recombination between the vector and viral genes, and was frequently detected in early versions of vector packaging cell lines in which all the viral genes (*gag/pol/env*) were expressed from a single plasmid *(16–19)*. By segregating *gag-pol* and *env* genes onto separate plasmids and minimizing homology between vector and packaging sequences, the rate of RCR development has been substantially decreased, but not eliminated *(20–25)*. In addition, the marked decrease in RCR generation resulting from decreased homologous recombination has resulted in rare recombinations between vector and cellular sequences leading to RCR, or rescue of endogenous viruses (*see* **Note 1**) especially when murine-based packaging cell lines are utilized *(26–28)*.

While murine oncoretroviruses have been known to develop recombinations leading to RCR, a number of factors complicate the detection of vector-associated RCL. First, RCL has not been reported with the commonly used lentiviral vector systems. Therefore, the RCL detection system must anticipate a currently theoretical virus, which will presumably arise through recombination between transfer vector and packaging construct sequences. Lentiviral vectors are generally produced by transient transfection of three or more plasmids: a plasmid expressing the transfer vector (containing the gene of interest in a lentiviral vector backbone), the packaging plasmid(s) (containing gag and pol), and an envelope-expressing plasmid (e.g., vesicular stomatitis virus glycoprotein (VSV-G)) *(29–32)*. Although areas of homology between the transfer and the packaging plasmids will greatly increase the likelihood of such recombination, nonhomologous recombination is also possible. Another concern is recombination with human endogenous retroviral (HERV) sequences *(33)*. RCL testing is also complicated by the transient transfection methods of lentiviral vector generation, which result in substantial contamination of supernatants with packaging plasmid DNA that contain the same viral sequences likely to be present in a RCL *(34, 35)*. Finally, a question that is still debated is whether vector product should be screened for a true RCL, or should signs of recombination events between vector and packaging plasmids (without complete generation of RCL) be grounds for rejecting a vector product.

**1.3. Testing Methodology**

Screening vector products for RCR has generally utilized biologic assays, while monitoring of patients after gene transfer rely on molecular or serologic testing *(36, 37)*. Molecular and serologic testing is rapid and less resource intensive than biologic assays but is prone to false positives. Particular care must be taken when performing PCR shortly after vector exposure since contaminating plasmids carried over from transient production methods or DNA from producer cell line can yield a false positive result suggesting RCR *(38)*. Therefore, screening of vector prior to clinical use has generally relied on biologic assays aimed at documenting a replication competent virus, rather than surrogate markers of virus.

The biologic assays most commonly used are the extended $S^+/L^-$ assay (**Fig. 1**) and the marker rescue assay *(13, 39–41)*. In these assays, test material is placed on a permissive cell line, and the cells are passaged for a minim um of 3 weeks (amplification phase). In the marker rescue assay, the permissive cell line contains a retroviral vector with an easily identifiable transgene (for example, a "marker" gene such as the neomycin phosphotransferase gene). After three weeks, the cell media from the amplification phase is collected. If RCR is present, it will package both the

RCR genome and "rescue" the marker vector. Cell free media from the amplification phase cells is then used to transduce a naïve cell line, which is then subjected to drug selection. Demonstration of drug resistance in the transduced cell population is indicative of RCR. In the extended $S^+/L^-$ assay, the amplification cells do not contain vector. Instead, virus is detected using indicator cell lines, such as the cat cell line PG-4 (42). The PG-4 cell line is referred to as a $S^+/L^-$ cell line, as it contains the murine sarcoma virus genome ($S^+$) but lacks the murine leukemia virus genome ($L^-$). Cells that express the murine sarcoma virus induce a transformed phenotype but only in cells coexpressing a murine leukemia virus. In this assay, media from the amplification phase is placed on PG-4 cells and transformation indicates the present of RCR. The $S^+/L^-$ assay can also be performed without the amplification phase (a direct $S^+/L^-$ assay), which allows one to titer the number of RCR (expressed as focus forming units per mL).

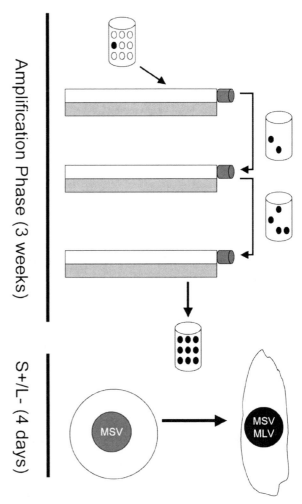

Fig. 1. The detection of replication competent retrovirus (rcr) using biologic assays. Amplification Phase: Test material depicts retroviral vector supernatant in which a small portion of the replication defective vector material (*open ovals*) is contaminated with RCR (*filled ovals*). Biologic assays often utilize a 3-week amplification phase in which a permissive cell line is used to increase the titer of any RCR present in the test material. In the Marker Rescue Assay, the cell lines used in the amplification phase contains an integrated retroviral vector that expresses a marker gene (such as a drug resistance gene). If RCR is present, it will "rescue" the marker vector and the cell supernatant will contain RCR along with virions capable of conferring drug resistance to naïve cells in the Indicator Phase. The $S^+/L^-$ assay also has an amplification step, but in this case the indicator cell line detects RCR directly. The indicator cell is termed an $S^+/L^-$ cells since it contains the murine sarcoma virus (MSV), which will transform the cell phenotype but only in the presence of a murine leukemia virus (MSV) RCR.

Molecular and biological assays for detection of RCL associated with lentiviral vectors have been developed *(43–52)*. The current experimental methodology for detection of RCL in vector preparations is outlined in **Fig. 2**, and was modeled after guidelines recommended by FDA for detecting RCR. In a typical assay for detecting RCL associated with HIV-1 based vectors, the test article is used to inoculate a cell line (C8166-45 derived from human umbilical cord blood lymphocytes) that is permissive for infection and growth of HIV-1, and the transduced cell line is cultured for 21 days to amplify any RCL in the test article *(49)*. After three weeks, cell free medium from amplified cells is used to infect naïve C8166 (Indicator) cells, which are cultured for an additional seven days and analyzed for the presence of viral markers. Sensitive assays have been developed for detection of RCL indicative markers including the $p24^{gag}$ antigen (by ELISA), viral reverse transcriptase (by product enhanced reverse transcriptase or PERT), psi-gag sequences that result from a recombination between vector and packaging constructs (by PCR) and the psuedotyping envelop VSV-G (by PCR) *(45, 48–52)*. In our current methodology for HIV-1 vectors, positive PCR for psi-gag sequences and presence of $p24^{gag}$ antigen by ELISA are indicative of RCL in vector test articles *(49)*. Biological end point assays (like the RCR detecting marker rescue assay) have been described for research grade vector product but have not been fully developed for testing clinical grade material.

PCR assays to detect other viral marker sequences like *tat* have been described to identify RCL associated with Jembrana disease virus (JDV) vectors *(44)*; however, the *tat* gene is not included in most of the later generation lentiviral vectors and therefore *tat* PCR has limited applicability for RCL testing. Most PCR assays described for RCL detection are designed with the assumption that the target sequences are present in the RCL being analyzed; however, as the structure of RCL associated with any of the current array of lentiviral vectors is unknown, it may be useful to have assays that are RCL structure/sequence independent.

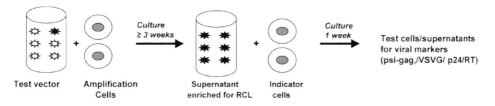

Fig. 2. RCL detection assay: Test article composed primarily of replication defective particles (*open circles*) is cultured with a permissive cell line (*shaded ovals*) to amplify any replication competent virus (*filled circles*). Supernatants from amplified cells enriched for RCL (if present in the test article) are subsequently incubated with naïve indicator cells. Infected indicator cells are analyzed for viral markers. Transfer of viral sequences from amplification cells to indicator cells is indicative of RCL in the test article.

The PERT assay detects reverse transcriptase activity associated with retroviruses and has shown to be as sensitive as the p24$^{gag}$ ELISA and psi-gag PCR for RCL detection when the RCL has been amplified *(52)*. The disadvantage of PERT is a high background found in certain cell types, but a major advantage is the enzymatic function (of RT) it detects is crucial for replication of any RCL, i.e., it can be used for detection of RCL associated with structurally distinct retroviruses. However, due to the hypothetical nature of RCL, it may be advisable to use at least two assays that detect different viral genes/functions for conclusive evidence of RCL in vector products. Ideally a molecular assay (PCR for viral sequence) can be used in conjunction with a functional (marker rescue) or protein expression (RT/p24 ELISA) assay for RCL testing.

For subject monitoring for evidence of RCR infection, positive results obtained by PCR or serologic methods should be confirmed by biologic assays to exclude false positives. Also, one must consider that complex recombinations may generate a RCR that lacks the sequence targeted in these assays. Clinical suspicion of RCR with a negative PCR or serologic result should prompt further testing for other target sequences, as well as testing in biologic assays. Subjects receiving lentiviral gene therapy should be monitored similarly for RCL infection by molecular and serological methods. Initial positive reactivity in any of these assays should be followed by biological amplification assays for vector derived RCL.

### 1.4. Regulatory Issues

The concern surrounding RCR and RCL has led US regulators to develop recommendations relevant to the clinical use of these vectors. As retroviral vectors generally utilized packaging cell lines the US FDA requires RCR testing on the Master Cell Bank, any Working Cell Banks, and the post-production cells (1% or a maximum of $10^8$ cells). Also, 5% of the final product volume must also be screened. In addition, vectors used in ex vivo transduction protocols in which the cells are cultured for greater than 4 days must be screened for RCR. A new guidance also defines screening that must occur for subjects treated with retroviral vectors. Patients treated with retroviral vectors must also be monitored for RCR pre-treatment, then at 3, 6, and 12 months. If all samples are negative for RCR the subject can be followed by yearly assessment of clinical status and archiving of blood or relevant tissue. Any suggestion of RCR exposure mandates extensive evaluation and close follow-up. There are similar expectations for RCL testing and follow-up.

To date, there have been no documented exposures of patients treated with gene therapy to RCR or RCL. Although this suggests current screening methods are sensitive, the known risk of insertional mutagenesis mandates continued vigilance and continued monitoring clinical trial subjects, especially given the limited experience to date with lentiviral vectors.

## 2. Materials

### 2.1. RCR Testing

#### 2.1.1. Media and Cell Culture Supplies

1. McCoys 5A Medium + 1% penicillin and streptomycin (all from GIBCO) containing 10% fetal bovine serum (FBS, Hyclone) (McCoys10); Trypsin-EDTA (GIBCO) and DPBS (Cambrex) for subculturing cells.

2. For cell expansion 75, 175, and 300, 450 cm² flasks, calibrated pipettes of various sizes, aspirating pipettes (1 and 10 mL). For dilution of samples 12 × 75 and 17 × 100 mm polypropylene tubes and 6-well plates for the assay. 1–200 µL sterile pipette tips for all standard procedures and 0.45 µm cellulose acetate syringe filters for filtering cell supernatants.

3. *Mus dunni* (CRL-2017) and PG4 (CRL-2032) cells available from the American Tissue Culture Collection (ATCC) at http://www.atcc.org.

4. Polybrene (Sigma). Final concentration will be 8 µg/mL throughout the procedure, it is helpful to prepare an 8 mg/mL stock.

#### 2.1.2. Positive Control for RCR Testing

1. 4070A positive control (VR-1450) available from the American Tissue Culture Collection (ATCC) at http://www.atcc.org.We infect cells with virus then grow up a high titer stock which is aliquoted into vials and used as a positive control for multiple assays. The stock is characterized for titer and the $TCID_{50}$ (determined by using infectious titer and Reed Muench formula) *(49)*. Stocks of virus are stored at –70°C.

### 2.2. RCL Testing

#### 2.2.1. Media and Cell Culture Supplies

1. RPMI 1640 + 1% penicillin and streptomycin (all from GIBCO) containing 10% fetal bovine serum (FBS, Hyclone) (subsequently referred to as RPMI 10); DMEM supplemented with 2 mM l-glutamine and 1 mM sodium pyruvate and containing 10% FBS (subsequently referred to as D10).

2. For cell expansion 75, 175, and 300 cm² flasks, calibrated pipettes of various sizes, aspirating pipettes (1 and 10 mL), centrifuge tubes (15 and 50 mL) from BD Falcon or equivalent; 10 cm²–50 mL flat bottom culture tubes for small scale transductions.

3. For RCL testing of HIV-1 vectors, C8166 cells are available from the NIH AIDS Research and Reference Reagents Program.

4. Polybrene (Sigma).

*2.2.2. Test Articles for RCL Testing*

Test articles can consist of lentiviral vector preparation or cell media from transduced cell populations.

*2.2.3. Positive Control Virus for RCL Testing*

For HIV-1 vectors, an attenuated HIV-1 virus lacking the accessory genes is used as a positive control in cGMP RCL assays (ex. R8.71) *(45, 48)*. Stocks of virus are stored at –70°C. The $TCID_{50}$ of the positive control virus is determined by using infectious titer and Reed Muench formula *(49)*.

*2.2.4. Reagents for RCL Detection Assays*

1. HIV-1 p24$^{gag}$ antigen ELISA kit (RETROtek or Perkin Elmer) for measuring p24$^{gag}$ protein in indicator cultures.
2. Puregene DNA extraction kit (Gentra) for isolating genomic DNA from indicator cells.
3. PCR:
   1. PCR primers for amplifying psi-gag sequences: GrecF1 (5-CAGGACTCGGCTTGCTGAA-3) and GrecR2 (5-TGTCTTATGTCCAGAATGCT-3), GrecP (5-AAGATTTAAACACCATGCTA-3)
   2. PCR reagents: HotStarTaq DNA polymerase, dNTPs, water (molecular biology grade).
   3. PCR reaction analysis: Agarose, 1 M Tris pH 7.6 (Sigma), ethidium bromide (Sigma), 1× TAE, Blue-orange loading dye (Promega), molecular standards (GIBCO).
4. Southern blotting and probing PCR products:

   Blotting: Nytran supercharged membrane (Midwest), NaOH
   Probe labeling: Probe GrecP, 10× phosphorylation buffer (Roche), Gamma $^{32}$P dATP (Perkin Elmer), 1× STE (0.1 M sodium chloride, 0.01 M Tris-HCl, 0.001 M EDTA), polynucleotide kinase (Roche), Scintillation counting fluid.
   Hybridization: 5 M NaCl, 1 M Tris, pH 7.6, 0.5 M EDTA (pH 8.0), 10% SDS, ExpressHyb hybridization solution, 20× SSC.

**2.3. Equipment**

1. Tissue Culture Incubator, Pipette Aid, Biological Safety Cabinet, Inverted Microscope.
2. Centrifuge (with aerosol protective cover).
3. Adjustable Pipettor: 1–200 µL.
4. Hybridization oven, Scintillation counter and protective barriers for radioactive work.
5. ELISA reader.
6. >–70°C Freezer.

# 3. Methods

### 3.1. RCR Detection

#### 3.1.1. Overview

The Extended S⁺/L⁻ Assay is composed of two stages (**Fig. 1**). The first stage is the amplification stage in which a test article/sample is inoculated into a culture of a cell line permissive for viral replication (for this assay Mus dunni cells). The second stage is the detection phase where media is obtained from the amplification cells at the three-week time point and analyzed on a cell line capable of detecting replication competent retrovirus. For this assay, the PG-4 S⁺/L⁻ cells are used. Foci of transformed cells indicate the presence of replication competent retrovirus. The assay described here is for detection of amphotropic virus, the assay may need to be modified for other viral envelopes (*see* **Note 2**).

#### 3.1.2. Amplification Phase

The size of the amplification phase depends on the volume of the sample being tested. As discussed above, for clinical material as much as 5% of the total batch volume must be tested. For research purposes, generally smaller amounts are tested. **Table 1** indicates the number of cells that will be needed based on the test volume.

1. Expand Mus dunni cells to the number required for plating in McCoy's10.

2. Day (-1): Count Mus dunni cells and seed into flasks using Table 17.1. Determine the size of flasks required for each test article according to the following table. For negative controls you will need three 25-cm² flasks with $2 \times 10^5$ Mus dunni cells per flask. For positive controls prepare five 25-cm² flasks with $2 \times 10^5$ Mus dunni cells per flask. As appropriate, additional "spiking control" may be required (*see* **Note 3**).

3. Day (0): Remove media from the three negative control flasks, add 1 mL of McCoy's media and POLYBRENE® to

### Table.1
### Cells needed for testing different quantities of test articles

| Test volume (mL) | Mus Dunni cells | Flask size (cm²) |
| --- | --- | --- |
| 1–4 | $2 \times 10^5$ | 25 |
| 10–15 | $5 \times 10^5$ | 75 |
| 30–50 | $1.5 \times 10^6$ | 175 |
| 60–90 | $2 \times 10^6$ | 300 |
| 60–120 | $3 \times 10^6$ | 450 |

a final concentration of 8 µg/mL. Remove media from the test article flasks, and add the test article and POLYBRENE®. Return flasks to incubator. Then remove media from the five 25-cm² flasks to be used as the positive control. Add 1 mL of 4070A virus stock that has been diluted to the $TCID_{50}$. Add POLYBRENE®. Incubate at 37°C, 5% $CO_2$ for 2–4 h. After infection remove media from all flasks and replace with the appropriate volume of fresh McCoy's medium. Return flasks to incubator.

4. Passage of cells during the three-week amplification. Mus dunni cells should be maintained in log phase growth through the 3-week period. To accomplish this, cells should be passaged a minimum of five times during the three-week period and at least two times per week. The amplification kinetics do allow scaling down of large flasks with later passages, as demonstrated in **Table 2.** Cell split ratios may be smaller than indicated if cells are less than confluent. Larger split ratios should not be used.

5. Day 14 or 15: Begin PG-4 cell expansion so that adequate number of cells will be available to assay amplification samples. Review the growth history of the PG-4 cells. If the cells

**Table. 2**
**Cell passaging during the amplification phase**

| Split no. and day | ORIGINAL FLASK SIZE | | | |
|---|---|---|---|---|
| | 450 or 300 cm² | 175 cm² | 75 cm² | 25 cm² |
| No. 1: Day 3–4 | 1:10 175 cm² | 1:10 75 cm² | 1:5 75 cm² | 1:5 25 cm² |
| No. 2: Day 7–8 | 1:10 75 cm² | 1:5 75 cm² | 1:5 75 cm² | 1:5 25 cm² |
| No. 3: Day 10–11 | 1:5 75 cm² | 1:5 75 cm² | 1:5 75 cm² | 1:5 25 cm² |
| No. 4: Day 14–15 | 1:5 75 cm² | 1:5 75 cm² | 1:5 75 cm² | 1:5 25 cm² |
| No. 5: Day 17–20 | 1:5 75 cm² | 1:5 75 cm² | 1:5 75 cm² | 1:5 25 cm² |
| No. 6: Day 21–22 (optional split) | 1:5 75 cm² | 1:5 75 cm² | 1:5 75 cm² | 1:5 25 cm² |

have been passaged greater than 15 times, discard and thaw new vial of PG-4 cells.

6. Harvest amplification media: Media can be harvested from amplification cultures $3 \geq 21$ days from culture initiation. Cultures should be confluent (usually 2–4 days after 5th or 6th split), then change media. Approximately 24 h after media change, collect supernatant and filter through a 0.45 μm filter. Supernatant may be inoculated directly onto PG-4 cells (see below) or frozen for later assay (store at –70°C). Do not leave supernatant out at room temperature for more than 2 h.

*3.1.3. Pg-4 Focus Forming Assay*

1. Determine the total number of 6-well plates required for the assay. Three wells will be needed for each test article. For the control samples 30 wells (6 plates) will be needed.

2. Day (-1): Plate $1 \times 10^5$ cells per well in 6-well plates. Add 4 mL of McCoy's10. Incubate overnight at 37°C, 5% $CO_2$.

3. Day (0): Prepare direct (nonamplified) negative control for the PG-4 Assay. Remove media from 3 wells and add 1 mL of McCoy's10 and add POLYBRENE®.

4. Day (0): Prepare amplification negative controls. Remove medium from each well of one plate. Add 1 mL of filtered supernatant from each of the three negative control flasks from the amplification phase to 2 wells and add POLYBRENE®.

5. Day (0): Prepare test articles. From each flask of the test articles, remove and filter supernatant; material will be tested undiluted and at $10^{-2}$ dilution (dilute in McCoys10, *see* **Note 4**). Remove media from designated wells. Add 1 mL of undiluted supernatant to two wells and 1 mL of supernatant diluted to $10^{-2}$ to the other well. Add POLYBRENE® to each well.

6. Day (0): Prepare amplification positive controls. From each of the 5 flasks of amplification positive control, remove and filter supernatant. Prepare the following dilutions for each flask: Flask #1: undiluted, $10^{-2}$, $10^{-4}$, $10^{-5}$, $10^{-6}$, and $10^{-7}$ Flasks #2–5: $10^{-2}$, $10^{-4}$, and $10^{-6}$. Remove media from designated wells and add 1 mL of the appropriate dilution, add POLYBRENE®.

7. Day (0): Prepare direct (nonamplified) positive control for the PG-4 assay. Thaw a vial of stock 4070A positive control virus. Infect cells with three dilutions that are more concentrated than the $TCID_{50}$, at the $TCID_{50}$ dilution, and two dilutions that are less concentrated than the $TCID_{50}$ (Example: If $TCID_{50} = 10^7$ test dilutions will be $10^{-4}$, $10^{-5}$, $10^{-6}$, the TCID ($10^{-7}$) and two dilutions less ($10^{-8}$ and $10^{-9}$). Prepare two wells per dilution, and then add POLYBRENE® to each well. Incubate for 2–4 h then remove medium from all wells and replace with 4 mL of McCoy's medium. Return plates to incubator.

8. Day (2): Remove medium and add 4 mL of fresh McCoy's medium. Return to incubator.

*3.1.4. Score Foci*

1. Day (4): Examine negative control wells for confluence using an inverted microscope with a 4× objective. If negative controls are not confluent, remove medium from all wells and replace with 4 mL of McCoy's medium. Check daily until negative controls show a confluent lawn of cells. If >2 additional days are required for cells to become confluent, suspect complications.

2. Each well should be examined for the presence of foci using a 4× objective. When the number of foci is between 0 and 20 in a well, record the actual number observed. If there are greater than 20 foci, record a qualitative assessment (+ if foci are grown together but normal cells observed, ++ if near or total ablation of the cell lawn is observed).

3. Examine the negative control wells. The assay is acceptable if no foci are observed.

4. Next examine the amplified positive controls. Foci should be present in at least one of the dilutions from the five amplified positive controls.

5. Examine the direct positive control. Foci should be present in the dilution which is two log more concentrated than the $TCID_{50}$ (Example: if $TCID_{50} = 10^7$, foci should be present in the $10^{-5}$ dilution). Foci in less concentrated dilutions are acceptable, if foci are not seen in dilutions 2 log greater than the $TCID_{50}$ consider complications. Count and record the number of foci in each well.

6. Examine the test article wells and record the number of foci. For cGMP work, we have two trained technicians read each well independently.

**3.2. RCL Detection in HIV-1 Vectors**

In this section, a detailed protocol for detecting RCL in HIV-1 vectors is described.

*3.2.1. Determine P24 Content of Test Articles*

1. If the test article is a lentiviral vector, the concentration of p24 must be determined as concentrated lentiviral vectors can inhibit the growth of the C8166 cell line. Determine the p24 value of test vector using p24$^{gag}$ ELISA as per manufacturers instructions.

2. For test articles with p24 values ≤1,000 ng/mL, use the material undiluted. For materials that exceed this amount, the test article should be diluted to final concentration of 1,000 ng/mL. Calculate the final volume of diluted test article and determine the amount of C8166 cells that will be required according to **Table. 3.** Propagate the required number of C8166 cells in RPMI10.

*3.2.2. Amplification Phase*

1. Day (-1) Preparation of C8166 cells (*see* **Note 5**): Harvest C8166 cells and count. Label six 50 mL culture tubes (3 for negative control and 3 for positive control). Add 12 mL RPMI to each culture tube. Aliquot $5 \times 10^6$ C8166 cells/culture tube. Label appropriate number of flasks for the test article. Add RPMI10 to each flask (12 mL/75 cm² flask or 50 mL culture tube, 30 mL/175 cm² flask, 60 mL/300 cm² flask). Add appropriate number of C8166 cells for the test article according to calculations from **Table. 3**. Place all culture tubes/flasks in the incubator (37°C, 5% $CO_2$).

2. Day (0) Inoculation of cultures: Centrifuge negative control tubes at $660 \times g$ for 3 min. Remove the media and add 1 mL of RPMI10. Add POLYBRENE® to a final concentration of 8 µg/mL.

3. Day (0) Prepare test article for inoculation into culture. If starting p24 concentration is >1,000 ng/mL dilute test article in RPMI medium to obtain 1,000 ng of p24. Centrifuge C8166 cells at $660 \times g$ for 3 min then add the appropriate amount of test article using the ratios defined in **Table. 1**. Add POLYBRENE®.

4. Day (0) Preparation of positive control cultures (*see* **Note 6**): Centrifuge C8166 cells as described above for negative control, but replacing media with replication competent virus (R8.71). The level of sensitivity will be based on the dilution of the virus. For example, use of virus at the $TCID_{50}$ would suggest maximal sensitivity of the assay. To insure that the test article is not inhibitory, additional positive controls may be added in which the test article is spiked with R8.71 (*see* **Note 3**). Add POLYBRENE®. Mix all cultures and return them to the (37°C, 5% $CO_2$). Incubate for 4 h.

## Table. 3
### Cells needed for testing different amounts of test vector for RCL

| Test volume (after appropriate dilution to obtain p24 levels of ≤1,000 ng/mL) | C8166 amplification cells | Flask (tube) size |
|---|---|---|
| 1–5 mL | $5 \times 10^6$ | 50 mL flat bottom culture tubes |
| 6–30 mL | $3 \times 10^7$ | 75 cm² |
| 31–60 mL | $6 \times 10^7$ | 175 cm² |
| 61–100 mL | $1 \times 10^8$ | 300 cm² |

5. 5 Day (0). After incubation, centrifuge the cells in all cultures, remove media and add appropriate amount ofRPMI10 (**Table. 4**) and transfer to appropriate sized flasks and transfer to the incubator. The positive and negative control samples are resuspended in 12 mL RPMI and transferred 75 cm² flasks.

6. Passage of cells during the 3-week amplification. Amplification cells should be maintained in log phase growth through the 3-week period by passing (ratio 1:5 to 1:10) a minimum of five times during the three-week period (least two times per week).

7. Day 17–19. Prepare for the Indicator Phase by expanding naïve C8166 cells 17–19 days after the start of the amplification phase.

8. Three to four days after 5th split and 24 h before collecting the amplification supernatants, transfer all cultures to conical tubes and spin at 660 × g for 3 min. Remove media and add 12 mL of fresh RPMI medium. Transfer to appropriate sized flasks and return cultures to the incubator.

9. Twenty-four hours after media change, transfer all cultures to 15 mL conical tubes, centrifuge and harvest supernatants and cells. Filter supernatants through a 0.45 μm filter, reserve 1 mL for p24$^{gag}$ determination. Use 2 mL of test article, and 2 mL from each of the negative and positive control flasks for the indicator phase testing. Reserve unused supernatant and store at –70°C.

10. OPTIONAL – Collect approximately $6 \times 10^6$ cells from each test article and controls for further analysis if RCL is detected after the indicator phase.

## Table. 4
## Processing of cultures after transduction

| Original sample volume (mL) | Resuspension volume after transduction (mL) | Flask size after transduction (cm²) |
|---|---|---|
| 1–5 | 12 | 75 |
| 6–30 | 30 | 75 |
| 31–60 | 60 | 175 |
| 61–100 | 100 | 300 |

*3.2.3. Indicator Phase (Begin at Least 21 Days after Initiation of the Amplification Phase)*

1. On the basis of the amplification phase, $2 \times 50$ mL culture tubes will be required for each test article, and 6 for the negative control and 6 for the positive controls. Plate $1 \times 10^6$ C8166 cells per 50 mL culture tube with 4 mL RPMI10 and incubate overnight.

2. After the overnight incubation, centrifuge cells. There will be three negative control supernatants: add 1 mL from each to duplicate culture tubes. For each test article, duplicate culture tubes should be inoculated with 1 mL. The three positive controls should be tested by adding 1 mL of each into duplicate culture tubes. Polybrene® should be added to each tube. Incubate for 4 h, centrifuge at $660 \times$ g for 3 min. Remove medium, and replace with fresh RPMI10, then return to the incubator and culture for two days.

3. Day (2): Transfer all cultures from tubes to a 75 cm² flask with 12 mL of fresh RPMI10 and return to incubator.

4. Day (5): Remove cells from flask, centrifuge and resuspend cell pellet in fresh RPMI10. Return to incubator.

5. Day (6): At least 24 h after media change, supernatant and cells are ready to be harvested. For each sample, transfer media to a 15 mL conical tubes, centrifuge at $660 \times$ g, for 3 min and filter separately through a 0.45 μm filter. Transfer an aliqout of filtered supernatant to a labeled tube for p24 determination. Then collect approximately $6 \times 10^6$ cells for DNA isolation in a 2.0 mL microcentrifuge tube and isolate as directed by manufacturer using a commercial kit (ex. Puregene DNA extraction kit, Qiagen,Valencia, CA). Freeze unused supernatants as a reserve samples at $\leq -70°C$.

*3.2.4. RCL Detection Assays*

RCL can be detected by a variety of methods and the stringency will relate to the level of assurance required. The easiest method is detection of p24[gag] antigen by ELISA *(43, 48–49)*. Additional tests can include psi-gag PCR and PCR for the VSV-G envelope *(49)*. The ELISA method is rapid and reproducible but the kits are expensive. The PCR methods assume that the RCL has generated from vector and packaging sequence (psi-gag), or that the RCL has not incorporated another envelope besides VSV-G. For clinical grade material, we perform both the p24 ELISA and psi-gag. For nonclinical material, we utilize the p24 ELISA and reserve the psi-gag if there is concerns about false positive ELISA results.

1. p24[gag] antigen estimation: Perform p24 ELISA on amplification and indicator phase cultures using a commercially available ELISA kit according to manufacturers specifications. The kit positive control standards (purified p24 gag protein) are used to establish a standard curve for the assay; a cut off value is established using buffer and negative controls from the RCL

assay to set a lower detection limit. All samples are analyzed in duplicate and are scored positive if they are above the cut off value of the assay.

2. Psi-gag PCR: Extract DNA from indicator cell samples according to manufacturer's specifications (Puregene Kit). Amplify 1 μg of DNA from indicator cell samples in a standard 50 μL PCR assay containing 0.5 μM psi-gag specific primers (Grec F1/ GrecR2), 200 μM dNTPs, 1× HotStart buffer, and 0.05 U of HotStart Taq. Amplification conditions are 95°C, 10 min and 40 cycles of 95°C for 1 min; 55°C, 1 min; 72°C, 1 min and a final extension of 72°C for 7 min. As a positive control for PCR use 100 copies of plasmid DNA containing recombinant HIV-1 psi-gag sequences (R8.71) in a background of C8166 genomic DNA; no template control is included as a negative control for the PCR assay. All samples are run in duplicate in the PCR assay and are analyzed by agarose gel electrophoresis followed by transfer of PCR products to a nylon membrane and probing with a radio-labeled probe (GrecP). Test articles are scored as positive if they exhibit a 953 bp band in the Southern analysis; negatives from the RCL assay should exhibit no signal and positive samples from the RCL assay should be positive. The PCR positive control (R8.71) should show the expected sized band and the assay negatives should show no signal for the assay to be valid.

*3.2.5. Interpretation of Assays*

1. Although p24 is normally below the detectable limit at the end of the amplification phase, the high concentration of p24 in lentiviral vectors occasionally results in low levels still detectable at 21 days. Also, low level transfer of sequences (such as psi-gag or VSV-G) can occur in amplification phase cells (possibly due to plasmid contamination) in the absence of RCL. Therefore, we consider a true RCL to be one that can be transferred to the indicator phase.

2. Positive controls from the indicator phase (at least 1/3) should be positive for p24 antigen (and psi-gag sequences if tested) if vector was inoculated at the $TCID_{50}$.

3. Negative controls from the indicator phase should be negative for p24 antigen (and psi-gag sequences if tested).

4. If test articles are negative for both p24 and psi-gag sequences at the indicator phases, no RCL is present in the test article. If any test articles from the indicator phase tested positive for p24 antigen alone or psi-gag sequences alone, repeat p24 ELISA and psi-gag PCR to confirm results. Consider analyzing saved material from the amplification phase.

5. If test articles are positive for both p24 antigen and psi-gag they may contain RCL.

# 4. Notes

1. As an example, Miller and colleagues identified a novel retrovirus, the *Mus dunni* endogenous virus (MDEV), which was present in the *Mus dunni* cell line used in a marker rescue assay. The virus was activated by hydrocortisone in the medium being tested for RCR *(41, 53)*. Therefore, cell lines used for amplification and assay must be considered as a possible source of RCR unrelated to the vector product.

2. The selection of the amplification phase cell line and the indicator cell assay will depend in part on the vector pseudotype. Pseudotype refers to the viral envelope selected for expression on the surface of the vector particle. The MoMLV from which many vectors are derived normally expressed the ecotropic envelope. As the ecotropic receptor is limited to rodent cells, vectors for human cell transduction were initially pseudotyped with the envelope from the amphotropic 4070A virus whose receptor is present on most mammalian cells. Methods detecting RCR pseudotyped with the 4070A envelope were developed, and the *Mus dunni* cell line is commonly used in the amplification phase due to the susceptibility to and amplification of a wide variety of murine leukemia viruses *(54, 55)*. Although *Mus dunni* propagates many RCRs, viruses enveloped with the ecotropic MoMLV glycoprotein are a notable exception and alternate cell lines must be used for their detection *(56)*. Also, the recent use of non-murine retroviral envelopes to pseudotype retroviral vectors has complicated the screening for RCRs. One that has now been used in a variety of clinical applications is the envelope derived from the Gibbon Ape Leukemia Virus (GALV) *(25, 57)*. The GALV envelope has demonstrated improved transduction efficiency in a number of target cells, in part, due to the increased expression of the GALV receptor on many target cells *(58–61)*. Although GALV acts as a xenotropic virus in that it infects primate, and other mammalian cells, it cannot infect murine cells *(25)*. Therefore, *Mus dunni* cells are not suitable for GALV amplification. To address this, 293 cells have been substituted for *Mus dunni* during the amplification phase of the extended S⁺/L⁻ assay with similar levels of virus detection *(62)*. Another retroviral envelope being developed for clinical trial use was cloned from the RD114 virus, which also displays properties of xenotropic viruses *(63–65)*. The 293 cell line is also useful for amplifying RD114 pseudotyped RCR in an extended S⁺/L⁻ assay *(66)*.

3. For testing clinical samples "spiking" or mixing of the test article with a known concentration of the positive control is

prudent to determine whether the test article is inhibitory. In this situation, use a concentration of virus above the $TCID_{50}$ since some inhibition is expected when the vector and RCR/RCL contain the same envelope (due to receptor interference). At present the FDA has not set guidelines as to the level of inhibitory effect that is allowable.

4. For RCR testing on PG-4 cells we test both undiluted and a $10^{-2}$ dilution. Generally foci can be seen only at low (<100 infectious virus) per well. Above this level, the majority of cells will be transformed and the lawn will be disrupted (generally when we titer virus after amplification it is >$10^4$ infectious viruses per mL). We use the $10^{-2}$ dilution because we have occasionally seen very high concentration of virus inhibit transformation (possibly through receptor interference by defective particles). In this case, the lawn of PG-4 cells will appear abnormal but not disrupted. At the $10^{-2}$ dilution the cells will be clearly transformed and the lawn will be disrupted, confirming the presence of RCR.

5. The choice of amplification cell line for RCL may vary depending on the type of lentiviral vector being tested, the pseudotyping envelop and on the positive control virus being used for the assay. For HIV-1 vectors, the C8166 cell line was used as it has high transduction efficiency for these vectors, can be infected efficiently by the positive control virus with a natural HIV-1 envelope, and can amplify the positive control over 100-fold *(49)*.

6. The choice of a positive control for the RCL assay is challenging as the structure of a RCL is not known. As the vectors are VSV-G pseudotyped, it may be argued that an HIV-1 virus pseudotyped with VSV-G is a better positive control for the RCL assay than an attenuated HIV-1 virus carrying its natural envelop. We chose the attenuated virus as a VSV-G pseudotyped HIV-1 presents an unacceptable risk to laboratory workers *(49)*. For RCL detection in EIAV vectors, 4070A amphotropic MLV has been used as a positive control *(51)*; however, whether an EIAV-RCL behaves like 4070A MLV in experimental situations is debatable.

## References

1. Miller, A.D., Eckner, R.J., Jolly, D.J., Friedman, T., Verma, I.M. (1984) Expression of a retrovirus encoding human HPRT in mice. *Science* 225, 630–632.

2. Williams, D.A., Orkin, S.H., and Mulligan, R.C. (1986) Retrovirus-mediated transfer of human adenosine deaminase gene sequences into cells in culture and into murine hematopoietic cells in vivo. *Proc Natl Acad Sci USA* 83, 2566–2570.

3. Eglitis, M.A., Kantoff, P., Gilboa, E., and Anderson, W.F. (1985) Gene expression in mice after high efficiency retroviral-mediated gene transfer. *Science* 230, 1395–1398.

4. Rosenberg, S.A., Aebersold, P.M., Cornetta, K., Kasid, A., Morgan, R.A., Moen, E.M., et al. (1990) Gene transfer into humans-immunotherapy of patients with advanced melanoma, using tumor infiltrating lymphocytes modified by retroviral gene transduction. *N Engl J Med* 323, 570–578.

5. Rosenberg, N. and Joelicoer, P. (1997) Retroviral Pathogenesis, in *Retroviruses* (Coffin, J.M., Hughes, S.H., and Varmus, H.E., ed.), Cold Spring Harbor Laboratory Press, Cold Spring Harbor, NY. pp. 475–586.

6. Li, Z., Dullman, J., Schiedlmeier, B., Schmidt, M., von Kalle, C., Meyer, J., et al. (2002) Murine leukemia induced by retroviral gene marking. *Science* 296, 497.

7. Cavazzana-Calvo, M., Hacein-Bey, S., de Saint Basile, G., Gross, F., Yvon, E., Nusbaum, P., et al. (2000) Gene therapy of human severe combined immunodeficiency (SCID)-X1 disease. *Science* 288, 669–672.

8. Hacein-Bey-Abina, S., von Kalle, C., Schmidt, M., Le Deist, F., Wulffraat, N., McIntyre, E., et al. (2003) A serious adverse event after successful gene therapy for X-linked severe combined immunodeficiency. *N Engl J Med* 348, 255–6.

9. Hacein-Bey-Abina, S., von Kalle, C., Schmidt, M., McCormack, M., Wulffraat, N., Leboulch, P., et al. (2003) LMO2-associated clonal T cell proliferation in two patients after gene therapy for SCID-X1. *Science* 302, 415–419 [erratum appears in *Science*, 302, 568].

10. Hacein-Bey-Abina, S., Le Deist, F., Carlier, F., Bouneaud, C., Hue, C., De Villartay, J.P., et-al. (2002) Sustained correction of X-linked severe combined immunodeficiency by ex vivo gene therapy. *N Engl J Med* 346,1185–1193.

11. Berns, A. (2004) Good news for gene therapy. *N Engl J Med* 350, 1679–1680.

12. Dave, U.P., Jenkins, N.A., and Copeland, N.G. (2004) Gene therapy insertional mutagenesis insights. *Science* 303, 333.

13. Cornetta, K., Nguyen, N., Morgan, R.A., Muenchau, D.D., Hartley, J., and Anderson, W.F. (1993) Infection of human cells with murine amphotropic replication-competent retroviruses. *Hum Gene Ther* 4, 579–588.

14. Donahue, R.E., Kessler, S.W., Bodine, D., McDonagh, K., Dunbar, C., Goodman, D., et-al. (1992) Helper virus induction T cell lymphoma in nonhuman primates after retroviral mediated gene transfer. *J Exp Med* 176, 1125–1135.

15. Miller, A.D. (1990) Retrovirus packaging cells. *Hum Gene Ther* 1, 5–14.

16. Muenchau, D.D., Freeman, S.M., Cornetta, K., Zwiebel, J.A., and Anderson, W.F. (1990) Analysis of retroviral packaging lines for generationof replication-competent virus. *Virology* 176, 262–265.

17. Bodine, D.M., McDonagh, K.T., Brandt, S.J., Ney, P.A., Agricola, B., Byrne, E., et al. (1990) Development of a high-titer retrovirus producer cell line capable of gene transfer into rhesus monkey hematopoietic stem cells. *Proc Natl Acad Sci USA* 87, 3738–3742.

18. Scarpa, M., Cournoyer, D., Muzny, D.M., Moore, K.A., Belmont, J.W., and Caskey, C.T. (1991) Characterization of recombinant helper retroviruses from Moloney-based vectors in ecotropic and amphotropic packaging cell lines. *Virology* 180, 849–852.

19. Otto, E., Jones-Trower, A., Vanin, E.F., Stambaugh, K., Mueller, S.N., Anderson, A.W., et al. (1994) Characterization of a replication-competent retrovirus resulting from recombination of packaging and vector sequences. *Hum Gene Ther* 5, 567–575.

20. Bosselman, R.A., Hsu, R.-Y., Bruszewski, J., Hu, S., Martin, F., and Nicolson, M. (1987) Replication-defective chimeric helper proviruses and factors affecting generation of competent virus: expression of Moloney Leukemia Virus structural genes via the metallothionein promoter. *Mol Cell Biol* 7, 1797–1806.

21. Markowitz, D., Goff, S., and Bank, A. (1988) A safe packaging line for gene transfer: seperating viral genes on two different plasmids. *J Virol* 62, 1120–1124.

22. Markowitz, D., Goff, S., and Bank, A. (1988) Construction and use of a safe and efficient amphotropic packaging line. *Virology* 167, 400–406.

23. Danos, O. and Mulligan, R.C. (1988) Safe and efficient generation of recombinant retroviruses with amphotroic and ecotropic host ranges. *Proc Natl Acad Sci USA* 85, 6460–6464.

24. Miller, A.D. and Rosman, G.J. (1989) Improved retroviral vectors for gene transfer and exprssion. *BioTechniques* 7, 980–990.

25. Miller, A.D., Garcia, J.V., Von Suhr, N., Lynch, C.M., Wilson, C., and Eiden, M.V. (1991) Construction and properties of retrovirus packaging cells based on gibbon ape leukemia virus. *J Virol* 65, 2220–2224.

26. Chong, H., Starkey, W., and Vile, R.G. (1998) A replication-competent retrovirus arising from a split-function packaging cell line was generated by recombination events between the vector, one of the packaging constructs, and endogenous retroviral sequences. *J Virol* 72, 2663–2670.

27. Garrett, E., Miller, A.R.-M., Goldman, J., Apperley, J.F., and Melo, J.V. (2000) Characterization of recombinant events leading to the production of an ecotropic replication-competent retrovirus in a GP + envAM12-derived producer cell line. *Virology* 266, 170–179.

28. Patience, C., Takeuch, Y., Cosset, F.L., and Weiss, R.A. (2001) MuLV packaging systems as models for estimating/measuring retrovirus recombination frequency. *Dev Biol* 106, 169–179.

29. Kim, V.N., Mitrophanous, K., Kingsman, S.M., and Kingsman, A.J. (1998) Minimal requirement for a lentivirus vector based on human immunodeficiency virus type 1. *J Virol* 72, 811–6.

30. Mochizuki, H., Schwartz, J.P., Tanaka, K., Brady, R.O., and Reiser, J. (1998) High-titer human immunodeficiency virus type 1-based vector systems for gene delivery into nondividing cells. *J Virol* 72, 8873–83.

31. Gasmi, M., Glynn, J., Jin, M.J., Jolly, D.J., Yee, J.K., and Chen, S. (1999) Requirements for efficient production and transduction of human immunodeficiency virus type 1-based vectors. *J Virol* 73, 1828–34.

32. Dull, T., Zufferey, R., Kelly, M., Mandel, R.J., Nguyen, M., Trono, D., et al. (1998) A third-generation lentivirus vector with a conditional packaging system. *J Virol* 72, 8463–71.

33. Urnovitz, H.B., and Murphy, W.H. (1996) Human endogenous retroviruses: nature, occurrence, and clinical implications in human disease. *Clin Microbiol Rev* 9, 72–99. Review.

34. Naldini, L., Blomer, U., Gallay, P., Ory, D., Mulligan, R., Gage, F.H., et-al. (1996) In vivo gene delivery and stable transduction of nondividing cells by a lentiviral vector. *Science* 12, 263–7.

35. Sastry, L., Johnson, T., Hobson, M.J., Smucker, B., and Cornetta, K. (2002) Titering lentiviral vectors: comparison of DNA, RNA and marker expression methods. *Gene Ther* 9, 1155–62.

36. Long, Z., Li, L.-P., Grooms, T., Lockey, C., Nader, K., Mychkovsky, I., et al. (1998) Biosafety monitoring of patients receiving intracerebral injections of murine retroviral vector producer cells. *Hum Gene Ther* 9, 1165–1172.

37. Martineau, D., Klump, W.M., McCormack, J.E., DePolo, N.J., Kamantigue, E., Petrowski, M., et al. (1997) Evaluation of PCR and ELISA Assays for screening clinical trial subjects for replication-competent retrovirus. *Hum Gene Ther* 8, 1231–1241.

38. Chen, J., Reeves, L., Sanburn, N., Croop, J., Williams, D.A., and Cornetta, K. (2001) Packaging cell line DNA contamination of vector supernatants: Implication for laboratory and clinical research. *Virology* 282, 186–197.

39. Printz, M., Reynolds, J., Mento, S.J., Jolly, D., Kowal, K., and Sajjadi, N. (1995) Recombinant retroviral vector interferes with the detection of amphotropic replication competent retrovirus in standard culture assays. *Gene Ther* 2, 143–150.

40. Forestell, S.P., Dando, J.S., Bohnlein, E., and Rigg, R.J. (1996) Improved detection of replication-competent retrovirus. *J Virol Methods* 60, 171–178.

41. Miller, A.D., Bonham, L., Alfano, J., Kiem, H.P., Reynolds, T., and Wolgamot, G. (1996) A novel murine retrovirus identified during testing for helper virus in human gene transfer trials. *J Virol* 70, 1804–1809.

42. Bassin, R.H., Ruscetti, S., Ali, I., Haapala, D., and Rein, A. (1982) Normal DBA/2 mouse cells synthesize a glycoprotein which interferes with MCF virus infection. *Virology* 123, 139–151.

43. Chang, L.J., Urlacher, V., Iwakuma, T., Cui, Y., and Zukali, J. (1999) Efficacy and safety analyses of a recombinant human immunodeficiency virus type 1 derived vector system. *Gene Ther* 6, 715–728.

44. Metharom, P., Takyar, S., Xia, H.H., Ellem, K.A., Macmillan, J., Shepherd, R.W., et al. (2000) Novel bovine lentiviral vectors based on Jembrana disease virus. *J Gene Med* 2, 176–185.

45. Farson, D., Witt, R., McGuinness, R., Dull, T., Kelly, M., Song, J., et al. (2001) A new-generation stable inducible packaging cell line for lentiviral vectors. *Hum Gene Ther* 12, 981–997.

46. Kappes, J.C., and Wu, X. (2001) Safety considerations in vector development. *Somat Cell Mol Genet* 26, 147–58. Review.

47. Delenda, C., Audit, M., and Danos, O. (2002) Biosafety issues in lentivector production. *Curr Top Microbiol Immunol* 261, 123–141. Review.

48. Escarpe, P., Zayek, N., Chin, P., Borellini, F., Zufferey, R., Veres, G., et al. (2003) Development of a sensitive assay for detection of replication-competent recombinant lentivirus in large-scale HIV-based vector preparations. *Mol Ther* 8, 332–341.

49. Sastry, L., Xu, Y., Johnson, T., Desai, K., Rissing, D., Marsh, J., et al. (2003) Certification assays for HIV-1-based vectors: frequent passage of gag sequences without evidence of replication-competent viruses. *Mol Ther* 8, 830–839.

50. Segall, H.I., Yoo, E., and Sutton, R.E. (2003) Characterization and detection of artificial replication-competent lentivirus of altered host range. *Mol Ther* 8, 118–29.

51. Miskin, J., Chipchase, D., Rohll, J., Beard, G., Wardell, T., Angell, D., et al. (2006) A replication competent lentivirus (RCL) assay for equine infectious anaemia virus (EIAV)-based lentiviral vectors. *Gene Ther* 13, 196–205.

52. Sastry, L., Xu, Y., Duffy, L., Koop, S., Jasti, A., Roehl, H., et al. (2005) Product Enhanced Reverse Transcriptase Assay for Replication-Competent Retrovirus and Lentivirus Detection. *Hum Gene Ther* 16, 1227–1236.

53. Bonham, L., Wolgamot, G., and Miller, A.D. (1997) Molecular cloning of Mus dunni Endogenous Virus: an unusual retrovirus in a new murine viral interference group with a wide host range. *J Virol* 71, 4663–4670.

54. Lander, M.R., and Chattopadhyay, S.K. (1984) A Mus dunni cell line that lacks sequences closely related to endogenous murine leukemia viruses and can be infected by ecotropic, amphotropic, xenotropic and mink cell focus-forming viruses. *J Virol* 52, 695–696.

55. Wilson, C.A., Ng, T., and Miller, A.E. (1997) Evaluation of recommendations for replication-competent retrovirus testing associated with use of retroviral vectors. *Hum Gene Ther* 8, 869–874.

56. Reeves, L., Duffy, L., Koop, S., Fyffe, J., and Cornetta, K. (2002) Detection of ecotropic replication-competent retroviruses: Comparison of S+/L– and marker rescue assays. *Hum Gene Ther* 13, 1783–1790.

57. Wilson, C., Reitz, M.S., Okayama, H., and Eiden, M.V. (1989) Formation of infectious hybrid virions with gibbon ape leukemia virus and human T-cell leukemia virus retroviral envelope glycoproteins and the gag and pol proteins of Molony Murine Leukemia Virus. *J Virol* 63, 2374–2378.

58. Bayle, J.Y., Johnson, L.G., St. George, J.A., Boucher, R.C., and Olsen, J.C. (1993) High-efficiency gene transfer to primary monkey airway epithelial cells with retrovirus vectors using the gibbon ape leukemia virus receptor. *Hum Gene Ther* 4, 161–170.

59. von Kalle, C., Kiem, H.P., Goehle, S., Darovsky, B., Heimfeld, S., Torok-Storb, B. et al. (1994) Increased gene transfer into human hematopoietic progenitor cells by extended in vitro exposure to a pseudotyped retroviral vector. *Blood* 84, 2890–2897.

60. Bauer, T.R. Jr., Miller, A.D., and Hickstein, D.D. (1995) Improved transfer of the leukocyte integrin CD18 subunit into hematopoietic cell lines by using retroviral vectors having a Gibbon Ape Leukemia Virus envelope. *Blood* 86, 2379–2387.

61. Bunnell, B.A., Muul, L.M., Donahue, R.E., Blaese, R.M., and Morgan, R.A. (1995) High-efficiency retroviral-mediated gene transfer into human and non-human primate peripheral blood lymphocytes. *Proc Natl Acad Sci USA* 92, 7739–7743.

62. Chen, J., Reeves, L., and Cornetta, K. (2001) Safety testing for replication-competent retrovirus (RCR) associated with Gibbon Ape Leukemia Virus pseudotyped retroviral vectors. *Hum Gene Ther* 12, 61–70.

63. Cosset, F., Takeuchi, Y., Battini, J., Weiss, R.A., and Collins, M.K.L. (1995) High-titer packaging cells producing recombinant retroviruses resistant to human serum. *J Virol* 69, 7430–7436.

64. Goerner, M., Horn, P.A., Peterson, L., Kurre, P., Storb, R., Rasko, J.E.J. et al. (2001) Sustained multilineage gene persistence and expression in dogs transplanted with CD34(+) marrow cells transduced by RD114-pseudotyped oncoretrovirus vectors. *Blood* 98, 2065–2070.

65. Kelly, P.F., Vandergriff, J., Nathwani, A., Nienhuis, A.W., and Vanin, E.F. (2000) Highly efficient gene transfer into cord blood nonobese diabetic/severe combined immunodeficiency repopulating cells by oncoretroviral vector particles pseudotyped with the feline endogenous retrovirus (RD114) envelope protein. *Blood* 96, 1206–1214.

66. Duffy, L., Koop, S., Fyffe, J., and Cornetta, K. (2003) Extended S+/L– assay for detecting replication competent retroviruses (RCR) pseudotyped with the RD114 viral envelope. *Preclinica* 53–59.

# Chapter 18

# Release Testing of Retroviral Vectors and Gene-Modified Cells

## Diana Nordling, Anne Kaiser, and Lilith Reeves

## Summary

This chapter will review the design and execution of release testing requirements for retroviral vectors and gene-modified cells consistent with ensuring the success of the clinical trial on the basis of current US regulatory requirements. It is the ethical and legal responsibility of the clinical trial sponsor(s) to ensure safety of the patients through proper evaluation of the drug products prior to use. Any clinical trial drug product used in human subjects must be produced and evaluated for safety, quality, purity, and effectiveness according to Current Good Manufacturing Practices appropriate for the stage of clinical development.

**Key words:** Gene Therapy, Cellular Therapy, Certification, Release Testing, Quality Control Testing, Retroviral Vectors, Gene-modified Cells.

## 1. Introduction

Innovative hematopoietic therapies based on gammaretroviral (retroviral) vectors have shown increasing promise as therapeutic agents in genetic, malignant, and infectious diseases *(1–5)*. Primarily, these are early clinical phase trials conducted under an Investigational New Drug (IND) application and many are investigator sponsored. Manufactured products for retroviral vector gene transfer include the gene transfer agent and insertion (transduction) of the genetic sequence into the target cells. Vector production and gene transfer are complicated undertakings and are extensively discussed in separate chapters within this text.

Christopher Baum (ed.), *Methods in Molecular Biology, Methods and Protocals, vol. 506*
© Humana Press, a part of Springer Science + Business Media, LLC 2009
DOI: 10.1007/978-1-59745-409_4_18

This chapter serves as a guide for release testing for retroviral vector products and the gene-modified cellular therapy products; however, final plans will be protocol specific and must be approved by the appropriate regulatory agency. Gene and cellular therapy is an evolving field, and changes to requirements in methodology and specifications are expected. The FDA's Office of Cellular, Tissue, and Gene Therapy (OCTGT) encourages early guidance in determining the approach to certification criteria *(6)*. Release and control specifications associated with these products are suggested in OCTGT Guidances and must be defined in the Chemistry Manufacturing and Control (CMC) section of IND application *(7, 8)*. Issues and concerns identified with the specifications during the FDA review must be adequately addressed prior to initiation of the study.

To ensure the safety, quality, purity, and effectiveness of the drug products produced for clinical trials, it is key to develop a set of specifications for release. Release criteria are defined during drug development, and are integral to the clinical production process. On the basis of inherent limitations of certification and release assays, it cannot be over-emphasized that certification does not alleviate the necessity of having quality built into the process from the beginning of development and assuring that it continues through final manufacture *(9)*. During the drug development processes, specifications are defined and evaluated to ensure that the clinical products will provide the quality, purity, and effectiveness required for the therapeutic product. An integral part of the mandatory quality program is ensuring that a clinical product is manufactured via aseptic processes in a cleanroom environment. Both aseptic processes and cleanroom environments are required to have a level of validation consistent with the phase of clinical trial *(10, 11)*. Likewise, it is essential that the quality of components intended for clinical production is appropriate for human clinical trials *(12, 13)*. Details for production of retroviral vector products are outside the scope of this chapter and are included elsewhere in this book.

Laboratories that perform the analysis of clinical products are a vital quality element in the overall schema. In early phase clinical production, functional testing of the vectors and gene-modified cells may be performed at a research level providing there is sufficient evidence demonstrated during method developmental to indicate proof of principal including suitability of selected methodology, accuracy, linearity, and sensitivity. Alternately, a laboratory with increased controls and validated methodology should perform safety testing for early phase clinical products. In both cases, testing of clinical products must be accompanied by an approved protocol or standard operating procedure (SOP), appropriate standards and controls, and documented evidence that the testing was properly conducted *(6, 14, 15)*.

Certification and release for retroviral vector gene therapy products are governed by Good Manufacturing Practices and require that the certification criteria and test methodology be formally defined and approved by the Quality Control Unit. In current industry practice, the Quality Control Unit includes a Quality Assurance discipline that must authorize the preliminary and final release of both the retroviral vector and gene-modified cells based on documented evidence that requirements are met *(9, 16, 17)*.

This chapter provides a stepwise approach to retroviral vector and transduced cell certification and release testing. It begins with steps for determining the assays required, follows with directions for establishing a sampling plan for the product and identifying laboratories to conduct the tests, and concludes with guidance for establishing acceptance and release criteria and ensuring adequate quality controls are implemented. The reference section provides a list of essential reference documents. This chapter is not intended to provide methodologies for performing the assays.

## 2. Materials

1. Protocols.
2. Sample plans.
3. Release specifications.
4. Standard operating procedures.

## 3. Methods

### 3.1. The Certification Schema

1. Develop a plan for product release to form the basis for both the product specifications and the sampling plan. Utilize a testing matrix to ensure that all relative quality attributes are incorporated in the design. See **Table 1** for an outline of a testing matrix for both retroviral vector products and gene-modified cellular therapy products.

2. Define an appropriate *identity* (ID) test to serve as a fingerprint of the specific product to be released and also to distinguish it from other similar products.
   a. *Retroviral vector ID tests* typically consist of sequencing data from the appropriate portions of vector (*see* **Note 1**).

**Table 1**
**Certification Matrix (Derived from USP < 1046>)**

| Quality Attribute | Cell Therapy Product | Retroviral Gene Therapy Product |
|---|---|---|
| Identity | Species phenotype and transgene detection by PCR and expression | Vector sequencing and transgene expression |
| Dose | Viable cell number | Infectious or genomic titer (particle number); Multiplicity of Infection (MOI) |
| Potency function of transduced cells | Total cell number posttransduction (e.g., colony-formation assay and function of expressed gene determined via CFU with PCR for vector genome) | Function of expressed gene determined via infectious titer and transduction rate of lineage specific target cells |
| Purity | Percentage of transduced target cells | Residual DNA (plasmid and host cell) |
| Safety | Mycoplasma<br>Sterility<br>Endotoxin<br>RCR (archived only if cells are in culture less than four days, e.g., residual ancillary products) | Mycoplasma PTC<br>Sterility<br>Endotoxin<br>RCR |
| | Additional trial specific assays may be requested | In vitro adventitious viruses |

b. Gene-modified *cellular therapy ID tests* include immunophenotypic surface markers to indicate cell lineage and maturity and functional and PCR assays to confirm insertion of the correct transgene.

3. Define the requirements for the *dose* or concentration of the drug product to be administered to the patient or to the patient cells for retroviral therapy products.

   a. *For retroviral vector products*, the total particle concentration is commonly used to express the dosage to be administered to the target cells, also expressed as multiplicity of infection (MOI). The measurement of particle concentration in the vector supernatant is quantified by controlled serial dilution transductions of an appropriate target cell and determination of rate of transduction by functional assays or PCR. Infectious unit number is calculated by the viral supernatant volume, dilution, target cell number, and percentage of cells transduced (*see* **Note 2**).

b. Gene-modified *cellular therapy product* dose is calculated on enumeration of target cells to be administered per kilogram body weight. The cell number is determined by counting viable cell number and immunophenotype, if indicated, and calculating total cell number per kilogram patient weight. Optimal cell dose for hematopoietic retroviral vector gene transfer protocols has not been defined. The investigator must consider that the number of cells required for engraftment of transduced cells in patients who have not received myeloablative conditioning may be in the range of $1 \times 10^6$ cells per kilogram of body weight or higher. The cell dose is highly study specific and should be discussed with the appropriate regulatory reviewers.

4. Define *potency* based on the bioactivity (expression) of the drug product when compared with accepted reference standards.

a. Potency of *retroviral gene therapy products* is measured by transduction of a target cell line in vitro, followed by a measure of the expressed gene of interest. The suitability of the potency assay depends on the relationship between the in vitro cell used and the intended human target cell in vivo, the transduction efficiency, typically the protein expression levels, and duration of expression necessary for the desired therapeutic effect. Final potency data must be shown to represent what is reasonably expected to be applicable to the clinical study.

b. Potency assays performed on gene-modified *cellular therapy products* include quantification of the viable cell number post-transduction and a target cell functional assay, such as a colony-forming assay for transduced hematopoietic cells and an assay demonstrating the desired functional expression.

5. Define assays to assess *purity* that measure the process and product impurities carried through to the final product, which pose a substantial risk to either the patient or the product itself. Methodologies that separate, isolate, and specifically quantify the intended active product components determine product purity. Assays for impurities must account for bioactive and immunogenic agents and other compounds that may have deleterious effects.

a. For *retroviral vectors,* the purity assay includes evaluation of residual plasmid and cellular DNA and may include assays to assess the level of ancillary products and degradation products in the final product.

b. For gene-modified *cellular therapy products,* the measure of purity is generally limited to the percentage of viable cells in the total cell population; however, additional tests may be required to assess the level of residual ancillary products.

6. *Define safety tests to meet United States Pharmacopoeia (USP) General Chapter 1046, stated requirements focusing on three issues: (1)* preventing use of contaminated cells, tissues, or gene therapy agents with the potential for transmitting infectious diseases; *(2)* preventing use of improperly handled or processed, and consequently contaminated, products; and *(3)* ensuring safety when cellular and gene therapies are adapted for use other than in their normal functions or setting.

   a. Safety testing requirements for the *retroviral vector* are clearly defined in the regulatory guidances and include the following:

      • Sterility testing for aerobic and anaerobic bacteria and fungi (21 CFR 610.12).

      • Mycoplasma testing, specified in the Points to Consider (PTC) in the Characterization of Cell Lines Used to Produce Biologicals (1993).

      • Endotoxin by LAL per method defined in the "Guideline on validation of the limulus amebocyte lysate (LAL) test as an end-product Endotoxin test for human and animal parenteral drugs, biological products, and medical devices," (1987).

      • In vitro adventitious viral agents, specified in the Points to Consider in the Characterization of Cell Lines Used to Produce Biologicals (1993).

      • Toxicological General Safety testing, (21 CFR 610.11) *Note: General Safety may not be required for early clinical phase products, but is expected for licensure.*

      • Replication Competent Retrovirus (RCR) for both the vector supernatant and end-of-production cells using the S + L⁻ method consisting of amplification on a permissive cell line and foci detection on PG4 cells.

   b. Safety testing requirements for the gene-modified *cellular therapy products* include both a preliminary and final release schema.

      • Sterility: A gram stain or other colorimetric detection for bacteria and fungi may be used for the preliminary release and the 14-day Sterility Test is required for final certification.

      • Mycoplasma may be assayed by a methodology other than the PTC culture assay (e.g., PCR) and confirmation that a sample has been submitted for analysis is satisfactory for preliminary release.

      • Endotoxin by a USP level assay is required prior to infusion.

      • RCR testing is required only when cells have been in transduction culture for more than four days and, otherwise, product samples are archived for future analysis if indicated. Confirmation that a sample has been submitted for analysis is satisfactory for preliminary release.

**3.2. The Sample Plan**

1. Develop a comprehensive sampling plan to assure that the samples required for release testing of each lot of drug product are obtained at appropriate steps during production processes. Assays may be required to be from supernatant and/or end of production (EOP) cells or, for retroviral vectors supernatant, may be from final harvest only or from all harvests. An example sample plan is provided in **Table 2** for a retroviral vector product. The sample plan for gene-modified cellular therapy products must be protocol specific and must include sufficient sample collection for preliminary release screening as well as the longer lead-time analysis. Discuss all sample plans with the appropriate regulatory agency prior to manufacture.

   a. Include quantity of sample required for each analysis (*see* **Note 3**).

## Table 2
## Example Sample Plan for Retroviral Vectors for Early Phase Human Gene Transfer

| Quality Attribute | Sample Size | Sample Location | Sample Prep | Storage |
|---|---|---|---|---|
| Sterility | 10 mL supernatant from begin, mid and end of harvest (Equivalent to represent 10% of final containers) *(see* **Note 8***)* | Each harvest | None | ≤ 70°C |
| Retest sterility samples | 20 mL supernatant from begin, mid and end of harvest (equivalent to represent 10% of final containers available for double volume testing.) *(see* **Note 8**) | Each harvest | None | ≤ 70°C |
| Preliminary sterility screen (optional but recommended) | 1 mL supernatant from begin, mid and end of harvest. (Equivalent to represent 10% of final containers). | Each harvest | None | ≤ 70°C |
| Endotoxin inhibition and enhancement | 0.5 mL supernatant | Each harvest | None | ≤ 70°C |
| Mycoplasma | 1 × 10e7 EOP cells | EOP cells | None | ≤ 140°C |

(continued)

**Table 2**
**(continued)**

| Quality Attribute | Sample Size | Sample Location | Sample Prep | Storage |
|---|---|---|---|---|
| In vitro adventitious viruses | 30 mL supernatant (*see* **Note 9**) | Final harvest | None | ≤70°C |
| In vitro adventitious viruses retain | 60 mL supernatant (*see* **Note 9**) | Final harvest | None | ≤70°C |
| RCR supernatant | 5% supernatant | Final harvest | None | ≤70°C |
| Retain RCR supernatant | 10% supernatant | Each harvest | None | ≤70°C |
| RCR EOP cells | 1% or 1 × 10e8 EOP cells, whichever is less | EOP cells[a] | Cryopreserve | ≤140°C |
| Retain RCR EOP cells | 2% or 2 × 10e8 EOP cells, whichever is less | EOP cells | Cryopreserve | ≤140°C |
| Infectious titer by qPCR | 1.2 mL supernatant | Each harvest | Submit 1.2 mL for transduction of permissive target cell | Supernatant at ≤70°C prior to transduction |
| Infectious titer by gene expression | 1.2 mL supernatant | Each harvest | Submit 1.2 mL for transduction of permissive target cell | Supernatant at ≤70°C prior to transduction |
| Sequencing | 1.2 mL supernatant | Final harvest | Submit 1.2 mL for transduction of permissive target cell | Supernatant at ≤70°C prior to transduction |
| Residual plasmid DNA | 1.2 mL supernatant | Final harvest | None | ≤70°C |
| Potency, identity, and purity retain | 8 vials of 1.2 mL of supernatant per vial | Each harvest | None | ≤70°C |
| EOP cells retain | Equal to two times the volume required to repeat all certification testing except sterility and endotoxin | Final harvest | Cryopreserve (in multiple vials) | ≤140°C |
| General supernatant reserve | Various (*see* **Note 10**) | Each harvest | None | ≤70°C |
| General EOP cells reserve | 5% EOP cells (*see* **Note 10**) | Final harvest | Cryopreserve (in multiple vials) | ≤140°C |

[a]End of production (EOP) Cells

b. Define proper storage and handling of the samples, including the physical storage location of the samples and duration of required storage.

c. Direct preparation steps prior to storage or submission for analysis.

2. Consider and include in-process sampling and testing and surrogate samples as required.

a. Incorporate essential information from the sample plan into the Master Production Record (MPR) for in-process or surrogate sampling (*see* **Notes 4** and **5**).

3. Include retention samples in the sampling plan (*see* **Note 6**).

4. Incorporate a preliminary screening for potency and specific safety tests prior to submission for final release testing for retroviral vector supernatants prior to submission of samples for all assays since most product failures are caused by contamination or subpotency (*see* **Note 7**).

***3.3. Selection of Laboratories***

1. Identify a laboratory to conduct each required assay according to the following basic criteria:

a. Experience in handling viral vectors or gene-modified products.

b. Adequate biosafety level containment controls.

c. Analytical methods developed to accommodate minimal testing sample volumes.

2. Assure that the laboratory testing is appropriate for the clinical phase of the gene or cellular therapy product.

a. Early phase products:

- GMP level testing is preferred for all assays; however, may not be readily available for many of the assays. GMP sterility and endotoxin are typically required.

- Good Laboratory Practice (GLP) level testing is the next preferred level, and is generally acceptable for safety assays for which GMP testing is not available. (*see* **Note 11**).

- Functional testing for early phase I clinical production may be acceptable if performed at a research level provided that a suitable method has been established with appropriate standards and controls. Additionally, research level development studies should be available to provide evidence of proof of principal to include suitability of selected methodology, accuracy, linearity, and sensitivity.

b. Late clinical phases and subsequent licensure:

Test methods will require full method validation, qualification of instrumentation and computer controls, use of recognized standards and controls (either compendial or industry qualified), and a fully compliant GMP compliance program.

3. Verify that each test laboratory has a quality unit to critically review processes and reports of analysis and maintain appropriate document control. This is necessary even for early phase clinical product certification.

**3.4. Release Specifications**

1. Establish release specifications for each certification test required for the respective certification of the retroviral vectors and gene-modified cells. Examples for preliminary and final release of gene-modified cells are provided in **Table 3** (*see* **Notes 12** and **13**.)

    a. Safety assay specifications are based on the regulatory guidelines.

    b. Functional assay specifications are based on results of developmental data, typically specific for the method of analysis and requirements of the clinical study.

**Table 3**
**Example Specification for Release of Gene-modified Cells (Cellular Therapy)**

| Assay | Method | Preliminary Acceptance Criteria | Final Acceptance Criteria |
|---|---|---|---|
| Endotoxin by LAL Endotoxin inhibition/enhancement | LAL Gel Cot (*LOD 0.03 EU*) (*See***Note 14**)[a] | <0.12 EU/mL No inhibition or enhancement at greatest dilution below MVD demonstrated | <0.12 EU/mL No inhibition or enhancement at greatest dilution below MVD demonstrated |
| Gram stain | Crystal Violet | Negative | Negative |
| Viability | Trypan Blue | >70% viable | >70% viable |
| CFU with PCR for vector genome | qPCR | Initiated | Report result; acceptability determined by sponsor |
| Mycoplasma by PCR | qPCR | Initiated | No evidence of mycoplasma contamination |
| Sterility | Direct inoculation | Initiated | Free from viable bacterial and fungal contamination |
| Transduction percent | Bulk transduced cells analyzed for vector expression | Initiated | Report result; acceptability determined by sponsor |

[a]Limit of detection (LOD); endotoxin units (EU)

2. Define specifications prior to analysis of the samples. Include the following:

   a. A brief description of the methodology used.

   b. The sensitivity or limit of detection for the assay (if known).

   c. Predefined acceptance criteria.

**3.5. Quality Controls**

1. Develop and implement standard operating procedures (SOPs) for all aspects of GMP operations, regardless of the phase of clinical trial (*see* **Note 15**).

2. Assure that the manufacturing Quality Unit is independent of the production and research group. In smaller organizations involved in the production and release of clinical trial materials, responsibility for both quality control (QC) and quality assurance (QA) functions may be a combined role provided that the person performing the analysis is not the same person who reviews the associated documentation and approves the final decision to release.

3. Ensure adequate QA staff within the organizational Quality Unit, which is assigned to the clinical product manufacturing area to ensure compliance and to oversee each gene transfer product release (*see* **Note 16**). This person(s) must be qualified to:

   a. Approve sample plans and specifications to ensure the process designed for making the drug product incorporates appropriate quality controls.

   b. Oversee the quality component in clinical manufacturing to ensure quality is built into the process.

   c. Critically review the assay results and certify that the product is acceptable for use in clinical trials.

   d. Compile results of testing and Certificates of Analysis and assure that the results are consistent with acceptance criteria and other quality requirements.

   e. Formally release the product including approval of the preliminary and final test results and associated Certificates of Analysis.

# 4. Notes

1. When sequencing a vector for identity testing, limitations due to the size of the vector construct may be overcome by sequencing the genetic insert in addition to the flanking regions, plus any significant modifications to the vector backbone or sites known to be at risk to alteration during manufacture.

2. Calculation of viral vector titer should be done from a dilution yielding 10–20% positive cells. Since a cell may be transduced by multiple viral particles, using a low level of transduction helps to minimize the underestimation of the particle number caused by having multiple viral particles transduce a single cell.

3. Sample requirements for sterility are defined in 21 CFR 610.12. Where the product lot size is between 20 and 200 containers, sample 10% of the containers; where the lot size is less than 20 containers, sample not less than 2 samples and ensure that these samples are taken at sequences representative of the beginning, middle, and end of the process. The volume for testing of the bulk supernatant is 10 mL and is obtained by surrogate sampling to minimize disruption of the final filled containers.

4. Development of the sample plan precedes development of the MPR. During development of the Master Production Record, the requirements to fill containers for release testing are incorporated at the correct production steps. By making this a stepwise process incorporated in the MPR prior to execution, there is reduced probability of missing a sample, thereby increasing the success of the overall product campaign.

5. Surrogate samples are representative of the actual production, and are obtained using the same methodology that specifically meets assay requirements, except that the volumes of surrogate samples are smaller than the final product containers. Where final product containers are pulled for analysis, the sample plan is an appropriate stand-alone document. Where surrogate samples are obtained or in-process sampling is performed, the batch production record provides the documented evidence that sampling occurred at appropriate steps in the process.

6. Retention samples are aliquots of sufficient quantity to be retained to repeat all analyses twice in the event and a safety or quality issue arises with the lot post-release. Additional criteria for investigational drugs require retention samples to be kept for at least 5 years after the application is approved or following the completion date of the investigational study.

7. Preliminary screening assays for potency and sterility require minimal product volume and incur minimal costs. Considerable expense may be avoided by early determination that a production run does not meet initial acceptance criteria. Consideration should be given to pooling of samples from individual harvests for sterility and ensuring that adequate surrogate samples are available for retesting if needed, so that the required retain quantities are not depleted investigating sterility test failures.

8. If the manufacturing process does not include the use of antibiotics, bacteriostasis and fungistatis testing are not required.

9. Not all guidances for these therapeutic products provide a consistent approach to determining of where samples should be pulled. For the in vitro adventitious viral assay, the Chemistry, Manufacturing, and Control (CMC) Guide for Gene Therapy specifies testing from end of production (EOP) cells and vector, and the International Conference on Harmonisation (ICH) Guideline specifies testing from bulk harvest or lysate or cell culture media only. Clarification from the appropriate regulatory bodies while developing the sampling plan is necessary.

10. Having adequate samples from each harvest of retroviral vector supernatant for final protocol development work is essential. A minimum of 5–10 samples of quantities varying from 1 to 5 mL is recommended. Discussion with the protocol sponsor to determine the remaining development work prior to supernatant harvest will be helpful for optimizing supernatant use. Research samples held for cellular therapy products are generally defined in the protocol and typically include 10% of the cells and at least 50% of the final wash supernatant.

11. GLP laboratories require approval of the test protocol, provide a comprehensive report on the design and execution of the testing as well as a quality statement of work, whereas GMP laboratories typically do not.

12. Acceptance criteria for phase 1 clinical materials should be based on data from lots used in preclinical trials; whereas the FDA will expect that the acceptance criteria for phase 2 clinical materials will have been refined and tightened on the basis of the data generated in phase 1. For phase 1, the FDA may permit acceptance criteria of *report results* on a limited basis, specifically if there is insufficient preclinical data to justify a specific range of acceptability. Alternately for phase 1, the ranges of acceptability may be wide, as variances in methodology for the bioassays are not yet defined. It is important to note however, that as the clinical studies move forward to later phase production, regulatory expectations include establishment of scientifically valid acceptance criteria based on results from information collected during both phase 1 and 2 clinical trials, known assay variability, and fully characterized processes and test methodology.

13. The release specifications for both retroviral vector supernatants and transduced cellular therapy products are part of the IND submission, and the Agency will provide feedback on acceptability prior to approval of the submission. Sponsors need to ensure that specifications developed for clinical product release at the manufacturing site are consistent with the IND submission.

14. The endotoxin limit of detection and release specifications are based on the USP Guidance limit of 5 EU/Kg/Hr for parenteral products. Therefore, the limit of detection of the assay must be correlated with the maximum valid dilution to assure that the overall sensitivity of the result clearly distinguishes products that exceed the endotoxin unit (EU) limit from those that are within the limit. A second consideration for establishing endotoxin limits is based on a historical perspective, i.e., if an endotoxin level of a production lot clearly exceeds the norm for that production process, the manufacturer should be alerted to the possibility of a bacterial or endotoxin contamination that may have occurred during the process.

15. SOPs should be written in a standard format to include specific sections defining the purpose, scope, materials, instrumentation, methods, and references. SOPs are assigned control numbers and are issued effective dates. They are subject to approval by the author, management, and QA. SOPs require change control to document change and the justification for such change. They are also subject to document control practices, where only the current version of an SOP is available for use and previous versions are removed from potential inadvertent use. SOPs are not effective without the implementation of a proper training program. A comprehensive training program is also required for all GMP operations. An effective training program will encompass reading and understanding the procedure, demonstration that the procedure can be followed properly, and for SOPs describing test methods, a verification of technique. Training records will need to be available for all team members assigned responsibility for execution of SOPs. Typical SOPs associated with release testing include the following:

    a. Personnel training.

    b. Change control.

    c. Document control.

    d. Development and approval of specifications and sample plans.

    e. Record review and product disposition.

    f. Selection and qualification of contract laboratories.

    g. Operation, calibration, and preventative maintenance for the facility and instrumentation.

    h. Test methods for each assay.

16. Final GMP manufacturing must have included Quality Unit personnel involved in the following:

    a. Sampling, testing, and inspecting raw materials.

b. Review of in-process testing and lot release testing, including administration of contract laboratory analysis.

c. Review of batch production records and support documentation.

d. Ensuring that manufacturing excursions are properly justified.

e. Performing audits and evaluating trends.

17. The materials referenced in the introduction of this chapter and the Center for Biologics Evaluation and Research (CBER) website (http://www.fda.gov/cber/) provide valuable information for guiding release testing.

## References

1. Abonour R, et al. Efficient retrovirus-mediated transfer of the multidrug resistance 1 gene into autologous human long-term repopulating hematopoietic stem cells. *Nat Med.* 2000;6:652–658.

2. Hacein-Bey H, et al. Gamma-c gene transfer into SCID X1 patients' B-cell lines restores normal high-affinity interleukin-2 receptor expression and function. *Blood.* 1996;87:3108–3116.

3. Kang EM, et al. Gene therapy-based treatment for HIV-positive patients with malignancies. *J Hematother Stem Cell Res.* 2002;11:809–816.

4. Kang EM, Nonmyeloablative conditioning followed by transplantation of genetically modified HLA-matched peripheral blood progenitor cells for hematologic malignancies in patients with acquired immunodeficiency syndrome. *Blood.* 2002;99:698–701.

5. Kelly PF, et al. Stem cell collection and gene transfer in fanconi anemia. *Mol Ther.* 2007;15:211–219.

6. Gombold J, Peden K, Gavin D, Wei Z, Baradaran K, Mire-Sluis A, and Schenerman M. (2006) Lot release and characterization testing of live-virus-based vaccines and gene therapy products, Part 1, factors influencing assay choices. *Bioprocess International.* 46–56.

7. Guidance for FDA Review Staff and Sponsors, Content and Review of Chemistry, Manufacturing, and Control (CMC) Information for Human Gene Therapy Investigational New Drug Applications (INDs). US Department of Health and Human Services, FDA. November 2004

8. Guidance for Industry, Guidance for Human somatic Cell Therapy and Gene Therapy. US Department of Health and Human Services, FDA. March 1998.

9. Guidance for Industry, Quality Systems Approach to Pharmaceutical CGMP Regulations. US Department of Health and Human Services, FDA. September 2006

10. Guidance for Industry, Sterile Drug Products Produced by Aseptic Processing – Current Good Manufacturing Practices CGMP Regulations. US Department of Health and Human Services, FDA. September 2004

11. 21 CFR 610. General Biological Products Standards. US Department of Health and Human Services, FDA.

12. USP 29/NF 24 < 1046 > Cell and Gene Therapy Products.

13. ICH Q5A (R1) Viral Safety Evaluation of Biotechnology Products Derived from Cell Lines of Human or Animal Original. Requirements for registration of Pharmaceuticals for Human Use. September 1999.

14. Guidance for Industry, INDs – Approaches to Complying with CGMP During Phase 1. US Department of Health and Human Services, FDA. January 2006

15 Gavin, Denise. Perspectives on Potency Assays for Complex Biological Products. Well Characterized Biotechnology Pharmaceutical (WCBP), Chemistry, Manufacturing, and Control (CMC) Forum on Bioassays. January 28, 2007.

16. 21 CFR 211. Current Good Manufacturing Practices for Finished Pharmaceuticals. US Department of Health and Human Services, FDA.

17. Cornetta, K, (2003) Regulatory issues in human gene therapy. *Blood Cells, Molecules, and Diseases.* 31:51–56.

# Chapter 19

# Copy Number Determination of Genetically-Modified Hematopoietic Stem Cells

## Todd Schuesler, Lilith Reeves, Christof von Kalle, and Elke Grassman

## Summary

Human gene transfer with gammaretroviral, murine leukemia virus (MLV) based vectors has been shown to effectively insert and express transgene sequences at a level of therapeutic benefit. However, there are numerous reports of disruption of the normal cellular processes caused by the viral insertion, even of replication deficient gammaretroviral vectors. Current gammaretroviral and lentiviral vectors do not control the site of insertion into the genome, hence, the possibility of disruption of the target cell genome. Risk related to viral insertions is linked to the number of insertions of the transgene into the cellular DNA, as has been demonstrated for replication competent and replication deficient retroviruses in experiments. At high number of insertions per cell, cell transformation due to vector induced activation of proto-oncogenes is more likely to occur, in particular since more than one transforming event is needed for oncogenesis. Thus, determination of the vector copy number in bulk transduced populations, individual colony forming units, and tissue from the recipient of the transduced cells is an increasingly important safety assay and has become a standard, though not straightforward assay, since the inception of quantitative PCR.

Key words: Viral insertion; Insertional mutagenesis; Vector copy number; Taqman q-PCR; Real time PCR.

# 1. Introduction

Side effects related to viral insertions after treatment of hematopoietic cells with replication deficient gammaretroviral vectors have been repeatedly reported (1–3). As the risk is linked to the number of vector insertions per cell (4–6), vector sequences need to be measured quantitatively for risk assessment in gene therapy with integrating vectors.

Christopher Baum (ed.), *Methods in Molecular Biology, Methods and Protocols, vol. 506*
© Humana Press, a part of Springer Science + Business Media, LLC 2009
DOI: 10.1007/978-1-59745-409-4_19

Taqman q-PCR, also called real time PCR, 5′ exonuclease-based PCR or kinetic PCR, exploits the 5′–3′ exonuclease activity of Taq polymerase *(7)* and uses a PCR buffer with dNTPs, magnesium, primers, and Taq polymerase as in conventional PCR, but also contains a fluorescent-labeled probe. The probe that is target sequence specific and nonextendable is labeled with a 5′ reporter dye and a 3′ quencher dye. As long as the probe is intact, the fluorescence emission is being absorbed by the quencher. As conventional PCR, real-time PCR has a denaturing, an annealing, and an extension phase. The probe hybridizes to the target sequence between the forward and reverse primers at the annealing phase. During the extension phase, Taq polymerase binds, then extends from the forward primer and cleaves the probe, separating the quencher from the reporter dye, which releases a permanent fluorescent signal. This signal is then detected and reported graphically in real time. When the signal multiplies enough to cross the threshold, the line that distinguishes positive signal from background, a data point called "Ct" (cycle threshold) is given. A low Ct value corresponds to a high number of target sequences. These values are used to quantify the amount of target sequence in the reaction. A key tool in using Taqman q-PCR is multiplexing. By using a probe labeled with a VIC dye and another probe labeled with FAM, the same quencher can be used on both dyes, and it is possible to detect two target sequences simultaneously. This allows for simultaneous quantitation of two different sequences in the same well, typically a target sequence (TS) specific for the integrated vector and a "normalizing" sequence present in each cell. We use the endogenous gene ApoB as the "normalizing" sequence to determine the number of genomes tested in each well. In the procedure described later, the probe specific for ApoB is labeled with VIC and the probe specific for vector sequences is labeled with FAM. A correlation between Ct value and copy number is established based on standards that are tested concurrently in each individual assay. The protocol below is written for the ABI 7900 Fast Real time system and has to be adapted if using other real time systems.

Standards need to be tested simultaneously to correlate the measured Ct value and copy numbers. Standards are serial dilutions of genomic DNA from a clonal cell line, derived from the same species as the test samples, with one target sequence insertion per cell. This way, the target sequence standard curve and the normalizing genes (ApoB) standard curve is produced simultaneously. A cell number-Ct value correlation is calculated based on the assumption that 1 μg DNA tested in the undiluted standard corresponds to 150,000 cells.

The controls are composed of DNA from a clone with single copy insertion of the target sequence (TS), which is diluted into genomic DNA from the same species. In contrast to the standards, the final DNA concentration of all TS dilutions is the same. This gives an accurate representation of different copy numbers of the target sequence present in a constant concentration of genomic background. These controls can be monitored over time to determine the performance, sensitivity, and reproducibility of the assay.

Special consideration must be taken when quantitating vector copies in cells that have been exposed to viral vector supernatant. The supernatant from stable producer cells contain varying but significant amounts of genomic transgene DNA from disrupted cells, and transiently produced supernatants contain plasmid DNA sequences in addition to packaging cell DNA *(8)*. The extracellular DNA can be removed by treating vector supernatant or the cells post transduction with a DNase enzyme such as benzonase as descibed in the literature *(9)* and in the **Subheading 3.1**, respectively. However, we found that DNase treatment is only marginally effective in bulk transduced cell populations and does not address the problem of detecting nonintegrated intracellular vector cDNA. Therefore, estimation of copy number per cell population in bulk transduced cells is not a specific tool for estimating the average inserted copy number. Furthermore, considering that bulk measurement cannot account for multiple insertions per cell, this technique is not effective for estimating percent transduced cells. However, when the transduced cell population is plated into colony forming unit assays, the treatment with benzonase effectively removes the extracellular DNA and subsequent cell doublings reduce nonintegrated sequences, the copy number per colony forming unit can be determined when stable integrations of the target sequence are quantitatively compared with a housekeeping gene such as species-specific ApoB. Comparison of the number of transgene positive colonies to total colony number is a reliable measure for estimating percent transduced progenitor cells, provided that the gene(s) encoded by the vector have no strong effect on cell survival and proliferation.

Tissues from recipients of the stably transduced cells do not pose the problem of nonintegrated DNA, and copy number per cell number may be clearly measured. One limitation, however, is that this does not provide specific copy number in a given cell but merely average copy number in the bulk of cells analyzed. Care must be taken to discriminate interpretation of insertions if the tissue of interest from those of peripheral blood that is captured with the tissue. This may be done by comparing transgene cDNA/ApoB copy number ratio in peripheral blood to the copy number ratio in the respective tissues. DNA for qPCR can be extracted from fresh or fixed tissue samples. Fixed tissue samples are also amenable to a technique known as laser capture microdissection (LCM). This procedure allows the target cells to be morphologically identified, laser captured, and analyzed for copy number. This technique is especially effective for determining potential for vector insertion caused malignancies *(10)*. This chapter describes the DNA extraction and PCR analysis of laser captured samples. The protocol described in **Subheading 3.3** is a modified version of the standard protocol recommended by the manufacturer (*see* **Subheading 2**) and covers the steps on DNA isolation after obtaining the captured tissue section via LCM.

This chapter will focus on techniques for quantitatively measuring copies of inserted vector. It will address selection of probes and primers, tissue collection and preparation, analysis by qPCR, standard and control materials, and interpretation of results.

## 2. Materials

***2.1. DNA Sample Preparation from Clonogenic Progenitors: DNase Treatment and Collection of Colonies***

1. IMDM/2%FBS; Cat. no. 7700; StemCell Technologies, Inc. (Vancouver, Canada). Store bulk at –20°C. After thawing it is stable at 4°C for two-month period.

2. Benzonase ® Nuclease, cat. no. 7066–4, Novagen (San Diego, CA), store at –20°C.

3. Magnesium Chloride 100 mM, sterile, store in aliquots at –20°C.

4. MethoCult GF M3434 methylcellulose medium with recombinant cytokines for colony assays of mouse cells; cat. no. 03434; StemCell Technologies, Inc. Store at –20°C.

5. CFU Lysis Buffer: 0.91 mg/ml Proteinase K, 0.5% Tween 20, 0.5% Nonidet P40 Substitute, 1× PCR buffer (cat. no. N8080189, Applied Biosystems, Foster City, CA) in nuclease free water.

6. 96-well Optical reaction plate; cat no. 4306737, Applied Biosystems.

7. Optical adhesive cover, cat no. 4311971, Applied Biosystems.

8. Binocular microscope.

***2.2. DNA Sample Preparation from Fresh Tissue or Cells***

1. Puregene DNA Purification System Blood Kit, cat no. D-5500, Gentra Systems (Valencia, CA), the Kit contains RBC lysis, cell lysis, protein precipitation, and DNA hydration solutions.

2. Puregene DNA Purification System Cell and Tissue Kit, cat no. D-5500A, Gentra Systems, the Kit contains cell lysis, protein precipitation, and DNA hydration solutions.

**2.3. DNA Sample Preparation from Tissue After Laser Capture Microdissection (LCM)**

1. PicoPure DNA Extraction Kit; Molecular Devices.

2. CapSure HS LCM Caps; Molecular Devices.

3. Alignment Tray; Molecular Devices cat no. LCM0504.

4. Incubation Block; Molecular Devices cat no. LCM 0505.

5. 0.5 ml microentrifuge tubes, sterile; Applied Biosystems cat no. N8010611.

**2.4. Preparation of Probes and Primers**

1. Order oligonucleotides and probes HPLC grade. The following primers and probes are used to quantify human (H-ApoB) and mouse (M-ApoB) genomic DNA.

   H-ApoB Forward 5′ > TGAAGGTGGAGGACATTCCTCTA

   H-ApoB Reverse 5′ > CTGGAATTGCGATTTCTGGTAA

   H-ApoB Probe 5′ > VIC-CGAGAATCACCCTGCCAGACT-TCCGT/Tamra

   M-ApoB Forward 5′ > CGTGGGCTCCAGCATTCTA

   M-ApoB Reverse 5′ > TCACCAGTCATTTCTGCCTTTG

   M-ApoB Probe 5′ > VIC-CCTTGAGCAGTGCCCGACCA-TTC/Tamra

2. Primers and probes binding a unique sequence present in the vector, the target sequence (TS). Refer to **Notes 1–2** for design of primers and probes for the target sequence. The probe is to be labeled 5′ with FAM and 3′ linked to the quencher Tamra.

**2.5. Preparation of Standards**

Standard TS (target sequence) DNA: Genomic DNA isolated from a single cell derived clonal cell line of the same species as the test samples, containing a single copy of the vector sequence of interest (*see* **Notes 3–5**). Store at –20°C. The isolated TS DNA must have a 260/280 ratio between 1.65 and 2.1 and a minimum concentration of $0.2\mu g/\mu l$.

**2.6. Preparation of Controls**

1. Genomic DNA isolated from a single cell derived clonal cell line of the same species as the test samples, containing a single copy of the vector TS of interest. It can be identical to the TS DNA used for the standards (see above) Store at –20°C. The isolated TS DNA must have a 260/280 ratio between 1.65 and 2.1 and a minimum concentration of $0.4\mu g/\mu l$.

2. Genomic DNA isolated from a HeLa, HT1080, 293, 3T3 or other Genomic DNA without the target sequence. The DNA must be of the same species than the test samples. Store at –20°C. The DNA must have a 260/280 ratio between 1.65 and 2.1 and a minimum concentration of $0.4\mu g/\mu l$.

**2.7. Taqman PCR: Real Time Acquisition of the Polymerase Chain Reaction**

1. ABI 7900 HT Fast Real-Time System.

2. To run in fast mode on the ABI 7900 HT Fast Real-Time System use the following reagent: TaqMan Fast Universal PCR Mix 2×, 5,000 Rxn cat no. 4367846, Applied Biosystems; store at 4°C.

3. To run q-PCR in standard mode use the following master mix (see **Note 6**): TaqMan® Universal PCR Master Mix, 2,000 Reactions 4318157, Applied Biosystems; store at 4°C

4. Real-time Fast PCR Plates, cat no. 3890, CLP.

5. Nuclease free water.

6. Optical adhesive covers.

7. DNA away, cat no. 7010, Molecular BioProducts (San Diego, CA).

# 3. Methods

**3.1. DNA Sample Preparation from Clonogenic Progenitors: DNase Treatment and Collection of Colonies**

1. DNase treatment (see **Note 7**)

    (a). Wash $10^5$–$10^6$ cells twice with 1ml IMDM/2% FBS to remove any possible inhibitors of Benzonase, e.g. EDTA.

    (b). Resuspend cells in 0.5ml IMDM/2%FBS.

    (c). Add 50μl of 100mM $MgCl_2$ to each cell suspension.

    (d). Add 1μl of Benzonase to each cell suspension. Mix gently.

    (e). Incubate cell suspension(s) at room temperature for 30min.

    (f). Centrifuge cell suspension(s) at 350× g, for 8min, at room temperature.

    (g). Aspirate the supernatant from the cell pellet.

    (h). Resuspend each cell pellet in 200μl IMDM/2%FBS.

2. Plate cells in Methocult, 1ml per 35×10mm dish. Use several dilutions of cells (between 500 and $10^5$ cells per dish, dependent on cell source) to obtain well separated colonies (ideally 30–100 colonies per 35x10mm petri dish). Incubate at 37°C and 5% $CO_2$ in humidified air for 7–14 days.

3. Collect colonies as follows

    (a). Prepare 96 well plate with 50μl of CFU lysis buffer per well.

    (b). Place CFU plate under microscope. Use 40× magnification power.

    (c). Set micropipettor to 3μl. Using micropipettor, harvest single clonogenic colonies growing in the methylcellulose.

    (d). Place single colony directly into one well containing lysis buffer, one sample per well. Pipette up and down to rinse pipette with lysis buffer.

(e). If possible, collect a total of 36 colonies per sample.

(f). Collect up 36 background samples containing Methocult as well as single cells, which have not grown into colonies.

4. After collecting all colonies and background samples seal the PCR plate with adhesive cover and store at –20°C freezer or proceed immediately to next step.

5. Incubate the plate in thermal cycler for 1h at 60°C, followed by 10-min incubation at 95°C.

6. The plate can be stored at –20°C freezer until PCR is performed. Use 5µl of each well as DNA sample for Taqman PCR.

**3.2. DNA Sample Preparation from Fresh Tissue or Cells**

1. DNA isolation from peripheral blood or bone marrow (*see* **Note 8**)

(a). Starting with a <300µl sample starting volume, add 3× the original sample volume of RBC Lysis Solution. Invert to mix and incubate at room temperature for 5min. Invert at least once during the incubation.

(b). Centrifuge for 2min at 16,000×g.

(c). Decant RBC Lysis supernatant leaving <5% residual liquid.

(d). Vortex the tube vigorously for 20s to resuspend the cells.

(e). Add 1× the original sample volume of Cell Lysis solution. Vortex vigorously for 10s. This solution is stable at room temperature for 2 years.

(f). Add 1/3 the original sample volume of protein precipitation solution. Vortex vigorously for 20s.

(g). Centrifuge at 16,000×g for 5min.

(h). Decant the supernatant into a new labeled tube containing equal volume as the original sample volume of 100% isopropanol.

(i). Invert the tube 50×. White threads of DNA will appear.

(j). Centrifuge at 16,000×g for 3min. Decant the supernatant and drain briefly on clean absorbent paper.

(k). Add equal volume of the original sample volume of 70% ethanol and invert.

(l). Centrifuge at 16,000×g for 1min. Carefully decant the ethanol, paying attention to not disturb the white DNA pellet. Drain briefly on clean absorbent paper.

(m). Air dry for 15min in a PCR hood.

(n). Dissolve DNA in DNA Hydration Buffer. Use 33% of the original sample volume (*see* **Note 9**).

(o). Incubate the DNA in a 65°C heat block for 1h. Gently vortex every 15min to facilitate dissolution.

(p). Store samples at –20°C for short term (<1month) or –80°C for long term (>1month).

2. DNA isolation from cell suspensions/cell pellets (*see* **Note 10**):

(a). Starting with $1-2 \times 10^6$ cells, centrifuge cells for 5min at 500×g. Aspirate and discard the supernatant, leaving ~20µl remaining. Cell pellets can be stored at –20°C for up to 1month.

(b). Vortex the tube vigorously for 20s to resuspend the cells.

(c). Add 0.3ml of cell lysis solution. Vortex vigorously for 10s. This solution is stable at room temperature for 2 years, if desired.

(d). Add 0.1ml protein precipitation solution. Vortex vigorously for 20s.

(e). Centrifuge at 16,000×g for 5min at 16,000×g.

(f). Decant the supernatant into a new labeled tube containing 0.3ml 100% isopropanol.

(g). Invert the tube 50×. White threads of DNA will appear.

(h). Centrifuge at 16,000×g for 3min. Decant the supernatant and drain briefly on clean absorbent paper.

(i). Add 70% ethanol according the chart and invert.

(j). Centrifuge for 1min. Carefully decant the ethanol, paying attention to disturb the white DNA pellet. Drain briefly on clean absorbent paper.

(k). Air dry for 15min in a PCR hood.

(l). Dissolve DNA in 50µl DNA Hydration Buffer.

(m). Incubate the DNA in a 65°C heat block for 1h. Gently vortex every 15min to facilitate dissolution.

(n). Store samples at –20°C for short term (<1month) or –80°C for long term (>1month).

*3.3. DNA Sample Preparation from Tissue after Laser Capture Microdissection (LCM)*

1. Generate an LCM cap containing the captured tissue section using Laser capture microdissection as described in detail in Chapter "Laser dissection of biopsy material."

2. To make the Extraction Solution pipette 155µl of Reconstitution Buffer in a vial of Proteinase K, both provided by the PicoPure DNA extraction Kit. Vortex and place on ice. Be sure to dissolve completely.

3. Pipette 30µl of the Extraction Solution into a sterile 0.5ml microcentrifuge tube (*see* **Note 11**).

4. Using aseptic technique with a set of sterile tweezers, invert the LCM cap and place it on top of the microcentrifuge tube. Press firmly (*see* **Note 12**).

5. Invert the tube and verify the solution is completely covering the surface of the cap.

6. Incubate inverted at 65°C for at least 16h (*see* **Note 13**).

7. After incubation centrifuge tubes at 1,000×g for one min.

8. Remove the cap, close the lid and incubate at 95°C for ten min to inactivate the Proteinase K.

9. Samples are ready for Q-PCR.

**3.4. Preparation of Probes and Primers**

1. Add appropriate volume of nuclease free water to each primer to make a 10µM working stock probe solution and a 50µM working stock primer solution. Leave the vial for 5min at room temperature to allow the powder to dissolve. Pipette up and down several times to mix.

2. Store in 50µl aliquots. When protected from light and stored at –20°C, the primers and probes are stable for up to 2 years. Thawing and refreezing up to three times is acceptable for these reagents.

3. Refer **Note 14** for options for resuspension.

**3.5. Preparation of Standards (see Note 15)**

1. Bring DNA from Standards (TS DNA) to a concentration of 0.2µg/µl with nuclease free water.

2. Label five 1.5ml reaction tubes with TS Std. 1–5

3. Pipette 450µl of nuclease free water into tubes 2–6.

4. Pipette 500 µl of Standard DNA (0.2 µg/µl) into tube 1.

5. Vortex tube 1 and spin briefly to collect the DNA.

6. Pipette 50µl from tube 1 to tube 2.

7. Vortex and spin briefly.

8. Repeat previous two steps for tubes 2–5. Always change pipette tips between steps. Dilution of standards are summarized in **Table 1.**

9. Pipette 990µl of nuclease-free water to tube 6.

10. Pipette 10µl from DNA tube 4 into tube 6.

11. Vortex and spin briefly.

12. Aliquot Standards 40µl per tube. Aliquoted standards should be thawed no more than two times.

**3.6. Preparation of Controls (see Note 15)**

1. Label tubes "TS DNA", numbers 1 through 5.

2. Add 375µl of stock TS DNA (0.4µg/µl) to tube 1.

3. Add 450µl of nuclease-free water to tubes 2–5.

4. Pipette 50µl from tube 1 into tube 2.

5. Vortex and spin briefly.

6. Repeat this for tubes TS DNA2–5.

7. Label appropriate size tubes "TS Controls," numbers 1–6

8. Pipette 250µl from "TS DNA 1" into "TS Control 1."

## Table 1
## Standard dilutions

| TS Std. No. | No. TS copies/5 μl[a] | Standard DNA (0.2μg/μl) | | Water (μl) | Final mass (μl) | Final volume (μl) | Final conc (μg/μl) |
|---|---|---|---|---|---|---|---|
| | | Volume (μl) | Mass (μg) | | | | |
| 1 | $1.5\times10^5$ | 500 | 100 | 0 | 100 | 500 | 0.2 |
| 2 | $1.5\times10^4$ | 50 | 10 | 450 | 10 | 500 | 0.02 |
| 3 | $1.5\times10^3$ | 50 | 1 | 450 | 1 | 500 | 0.002 |
| 4 | $1.5\times10^2$ | 50 | 0.1 | 450 | 0.1 | 500 | 0.0002 |
| 5 | $1.5\times10^1$ | 50 | 0.01 | 450 | 0.01 | 500 | 0.00002 |

[a]The copy numbers in the standards are based on the assumption that there are 150,000 copies in 1μg of DNA

## Table 2
## Preparation of controls

| TS Ctl. No. | No. TS copies/ 5μl | TS DNA (0.4μg/μl) | | Genomic DNA (TS negative) (0.4μg/μl) | | Water (μl) | Final mass (μg) | Final volume (μl) | Final conc (μg/μl) |
|---|---|---|---|---|---|---|---|---|---|
| | | volume (μl) | mass (μg) | volume (μl) | mass (μg) | | | | |
| 1 | $1.5\times10^5$ | 250 | 100 | 0 | 0 | 250 | 100 | 500 | 0.2 |
| 2 | $1.5\times10^4$ | 250 | 10 | 225 | 90 | 25 | 100 | 500 | 0.2 |
| 3 | $1.5\times10^3$ | 250 | 1 | 247.5 | 99 | 2.5 | 100 | 500 | 0.2 |
| 4 | $1.5\times10^2$ | 250 | 0.1 | 249.75 | 99.9 | 0.25 | 100 | 500 | 0.2 |
| 5 | $1.5\times10^1$ | 250 | 0.01 | 249.975 | 99.99 | 0.025 | 100 | 500 | 0.2 |

9. Pipette 250μl from "TS DNA 2" into "TS Control 2."

10. Repeat this for tubes 3–5. Add to each TS Control Tube the appropriate amount of genomic, TS negative DNA and nuclease-free water as listed in **Table 2**.

11. Label a 14ml round-bottom tube as TS Control 6.

12. Pipette 2,475μl of nuclease-free water into the tube.

13. Pipette 25µl of TS DNA #4 into the same tube.

14. Pipette 2,500µl of genomic DNA (0.4µg/µl) into the same tube to create TS Control 6. Mix well by vortexing.

15. Aliquot 500µl of TS Control 6 to each tube.

16. Aliquot TS Controls 1–5 in 50µl per tube.

17. Dilute TS negative genomic DNA to a final concentration of 0.2µg/µl to use as negative control in the PCR. Aliquot and store at −20°C.

*3.7. Taqman PCR: Real Time Acquisition of the Polymerase Chain Reaction*

1. Set up the PCR reactions in a PCR workstation, if available. Prior to starting the procedure, clean the PCR workstation and micropipette barrels with DNA Away.

2. Use aseptic technique for open samples and reagent manipulations. Change gloves often, and whenever contamination of gloves is suspected. Use powder free gloves, as powder from the gloves could potentially enter a well on the plate and give false positive fluorescence. (*see* **Notes 16–21** for additional practice hints).

3. Prepare the master mix (MM) in a 1.5ml tube according to **Table 3**. Standardly use a reaction volume of 20µl (*see*

**Table 3**
**Contents of reaction mix for Taqman PCR**

| Reagent | Volume(µl) per/ sample | Final concentration |
|---|---|---|
| TaqMan Fast Universal PCR Mix 2× | 10 | 2× |
| Nuclease free water | Varies | N/A |
| TS forward primer [50µM] | 0.36 | 900nM |
| TS reverse primer [50µM] | 0.36 | 900nM |
| TS-FAM probe [10µM] | 0.5 | 250nM |
| ApoB(human or murine) forward primer [50µM] | 0.1 | 250nM |
| ApoB (human or murine) reverse primer [50µM] | 0.1 | 250nM |
| ApoB-(human or murine) VIC Probe [10µM] | 0.5 | 250nM |
| Sample DNA (0.01–1µg or 5µl from colony lysis) | Up to 8µl | |
| Total volume | 20 | N/A |

Notes **22** and **23**). Calculate total volumes based on the number of samples plus two extra, to allow for pipetting loss. DNA is to be added after MM is placed in the wells of the reaction plate.

4. Add appropriate amount (dependent on sample volume) of master mix to each well of a new Fast PCR optical reaction plate.

5. Add the DNA samples to wells. Use 0.01–1µg of DNA template per reaction in a total of up to 8µl (*Example 1*: If 8.0µl DNA sample is to be added, add 0.28µl of water per reaction to the MM. Add 12µl of MM to each well. Since only 5µl of template is used for the Controls and Standards, an additional 3µl of water needs to be added to those wells to bring the total reaction volume to 20µl). Use at least 12 wells for controls and standards.

6. Add controls: no template control (add water only or MM only), negative control (TS negative genomic DNA, 1µg, 5µl) and controls 1–5 as defined in **Subheading 3.6** (*see* **Note 24**).

7. Add standards in at least 5 dilutions as prepared in **Subheading 3.5**. Standards and controls may be run in triplicates (*see* **Note 25**).

8. Apply optical adhesive cover, and briefly spin in centrifuge to collect the DNA at the bottom of the wells.

9. Set Fast Thermal Cycler Conditions as follows (*see* **Note 26** for other real-time scenarios). Denaturation: 95°C – 20s; then 40 cycles of 95°C – 1s, 60°C – 30s (*see* **Note 27**).

10. Save file and select run.

***3.8. Taqman PCR: Data Analysis***

1. When run is complete, open the run file and assign values to the standards

   a. Under analysis options set threshold to manual and click the analysis button. The ApoB and vector copy numbers are calculated automatically from the Ct values.

   b. Adjust threshold to obtain optimal correlation coefficient ($R^2$>0.99), verify the slope is 3.20–3.70.

2. Under "file," "export" results to a text file. Copy and paste text file to excel spreadsheet.

3. Divide the vector copy number value (FAM) by the cell number calculated based on the ApoB value (VIC) to arrive at copy number/cell

**3.9. Determine Validity of the Assay**

1. Review controls

   (a). No Template Control and Negative Control have a Ct value for the target sequence that is undetermined.

   (b). All controls perform within expected range (*see* **Note 28**).

   (c). The mean is established from at least 5 runs with the standards. These should be charted over time. If they deviate remake controls.

2. Review standards

   (a). Standards produce a linear regression curve with a correlation coefficient $R^2 > 0.99$ and a slope ranging from 3.20 to 3.70.

   (b). Ct for the standards should range from 19 to 39.9

   (c). If the standards have deviated from above stated range over time, remake standards, rerun assay.

   (d). If high sensitivity is requested, standard 6 (1.5 copies) needs to be detected in a minimum of 1 out of 10 wells.

**3.10. Interpretation**

1. Interpretation criteria are dependent on sample type

2. Taqman PCR of colony derived DNA may be interpreted as follows (*see* **Note 29**)

   (a). Exclude colonies from analysis where the ApoB Ct is higher than the second highest background Ct value or where less than 100 cells (as calculated based on ApoB Ct value) were present.

   (b). Colonies with <0.5 copies per cell are considered vector negative.

   (c). The calculated copy number per cell of colonies with >0.5 copies per cell is rounded to the closest full number.

3. Taqman PCR of DNA derived from non-clonal fresh or LCM derived tissue may be interpreted as follows

   (a). The measured copy number per cell is a mean value for all present cells.

4. If the PCR needs to determine whether a tumor found in an experimental animal is vector positive or vector negative: The copy number of fresh or microdissected (e.g., by LCM) tumor tissue needs to be compared with the copy number per cell in not affected hematopoietic tissues of the same experimental animal. Also, the contribution of the tumor cells to the particular sample need to be considered. For example, a copy number of <0.3 copies per cells in a tissue sample where the contribution of tumor cells to the sample is >70% indicates that the tumor is vector-negative, and the signal is derived from nonmalignant blood cells present in the tissue sample.

## 4. Notes

1. If ABI primer express is not available, there are online resources that assist with primer design. If neither of these programs can suggest a primer/probe set here are some design tips: A good "rule of thumb," for every "GC" pair add 4°C to the Tm and for every "TA" pair add 2°C to the total Tm. The Tm for the primers should be around 60°C, while the Tm for the probe should be 10° higher than the primer, around 70°C. Be sure to design the forward and reverse primer to span an intron if vector cDNA is targeted so that the potentially endogenous gene is not amplified. Be sure to perform a BLAST on the new primer sequences at the NCBI Website. This should check the sequence against the organism of choice (in our case mouse and human genome), and displays any possible cross-matches the primers will have with the selected genomic DNA. Cross-matches should be avoided to minimize unwanted binding of primers to background genomic DNA.

2. Other helpful Websites for primer design assistance and calculations are the following:

    a. Integrated DNA Technologies (http://www.idtdna.com) has a "SciTools" link.

    b. Roche has an assay design center (http://www.roche-applied-science.com).

    c. Identification of exons and introns within known genes can be found at Ensembl Genome Browser (http://www.ensembl.org) along with vast information related to specific eukaryotic genomes.

3. The ideal standards are topics of ongoing discussion. In general, we have chosen to use a clonal cell line with a single integration of the target sequence because we believe that this is as close as possible to the clinical samples we expect to assay. The clonal cell lines, if not available, must be generated by transducing the cell line with a retroviral vector containing the target sequence, performing single cell deposition and isolating clones with a single insertion. Great attention must be paid to the starting cell line. The cell line must be of the same species as the samples and must have a near diploid karyotype to appropriately represent the "normal" copy number of ApoB and total DNA per cell. HT1080 cells can be used for human and NIH/3T3 cells for murine background.

4. Another valid consideration is to use the "controls" as they are described in this chapter as the standards because they contain the target sequence within a larger background of genomic DNA. This more closely mimics the composition of

the samples. For instance, at low TS copy numbers (e.g., 15 copies), the reaction frequently shows a lower copy number for the controls (TS DNA diluted in genomic DNA) than for standards containing the same number of TS copies (TS DNA diluted with water). This is hypothesized to be due to template competition.

5. Standards can also be generated by diluting plasmid DNA containing the target sequence into genomic DNA. However, manipulations involving plasmid DNA are avoided to minimize opportunity for plasmid contamination.

6. The "FAST" protocol allows a significantly faster turnover and can be used once new target sequence specific primers are validated by running with the "standard" protocol.

7. The DNase treatment step can be omitted if the hematopoietic cells are derived from test animals or individuals which had been transplanted with vector transduced cells or if the cells were growing in culture for >10 days after vector exposure.

8. The volume of reagents and tubes used is dependent on the starting volume of the *original sample*. For samples <300μl, use 2.0ml tubes. For samples >300μl, use 15ml conical tubes. All centrifugation for 2.0ml tubes is at 16,000×g. All centrifugation for 15ml tubes is at 2,000×g.

9. If the original volume is >300μl, use 10% of the original sample volume.

10. The volume of reagents and tubes used is dependent on the starting cell number of the *original sample*. Use **Table 4** to determine how much of each reagent and which tube to use. All centrifugation for 2.0ml tubes is at 16,000×g, except when noted. All centrifugation for 15ml tubes is at 2,000×g.

11. With over-dilution of a limited sample as a concern, 30μl allowed for enough template at a sufficient concentration. The Sample Extraction Devise that comes with the kit was not used as it often would separate from the tube during centrifugation causing mass contamination of all samples.

12. During transport of the caps from capture to prep bench the Alignment tray was used. This held the caps well during transport. Tweezers were found to be helpful when removing the caps from the alignment tray. Be sure to spray the tray with 70% ethanol and DNase before using it again.

13. This is best done overnight. Preheat the incubation block, then place the block over the alignment tray containing the inverted tubes and place back in the incubator.

14. When receiving the primers and probes, some companies measure the dry primer/probe in OD260, nmol, or mg.

**Table 4**
**Chart for DNA isolation with Puregene DNA Isolation Kit (Gentra)**

| | | | | | Solution (µλ) | | |
|---|---|---|---|---|---|---|---|
| No. of cells | Expected yield range (µg) | Tube size (ml) | Cell lysis | Protein ppt. | Isopropanol | 70% ethanol | DNA hydration |
| 0.5–1×10⁶ | 2–6 | 2 | 0.15 | 0.05 | 0.15 | 0.15 | 25µl |
| 1–2×10⁶ | 5–10 | 2 | 0.3 | 0.1 | 0.3 | 0.3 | 50µl |
| 2–6×10⁶ | 15–30 | 2 | 0.6 | 0.2 | 0.6 | 0.6 | 0.1 |
| 6–9×10⁶ | 28–43 | 2 | 0.75 | 0.25 | 0.75 | 0.75 | 0.1 |
| 9–20×10⁶ | 40–120 | 15 | 3 | 1 | 3 | 3 | 0.25 |

There are online resources that will perform the resuspension calculations automatically from various measurements. The forward and reverse primers can also be combined into one tube. If the final concentration desired is 50µM, resuspend the forward and reverse primers at a concentration of 100µM and then combine equal parts into one tube.

15. Dilutions of standards and controls can be prepared ahead of time, aliquoted and stored at –20°C. This ensures that the same standards/control is used in each run and allows quality assurance. To establish an acceptance range for the control, the controls must be run at least five times with the standards, to establish a mean and a standard deviation (SD). The acceptance range for the controls is 1 SD. The controls are measured in percentages based on TS copies/cell.

16. It is critical to maintain a separate area for all pre-PCR work. This is best accomplished in a PCR Workstation in a room segregated from plasmid and post-PCR materials. Maintain controlled and limited access.

17. HEPA filtered supply and positive differential pressure relative to adjacent areas is suggested.

18. Use lab coats designated for the PCR area only. Hair and beard covers are suggested.

19. Maintain equipment designated for the PCR area only.

20. When moving samples into the pre-PCR area, transfer to designated tube racks.

21. To reduce the risk of contamination when setting up the PCR plate, first add the Negative Control, then the Standards and the (Positive) TS Controls, change gloves, then add the

samples. When pipetting samples, standards, and controls, make sure to open the tubes, and draw the sample away from the top of the open plate. This reduces the chance of the samples DNA dropping onto the plate. A repeat pipettor is ideal for aliquoting the master mix into the PCR plate, and also reduces the risk of mis-aliquoting.

22. Although 20μl is suggested, the reaction volume can range from 10 to 25μl. If sensitivity is paramount a reaction of 20–25μl is recommended. This allows for more reagents and primers per potential cDNA. However, 15μl has shown to be an adequate amount for many q-PCR assays. Mix the master mix by pipetting up and down.

23. When setting up a multiplex reaction start with forward and reverse primer concentrations of 900nM for both genes, and 250nM for the probes. Because of the improved technology of the newer PCR machines, it is not necessary to optimize forward and reverse primers individually. The two recommended reporter dyes for the probes are FAM and VIC. If the resulting curves show that one gene (A) is outperforming the other (B), meaning curves from gene A has a longer geometric phase and curves from gene B plateau early or do not amplify properly, run a primer limiting assay. The primer limiting assay consists of reducing the concentration of primers of the better performing gene. Try a few different reaction mixes. Keep 900nM primer concentration for gene B and 750, 500, and 250nM primer concentrations for gene A. Use 250nM concentration for both probes. Test these mixes on the standards and controls, as there could be a balance needed between sensitivity and reproducibility of the amplification kinetics of target sequence and normalizing gene. Also, it is possible to go well below a primer concentration of 250nM and still be effective.

24. 10 wells with Control 6 (1.5 copies in 150,000 cells) may be added if sensitivity at very low copy number needs to be demonstrated.

25. When setting up an assay running the Standards in triplicate is highly recommended. If, over time reproducibility becomes evident, one set of standards should suffice. When new standards are made running in triplicate is again recommended.

26. If using a real-time machine without the fast cycling capabilities Standard Thermal Cycler Conditions for Taqman q-PCR are set as follows: 50°C – 2min, 95°C – 10min, 40 Cycles: 95°C – 15s, 60°C – 1min

27. Long annealing (30s) increases sensitivity, whereas shorter annealing (20s) increases specificity.

28. If apoB is not detected for the sample and the standards and controls worked properly, check the DNA spectrophotometer reading. If the reading is acceptable, reisolate the DNA and rerun. There is a possibility that an inhibitor was either not removed or was introduced during isolation of the sample. An example of this would be when rinsing the DNA in ethanol, and not properly drying the DNA pellet. This would potentially inhibit the reaction.

29. An alternative reported method is to base interpretation of q-PCR data from colonies on the background values only *(11)*. If 36 background samples were collected, the second highest Ct value will serve as a cutoff value. All samples with a Ct values higher than the Cutoff value are considered negative.

## References

1. Baum, C., Dullmann, J., Li, Z., Fehse, B., Meyer, J., Williams, D.A. and von Kalle, C. (2003) Side effects of retroviral gene transfer into hematopoietic stem cells. *Blood*, 101, 2099–2114.

2. Hacein-Bey-Abina, S., Von Kalle, C., Schmidt, M., McCormack, M.P., Wulffraat, N., Leboulch, P., Lim, A., Osborne, C.S., Pawliuk, R., Morillon, E.et-al. (2003) LMO2-associated clonal T cell proliferation in two patients after gene therapy for SCID-X1. *Science*, 302, 415–419.

3. Li, Z., Dullmann, J., Schiedlmeier, B., Schmidt, M., von Kalle, C., Meyer, J., Forster, M., Stocking, C., Wahlers, A., Frank, O.et-al. (2002) Murine leukemia induced by retroviral gene marking. *Science*, 296, 497.

4. Du, Y., Spence, S.E., Jenkins, N.A. and Copeland, N.G. (2005) Cooperating cancer-gene identification through oncogenic-retrovirus-induced insertional mutagenesis. *Blood*, 106, 2498–2505.

5. Du, Y., Jenkins, N.A. and Copeland, N.G. (2005) Insertional mutagenesis identifies genes that promote the immortalization of primary bone marrow progenitor cells. *Blood*, 106, 3932–3939.

6. Modlich, U., Kustikova, O.S., Schmidt, M., Rudolph, C., Meyer, J., Li, Z., Kamino, K., von Neuhoff, N., Schlegelberger, B., Kuehlcke, K.et-al. (2005) Leukemias following retroviral transfer of multidrug resistance 1 (MDR1) are driven by combinatorial insertional mutagenesis. *Blood*, 105, 4235–4246.

7. Holland, P.M., Abramson, R.D., Watson, R. and Gelfand, D.H. (1991) Detection of specific polymerase chain reaction product by utilizing the 5--3 exonuclease activity of Thermus aquaticus DNA polymerase. *Proc Natl Acad Sci USA*, 88, 7276–7280.

8. Chen, J., Reeves, L., Sanburn, N., Croop, J., Williams, D.A. and Cornetta, K. (2001) Packaging cell line DNA contamination of vector supernatants: implication for laboratory and clinical research. *Virology*, 282, 186–197.

9. Sastry, L., Xu, Y., Cooper, R., Pollok, K. and Cornetta, K. (2004) Evaluation of plasmid DNA removal from lentiviral vectors by benzonase treatment. *Hum Gene Ther*, 15, 221–226.

10. Will, E., Bailey, J., Schuesler, T., Modlich, U., Balcik, B., Burzynski, B., Witte, D., Layh-Schmitt, G., Rudolph, C., Schlegelberger, B.et-al.. (2007) Importance of murine study design for testing toxicity of retroviral vectors in support of phase I trials. *Mol Ther*, 15, 782–791.

11. Villella, A.D., Yao, J., Getty, R.R., Juliar, B.E., Yiannoutsos, C., Hartwell, J.R., Cai, S., Sadat, M.A., Cornetta, K., Williams, D.A.et-al. (2005) Real-Time PCR: An Effective Tool for Measuring Transduction Efficiency in Human Hematopoietic Progenitor Cells. *Mol Ther*, 11, 483–491.

# Chapter 20

# Tissue Procurement for Molecular Studies Using Laser-Assisted Microdissection

## Ulrich Lehmann and Hans Kreipe

## Summary

Properly collected and stored human specimens offer the unique opportunity to study human diseases at the molecular level in the real in vivo situation. The intention of this chapter is, first, to raise the awareness for important points, which have to be clarified before human tissue samples are collected and stored for molecular studies. Second, detailed protocols are provided for the fixation and processing of bone marrow trephines and the isolation of morphologically and immunohistochemically defined pure cell populations from bone marrow trephine sections using laser-assisted microdissection.

**Key words:** Laser assisted-microdissection, Tissue banking, Formalin-fixed paraffin-embedded (FFPE), Bone marrow trephines.

## 1. Introduction

The collection and long-term storage of human samples offer the unique opportunity to study molecular alterations underlying a certain disease in the context of the real in vivo situation in a retrospective or prospective manner. This allows the correlation of long-term clinical follow-up data with molecular alterations in the very same patients. In addition, it will enable in the future the application of analytical tools for the patient's benefit, which are not yet developed. However, collection and storage of human samples for molecular studies requires careful consideration of the following points before biopsies can be collected and experiments can be started:

1. Legal questions (informed consent of patients necessary/ available? data protection ensured?).

Christopher Baum (ed.), *Methods in Molecular Biology, Methods and Protocals, vol. 506*
© Humana Press, a part of Springer Science+Business Media, LLC 2009
DOI: 10.1007/978-1-59745-409_20

2. Interference of sample collection with necessary routine diagnostic procedures.

3. Rapid preservation by freezing necessary or formalin-fixation sufficient or even better?

4. Dedicated personnel for collection, processing, and management of the archive available?

5. Equipment for every step on-hand?

6. Data management set up? (collection, storage, and connection of patient data and experimental results)

7. Instruction of all health care professionals involved.

8. Sample processing (e.g., decalcification) compatible with subsequent molecular studies?

A comprehensive discussion of all points raised is beyond the scope of this chapter. In addition, many points concerning the "infrastructure" (and not the specimen processing itself), such as interference with routine diagnostics, personnel involved, cooperation between departments, data storage and management require individual "local" solutions (and might differ substantially between different countries). Therefore, we intend to raise the awareness for several problems, which have to be addressed (and solved!) in advance and to focus on the collection, processing, and molecular analysis of human bone marrow trephines **(Fig. 1)**.

The collection of human specimens and patient data as well as the use of human material for (future) molecular studies raises several ethical and legal questions which require thorough consideration and lead to the development of a variety of legal and organizational frameworks with differences from country to country *(1, 2)*.

A major hurdle that needs considerable time and effort is the involvement of personnel from different laboratories and/or institutions and the allocation of sufficient resources in terms of space and equipment required for the processing and storage of biopsies. One of the most important questions concerning the procurement of human specimens that has to be answered as early as possible is: "fresh frozen or archival?". Rapid preservation by freezing requires considerable efforts in terms of planning, infrastructure (e.g., constant supply of liquid nitrogen and appropriate storage facilities), and personnel involved. It interferes also more or less with routine diagnostic procedures because fresh or fresh frozen material is under many circumstances not suitable for routine diagnostics. Further disadvantages are reduced morphological quality of cryo sections, after thawing autolytic destruction ensues immediately (making staining of frozen sections and subsequent RNA analysis a considerable challenge).

In contrast to this, formalin-fixation and paraffin-embedding is widely used in pathology and has several advantages: excellent preservation of morphological details, compatible with all standard

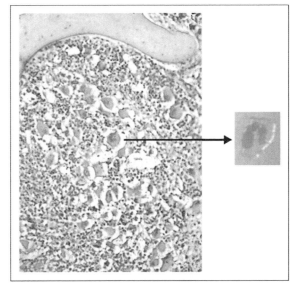

## Lehmann, Kreipe

Fig. 1. Isolation of megakaryocytes from a bone marrow trephine section. The reduced optical quality of the *left panel* is due to the fact that the tissue section is dried and not coverslipped (original magnification: ×100). Isolated megakaryocyte in the lid of a reaction tube (original magnification: ×400). To improve the optical quality of the microscopic picture, a small drop of absolute ethanol can be applied. The appearance of the section will improve temporarily until the ethanol has evaporated. But too much liquid will interfere with the laser cutting and in particular with the catapulting of single cells. Therefore, a compromise between optical quality and efficient cutting and recovery of cells has to be found (Original photographs kindly provided by Oliver Bock).

histological and the vast majority of immunohistochemical stains, very easy to apply and quite cheap, container with fixative can be stored at room temperature until use, fixed and embedded specimens can be stored at room temperature virtually for decades, nuclease activity completely abolished due to cross-linking of all proteins, killing of all potential infectious hazards. However, formalin-fixation leads to extensive fragmentation of nucleic acids and the destruction or masking of several interesting antigens *(3, 4)*. Therefore, molecular studies requiring high-molecular weight DNA or full length RNA molecules or aiming at antigens destroyed by formalin-fixation have to be performed with rapidly frozen tissue.

Bone marrow trephines, which enable the morphological evaluation of pathological changes affecting hematopoiesis in the bone marrow (e.g., development of bone marrow fibrosis), represent a peculiar type of specimen, because due to the presence of bone fragments only formalin-fixation followed by extensive decalcification and paraffin-embedding prior to sectioning will produce adequate morphological preservation. But we and others could show (3, 4) that DNA as well as RNA extracted from these formalin-fixed and paraffin-embedded specimens is very well suitable for qualitative and quantitative molecular analysis if adequate tissue processing is performed and subsequent molecular assays are optimized accordingly. Very recently it was also shown that even the extraction of intact full-length proteins from formalin-fixed paraffin-embedded biopsies is possible (5). In addition, laser-assisted microdissection of bone marrow sections allows the cell-type-specific molecular analysis of different cellular compartments in the bone marrow. This approach enables the direct comparison of molecular data with morphological findings and is under many circumstances (e.g., marrow fibrosis) the only way of accessing the cells of interest. Since laser-assisted microdissection can be combined with prior immunohistochemical labeling, direct correlation of DNA data (e.g., mutation), mRNA expression and protein expression data can be performed (6). We have used the microdissection methodology in hematopathology for mutation detection and mRNA quantification in defined cell populations of the bone marrow (see for example references (7–9)). Additional applications in our group include the exact quantification of gene copy numbers in intraductal and invasive components of breast carcinomas (10, 11), the quantitative assessment of promoter hypermethylation during breast cancer development (12), and the detection of in situ chimerism after organ transplantation (13, 14). For an overview of commercially available microdissection systems, see **Table 1**.

## Table 1
## This gives a list of commercially available microdissection devices currently in use

| **Arcturus** | **(http://www.moleculardevices.com)** |
|---|---|
| Bio-Rad Laboratories | (http://www.bio-rad.com) |
| Cell Robotics | (http://www.cellrobotics.com) |
| Eppendorf | (http://www.eppendorf.com) |
| Leica Microsystems | (http://www.leica-microsystems.com) |
| MMI AG | (http://www.molecular-machines.com) |
| P.A.L.M. | (http://www.PALM-microlaser.com) |
| SEL (Nikon AG) | (http://www.nikonusa.com) |

The interested reader is referred to the specified Web addresses for further information. Also more or less comprehensive lists of references using the different systems are available on the Web sites of the manufacturers. We are working with the P.A.L.M-system since several years and most of the protocols described inhere are specifically adapted to this system

***1.1. Organisation of the Laboratory***

The enormous amplification power of the polymerase chain reaction (up to $10^{13}$ times!), which is the basis for the exquisite sensitivity of this technology, creates also a serious problem: risk of contamination due to the introduction of PCR product from a reaction tube after amplification into the reaction mixture prepared for the analysis of the same target in different samples. This problem is especially prevalent in the case of analysing minute amounts of starting material isolated by laser-assisted microdissection and if the same assay is performed on a regular base over a long period of time (as it is the case for routine diagnostics or comprehensive prospective molecular studies). Since every PCR produces vast amounts of amplifiable molecules (usually much more than $10^9$ molecules), the strict physical separation of the analysis of reaction products (postamplification) from all stages of sample preparation (preamplification) has to be implemented.

For these reasons, strictly enforced guide-lines concerning the handling of samples before and after amplification and the cleaning of all instruments used need to be followed by all personnel involved. We perform all preamplification steps, including sectioning, staining, and laser-assisted microdissection in a separate laboratory ("pre-PCR-area"). Everything used in this laboratory (including lab coats, pens, and notepads) is dedicated exclusively to this room and strictly separated from the post-PCR-area. Plastic labware and the benches are cleaned using a 3% hypochlorite solution. PCR products are analysed in a separate laboratory ("post-PCR-area"). Under no circumstances should amplified samples or equipment from this working area be brought back to the pre-PCR area. Great care has also to be taken to ensure that the "primary sample processing" in the institution (registration, labelling, decalcification, and embedding) is well separated from the post PCR-laboratory.

The protocols described below concentrate on the following steps:

1. Fixation and decalcification of bone marrow trephines.

2. Cutting and staining of histological tissue sections.

3. Laser microdissection and specimen recovery.

4. Isolation of DNA and/or RNA from microdissected cells.

## 2. Materials

The manufacturers or distributors are only specified if reagents or laboratory equipment might be important for the outcome, or if a source might be difficult to identify. All chemicals were purchased in p.a. quality from MERCK, Roth, or SIGMA and kept strictly separate from the postamplification area in our institute.

1. Bone marrow fixative: 64% methanol (v/v), 4.48 M formaldehyde, 1.6 mM sodium hydrogenphosphate pH 7.4, 7.4 mM glucose.

2. Decalcification solution: 270 mM Tris-HCL, 270 mM EDTA pH 7.4.

3. 60 mM sodium hydrogenphosphate, pH 7.0–7.2.

4. Poly-propylene foil-coated glass slides (P.A.L.M., Bernried, Germany).

5. "Xylol-Ersatz", a xylene substitute, which is less toxic and smells less unpleasant (Vogel, Karlsruhe, Germany).

6. Ethanol (100%, 96%, 70%).

7. Glass cuvettes.

8. ABC Vectastain-Kit (Vector Laboratories, Burlingame, CA, US).

9. Methylene blue (Loeffler's Methylene blue, MERCK, Darmstadt, Germany).

10. Methyl green (MERCK, Darmstadt, Germany); Staining solutions are stored at room temperature in the dark.

11. Liquid wax (MJ Research, Boston, MA, USA).

12. Proteinase K-buffer: 50 mM Tris-HCl pH 8.1, 1 mM EDTA, 0.5% Tween 20.

13. Digestion solution: 4 M guanidinium isothiocyanate, 30 mM Tris-HCl pH 8, 0.5% sarcosyl, 0.1 M β-mercaptoethanol.

14. Linear polyacryl amide ("LPA" from SIGMA, Taufkirchen, Germany).

15. Proteinase K, stock solution: 20 mg/ml in water, aliquots stored at –20°C (Merck, Darmstadt, Germany).

16. TE buffer: 10 mM Tris-HCl pH 8.1, 1 mM EDTA.

17. Sterile canula (26 GA 3/8, 0.45 × 10), for picking dissected cells.

18. 0.5 ml tubes with transparent lid and lowered inner lid, for collecting dissected and catapulted cells (P.A.L.M., Bernried, Germany).

19. Tips with aerosol protection, DNAse-, RNAse-free (Sarstedt, Nümbrecht, Germany).

20. Sterile water.

21. 3 M Sodium acetate pH 7.0 containing 100 µg/ml Dextran T500 (SIGMA, Taufkirchen, Germany).

22. Hypochlorite-solution, for use diluted 1:4 with tab water (Roth, Karlsruhe, Germany).

23. PCR-bench with UV-lamp, for decontamination of racks and pipettes (in-house construction).

24. Refrigerated table-top centrifuge for 0.2–2.0 ml tubes (max. 14,000 × g).

25. Vortex.

26. 40°C-incubator.

27. Thermoshaker (Eppendorf, Hamburg, Germany).

28. Thermoshaker with heated lid (CLF, Emersacker, Germany).

# 3. Methods

## 3.1. Fixation and Decalcification

The optimal fixation procedure using buffered formalin is 24 h at 4°C in the dark. This cannot always be achieved under routine conditions but some deviation from this optimal conditions (longer duration, higher temperature i.e., room temperature) seems to be tolerable. The volume of the fixative should be much greater (10 times or more) than the volume of the specimen. But this is no problem dealing with bone marrow trephines, which are quite small.

The pH of the EDTA solution is adjusted to pH 7.4 using 60 mM sodium hydrogenphosphate pH 7.0–7.2. The solution is stirred constantly and changed every day. The decalcification in the neutral EDTA-buffer can take several days. But conservation of nucleic acids is guaranteed under these conditions. By using an ultrasonic bath this time can be shortened considerably (*see* **Note 1**).

## 3.2. Sectioning and Staining

### 3.2.1. Sectioning

Formalin-fixed paraffin-embedded biopsies are cut using a conventional microtom, and sections are mounted on foil-covered slides. To improve the adhesiveness, the slides with sections are stored over night at 45°C. Afterwards the sections are deparaffinised and rehydrated using "Xylol-Ersatz" and ethanol (2× xylene-substitute for 10 min, 2× 100% ethanol for 5 min, 2× 96% ethanol for 5 min, 1× 70% ethanol for 5 min, sterile water for 5 min).

The sections are stained with methylene blue or methyl green (depending on personal preferences and the combination with immunohistochemical stains, see below) for 10–30 s, rinsed with distilled water and absolute ethanol and dried thereafter at 40°C over night (*see* **Note 2**).

*3.2.2. Immuno-
histochemical Stains*

For labeling cells with antibodies before microdissection, we use the Vectastain kit from Vector Laboratories (Burlingame, CA, USA). As outlined earlier, all reagents brought into contact with samples before amplification (pre-PCR) have to be strictly separated from all other reagents used in the laboratory (post-PCR). The great advantage of ready-to-use kits is that the components are free of any potentially contaminating PCR products and completely separated from all reagents normally used in the laboratory. This justifies the higher costs (*see* **Note 3**).

### 3.3. Microdissection Using the P.A.L.M.™ Laser Microdissection Microscope

The width of the laser cut can be trimmed by adjusting the laser energy and/or the focus of the laser. In our setting, the optimal focus for using the 40× long distance objective from Zeiss (40×/0.60 Korr ∞/0–2) is around 13 (arbitrary units), for using the 10× objective from Zeiss (10×/0.50 ∞/0.17) the optimal value is 68 (arbitrary units). Depending on the thickness of the section and the tissue type, the energy for cutting is between 70 and 85 (arbitrary units), for catapulting greater than 85. The actual numbers for the energy and focus setting may vary a bit for different instruments calibrated and adjusted in a slightly different way (*see* **Note 4**).

For recovery of dissected cells using the Laser Pressure Catapulting technology of the P.A.L.M. system, the 0.5 ml-tubes distributed by P.A.L.M. itself are most suitable because of the lowered inner lid, which shortens the distance the catapulted cells have to fly. Also the visibility of the specimens in the transparent lid is quite well. For catapulting a dissected cell or a group of cells, the laser is slightly focused below the section and the laser energy is increased. A single short laser impulse is sufficient for catapulting the specimen into the lid of a reaction tube placed directly above the section. The reaction tube is conveniently positioned in the holder of the micromanipulator (*see* **Note 5**).

Despite having the very sophisticated (and also often necessary) laser pressure catapulting technology, we recover dissected pieces of tissue with a sterile needle if they are large enough. This can be done easily "by hand" without any technical support. After some exercising this turns out to be unsurpassed fast and easy for every person working with the laser microscope.

### 3.4. Isolation of Nucleic Acids

*3.4.1. Isolation of DNA*

Very few cells *(1–30)* are lysed in the lid of the reaction tube by adding 10–30 µl TE-buffer containing 40 µg Proteinase K. The closed tubes are incubated in a small incubator in an inverted position at 45°C over night. More than ~50 cells are lysed in Proteinase K-buffer (*see* **Subheading 2**) containing 500 µg/ml of Proteinase K. The samples are incubated in a thermoshaker with a heated lid at 56°C over night. If several hundred or even thousands of cells are isolated, the samples are lysed in a larger volume of Proteinase K-buffer (100–300 µl) containing 500 µg/ml of Proteinase K. The samples are incubated in a vigorously shaking thermoshaker at 56°C over night. The next day the lysate is transferred to a new tube (*see* **Note 6**), and the DNA is precipitated by adding sodium acetate (pH 7) containing Dextrane T500 as a carrier (100 µg/ml) and ethanol. This precipitation steps removes contaminating dyes and cell debris nearly completely.

*3.4.2. Isolation of RNA*

Isolation of RNA from Formalin-Fixed Paraffin-Embedded Specimens

For the isolation of RNA from formalin-fixed paraffin-embedded specimens, we use a protocol described elsewhere in detail *(15)* with a few modifications:

1. The dissected specimens are incubated overnight at 55°C in a vigorously agitating thermoshaker (Eppendorf, Hamburg, Germany) in the digestion solution (*see* **Subheading 2**) containing 0.5 mg of Proteinase K (total volume: 100 µl) and 1 µl LPA (25 µg).

2. Extraction of the lysate with water-saturated phenol and chloroform in the presence of LPA as a carrier.

3. Precipitation of RNA from the aqueous phase with isopropanol.

   If only a few cells are collected, the lysis buffer (20–50 µl) is directly added to the lid of the reaction tube, and the closed tube is incubated in an inverted position at 45°C over night (*see* **Note 7**).

---

# 4. Notes

1. All "rapid decalcifiers" destroy more or less the nucleic acids due to a low pH (acid-catalyzed hydrolysis of nucleic acids). Also histotechnicians like to apply formic or acetic acid directly onto the block immediately prior to sectioning to soften the tissue. This also severely damages nucleic acids contained within the tissue and has to be avoided under any circumstances.

   The widely used fixatives "B-5" and "Bouin's" do interfere with subsequent molecular studies *(16, 17)*. In our hands B-5 fixed bone marrow trephines are "lost" for DNA or RNA studies. Therefore, we strongly discourage the use of these solutions. In addition, some components of these fixatives pose serious safety problems concerning preparation, storage, and waste disposal (mercuric cloride, picric acid).

2. Duration of staining and washing with water and ethanol depends very much on the tissue under study and personal

preferences concerning staining intensity. It has to be figured out individually.

3. We did not perform a comprehensive comparison of commercially available staining kits. Therefore other kits might work as well.

4. The width of the laser cut depends very much on the thickness of the section and the type of tissue structure which has to be cut (e.g., fat tissue is very easily cut, connective tissue often quite resistant). That means, the energy sufficient for a fine cut through fat tissue will not be sufficient for cutting connective tissue structures and the energy adjusted to the latter tissue type will create a quite broad, irregular cut through fat tissue. However, adjustment of the appropriate laser focus and energy has to be learned by trial-and-error, and the numbers given in the text give only a rough orientation.

5. Dissected cells can also be catapulted into the lid of a reaction tube without changing the focus of the laser but this will create a "bullet hole" in the specimen. This is no problem if larger structures are dissected and the laser "bullet hole" can be placed in an irrelevant part of the specimen (e.g., the lumen of a vessel dissecting the vessel lining endothelial cells). Dissecting single cell or very small cell clusters, it is absolutely necessary to change the focus of the laser. This adjustment of the laser focus for catapulting has to be learnt by trial-and-error, and the right adjustment is a delicate balance between the size of the specimen and the laser focus and the laser energy. A tiny drop of liquid wax from MJ Research (Boston, MA, USA) is distributed in the lid of the reaction tube. This wax film ensures that the catapulted specimens will firmly adhere to the lid and facilitates the visual inspection of the catapulted cells.

6. The transfer of the lysate of larger groups of cells to a new tube before precipitation is necessary to separate the pieces of supporting membranes from the cell lysate. These pieces are isolated together with the dissected cells. They are not lysed and interfere physically with precipitation of nucleic acids by preventing the formation of a compact pellet at the bottom of the tube during centrifugation.

7. The average fragment length of DNA or RNA isolated from formalin-fixed biopsies ranges from 200 to 400 bp. Therefore, the PCR products (amplicons) have to be as short as possible to achieve maximal amplification efficiency and to reduce the influence of nucleic acid fragmentation due to fixation.

## Acknowledgments

The authors thank Britta Hasemeier for expert assistance in preparing Figure 1 and Oliver Bock for providing original photomicrographs. The work of the authors is supported by grants from the Deutsche Forschungsgemeinschaft and the Deutsche Krebshilfe.

## References

1. Bauer, K., Taub, S., and Parsi, K. (2004) Ethical issues in tissue banking for research: a brief review of existing organizational policies. *Theor Med Bioeth* 25, 113–142.

2. Caulfield, T. (2004) Tissue banking, patient rights, and confidentiality: tensions in law and policy. *Med Law* 23, 39–49.

3. Lehmann, U. and Kreipe, H. (2001) Real-time PCR analysis of DNA and RNA extracted from formalin-fixed and paraffin-embedded biopsies. *Methods* 25, 409–418.

4. Fend, F., Bock, O., Kremer, M., Specht, K., and Quintanilla-Martinez, L. (2005) Ancillary techniques in bone marrow pathology: molecular diagnostics on bone marrow trephine biopsies. *Virchows Arch* 447, 909–919.

5. Becker, K. F., Schott, C., Hipp, S., Metzger, V., Porschewski, P., Beck, R., Nahrig, J., Becker, I., and Hofler, H. (2007) Quantitative protein analysis from formalin-fixed tissues: implications for translational clinical research and nanoscale molecular diagnosis. *J Pathol* 211, 370–378.

6. Fend, F., Emmert-Buck, M. R., Chuaqui, R., Cole, K., Lee, J., Liotta, L. A., and Raffeld, M. (1999) Immuno-LCM: laser capture microdissection of immunostained frozen sections for mRNA analysis. *Am J Pathol* 154, 61–66.

7. Bock, O., Schlué, J., Lehmann, U., von Wasielewski, R., Länger, F., and Kreipe, H. (2002) Megakaryocytes from myeloproliferative disorders show enhanced nuclear bFGF expression. *Blood* 100, 2274–2275.

8. Bock, O., Loch, G., Schade, U., von Wasielewski, R., Schlue, J., and Kreipe, H. (2005) Aberrant expression of transforming growth factor beta-1 (TGF beta-1) per se does not discriminate fibrotic from nonfibrotic chronic myeloproliferative disorders. *J Pathol* 205, 548–557.

9. Hussein, K., Brakensiek, K., Buesche, G., Buhr, T., Wiese, B., Kreipe, H., and Bock, O. (2007) Different involvement of the megakaryocytic lineage by the JAK2 (V617F) mutation in Polycythemia vera, essential thrombocythemia and chronic idiopathic myelofibrosis. *Ann Hematol* 86, 245–253.

10. Lehmann, U., Glockner, S., Kleeberger, W., von Wasielewski, H. F., and Kreipe, H. (2000) Detection of gene amplification in archival breast cancer specimens by laser-assisted microdissection and quantitative real-time polymerase chain reaction. *Am J Pathol* 156, 1855–1864.

11. Glockner, S., Lehmann, U., Wilke, N., Kleeberger, W., Langer, F., and Kreipe, H. (2001) Amplification of growth regulatory genes in intraductal breast cancer is associated with higher nuclear grade but not with the progression to invasiveness. *Lab Invest* 81, 565–571.

12. Lehmann, U., Langer, F., Feist, H., Glockner, S., Hasemeier, B., and Kreipe, H. (2002) Quantitative assessment of promoter hypermethylation during breast cancer development. *Am J Pathol* 160, 605–612.

13. Kleeberger, W., Versmold, A., Rothamel, T., Glockner, S., Bredt, M., Haverich, A., Lehmann, U., and Kreipe, H. (2003) Increased chimerism of bronchial and alveolar epithelium in human lung allografts undergoing chronic injury. *Am J Pathol* 162, 1487–1494.

14. Bröcker, V., Langer, F., Fellous, T. G., Mengel, M., Brittan, M., Bredt, M., Milde, S., Welte, T., Eder, M., Haverich, A., Alison, M. R., Kreipe, H., and Lehmann, U. (2006) Fibroblasts of recipient origin contribute to bronchiolitis obliterans in human lung transplants. *Am J Respir Crit Care Med* 173, 1276–1282.

15. Bock, O., Kreipe, H., and Lehmann, U. (2001) One-step extraction of RNA from archival biopsies. *Anal Biochem* 295, 116–117.

16. Tbakhi, A., Totos, G., Pettay, J. D., Myles, J., and Tubbs, R. R. (1999) The effect of fixation on detection of B-cell clonality by polymerase chain reaction. *Mod Pathol* 12, 272–278.

17. Wan, X., Cochran, G., and Greiner, T. C. (2003) Removal of mercuric chloride deposits from B5-fixed tissue will affect the performance of immunoperoxidase staining of selected antibodies. *Appl Immunohistochem Mol Morphol* 11, 92–95.

## Suggested Readings

First description of the laser pressure catapulting technology from P.A.L.M.

Schütze, K. and Lahr, G. Identification of expressed genes by laser-mediated manipulation of single cells [see comments]. *Nat Biotechnol, 16:* 737–742, 1998.

First and comprehensive description of the Acturus system, which is based on different principals for dissection and specimen recovery and a very good early review.

Emmert-Buck, M. R., Bonner, R. F., Smith, P. D., Chuaqui, R. F., Zhuang, Z., Goldstein, S. R., Weiss, R. A., and Liotta, L. A. Laser capture microdissection [see comments]. *Science, 274:* 998–1001, 1996.

Simone, N. L., Bonner, R. F., Gillespie, J. W., Emmert-Buck, M. R., and Liotta, L. A. Laser-capture microdissection: opening the microscopic frontier to molecular analysis. *Trends Genet, 14:* 272–276, 1998.

A detailed practical guide introducing the Arcturus system can be found in "Current Protocols of Molecular Biology" (eds. Ausubel X et al., Wiley.)

First description of quantitative mRNA analysis in laser-microdissected cells.

Fink, L., Seeger, W., Ermert, L., Hanze, J., Stahl, U., Grimminger, F., Kummer, W., and Bohle, R. M. Real-time quantitative RT-PCR after laser-assisted cell picking. *Nat Med, 4:* 1329–1333, 1998.

First description of cDNA array based expression profiling of laser-microdissected cells.

Luo, L., Salunga, R. C., Guo, H., Bittner, A., Joy, K. C., Galindo, J. E., Xiao, H., Rogers, K. E., Wan, J. S., Jackson, M. R., and Erlander, M. G. Gene expression profiles of laser-captured adjacent neuronal subtypes [published erratum appears in Nat Med 1999 Mar;5(3):355]. *Nat Med, 5:* 117–122, 1999.

Two recent reviews about laser-microdissection.

Fink, L., Kwapiszewska, G., Wilhelm, J., and Bohle, R. M. Laser-microdissection for cell type- and compartment-specific analyses on genomic and proteomic level. *Exp Toxicol Pathol,* 57 Suppl 2: 25–29, 2006.

Ladanyi, A., Sipos, F., Szoke, D., Galamb, O., Molnar, B., and Tulassay, Z. Laser microdissection in translational and clinical research. *Cytometry A, 69:* 947–960, 2006.

# Chapter 21

## Leukemia Diagnosis in Murine Bone Marrow Transplantation Models

### Zhixiong Li, Ute Modlich, and Anjali Mishra

### Summary

The mouse is the most commonly used experimental animal, and a wide range of tumor types can arise in their hematopoietic system. Therefore, for research scientists and graduate students working in the field of experimental hematology, immunology, and cancer research, there is an urgent need for well-established protocols for the preparation of histology and cytology for leukemia diagnosis. Moreover, the criteria for the classification of hematopoietic neoplasms often vary between different laboratories. In this chapter, we describe diagnosis and analysis of leukemia in murine bone marrow transplantation models based primarily on the findings of the histology and cytology of hematopoietic and infiltrated tissues, peripheral blood smear, and immunophenotyping by FACS analysis.

**Key words:** Murine leukemia, Diagnosis, Bone marrow transplantation, Histology, Cytology.

## 1. Introduction

Leukemia diagnosis primarily relies on morphological, cytochemical, and immunophenotypic features of the neoplastic cells to define their lineage and degree of maturation. In human patients, the diagnosis of leukemia is based primarily on the findings from examination of peripheral blood smear, and bone marrow biopsy and aspirate. The criteria for the identification of myeloid and lymphoblastic leukemia and their subtypes are well described by the French-American-British (FAB) Cooperative Group, a classification that had been used during the last 3 decades (1, 2). Cytochemical studies (e.g., myeloperoxidase, nonspecific esterase, PAS) and/or immunophenotyping studies provide evidence that the neoplastic cells belong to one or more lineages. It was

Christopher Baum (ed.), *Methods in Molecular Biology, Methods and Protocals, vol. 506*
© Humana Press, a part of Springer Science + Business Media, LLC 2009
DOI: 10.1007/978-1-59745-409_21

also discovered that many cases of leukemia are associated with recurring genetic abnormalities that affect cellular pathways of stem/progenitor cell maturation and proliferation. In many leukemia, recurring genetic abnormalities (e.g., AML1/ETO) can be identified by reverse transcriptase-polymerase chain reaction (RT-PCR) or fluorescent in situ hybridization (FISH), but cytogenetic studies should be performed early on and at regular intervals throughout the course of the disease for establishing a complete genetic profile and for detecting genetic evolution. The World Health Organization (WHO) classification of hematopoietic and lymphoid neoplasms has recently been published (3), which includes specific genetic subcategories. One of the major differences between the FAB and WHO classifications is the threshold of the percentage of blasts for leukemia diagnosis (>30% vs. 20%).

The outstanding role of the laboratory mouse for modeling human development and disease has received further support by the mouse genome project (4). Nevertheless, the differences between the hematopoietic systems of mice and man must be carefully evaluated to diagnose with certainty reactive and neoplastic blood cell disorders and to improve the predictive value of the animal model. The murine hematopoietc system has a significantly higher daily cell turnover (especially of red blood cells and platelets) compared with humans. Since there are few or no fat cells interspersed in murine bone marrow, the complete bone cavities are used for hematopoiesis (5). In mice, extramedullary hematopoiesis continues in the red pulp of the spleen throughout life. Reactive or malignant increase of hematopoietic cells rapidly lead to extramedullary hematopoiesis, typically starting in the spleen. Thus, splenomegaly with expansion of red pulp caused by leukemic infiltrations (>20% blasts) and regression of the white pulp (peri-arteriolar lymphatic sheath and lymph follicles) are early characteristics of murine non-lymphoid leukemia (6–8). In advanced stages of the disease, normal architecture of the spleen is totally disturbed and widely replaced by masses of blasts. Amongst many other organs, leukemic infiltrates in periportal areas of the liver is a characteristics for acute myeloid leukemia (AML). Unlike human leukemia, in murine models bone marrow is only variably involved but spleen in almost all cases (6–8).

Mouse models may not be fully predictive for the development of human leukemia considering differences in HSC turnover. Leukemia development often involves genetic alterations of long term repopulating HSCs (9). However, it is unclear whether this also applies to oncogenesis related to retroviral manipulations. The pool size of murine HSCs is tightly regulated, although with considerable genetic and age-dependent variability (10). Abkowitz postulated a similar size of the HSC pool in mice and cats (~12,000 per animal), with the possible extension of this finding in humans (11). Her study also suggests a conservation of the replicative activity per lifetime between murine and human HSC. Thus, mice would present with a higher density and shortened cycling times of HSCs within the bone marrow.

Generally, terminology of hematopoietic neoplasms is inconsistently applied to lesions of mice. For instance, similar abnormalities may be described as myeloproliferative disease in one publication but as leukemia in a publication from a different laboratory. Acute leukemia has been very often incorrectly used in the literature describing neoplasms that occur very late in lifetime of the mouse i.e., after several months to upto a year *(8)*. A systematic classification of mouse hematopoietic neoplasms was first put forward by Dunn in 1954 *(12)*. Many more research groups involved in the identification of leukemia are now following recently published comprehensive Bethesda proposals *(6, 8, 13)*.

Unless a genetic predisposition is involved (such as endogenous replication competent retrovirus, RCRs), spontaneous leukemia occurs only sporadically in older animals. It can be induced with high incidence by irradiation and inoculation of newborn or immunodeficient animals with murine leukemia viruses (MLVs). Susceptibility of mice to develop leukemia varies according to strain and its contamination with MLVs. The murine bone marrow transplantation model using gene-modifed stem/progenitor cells (mainly by retroviral vectors) has been widely used in studies of gene marking, gene therapy, leukemogenesis, and stem cell biology *(6, 14–16)*. Animals can develop hematopoietic neoplasms with latencies ranging between several weeks to upto 2 years. However, many mouse studies were constructed in the way that observation periods rarely exceeded 6–12 months and visits of infected mice were not performed continually, particularly with lacks on weekend and holidays. Thus many animals might have been lost and reported as death of unknown reasons or even not mentioned in the publication due to the loss of animals by discontinuous observations. Along the same lines, we suspect that the death of a large proportion of such animals might have been caused by hematopoietic neoplasms. In addition, there is an urgent need for experienced scientists and pathologists to make appropriate diagnosis.

## 2. Materials

### 2.1. Peripheral Blood Analysis

1. EDTA-coated blood tubes, 1,3 ml (Sarstedt, Nürnbrecht).

2. BD PharM Lyse™ ammonium chloride erythrocyte lysing reagent, 10× concentrated, BD Biosciences Pharmingen.

3. Glass slides for blood smears (Menzel, Germany).

4. SCIL Vet abc™ blood counter or comparable machine.

### 2.2. Preparation of Mouse Tissues and Cells for Final Analysis (Fig. 1)

1. Sterile scissors and forceps.

2. Dulbecco's modified Eagle Medium (DMEM, Biochrom, Berlin, Germany), containing 10% fetal calf serum (Biochrom, Berlin, Germany), 2 mM glutamine (Biochrom, Berlin, Germany), 1% penicillin/streptomycin (PAA, Pasching, Austria).

3. Sterile syringes and needles.

4. Cell strainer (BD Falcon, 70 μm, Nylon).

5. Table top balance for weighing of the organs.

6. 90% fetal calf serum (FCS) (Biochrom, Berlin, Germany), 10% DMSO for freezing of cells.(**Fig. 1**)

### 2.3. Histology

1. 37% formaldehyde solution (Merck, Darmstadt, Germany).

2. For embedding of tissues, sectioning and staining a pathological unit should be available.

### 2.4. Cytospin

1. Suitable Cytospin centrifuge, e.g., Shandon Cytospin 4 (Thermo Electron Corporation, Pittsburgh, USA).

2. Sample Chambers (Cytofunnel., Thermo Electron Corporation, Pittsburgh, USA).

3. Shandon filter cards, thick, (Thermo Electron Corporation, Pittsburgh, USA).

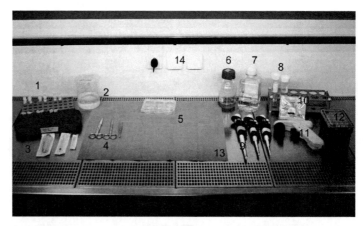

| | | |
|---|---|---|
| 1 cryotubes | 6 DMEM medium | 11 pipette boy |
| 2 ethanol | 7 PBS | 12 sterile tips |
| 3 syringes and needles | 8 sterile falcon tubes | 13 sterile drape sheet |
| 4 scissors and forceps | 9 pipettes | 14 power socket (possible |
| 5 tissue culture plate | 10 cell strainer | to plug in balance) |

Fig. 1. Typical set up for the endpoint analysis under a flow hood for sterile work. (*See Color Plates*)

4. Giemsa and May-Grünwald staining solutions (Sigma-Aldrich, Steinheim, Germany).

***2.5. FACS Phenotyping***

1. Lineage specific antibodies as indicated in the below tables, supplied by BD Biosciences or eBiosciences or any other suitable supplier.

2. FACS buffer: PBS with 4% FCS.

3. Flow cytometer that can at least measure four different colours, e.g., FACS Calibur, FACS Canto II, or BD™ LSR II (BD Biosciences).

# 3. Methods

***3.1. Visit After BMT***

1. It is very important to observe transplanted animals at least three times a week (including weekends) until the final analysis is done. This should start already a week after BMT. Seriously sick mice (with very obvious sign of mentioned illness, *see* 3.1.2) which are expected to die within 24 h, should be sacrificed and analysis should be performed as described in **Subheading 3.1**.

2. Parameters for observations:

   (a). Is the animal physically active? Does the animal have breathing problems (in most cases due to enlarged organs such as thymus, spleen or liver)?

   (b). Are there signs of anemia (by observation of paleness of toes of the mouse)?

   (c). Has normal posture of the mouse changed? Does mouse sits in hunched position (this could be possible sign of enlarged spleen and liver)?

   (d). Has the animal lost weight? Does the mouse have uneven and ruffled fur?

3. If the answer to any one of the above questions is affirmative, the animal may have a hematopoietic neoplasm, and should be observed more frequently at much shorter time intervals than usual (e.g. daily if possible). To be certain, peripheral blood analysis can be performed (*see* **Subheading 3.2**). A good clue for the progression of leukemia includes presence of blast cells in the peripheral blood (by blood smears) or leukocytes counts of more than >20,000/μl (by complete blood counts). In this case the mouse can be sacrificed for the endpoint analysis.

***3.2. Interval Analysis***

1. Time points for analysis

   While the mice are observed after the BMT we recommend interval analysis of the peripheral blood at certain time points at

6–9 weeks, 17 weeks, 28 weeks, and 54 weeks (if necessary) after BMT (*see* **Note 1**).

2. Peripheral blood analysis

Take 100–150 μl peripheral blood by retro-orbital bleeding or from the tail vein. Prepare and perform the following analysis:

(a) Blood smear and blood counts: For a blood smear use 1–2.5 μl of blood. Blood counts should be performed on a blood counter suitable to measure mouse blood cells, e.g. SCIL Vet abc™ blood counter. Only 11 μl of blood is needed for the Vet abc, however this is variable from instrument to instrument.

(b) FACS analysis of leukocytes (lysed blood): Pipette 30–50 μl blood into an Eppendorf tube and add 1 ml of erylysis buffer. Vortex and incubate for 10 min at room temperature (RT). Spin down the cells with 300 g for 5 min at RT, take off the supernatant, wash once with PBS (optional) and resuspend the cell pellet in 100 μl FACS buffer for further immunostaining. FACS panel: *(1)* GFP (transgene)/CD45-PE/CD11b-APC (Percentage of myeloid cells, transgene expression in total leukocytes); *(2)* GFP (transgene)/CD19-PE/CD3-APC (Expression of transgene in T cells and B cells); *(3)* for Ly5.2/Ly5.1 (donor/recipient) model: GFP (transgene)/CD45.2-PE/CD45-APC (Percentage of donor cells, i.e., CD45.2$^+$/CD45$^+$; expression transgene in donor cells).

**3.3. Endpoint Analysis**

**3.3.1. Blood Counts and Blood Smears**

Blood is harvested as follows: (a) Anesthesize mouse; (b) Cut axillary blood vessels, collect as much blood as possible, instantly transfer into EDTA tube; (c) Sacrifice the mouse immediately and soak it in ethanol; d) Perform blood counts and blood smears as described above.

**3.3.2. Dissection and Harvest of Organs (17)**

Place the mouse on its back and open the abdomen by a cut along the *linea alba*. Perform two further cuts toward the sides of the mouse to get a better view. After mouse has been opened, follow a routine on all mice with respect to the sequence in which organs are going to be harvested.

1. (a) Harvest inguinal lymph nodes; (b) Open the peritoneal cavity and harvest spleen located at posterior left of mouse body. Detach it very carefully from the omentum with the help of forcep; (c) Pull apart the small intestine by stretching it gently and pulling it away from the body. Excise approximately 3 cm of intestine with the help of scissor; (d) Carefully inspect the liver across diaphragm and smoothly remove any attachment to the body posterior with the help of scissors. Grab whole liver as a single block; (e) To visualize posterior of mouse pinup small intestine and stomach on the right side. Now kidney and ovary are clearly visible; (f) Remove both the kidneys with the help of forcep and scissors; (g) Open the thoracic cavity along lateral

axis to cut open sternum and carefully opening it upto over the thymus. Excise it completely with the help of scissor in the shape of 'V'; (h) Visualize thymus as a pale white compact mass of cells over the heart. Pull out the thymus by grabbing it in the middle (isthmus) and gently detaching it from mediastinum over the heart with the help of fine scissors; (i) Remove heart carefully by first lifting it up with the help of forceps and cutting off the artery with the help of fine scissors; (j) Remove lungs.

2. Report all observations regarding size, weight, color, consistency of organs, and any abnormalities on a report sheet. Work sterile!

3. To do further analysis of tissue sections and to ensure that the results of histopathology are accurate, proper formalin fixation of tissue samples is essential. Fix all organs for histopathology in 4% formalin in PBS (50 ml Falcon) Bone marrow: whole sternum and one tibia; spleen: >1/3 part, transverse direction; liver: half of left lateral lobe, right lateral lobe and optional caudate lobe, transverse direction; lung both whole left and right lobe; kidney: whole of one side; brain: whole; thymus (if enlarged): half, longitudinal direction; and other enlarged organs, e.g. lymph nodes.

4. For further processing of fixed tissues, which includes the embedding into paraffin, preparation of the paraffin blocks, sectioning and staining of the tissues, the co-operation with a pathological unit is advisable.

5. Sections of tissues are stained with hematoxillin/eosin (HE) (*see* **Note 2**).

*3.3.3. Preparation of Single Cell Suspensions for Cytospins and FACS Analysis*

1. Prepare single cell suspensions of the spleen, bone marrow (BM), possibly liver and thymus when enlarged for subsequent cytospins and FACS analysis:

    Grind spleen and liver tissues with cell strainers into a 50 ml Falcon tube. Wash the membrane of the cell strainer with DMEM medium, 10% FCS, 1% penicillin/streptomycin.

    BM: Wash the BM from the bones with DMEM by flushing with a syringe. For preparation of single cell suspensions gently draw up the BM trough a needle into a syringe twice.

2. Count cells and take $2 \times 10^5$ cells for cytospins and $5 \times 10^5$ cells per FACS staining

3. Work fast and store cells in fridge while counting etc.

4. Freeze the rest of the cells in 90%FCS/10% DMSO and store in liquid $N_2$ (*see* **Note 3**).

*3.3.4. Cytospins*

1. Cytospins are prepared using a Cytospin Centrifuge. Typically, $2 \times 10^5$ cells in 150 µl PBS are used for one cytospin. Centrifuge for 10 min at 800 rpm (in a Shandon Cytospin 4).

2. Cytospins are stained after Pappenheim (Giemsa/May-Grünwald). Incubate the slides 5 min in May-Grünwald followed by 20 min in Giemsa staining solution.

*3.3.5. FACS Analysis*

Flow cytometry (FACS) analysis is performed to check for transgene expression and immunophenotyping of leukemic cells in the following organs: spleen, liver, bone marrow, peripheral blood, and thymus (if enlarged). FACS panel for endpoint analysis (transgene expression in FL1 channel, e.g., green fluorescence protein as an example) (*see* **Note 4**):

1. Spleen is enlarged, but not the thymus (possibly myeloid or B cell lymphoblastic leukemia) (**Table 1**)

2. Thymus is enlarged (possibly T cell lymphoma/lymphoblastic leukemia) (**Table 2**)

**3.4. Time Schedule**

1. Histology: Tissues should be fixed in 4% Formalin immediately.

2. Cells should be frozen within 3 h.

3. Blood smears and differential blood counts should be performed within 6 h after blood sample was taken.

## Table 1
## FACS panel for endpoint analysis in animals with enlarged spleen, but normal size of thymus (examples for using FACS Calibur)

| Organs | FL1 | FL2 | FL3 | FL4 |
|---|---|---|---|---|
| Spleen: | GFP | CD11b-PE | PI | Gr1-APC |
| | GFP | F4/80-PE | PI | CD11b-APC |
| | GFP | CD61-PE | PI | Ter119-APC |
| | GFP | B220-PE | PI | C-Kit-APC |
| | GFP | CD19-PE | PI | CD45-APC |
| | GFP | CD3e-PE | PI | NK1.1-APC |
| | GFP | CD8-PE | PI | CD4-APC |
| Liver | GFP | CD11b-PE | PI | Gr1-APC |
| | GFP | CD19-PE | PI | CD3-APC |
| Peripheral blood (optional) | GFP | CD11b-PE | PI | Gr1-APC |
| | GFP | F4/80-PE | PI | CD11b-APC |
| | GFP | CD19-PE | PI | CD45-APC |
| | GFP | B220-PE | PI | CD3-APC |
| Bone marrow (optional) | GFP | CD11b-PE | PI | Gr1-APC |
| | GFP | CD19-PE | PI | CD3-APC |

*Note* Dead cells are excluded by propidium iodide staining (PI, concentration: 1 µg/ml)

**Table 2**
**FACS panel for endpoint analysis in animals**
**with enlarged thymus**

| Organs | FL1 | FL2 | FL3 | FL4 |
|--------|-----|-----|-----|-----|
| Thymus | GFP | CD3e-PE | PI | NK1.1-APC |
|        | GFP | CD8-PE | PI | CD4-APC |
|        | GFP | CD44-PE | PI | CD25-APC |
|        | GFP | CD11b-PE | PI | C-Kit-APC |
|        | GFP | CD19-PE | PI | CD45-APC |
| Liver | GFP | CD11b-PE | PI | Gr1-APC |
|       | GFP | CD8-PE | PI | CD4-APC |
|       | GFP | CD44-PE | PI | CD25-APC |
|       | GFP | CD19-PE | PI | CD45-APC |
| Peripheral blood (optional) | GFP | CD3e-PE | PI | NK1.1-APC |
|       | GFP | CD8-PE | PI | CD4-APC |
|       | GFP | CD44-PE | PI | CD25-APC |
|       | GFP | CD11b-PE | PI | Gr1-APC |
|       | GFP | CD19-PE | PI | CD45-APC |
| Spleen (optional) | GFP | CD11b-PE | PI | Gr1-APC |
|       | GFP | CD19-PE | PI | CD45-APC |
|       | GFP | CD8-PE | PI | CD4-APC |
|       | GFP | CD44-PE | PI | CD25-APC |
| Bone marrow (optional) | GFP | CD11b-PE | PI | Gr1-APC |
|       | GFP | CD19-PE | PI | CD45-APC |
|       | GFP | CD8-PE | PI | CD4-APC |
|       | GFP | CD44-PE | PI | CD25-APC |

4. FACS stain and acquisition should be done as fast as possible (within 24 h) after the mice were sacrificed. In the case this is not possible, the cells can be frozen and stained after thawing.

5. Perform the cytospin on the same day (12 h after sacrifice).

6. Blood smears and cytospin should be stained within 7 days.

7. Preparation of genomic DNA within 2 days after sacrifice if fresh cells are used.

8. Peripheral blood samples can be frozen for later preparation of genomic DNA.

### 3.5. Diagnosis of Hematopoietic Neoplasms (8, 13)

#### 3.5.1. Information Needed for Diagnosis and Classification (see Note 5)

1. Clinical features (e.g., sign of anemia, weight loss) and macroscopic findings of relevant organs (**Fig. 2**).

2. Histopathology of mouse tissues as assessed by microscopic examination of paraffin-embedded tissue sections is central to pathologic diagnosis of hematopoietic neoplasms.

3. Peripheral blood findings (including blood cell counts and cytology), morphological phenotyping of spleen and bone marrow cytology play a critical role in diagnosis.

   The percentage of blasts and assessment of degree of maturation and dysplastic abnormalities of the neoplastic cells should be determined, if possible, phenotyping should be performed on 300 leukocytes of spleen and bone marrow cytospins stained with May-Grünwald Giemsa.

4. Lineage assessment with cytochemical stains and flow cytometric immunophenotyping provides valuable information (**Fig. 3**). Immunohistochemical stains can provide additional information, but not absolutely necessary to our opinion.

5. Serial transplantation of transformed cells assess whether a lesion is rapidly fatal upon transplantation. This analysis provides additional information for the diagnosis of leukemia and provides good evidence for the exclusion of reactive forms of hyperproliferation of hematopoietic cells.

#### 3.5.2. Classification and Diagnosis of Hematopoietic Neoplasms

1. Hematopoietic neoplasms can be divided into *(1)* lymphoid neoplasms containing transformed cells that have fully or partially differentiated into T cells, B cells, or natural killer cells, *(2)* and non-lymphoid neoplasms, lesions of the other lineages that

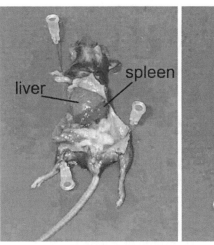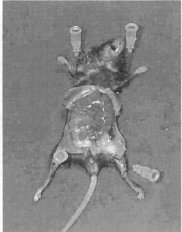

Fig. 2. Typical sectio of a mouse with enlarged liver and spleen. (*See Color Plates*)

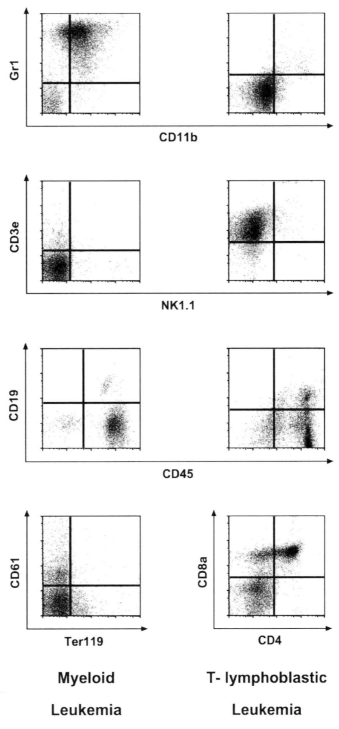

**Myeloid**

**Leukemia**

**T- lymphoblastic**

**Leukemia**

Fig. 3. Representative FACS analysis of murine leukemia. Myeloid leukemic cells isolated from diseased mice expressed CD11b, Gr1, and CD61, while T-lymphoblastic leukemic cells expressed CD3e, CD4, and CD8.

arise from hematopoietic stem/progenitor cells. For classification of non-lymphoid neoplasms, we recommend to consult the pertinent classification provided by the Bethesda proposals by Kogan et al. (**Table 3**) *(8)*.

2. Non-lymphoid leukemias, myeloid dysplasias, and myeloid proliferations (nonreactive) include diseases that arise primarily as increased numbers of non-lymphoid hematopoietic cells in the spleen and/or bone marrow. Leukemia is a disseminated disease that is rapidly fatal. Many leukemias are characterized by impaired differentiation, but myeloproliferative-disease-like (MPD-like) myeloid leukemia retain differentiation to mature forms. Myeloid dysplasias are characterized by cytopenias and abnormal differentiation. Myeloid proliferations (nonreactive)

**Table 3**
**Murine nonlymphoid hematopoietic neoplasms after Bethesda proposals** *(8)*

| **Nonlymphoid** |
| --- |
| -Myeloid leukemia |
| Myeloid leukemia without maturation |
| Myeloid leukemia with maturation |
| MPD-like myeloid leukemia |
| Myelomonocytic leukemia |
| Monocytic leukemia |
| -Erythroid leukemia |
| -Megakaryocytic leukemia |
| -Biphenotypic leukemia |
| **Nonlymphoid hematopoietic sarcoma** |
| -Granulocytic sarcoma |
| -Histiocytic sarcoma |
| -Mast cell sarcoma |
| **Myeloid dysplasias** |
| -Myelodysplastic syndrome |
| -Cytopenia with increased blasts |
| **Myeloid Proliferations (nonreactive)** |
| -Myeloproliferatiion (genetic) |
| -Myeloproliferative disease |

## Table 4
## Criteria for diagnosis of nonlymphoid leukemia (modified after the Bethesda proposals) *(8)*

| |
|---|
| (1) >20% nonlymphoid blasts in one of hematopoietic tissues (blood, spleen, or bone marrow) |
| (2) Mice exhibit anemia, neutropenia, and/or thrombocytopenia |
| Leukocytosis is present, and nonlympoid immature/blasts are seen in the peripheral blood |
| Nonlymphoid immature/blasts are increased in spleen or bone marrow (generally in both organs with enlarged spleen) |
| Nonlymphoid hematopoietic cells are increased in tissues other than blood, spleen, and bone marrow (e.g., liver) |
| Rapidly fatal to primary animal or serially transplanted recipients (median time from transplantation to moribund is 8 weeks or less) |

encompass subtle to marked expansions of genetically drived cells that do not meet criteria for either leukemia or myeloid dysplasia. Non-lymphoid hematopoietic sarcomas are cellular proliferations that arise primarily as solid tumors *(8)*.

3. Classification of lymphoid disorders can be ascertained in accordance with the nomenclature proposed by Morse et al. *(13)*, but in our opinions, in the majority of cases it is enough to make a diagnosis of B-cell lymphoma/leukemia or T-cell lymphoma/leukemia and the grade of maturation. Lymphoma/leukemia subtypes can be further characterized, but this is not absolutely essential for most studies.

4. A non-lymphoid leukemia diagnosis can be made if a disorder satisfies criteria *(1)* plus at least one of five criteria *(2)* in **Table 4 (Fig. 4)**. Similar criteria can also be applied for diagnosis of lymphoblastic leukemia.

# 4. Notes

1. Why is the first analysis of peripheral blood performed 6 weeks after BMT?

   Data from Weissman's group demonstrated that emergence of T cells occurs at day 34 after BMT, whereas other cell types (B-cells and myeloid cells) appear earlier in the peripheral blood *(18)*. The first analysis can be done early, e.g., 2 weeks after transplantation, if necessary.

Spleen
(x100)

Spleen
(x1000)

Liver
(x100)

Bone
marrow
(x1000)

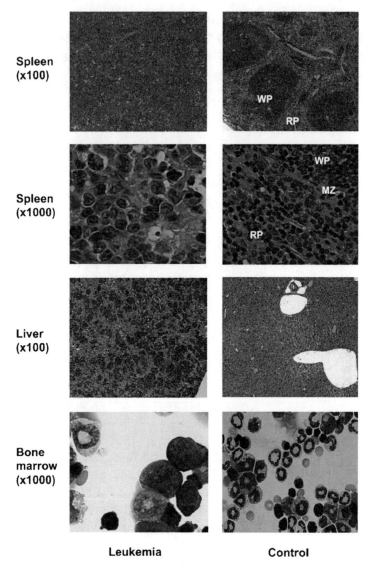

**Leukemia**                    **Control**

Fig. 4. Histopathology and cytology of mice with myeloid leukemia (*left panel*) and healthy controls (*right panel*). Histopathology of spleen (low and high power views): normal architecture of spleen in a leukemia mouse is totally abolished and replaced by myeloid blasts with poor differentiation (diagnosis: myeloid leukemia without maturation). In comparison, splenic architecture is normal in the control mouse. *WP* white pulp, *MZ* marginal zone, *RP* red pulp. Histopathology of liver (low power view): Extensive infiltration of leukemic cells in liver of a diseased mouse, while none of hematopoietic cells is present in the liver of the control mouse. Cytospin of bone marrow (high power view): There are >50% myeloblasts in bone marrow of a diseased mouse. Note that myeloblasts have fine chromatin with fairly distinct nucleoli. The control mouse shows a normal myelogram with approximately 60% myeloid cells, 20% erythroid cells, and 15% lymphoid cells. All hematopoitic cells, at all stages of maturation, are present together. (*See Color Plates*)

2. HE or Giemsa staining for histopathology?

The well-established place of the HE stain in general diagnostic pathology has assured it of much support amongst pathologists as the primary stain in bone marrow histology. However, many hematologists believe that a good Giemsa stain provides more information than its HE counterpart, e.g. in identifying cell lineage.

3. After final analysis, for each animal, single cell suspension from bone marrow, spleen, liver (at least one of them), or other enlarged organs must be sterilely frozen down for future experiments (for instance: serial transplantation).

4. Immunophenotyping for leukemic cells (8):

(a) Analyzing splenic cells and bone marrow cells is much more important than peripheral blood. In the cases of leukemic infiltrations, the liver is a very informative source for immunophenotyping of leukemic cells.

(b) Why using double Gr1/CD11b staining as myeloid markers?

CD14 is widely perceived as the best marker for the human macrophage/monocyte population. In mice, the level of CD14, as recognized by an anti-CD14 mAb, mC5–3, was low or undetectable on resting blood monocytes. CD11b (Mac-1), another commonly used marker for the macrophage/monocyte lineage, is expressed on NK cells, granulocytes, a T cell subset, and peritoneal B-1 B cells (19). Gr1 (Ly-6G) is a good marker for murine granulocytes (20); however, memory-type CD8(+)CD44(high) CD62L(high) T cells also express Gr1 (21). Using double Gr1/CD11b staining, we can distinguish granulocytes (Gr1 + /CD11b + population) and monocytes (Gr1-/CD11b + +).

(c) Why F4/80?

The F4/80 antigen is a macrophage-specific marker (22). We observed expression of F4/80 in histiocytic sarcoma and myeloid leukemia (23), but not T cell leukemia, thus suggesting that F4/80 may be a very specific marker for myeloid leukemia.

(d) Why CD41, CD61?

CD41 is the best marker for platelets in the mice (19). Other surface proteins expressed by platelets include CD61 and CD9 (19, 24). In the immunological classification of human acute leukemia, CD41 and/or CD61 are the best markers for megakaryocytic lineage (M7) (25)

(e) Ter119 or Glycophorin A?

The cell-surface sialoglycoproteins (glycophorins) have been shown to be selectively expressed in erythroid cells. Thus, mAbs against glycophorins have been considered as specific markers for erythroid lineage in humans. Although the expression of glycophorin molecules has been well studied in human erythroid cells, little is known about glycophorin expression in the mouse. In mice, several mAbs that react with the erythroid lineage have been reported. Expression of these determinants appears to be limited only to a certain stage of erythroid cells or to erythroleukemia cells. Ter119 is highly specific for erythroid cells *(26)*, at the stages from early proerythroblast to mature erythrocytes, but not expressed in four murine erythroleukemia lines analyzed *(27)*. However, Moreau-Gachelin et al. demonstrated expression of Ter119 in erythroleukemias in PU.1 transgenic mice *(28)*. So far there is no commercial antibody against glycophorin A available. Therefore, we recommend Ter119 as a marker for erythroleukemia.

(f)   Why CD117 (c-Kit)?

In human leukemia, CD117 (c-Kit) is a very good marker for classification of leukemia of the myeloid lineage *(25, 29, 30)*. Its specificity is significantly higher than some other markers for myeloid leukemia, e.g., CD13, CD33 *(31, 32)*. Expression of CD117 was found only in very few patients with ALL (<5%) *(30–32)*. So far, we only observed expression of CD117 in murine non-lymphoid leukemia *(6, 14)*.

(g)   Why CD19 as a marker for B lineage?

In the mouse, one of the most commonly used pan-B cell markers is identified by the CD45R/B220 *(33)*. However, this epitope is also expressed on T lymphoid progenitor cells in mouse fetal liver *(34)*, activated T and NK cells progenitors *(19)* and dendritic cell *(35)*. The CD19 antigen (Ag) appears to be more restricted to the B cell lineage and is for instance not expressed by NK progenitor cells *(19)*. In mouse bone marrow, the majority of all B-cell precursors coexpress B220 and CD19 *(36)*. In B cell development, B220 + /CD19-bone marrow cells are referred to as pro-B (Basel nomenclature), pre-pro B (Philadelphia nomenclature) or fraction A of B cell precursors (Hardy's nomenclature) *(37)*. In the immunological classification of human acute leukemia, CD19 is the most important marker for B lineage *(25)*. In one study, all of 180 patients with Precursor B-ALL showed expression of pan B-cell marker CD19 *(30)*.

(h)   Why both CD117/B220?

CD117/B220 double positive cells are referred to as Pro-B cells *(38, 39)*.

(i)   Why CD3 as marker for T lineage? Why not Thy-1?

While monoclonal antibodies (mAbs) to Thy-1 (CD90) and TCR complex antigens have commonly been used as pan-T cell markers in mice, Thy-1 is not restricted to T cells. The CD3e expression is correlated with T cell maturation *(19)*. Joosten M et al. reported the phenotype of Cas-Br-M MuLV-induced leukemias: T cell leukemias were CD3 + /Thy-1 + (11 cases), CD3 + /Thy-1- (1 case), and CD3-/Thy-1 + (1 case) *(40)*. In another study, CD3 was expressed in all of murine T type tumors analyzed *(41)*. In the immunological classification of human acute leukemias, cytoplasmic/membrane CD3 expression is the best marker for T lineage *(25)*

  (j)  In the case of erythoid leukemia immunophenotyping with CD71/Ter119 is very helpful to characterize the stage of maturation of the erythroid precursor cells *(42, 43)*.

5. Genetic characterization of leukemic cells plays a limited role in classification of mouse leukemia at present. There are some critical parameters for distinguishing reactive from malignant processes *(8)*. Investigators need to be alert to the possibility that an expansion of hematopoietic cells represents a reactive, as opposed to a neoplastic, process, and should be especially cautious in making a diagnosis of hematopoietic neoplasm whenever an infection or a non-hematopoietic tumor is present *(8)*.

## Acknowledgments

This study was supported by the Deutsche Krebshilfe (grant: 10–2090-Li I) and DFG SPP1230. We thank Mathias Rhein and Daniel Wicke for preparing figures.

## References

1. Bennett, J. M., Catovsky, D., Daniel, M. T., Flandrin, G., Galton, D. A., Gralnick, H. R., et al. (1976). Proposals for the classification of the acute leukaemias. French-American-British (FAB) co-operative group. *Br J Haematol.* 33, 451–8.

2. Bennett, J. M., Catovsky, D., Daniel, M. T., Flandrin, G., Galton, D. A., Gralnick, H. R., et al. (1985). Proposed revised criteria for the classification of acute myeloid leukemia. A report of the French-American-British Cooperative Group. *Ann Intern Med.* 103, 620–5.

3. Harris, N. L., Jaffe, E. S., Diebold, J., Flandrin, G., Muller-Hermelink, H. K., Vardiman, J., et al. (1999). World Health Organization classification of neoplastic diseases of the hematopoietic and lymphoid tissues: report of the Clinical Advisory Committee meeting-Airlie House, Virginia, November 1997. *J Clin Oncol.* 17, 3835–49.

4. Waterston, R. H., Lindblad-Toh, K., Birney, E., Rogers, J., Abril, J. F., Agarwal, P., et al. (2002). Initial sequencing and comparative analysis of the mouse genome. *Nature.* 420, 520–62.

5. Baum, C., Dullmann, J., Li, Z., Fehse, B., Meyer, J., Williams, D. A., et al. (2003). Side effects of retroviral gene transfer into hematopoietic stem cells. *Blood.* 101, 2099–114.

6. Modlich, U., Kustikova, O. S., Schmidt, M., Rudolph, C., Meyer, J., Li, Z., et al. (2005). Leukemias following retroviral transfer of multidrug resistance 1 (MDR1) are driven by combinatorial insertional mutagenesis. *Blood.* 105, 4235–46.

7. Perkins, A. S. (1989). The pathology of murine myelogenous leukemias. *Curr Top Microbiol Immunol.* 149, 3–21.

8. Kogan, S. C., Ward, J. M., Anver, M. R., Berman, J. J., Brayton, C., Cardiff, R. D., et al. (2002). Bethesda proposals for classification

of nonlymphoid hematopoietic neoplasms in mice. *Blood.* 100, 238–45.

9. Reya, T., Morrison, S. J., Clarke, M. F. and Weissman, I. L. (2001). Stem cells, cancer, and cancer stem cells. *Nature.* 414, 105–11.

10. Geiger, H. and Van Zant, G. (2002). The aging of lympho-hematopoietic stem cells. *Nat Immunol.* 3, 329–33.

11. Abkowitz, J. L., Catlin, S. N., McCallie, M. T. and Guttorp, P. (2002). Evidence that the number of hematopoietic stem cells per animal is conserved in mammals. *Blood.* 100, 2665–7.

12. Dunn, T. B. (1954). Normal and pathologic anatomy of the reticular tissue in laboratory mice, with a classification and discussion of neoplasms. *J Natl Cancer Inst.* 14, 1281–433.

13. Morse, H. C., 3rd, Anver, M. R., Fredrickson, T. N., Haines, D. C., Harris, A. W., Harris, N. L., et al. (2002). Bethesda proposals for classification of lymphoid neoplasms in mice. *Blood.* 100, 246–58.

14. Li, Z., Dullmann, J., Schiedlmeier, B., Schmidt, M., von Kalle, C., Meyer, J., et al. (2002). Murine leukemia induced by retroviral gene marking. *Science.* 296, 497.

15. Kustikova, O., Fehse, B., Modlich, U., Yang, M., Dullmann, J., Kamino, K., et al. (2005). Clonal dominance of hematopoietic stem cells triggered by retroviral gene marking. *Science.* 308, 1171–4.

16. Daley, G. Q., Van Etten, R. A. and Baltimore, D. (1990). Induction of chronic myelogenous leukemia in mice by the P210bcr/abl gene of the Philadelphia chromosome. *Science.* 247, 824–30.

17. Bahnemann, R., Jacobs, M., Karbe, E., Kaufmann, W., Morawietz, G., Nolte, T., et al. (1995). RITA--registry of industrial toxicology animal-data–guides for organ sampling and trimming procedures in rats. *Exp Toxicol Pathol.* 47, 247–66.

18. Forsberg, E. C., Serwold, T., Kogan, S., Weissman, I. L. and Passegue, E. (2006). New evidence supporting megakaryocyte-erythrocyte potential of flk2/flt3 + multipotent hematopoietic progenitors. *Cell.* 126, 415–26.

19. Lai, L., Alaverdi, N., Maltais, L., and Morse, H. C., 3rd. (1998). Mouse cell surface antigens: nomenclature and immunophenotyping. *J Immunol.* 160, 3861–8.

20. Fleming, T. J., Fleming, M. L. and Malek, T. R. (1993). Selective expression of Ly-6G on myeloid lineage cells in mouse bone marrow. RB6–8C5 mAb to granulocyte-differentiation antigen (Gr-1) detects members of the Ly-6 family. *J Immunol.* 151, 2399–408.

21. Matsuzaki, J., Tsuji, T., Chamoto, K., Takeshima, T., Sendo, F. and Nishimura, T. (2003). Successful elimination of memory-type CD8 + T cell subsets by the administration of anti-Gr-1 monoclonal antibody in vivo. *Cell Immunol.* 224, 98–105.

22. Hume, D. A., Robinson, A. P., MacPherson, G. G. and Gordon, S. (1983). The mononuclear phagocyte system of the mouse defined by immunohistochemical localization of antigen F4/80. Relationship between macrophages, Langerhans cells, reticular cells, and dendritic cells in lymphoid and hematopoietic organs. *J Exp Med.* 158, 1522–36.

23. Li, Z., Kustikova, O. S., Kamino, K., Neumann, T., Rhein, M., Grassman, E., et al. (2007). Insertional Mutagenesis by Replication-Deficient Retroviral Vectors Encoding the Large T Oncogene. *Ann N Y Acad Sci.*

24. Smyth, S. S., Tsakiris, D. A., Scudder, L. E. and Coller, B. S. (2000). Structure and function of murine alphaIIbbeta3 (GPIIb/IIIa): studies using monoclonal antibodies and beta3-null mice. *Thromb Haemost.* 84, 1103–8.

25. Bene, M. C., Castoldi, G., Knapp, W., Ludwig, W. D., Matutes, E., Orfao, A., et al. (1995). Proposals for the immunological classification of acute leukemias. European Group for the Immunological Characterization of Leukemias (EGIL). *Leukemia.* 9, 1783–6.

26. Ikuta, K., Kina, T., MacNeil, I., Uchida, N., Peault, B., Chien, Y. H., et al. (1990). A developmental switch in thymic lymphocyte maturation potential occurs at the level of hematopoietic stem cells. *Cell.* 62, 863–74.

27. Kina, T., Ikuta, K., Takayama, E., Wada, K., Majumdar, A. S., Weissman, I. L., et al. (2000). The monoclonal antibody TER-119 recognizes a molecule associated with glycophorin A and specifically marks the late stages of murine erythroid lineage. *Br J Haematol.* 109, 280–7.

28. Moreau-Gachelin, F., Wendling, F., Molina, T., Denis, N., Titeux, M., Grimber, G., et al. (1996). Spi-1/PU.1 transgenic mice develop multistep erythroleukemias. *Mol Cell Biol.* 16, 2453–63.

29. Buhring, H. J., Ullrich, A., Schaudt, K., Muller, C. A. and Busch, F. W. (1991). The product of the proto-oncogene c-kit (P145c-kit) is a human bone marrow surface antigen of hemopoietic precursor cells which is expressed on a subset of acute non-lymphoblastic leukemic cells. *Leukemia.* 5, 854–60.

30. Kaleem, Z., Crawford, E., Pathan, M. H., Jasper, L., Covinsky, M. A., Johnson, L. R., et al. (2003). Flow cytometric analysis of

acute leukemias. Diagnostic utility and critical analysis of data. *Arch Pathol Lab Med.* 127, 42–8.

31. Bene, M. C., Bernier, M., Casasnovas, R. O., Castoldi, G., Knapp, W., Lanza, F., et al. (1998). The reliability and specificity of c-kit for the diagnosis of acute myeloid leukemias and undifferentiated leukemias. The European Group for the Immunological Classification of Leukemias (EGIL). *Blood.* 92, 596–9.

32. Auewarakul, C. U., Lauhakirti, D., Promsuwicha, O. and Munkhetvit, C. (2006). C-kit receptor tyrosine kinase (CD117) expression and its positive predictive value for the diagnosis of Thai adult acute myeloid leukemia. *Ann Hematol.* 85, 108–12.

33. Coffman, R. L. and Weissman, I. L. (1981). B220: a B cell-specific member of th T200 glycoprotein family. *Nature.* 289, 681–3.

34. Sagara, S., Sugaya, K., Tokoro, Y., Tanaka, S., Takano, H., Kodama, H., et al. (1997). B220 expression by T lymphoid progenitor cells in mouse fetal liver. *J Immunol.* 158, 666–76.

35. Nikolic, T., Dingjan, G. M., Leenen, P. J. and Hendriks, R. W. (2002). A subfraction of B220(+) cells in murine bone marrow and spleen does not belong to the B cell lineage but has dendritic cell characteristics. *Eur J Immunol.* 32, 686–92.

36. Ghia, P., ten Boekel, E., Rolink, A. G. and Melchers, F. (1998). B-cell development: a comparison between mouse and man. *Immunol Today.* 19, 480–5.

37. Martin, C. H., Aifantis, I., Scimone, M. L., von Andrian, U. H., Reizis, B., von Boehmer, H., et al. (2003). Efficient thymic immigration of B220 + lymphoid-restricted bone marrow cells with T precursor potential. *Nat Immunol.* 4, 866–73.

38. Osmond, D. G., Rolink, A. and Melchers, F. (1998). Murine B lymphopoiesis: towards a unified model. *Immunol Today.* 19, 65–8.

39. Hoffmann, R., Seidl, T., Neeb, M., Rolink, A. and Melchers, F. (2002). Changes in gene expression profiles in developing B cells of murine bone marrow. *Genome Res.* 12, 98–111.

40. Joosten, M., Valk, P. J., Vankan, Y., de Both, N., Lowenberg, B. and Delwel, R. (2000). Phenotyping of Evi1, Evi11/Cb2, and Evi12 transformed leukemias isolated from a novel panel of cas-Br-M murine leukemia virus-infected mice. *Virology.* 268, 308–18.

41. Davies, J., Badiani, P. and Weston, K. (1999). Cooperation of Myb and Myc proteins in T cell lymphomagenesis. *Oncogene.* 18, 3643–7.

42. Hall, M. A., Slater, N. J., Begley, C. G., Salmon, J. M., Van Stekelenburg, L. J., McCormack, M. P., et al. (2005). Functional but abnormal adult erythropoiesis in the absence of the stem cell leukemia gene. *Mol Cell Biol.* 25, 6355–62.

43. Socolovsky, M., Nam, H., Fleming, M. D., Haase, V. H., Brugnara, C. and Lodish, H. F. (2001). Ineffective erythropoiesis in Stat5a(-/-)5b(-/-) mice due to decreased survival of early erythroblasts. *Blood.* 98, 3261–73.

# Chapter 22

# Humanized Mouse Models to Study the Human Haematopoietic Stem Cell Compartment

## Dominique Bonnet

## Summary

The ideal way to assess hematopoietic stem cells is to observe their growth in the endogenous microenvironment where they would receive the appropriate signals. With colonies of inbred mice, it is possible to myeloablate recipients and transplant hematopoietic cells from genetically similar mice and observe the growth of primitive hematopoietic cells in their endogenous environment for a significant proportion (10 months) of an organisms lifespan (29 months average). It is not possible to perform these experiments in humans, but xenotransplantation mouse models provide the closest paradigm for the human hematopoietic environment at the present time.

**Key words**: Hematopoietic stem cell, Xenotransplantation, Immunodeficient mice.

## 1. Introduction

The greatest challenge to the development of xenotransplantation models is the immune rejection of the transplanted cells. This may be overcome by transplanting cells before the immune system has developed or by modulation of the hosts' immune response. The injection of primitive cells into fetal sheep that have a naive immune system has allowed the growth of hematopoietic cells to be observed in an in vivo environment for over one year (1, 2). However, this model is technically difficult, expensive to manage, and hence is limited to small sample numbers (3).

The engraftment of normal human hematopoietic cells in immuno-deficient mice provides a suitable assay that measures

Christopher Baum (ed.), *Methods in Molecular Biology, Methods and Protocols, vol. 506*
© Humana Press, a part of Springer Science + Business Media, LLC 2009
DOI: 10.1007/978-1-59745-409-4_22

the repopulating capacity of human stem cells. Originally this model used the severe combined immune-deficient (SCID) mice, which, based on the SCID mutation, have a lack of functional T and B lymphocytes. Dr. Dick's group has shown that intravenous injection of human bone marrow or umbilical cord blood into these mice resulted in the engraftment of primitive cells that proliferate and differentiate in the murine bone marrow, producing large numbers of CFC, LTC-IC, immature CD34$^+$ CD38$^-$ as well as mature myeloid, erythroid, and lymphoid cells (4, 5). The primitive cells that initiated the graft were operationally defined as SCID repopulating cells (SRCs). Most of the earlier studies used the SCID mice, which were not ideal, as they still possessed significant innate immunity. As a result, high cell doses were required to overcome any residual host resistance, ruling out the development of any purification strategies. A new mouse strain, created by backcrossing the SCID gene onto the nonobese diabetic (NOD) background, proved to be a better recipient. These mice have an impaired natural killer (NK) and antigen-presenting cell function, and the growth of human cells in these mice is possible without cytokine stimulation (6–8). These nonobese diabetic/severe combined immuno deficient mice – NOD/Lt-Sz-Scid/Scid (NOD/SCID) – have proven to be reliable for detecting human hematopoietic repopulating cells that differentiate into multilineage mature cells and self-renew in mice. Overall, the NOD/SCID mice allowed high level engraftment of normal and leukemic human transplants and, more importantly, enabled engraftment with lower cell doses, rendering purification strategies possible (9). The NOD/SCID model can support multilineage (B-cell and myeloid) for over 3 months and can support secondary engraftment (10). In this way, the NOD/SCID assay can assess the self-renewal, proliferation, and differentiation of human hematopoietic stem cells.

The ideal demonstration of human stem cell potential would be the ability of a single cell to produce all the hematopoietic lineages for a significant period of time. Because of apparent inefficiencies in transplantation, human single cell, xenotransplantation studies are still elusive. Indeed, although transplantation in a mouse-mouse setting is very efficient (11), transplantation is not efficient in the human–mouse situation. This may be partially due to be an inefficiency in homing given that injection of cells directly into the bone increases the frequency of repopulating cells when compared with intravenous administration (12). In spite of this, it has been demonstrated via viral tracking experiments that a single cell is capable of multilineage differentiation in the NOD/SCID model (13).

Other mouse models have been developed to support the growth of human hematopoietic cells. The NOD/SCID-$\beta_2$ microglobulin null ($\beta$2 m$^{-/-}$) mouse has a further defect in NK cell activity and is more tolerant of human grafts than the NOD/SCID model *(14–16)*. However, in these mice limited T-lymphoid reconstitution is achieved. Furthermore, all these immunodeficient mice are problematic as lymphomas limit their lifespan, preventing long-term reconstitution assessment. These hurdles were recently overcome in three new strains: NOD/Shi-Scid IL2Rg$^{null}$ *(17, 18)*, NOD/SCID IL2Rg$^{null}$ *(19, 20)*, and BALB/c-Rag2$^{null}$ IL2Rg$^{null}$ *(21)*, which all lack the IL-2 family common cytokine receptor gamma chain gene. The absence of functional receptors for IL-2, IL-7, and other cytokines may prevent the expansion of NK cells and early lymphoma cells in NOD/SCID IL2Rg$^{null}$ mice, resulting in better engraftment of transplanted human cells and longer lifespan of the mice. It was reported recently that human HSCs and progenitor cells engraft successfully in these mice and produce all human myeloid and lymphoid lineages. T and B cells migrate into lymphoid organs and mount HLA-dependent allogeneic responses, and generate antibodies against T cell-dependent antigens such as ovalbumin and tetanus toxin *(19, 21)*. A recent review by Shultz et al. summarises brilliantly the important events in the development of humanized mice *(22)*.

The SRC assay as originally developed is based on intravenous injection, a complex process that requires for the candidate human HSC circulation through the blood, recognition and extravasation through bone marrow vasculature, and migration to a supportive microenvironment. This process called "homing" has been shown to be quite inefficient even in a syngeneic murine situation *(23, 24)*. Cashman and Eaves *(23)* reported that the proportion of total injected human CB competitive units in the marrow was 7%, as determined by limiting-dilution assays in NOD/SCID mice. Thus, this assay quite possibly underestimates human hematopoietic-repopulating cell frequencies. To exclude stem cell homing interference and focus on the intrinsic capacity of a cell to self-renew and give rise to multilineage engraftment, a few groups recently developed a highly sensitive strategy for SRC assay on the basis of direct intra-bone marrow (IBM) injection of the candidate human stem cell *(12, 25, 26)*. IBM injection was found to be a more sensitive and adequate means to measure human HSC capacity.

## 2. Materials

1. Immuno-deficient mice NOD/SCID, β2 m$^{-/-}$ or NOD/SCID IL2Rg$^{null}$ can be purchased from Jackson Laboratory (Bar Harbor, Maine, USA).
2. 29½ gauge Needle/Syringe from Kendall Monoject Insulin syringe (Tyco Healthcare, Basingstoke, Hampshire, UK).

3. Phosphate buffered saline (PBS): either commercially available or prepared as described. Prepare 10× stock with 1.37 M NaCl, 27 mM KCl, 100 mM Na$_2$HPO$_2$, 18 mM KH$_2$PO$_4$ (adjust to Ph 7.4 if necessary) and autoclave before storage at room temperature. Prepare working solution by dilution of one part with nine parts of water.

4. Acidified water: a solution of HCL at a final pH 2.8 to 3.2.

5. Ketaset solution (Fort Dodge Animal Health Ltd., Southampton, UK).

6. Rompun 2% solution (Bayer Plc, Newbury, UK).

7. Vetergesic (Alstoe Animal Health, Melton Mowbray, Leicestershire, UK).

8. Purified Rat Anti-mouse CD122 antibody (BD Pharmingen, Oxford, UK).

9. Antibodies against human anti-CD45, CD33, and CD19 (BD Pharmingen, Oxford, UK).

## 3. Methods

### 3.1. Preparation, Irradiation, Anaesthesia, and Anitibody Treatment of the Immunodeficient Mice

1. All the animals used have some impairment of their immune system and may succumb to infections not affecting normal mice. They thus should be kept in pathogen free status within barrier systems to protect them from inter-current infections.

2. These mice are sublethally irradiated before the adoptive transfer of cells. The dose of irradiation depends on the mouse strains used and also on the source and irradiator used. It usually varies from 300 to 375 cG (see **Notes 1–3**).

3. General anaesthesia of the mice may be required for intrabone injection and biopsy.

4. The mice are injected intraperitoneal with a dose of 0.2–0.25 ml of anaesthetic solution (see **Note 4**). General anaesthesia suppresses the heat regulating mechanisms of the body. This is overcome by intra and postoperative maintenance of body temperature in appropriate thermostatically-controlled incubators or by other heat sources. During recovery, animals should be kept under regular observation until full mobility is regained. At this stage, animals should receive at least one dose of postoperative analgesic following bone marrow injection (100 µl of analgesic solution (Vetergesic diluted 1/10; see **Note 5**).

5. Treatment of mice with anti-natural killer (NK) agents is useful to promote engraftment of human cells. Since natural killer cells were previously shown to be important factor for resisting early

engraftment *(27)*, NOD/SCID mice can be treated with anti-CD122 antibody directed against the IL-2β chain. This antibody targets several mature hematopoietic cells including NK cells and macrophages. Recipient mice will be gamma-irradiated as above and in addition receive injections of anti-NK antibodies. Anti-CD122 antibody is administered by a single intraperitoneal (IP) injection (dose 200 μg per mouse) at the time of the adoptive transfer of cells.

### 3.2. Adoptive Transfer of Cells

Mice that have been subjected to sublethal irradiation will receive cell preparations (unpurified or purified cell fraction, genetically modified or not). This may be performed on the same day as, or up to 72 h following, irradiation.

1. Cells will be injected intravenously via the tail vein (maximum volume 1% body weight) using a syringe with a 29½ gauge needle.

2. Intrabone injection is an alternative to intravenous injection. This technique is performed under a short general anaesthesia (*see* above) following a method described originally by Verlinden et al. *(28)*. A syringe with a 25-gauge needle (maximum) is inserted into the joint surface of the right or left tibia/femur, and cells to a maximum volume of 40 μl suspension are injected into the bone marrow cavity of the tibia or femur. After this procedure, animals receive at least one dose of postoperative analgesic.

3. Mice transplanted with human cells should be monitored daily (*see* **Note 6**).

### 3.3. Analysis of the Engraftment

1. *For bone marrow aspiration,* samples can be taken under general anaesthesia, as described previously, on up to 4 occasions per mouse at least 14 days apart, alternatively from the left and right femur. Bone marrow will be aspirated by puncture through the knee joint using a maximum 25-gauge needle. The maximum amount taken per sample will not exceed 30 μl. The first sample should be taken at least 16 h after adoptive transfer of cells.

2. Mice between 3 and 18 weeks after transplantation will be sacrificed either using cervical dislocation or terminal anaesthesia. When work under terminal anaesthesia is involved, the level of anaesthesia should be maintained at sufficient depth for the animal to feel no pain. Dissect the femurs, tibias, and pelvis from the mice and store at room temperature (RT) in PBS before flushing.

3. To prepare mouse bone marrow cells, place 1 ml of room temperature PBS in a 5 ml snap-top polystyrene tube. Cut both ends of each bone to provide an opening. To flush, insert a PBS-containing insulin syringe into one end of each bone and wash the lumen of the bone with a medium pressure. Repeat twice for both ends of the bone or until the bone appears relatively white.

4. To prepare the cells for FACS analysis, you need to lyse red blood cells. First, cool cell suspension for 5 min on ice and then add 3 ml of cold ammonium chloride solution (Stem Cell Technologies) to the 1 ml PBS/cell suspension, mix and leave for 5 min at 4°C. Add 0.5 ml of FCS, spin down cells at 1,200 rpm for 5 min, remove the supernatant and resuspend the cells in 1 ml of cold PBS containing 2% FCS. Count the cells using a hematocytometer. Store on ice ready for FACS analysis.

5. For antibody staining, prepare a mix of human-specific FITC-conjugated anti-CD19, PE-conjugated anti-CD33 and PerCP-conjugated anti-CD45 antibodies (5 µl/sample/antibody for all stains and compensation/isotype controls). Also prepare FITC, PE, and PerCP single color compensation control tubes (5 ml snap-top polystyrene as before) and a combined FITC/PE/PerCP matched isotype control tube. Distribute 15 µl of the antibody mix into each tube of a fresh set of tubes for antibody labeling. Dispense 40 µl of each cell suspension into the appropriate antibody labeling tube and leave to label for 30 min at 4°C. Wash cells in 2 ml PBS-2%FCS and resuspend in 500 µl of PBS-2% FCS supplemented with a cell impermeant DNA dye for live/dead discrimination, either 100 ng/ml 4′,6-diamidino-2-phenylindole (DAPI), or TOPRO-3.

6. For FACS analysis of this combination of fluorochromes, we require a 488 nm excitation source and either a UV or HeNe (633 nm) source depending on your choice of live/dead discriminator. For emission collection, we need a 440/40 band-pass (bp) filter for analysis of DAPI, a 530/30 bp filter for FITC, a 575/26 bp for PE, a 695/40 bp for PerCP, and a 660/20 bp for TOPRO-3. During FACS analysis, set the photomultiplier gains so that the background signal from the combined isotype control gives 2–5% positive cells in each collection channel. Set the compensation amount according to the detected spectral overlap. To analyze the engraftment, draw 4 dotplots as in A, B, C, and D in **Fig. 1** (440/40 nm vs. side-scatter (SSC), forward scatter (FSC) vs. SSC, 695/40 nm vs. SSC, and 530/30 nm vs. 575/26 nm). First, exclude dead cells from the analysis via a region (R1) around the live, unstained cells as in **Fig. 1a**. Next, display these cells on a FSC vs. SSC plot and select the lymphoid and myeloid cells for further analysis but exclude debris via a region (R2) as in **Fig. 1b**. Display cells that fall into the first 2 regions on a 695/40 nm vs. SSC plot and draw a generous region around the CD45-PerCP positive cells as in **Fig. 1c** (R3). Display these CD45$^+$ cells on a CD19-FITC vs. CD33-PE (530/30 nm vs. 575/26 nm) dotplot and draw a quadrant to define FITC$^+$/PE$^-$ cells and FITC$^-$/PE$^+$ cell subsets as in Fig. 1d. The number of events that fall within these regions may be

used to calculate the percentage of live, debris-free cells (R2) that are human cells. In addition, the scatter characteristics of cells may be confirmed as consistent with myeloid (high FSC and SSC, example in **Fig. 1e**) and lymphoid (low FSC and SSC, example in **Fig. 1f**).

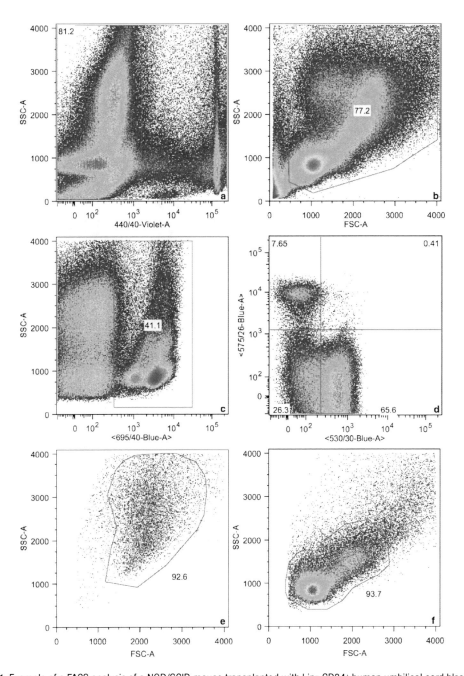

Fig. 1. Example of a FACS analysis of a NOD/SCID mouse transplanted with Lin⁻ CD34⁺ human umbilical cord blood cells and examined 10 weeks after the transplantation.

# 4. Notes

1. The mice should be treated for at least 8 days with acidified water before the irradiation.

2. The sublethal irradiation dose (between 320 and 375 cGy) should be adapted to the irradiator used. It indeed depends on the dose rate/minute of irradiation. More the dose rate is high less the irradiation should be lengthy as this impede on internal organs. If the dose rate is initially too high, the time should be increased (as should the shielding used), to reduce the impact on internal organs.

3. The gamma radiation damages primarily cells dividing at a high rate, such as gut epithelium or cells of the hemopoietic system. Any mouse receiving irradiation should be maintained on acidified water at least one week before the irradiation and two weeks after the irradiation dose to prevent diarrhoea or weight loss possibly arising through epithelial damage of the intestines. Any animal showing persistent weight loss exceeding 20% of body weight and/or other signs of illness (rough fur, inappetance, inability to groom, immobility, tympany) should be sacrificed. Experience shows that with these measures, the above side effects rarely arise.

4. Anaesthetic solution: Mix 1 ml of Ketaset with 0.5 ml of Rompun 2% and dilute with 8.5 ml of PBS. This solution can be kept at least one month at 4°C.

5. Analgesic: Vetergesic should be diluted 1/10 in PBS and injected at 100 µl subcutaneously per mouse.

6. General health monitoring: Mice should be monitored daily and might be culled if they show signs of ill health such as piloerection and hunched posture, inactivity or inappetance for a period of 48 h. In addition, any animal that loses 20% body weight or that develops more serious clinical signs such as diarrhoea or dyspnoea should also be sacrificed.

## Acknowledgments

The author would like to thank Mr. Christopher Ridler and Dr. Daniel Pearce for their assistance in the preparation of this manuscript.

# References

1. Zanjani, E. D. (1997). The human/sheep xenograft model for assay of human HSC (discussion). *Stem Cells* 15 Suppl 1, 209.

2. Zanjani, E. D., Flake, A. W., Almeida-Porada, G., Tran, N., and Papayannopoulou, T. (1999). Homing of human cells in the fetal sheep model: modulation by antibodies activating or inhibiting very late activation antigen-4-dependent function. *Blood* 94, 2515–2522.

3. Almeida-Porada, G., Porada, C., and Zanjani, E. D. (2001). Adult stem cell plasticity and methods of detection. *Rev Clin Exp Hematol* 5, 26–41.

4. Lapidot, T., Pflumio, F., Doedens, M., Murdoch, B., Williams, D. E., and Dick, J. E. (1992). Cytokine stimulation of multilineage hematopoiesis from immature human cells engrafted in SCID mice. *Science* 255, 1137–1141.

5. Vormoor, J., Lapidot, T., Pflumio, F., Risdon, G., Patterson, B., Broxmeyer, H. E., et al. (1994). Immature human cord blood progenitors engraft and proliferate to high levels in severe combined immunodeficient mice. *Blood* 83, 2489–2497.

6. Cashman, J. D., Lapidot, T., Wang, J. C., Doedens, M., Shultz, L. D., Lansdorp, P., et al. (1997). Kinetic evidence of the regeneration of multilineage hematopoiesis from primitive cells in normal human bone marrow transplanted into immunodeficient mice. *Blood* 89, 4307–4316.

7. Larochelle, A., Vormoor, J., Hanenberg, H., Wang, J. C., Bhatia, M., Lapidot, T., et al. (1996). Identification of primitive human hematopoietic cells capable of repopulating NOD/SCID mouse bone marrow: implications for gene therapy. *Nat Med* 2, 1329–1337.

8. Shultz, L. D., Schweitzer, P. A., Christianson, S. W., Gott, B., Schweitzer, I. B., Tennent, B., et al. (1995). Multiple defects in innate and adaptive immunologic function in NOD/LtSz-scid mice. *J Immunol* 154, 180–191.

9. Dick, J. E. (1996). Normal and leukemic human stem cells assayed in SCID mice. *Semin Immunol* 8, 197–206.

10. Hogan, C. J., Shpall, E. J., McNiece, I., and Keller, G. (1997). Multilineage engraftment in NOD/LtSz-scid/scid mice from mobilized human CD34 + peripheral blood progenitor cells. *Biol Blood Marrow Transplant* 3, 236–246.

11. Matsuzaki, Y., Kinjo, K., Mulligan, R. C., and Okano, H. (2004). Unexpectedly efficient homing capacity of purified murine hematopoietic stem cells. *Immunity* 20, 87–93.

12. Yahata, T., Ando, K., Sato, T., Miyatake, H., Nakamura, Y., Muguruma, Y., et al. (2003). A highly sensitive strategy for SCID-repopulating cell assay by direct injection of primitive human hematopoietic cells into NOD/SCID mice bone marrow. *Blood* 101, 2905–2913.

13. Guenechea, G., Gan, O. I., Dorrell, C., and Dick, J. E. (2001). Distinct classes of human stem cells that differ in proliferative and self-renewal potential. *Nat Immunol* 2, 75–82.

14. Christianson, S. W., Greiner, D. L., Hesselton, R. A., Leif, J. H., Wagar, E. J., Schweitzer, I. B., et al. (1997). Enhanced human CD4 + T cell engraftment in beta2-microglobulin-deficient NOD-scid mice. *J Immunol* 158, 3578–3586.

15. Glimm, H., Eisterer, W., Lee, K., Cashman, J., Holyoake, T. L., Nicolini, F., et al. (2001). Previously undetected human hematopoietic cell populations with short-term repopulating activity selectively engraft NOD/SCID-beta2 microglobulin-null mice. *J Clin Invest* 107, 199–206.

16. Kollet, O., Peled, A., Byk, T., Ben-Hur, H., Greiner, D., Shultz, L., et al. (2000). beta2 microglobulin-deficient (B2m(null)) NOD/SCID mice are excellent recipients for studying human stem cell function. *Blood* 95, 3102–3105.

17. Hiramatsu, H., Nishikomori, R., Heike, T., Ito, M., Kobayashi, K., Katamura, K., et al. (2003). Complete reconstitution of human lymphocytes from cord blood CD34 + cells using the NOD/SCID/gammacnull mouse model. *Blood* 102, 873–880.

18. Yahata, T., Ando, K., Nakamura, Y., Ueyama, Y., Shimamura, K., Tamaoki, N., et al. (2002). Functional human T lymphocyte development from cord blood CD34 + cells in nonobese diabetic/Shi-scid, IL-2 receptor gamma null mice. *J Immunol* 169, 204–209.

19. Ishikawa, F., Yasukawa, M., Lyons, B., Yoshida, S., Miyamoto, T., Yoshimoto, G., et al. (2005). Development of functional human blood and immune systems in NOD/SCID/IL2 receptor {gamma} chain(null) mice. *Blood* 106, 1565–1573.

20. Shultz, L. D., Lyons, B. L., Burzenski, L. M., Gott, B., Chen, X., Chaleff, S., et al. (2005). Human lymphoid and myeloid cell development in NOD/LtSz-scid IL2R gamma null mice engrafted with mobilized human hemopoietic stem cells. *J Immunol* 174, 6477–6489.

21. Traggiai, E., Chicha, L., Mazzucchelli, L., Bronz, L., Piffaretti, J. C., Lanzavecchia, A., et al. (2004). Development of a human adaptive immune system in cord blood cell-transplanted mice. *Science* 304, 104–107.

22. Shultz, L. D., Ishikawa, F., and Greiner, D. L. (2007). Humanized mice in translational biomedical research. *Nat Rev Immunol* 7, 118–130.

23. Cashman, J. D., and Eaves, C. J. (2000). High marrow seeding efficiency of human lymphomyeloid repopulating cells in irradiated NOD/SCID mice. *Blood* 96, 3979–3981.

24. van der Loo, J. C., and Ploemacher, R. E. (1995). Marrow- and spleen-seeding efficiencies of all murine hematopoietic stem cell subsets are decreased by preincubation with hematopoietic growth factors. *Blood* 85, 2598–2606.

25. Mazurier, F., Doedens, M., Gan, O. I., and Dick, J. E. (2003). Rapid myeloerythroid repopulation after intrafemoral transplantation of NOD-SCID mice reveals a new class of human stem cells. *Nat Med* 9, 959–963.

26. Wang, J., Kimura, T., Asada, R., Harada, S., Yokota, S., Kawamoto, Y., et al. (2003). SCID-repopulating cell activity of human cord blood-derived CD34- cells assured by intra-bone marrow injection. *Blood* 101, 2924–2931.

27. Shultz, L. D., Banuelos, S. J., Leif, J., Appel, M. C., Cunningham, M., Ballen, K., et al. (2003). Regulation of human short-term repopulating cell (STRC) engraftment in NOD/SCID mice by host CD122 + cells. *Exp Hematol* 31, 551–558.

28. Verlinden, S. F., van Es, H. H., and van Bekkum, D. W. (1998). Serial bone marrow sampling for long-term follow up of human hematopoiesis in NOD/SCID mice. *Exp Hematol* 26, 627–630.

# Chapter 23

## Canine Models of Gene-Modified Hematopoiesis

### Brian C. Beard and Hans-Peter Kiem

## Summary

Large animal models have played a crucial role in the development of gene therapy protocols. A significant advantage of large animal models over rodent models includes the ability to more easily translate protocols developed in large animals to humans. For gene therapy applications, nonhuman primates and canines have been the main large animal models. Canines have the advantage that there are disease models available, e.g., hemolytic anemia (pyruvate kinase deficiency), leukocyte adhesion deficiency, severe combined immunodeficiency (XSCID), storage diseases, and others. In addition, all three major integrating virus systems, i.e., gammaretrovirus-, HIV-derived lenti- and foamy virus vectors are able to efficiently transduce canine hematopoietic cells. Here we describe protocols developed for efficient transduction of canine hematopoietic repopulating cells.

**Key words:** Canine, CD34, Gammaretrovirus, HIV-derived lentivirus, Foamy virus, Transduction, Hematopoiesis, In vivo selection.

## 1. Introduction

Genetic modification of hematopoietic repopulating cells has significant potential for many diseases including malignant and infectious diseases and genetic disorders (for reviews see (1–4)). A major obstacle in the development of efficient transduction protocols for hematopoietic repopulating cells or stem cells (HSC) has been the inability to effectively test HSC transduction conditions in vitro, and studies in the mouse have not accurately predicted outcomes in humans. Thus, over the past decade, several research groups have focused on studies in large animal models, mainly nonhuman primates and canines, which generally are more predictive for human applications (5–8). Canines have

Christopher Baum (ed.), *Methods in Molecular Biology, Methods and Protocols, vol. 506*
© Humana Press, a part of Springer Science+Business Media, LLC 2009
DOI: 10.1007/978-1-59745-409-4_23

several other advantages over nonhuman primates, including the availability of canine disease models, such as metabolic (alpha-l-iduronidase deficiency) and hematologic (pyruvate kinase deficiency, X-linked severe combined immunodeficiency (XSCID), and leukocyte adhesion deficiency) diseases, which closely mimic the correlating human disease and can be used to study the clinical effectiveness of gene therapy. In addition, the canine also allows for the study of drug resistance genes such as MDR1 *(9)* and MGMT *(10, 11)* or in vivo selection strategies *(12)*. Drug resistance gene therapy or in vivo selection strategies will likely be necessary for most diseases in which the transgene alone will potentially not confer a substantial selective advantage.

## 2. Materials

1. *Cytokines and small molecules:* Canine granulocyte colony stimulating factor (cG-CSF Amgen, Thousand Oaks, CA), canine stem cell factor (cSCF Amgen, Thousand Oaks, CA), canine granulocyte/macrophage colony stimulating factor (cGM-CSF), human FMS-like tyrosine kinase 3 ligand, erythropoietin (EPO), human megakaryocyte growth and development factor (hMGDF), AMD3100 (Sigma-Aldrich St. Louis, MI), and CH-296 (RetroNectin® Takara Kyoto, Japan).

2. *Antibodies:* Biotinylated anti-canine CD34 antibody (1H6 FHCRC), CD3 (CA17.2A12, Dr. Peter Moore), CD21 (CA2.1D6, Serotec Raleigh, NC), CD14 (TüK4, Serotec Raleigh, NC), DM5 (Clone 1.2, Dr. Brenda Sandmaier), PE conjugated IgG goat anti mouse, PE conjugated murine $IgG_1$, and PE conjugated streptavidin.

3. *Chemotherapy:* $O^6$-benzylguanine, BCNU (*Carmustine*), and temozolomide (Schering-Plough Kenilworth, NJ).

4. *Antibiotics:* baytril, amikacin, and ceftazidime.

5. *Miscellaneous drugs:* diphenhydramine, Zofran® (GlaxoSmith-Kline London, United Kingdom), cyclosporine A, mycophenolate mofetil.

6. *Miltenyi buffer:* 4% (w/v) ultra pure BSA, 2 mM EDTA, pH: 7.0, in 1× PBS.

7. *Hemolytic buffer:* 155 mM $NH_4Cl$, 12 mM $NaCO_3$, 0.1 mM EDTA adjusted to pH: 7.4.

8. *DNase buffer:* 2% (v/v) fetal bovine serum (FBS) and 5.1 µg/mL DNase in 1× PBS

9. *CH-296 buffer I:* 2% (w/v) BSA fraction V in Hanks.

10. *CH-296 buffer II:* 25 mM HEPES, pH: 7.0, in Hanks.

11. *Human Dexter:* 12.5% (v/v) FBS, 12.5% (v/v) horse serum, 1 mM sodium pyruvate, 1 mM L-glutamine, 5 mL penicillin/streptomycin (5,000 U/mL stock), 10 μM 2-mercaptoethanol, 1 μM hydrocortisone hemisuccinate in IMDM.

12. StemSpan SFEM (StemCell Technologies Vancouver, Canada).

13. IMDM 10/1 FBS: 10% (v/v) FBS and 1% (v/v) penicillin/streptomycin (5,000 U/mL stock) in IMDM.

14. 2× CFU media: 50% (v/v) FBS, 40% (v/v) 3× Alpha MEM, 10 mL 2% (w/v) BSA Fraction V (resuspended in 3× MEM), 1 mL L-glutamine (100×), 1 mL Penicillin–Streptomycin (100×).

15. FACS buffer: 2% (v/v) cosmic calf serum in 1× PBS with propidium iodide (1 μg/mL).

16. QIAamp DNA Blood Mini Kit (Qiagen Valencia, CA).

17. Proteinase K.

18. Seaplaque low melting temperature agarose.

19. 70 μM nylon cell strainer.

20. MACSmix (Miltenyi Biotec Auburn, CA), MidiMACS Magnetic Holder (Miltenyi Biotec Auburn, CA), MACS Separation LS Columns (Miltenyi Biotec Auburn, CA), and streptavidin microbeads (Miltenyi Biotec Auburn, CA).

21. Steriflip 0.22 μM filter (Millipore Billerica, MA)

22. Hemocytometer and trypan blue

23. 50 mL conical tubes, 15 mL conical tubes, and 5 mL polystyrene round-bottom tubes.

24. 35 mm tissue culture dishes

25. Tissue culture treated and nontissue culture treated T-75 flasks.

26. Vacutainer K2 EDTA blood collection vials (Becton Dickinson, Franklin Lakes, NJ).

27. Taqman instrument and reagents needed for Taqman PCR.

## 3. Methods

The methods described below outline *(1)* isolation of cG-CSF/cSCF or AMD3100 mobilized canine CD34⁺ cells and retrovirus transduction, *(2)* posttransplantation gene marking analysis, and *(3)* in vivo selection strategies in canines that received cells gene-modified with drug resistance transgenes.

**3.1. Canine CD34⁺ Cell Isolation and Retrovirus Transduction**

The isolation and retrovirus transduction of canine CD34⁺ cells from peripheral blood or bone marrow is described in **Subheadings 3.1.1–3.1.5.** This includes (a) the description of mobilization/priming and selection of canine CD34⁺ cells from leukapheresis or bone marrow product, (b) the description of CD34⁺ staining of canine cells, (c) the description of canine CD34⁺ retrovirus transduction, (d) the description of retrovirus transduction analysis in bulk liquid cultures and colony forming units (CFUs), and (e) the description of gene-modified cell infusion into conditioned canine recipients.

*3.1.1. Canine White Blood Cell Mobilization/Priming and CD34 Selection*

Mobilization/priming of canine CD34⁺ cells is achieved with five consecutive days of cG-CSF (5 µg/kg body weight subcutaneously, twice daily) and cSCF (25 µg/kg body weight subcutaneously, once daily). On the sixth day prior to leukapheresis or bone marrow harvest, a single dose of cG-CSF (5 µg/kg body weight subcutaneously) is administered. As an alternative mobilization strategy to cG-CSF/cSCF, a single dose of AMD3100 (4 mg/kg body weight subcutaneously) is administered 6–8 h prior to leukapheresis (*see* **Note 1**). Bone marrow harvest and leukapheresis is carried out by a qualified animal technician, and from a 10 kg canine, the expected yield of total nucleated white blood cells ranges from $2 \times 10^9$ to $1 \times 10^{10}$ cells.

Selection of Canine CD34⁺ Cells from Leukapheresis or Bone Marrow Product

1. Freshly prepare Miltenyi buffer and filter sterilize (0.22 µM). De-gas a portion of the fresh Miltenyi buffer (50 mL) by attaching 50 mL conical to a Steriflip (wrap the junction with parafilm) and place under vacuum for ~20 min and leave on ice.

2. For efficient red blood cell lysis, split the cell product (bone marrow or apheresis) into 50 mL conical tubes (10–12 mL/tube) and add hemolytic buffer to 50 mL. Incubate several minutes at room temperature and centrifuge at 450 g for 5 min.

3. Aspirate supernatant from pellet(s) leaving enough hemolytic buffer for resuspension (3–5 mL) (*see* **Note 2**). Combine all cell pellets in hemolytic buffer, bringing volume to 50 mL with hemolytic buffer (to lyse residual red blood cells), incubate several minutes at room temperature, and centrifuge cells as above.

4. Aspirate supernatant and resuspend cells in 15–20 mL of DNase buffer (*see* **Note 3**) and filter through 70 µM nytex filter. Rinse filter and tube with DNase buffer to bring to 50 mL and centrifuge cells as above.

5. Aspirate supernatant and resuspend cell pellet in Miltenyi buffer (~10–15 mL). Make 1:10, 1:100, and 1:1,000 dilutions of cells in trypan blue and count cells on a hemocytometer.

6. Bring cells to a concentration of $1 \times 10^8$/mL in Miltenyi buffer and add biotinylated anti-canine CD34 antibody (1H6) at the volume required for $1 \times 10^6$ cells empirically tested prior to transplant (*see* **Note 4**), then put the tube on MACSmix rotator at 4°C for 30 min. After incubation bring up to 50 mL volume with Miltenyi buffer and centrifuge cells as above. Repeat resuspension and centrifugation once more.

7. Resuspend pellet with Miltenyi buffer that will give $1 \times 10^8$ cells/mL, but leave out 10% of the volume for addition of streptavidin microbeads and incubate on MACSmix rotator at 4°C for 30 min. Filter and rinse cell suspension through 70 µM nytex filter into fresh 50 mL conical tube.

8. Centrifuge cells as above and resuspend pellet in degased Miltenyi buffer (d-Miltenyi buffer) at $3 \times 10^8$ cells/mL.

9. Equilibrate Miltenyi LS columns (1 column for every $6 \times 10^8$ cells) attached to *Midi*MACS magnetic holder with 1 mL of d-Miltenyi buffer (*see* **Note 5**).

10. Add 2 mL of the cell suspension to each column and collect the flow through.

11. After the columns have stopped dripping add 2 mL of d-Miltenyi buffer to wash tubes and transfer to columns and continue to collect the flow through.

12. Add 4 mL d-Miltenyi buffer for the last wash and collect the flow through.

13. Prepare a clean 50 mL conical tube and remove the column from the *Midi*MACS magnetic holder. Add 5 mL IMDM 10/1 FBS and push plunger to collect CD34-selected cells.

14. Centrifuge positive and negative fractions as above, resuspend pellet(s) in 5 mL IMDM 10/1 FBS and count cells as before (*see* **Subheading 3.1.1**, step 5). Remove CD34-selected cells for nontransduced liquid culture (*see* **Subheading 3.1.4**) and colony forming unit plating (*see* **Subheading "Colony Forming Unit (CFU) Analysis of Transduced Canine CD34-Selected Cells"**).

*3.1.2. CD34+ Staining of Canine Cells*

CD34+ staining using the anti-canine CD34 antibody (1H6) should be carried out on peripheral blood samples during mobilization and compared with a control animal to confirm efficient mobilization. Staining is also carried out on all of the various fractions during CD34-selection (i.e., bone marrow or leukapheresis product, total nucleated white blood cell fraction, flow through fraction, and CD34-selected fraction).

1. Take 100 µL of total peripheral blood, bone marrow, or $5 \times 10^5$ to $1 \times 10^6$ selected cell fractions and transfer to separate 5 mL

polystyrene round-bottom tubes (FACS tube), add 2 mL of hemolytic buffer, and incubate at room temperature for 5 min.

2. Centrifuge cells at 450 g for 5 min, aspirate the majority of supernatant, discard and resuspend the pellet in ~2 mL DNase buffer. Centrifuge cells again as above and then aspirate all but ~100 μL of DNase buffer. Add 10 μL of IgG-PE isotype control antibody or 10 μL of diluted 1H6 antibody (dilution: 50 μL of stock biotinylated 1H6 plus 950 μL DNase buffer) to appropriate tubes (*see* **Note 6**). Incubate samples at 4°C for 20 min then add ~2 mL of DNase buffer and centrifuge cells as above.

3. Aspirate all but 100 μL and add 2 μL of PE conjugated streptavidin to 1H6 antibody tubes and incubate for 20 min at 4°C. After incubation bring all tubes to volume with DNase buffer and centrifuge cells as above. Aspirate supernatant and resuspend with 250 μL DNase buffer with propidium iodide (2 μg/mL).

4. Analyze flow cytometric quantification of at least 20,000–100,000 events (gated by forward and right-angle light scatter and excluded for propidium iodide positive cells) on a FACSCalibur (Becton Dickinson, Franklin Lakes, NJ) or similar instrument. Perform analysis of flow cytometric with CELLQuest v3.3 software (BD Biosciences, San Jose, CA) or similar program (i.e. FlowJo). Generally, the CD34⁺ population is ~60–90% positive based on the above gating criteria (*see* **Fig. 1**).

*3.1.3. Retrovirus Transduction*

Described below are general transduction protocols for retrovirus (gammaretrovirus, HIV-derived lentivirus, and foamy virus) vectors (*see* **Fig. 2**). All retroviral vector transductions are carried out on CH-296-coated nontissue culture-treated T75 flasks with (gammaretrovirus) and without (lentivirus and foamy virus) prestimulation. Also, in the case of transductions using unconcentrated

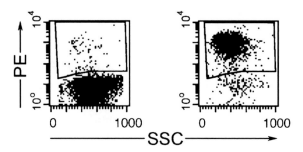

Fig. 1. Canine CD34 Selection: Flow cytometric profiles (side scatter [SSC] and PE) of canine bone marrow cells before (*left panel*) and after (*right panel*) CD34 selection.

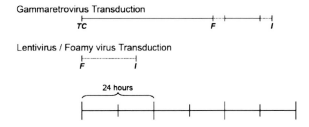

Fig. 2. Retrovirus Transduction: Schematic representation of gammaretrovirus, lentivirus, and foamy virus transduction protocols. *Solid lines* represent prestimulation or standard ex vivo culture without virus and *dashed lines* represent retrovirus transduction. **TC** = tissue culture treated flasks without CH-296 coating; **F** = CH-296 coated non-tissue culture treated flasks; and **I** = end of transduction/ex vivo culture and infusion of gene modified cells into conditioned canine recipients.

gammaretrovirus vector, preloading the CH-296-coated flasks improves transduction.

To prepare nontissue culture-treated T75 flasks for retrovirus transduction, the protocol is identical regardless of the retrovirus that will be used. The CH-296 protein fragment is stored long-term as a lyophilized powder at –70°C. Dissolve the powder in sterile water at a concentration of 1 mg/mL and filter it through a 0.22 μm syringe filter into a 50 mL conical tube. Rinse vial and syringe/filter twice with 2.5 mL sterile Hanks and add to 50 mL conical. Add sterile Hanks to bring to a final volume of 50 mL. The final stock concentration is 50 μg/mL, and the solution can be stored at 4°C for up to a month. Flasks are freshly CH-296 coated the day of transduction.

1. To coat T-75 nontissue culture-treated canted-neck flasks, aliquot the appropriate amount of CH-296 stock to each dish (2 μg/cm$^2$) and add 7 mL sterile Hanks to completely cover the bottom of the flask. Allow the dishes to sit undisturbed at room temperature in a laminar flow hood for 2 h.

2. At the end of 2 h, aspirate off the CH-296 solution and replace it with a similar volume of CH-296 buffer I and incubate at room temperature for 30 min. Aspirate CH-296 buffer I and replace it with a similar volume of CH-296 buffer II.

3. Remove the CH-296 buffer II solution just prior to adding CD34-selected cells or virus containing media (VCM) for preloading. Do not let the dishes dry at any time during the protocol.

Gammaretrovirus Prestimulation

1. CD34-selected cells (*see* **Subheading "Selection of Canine CD34$^+$ Cells from Leukapheresis or Bone Marrow Product", step 14**) are centrifuged at 450 g for 5 min and resuspended at a concentration of $1.0 \times 10^6$ cells/mL in Human Dexter media supplemented with cG-CSF, cSCF, hMGDF,

and hFlt3-L all at 50 ng/mL. As an alternative media for serum-free conditions, CD34-selected cells can be resuspended in StemSpan SFEM without serum and supplemented with the same growth factors as mentioned above (*see*Note 7). Ten milliliters (1 × 10$^7$ cells) are added to each tissue culture-treated T-75 flask and incubated at 37°C in 5% $CO_2$ and 95% air in a humidified incubator for ~36–48 h.

2. After prestimulation, transfer media to 50 mL conical tubes (use separate tubes for each T75 flask) and rinse flasks 1–2 times with 10 mL Hanks and transfer to 50 mL conical tubes. Mobilize any remaining adherent cells with Cell Dissociation Buffer (2 mL/flask, incubate 5 min at 37°C) and transfer to 50 mL conical tubes.

**Gammaretrovirus Transduction (Unconcentrated Virus Stocks)**

1. Preload CH-296-coated T75 flasks (*see* **Subheading 3.1.3.**, step 3) with 10 mL of unconcentrated VCM (*see*Note 8) and incubate for 15 min at room temperature. Aspirate the VCM, repeat the procedure, and remove media just prior to adding CD34-selected cells.

2. Following preloading centrifuge cells (*see* **Subheading "Gammaretrovirus Prestimulation," step 2**) as above, aspirate the media from the 50 mL conical tubes and add 7 mL of VCM supplemented with growth factors (cG-CSF, cSCF, hMGDF, and hFlt3-L at 50 ng/mL) and 8 μg/mL protamine sulfate. Transfer CD34-VCM solution to preloaded CH-296-coated T-75 flasks. Wash the 50 mL conical tubes with 3 mL of the above VCM media, transfer to the respective T-75 flask, and incubate for 4 h at 37°C in 5% $CO_2$ and 95% air in a humidified incubator.

3. Following the first transduction, remove the media and transfer to 50 mL conical tubes and rinse flasks with ~10 mL of sterile Hanks and transfer that solution to the appropriate conical tubes and centrifuge as above. Add ~3 mL of fresh Human Dexter supplemented with the above growth factors to keep cells that adhere to the plate covered. Aspirate the media from the 50 mL conical tubes and resuspend in ~50 mL of Hanks and repeat the centrifugation. Aspirate the media from the 50 mL conical tubes and resuspend the cell pellets in 4 mL of the supplemented Human Dexter media and transfer to the appropriate T-75 flask and repeat with 3 mL of supplemented Human Dexter media. Incubate overnight at 37°C in 5% $CO_2$ and 95% air in a humidified incubator.

4. To begin the second transduction following the overnight incubation, remove the cells, rinse the flasks, and replace the media as described in **step 3**. The only modification is that the media is the VCM media described in **step 2**. Incubate for 4 h at 37°C in 5% $CO_2$ and 95% air in a humidified incubator. If desired additional transductions can be carried out *(13)*.

**Gammaretrovirus Trans-
duction (Concentrated
Virus Stocks)**

1. Centrifuge cells (*see* **Subheading "Gammaretrovirus Prestimulation," step 2**) as before and aspirate the media from
the 50 mL conical tubes and add 7 mL of IMDM 10/1 FBS
VCM or, for serum-free conditions, StemSpan SFEM VCM
supplemented with growth factors (cG-CSF, cSCF, hMGDF,
and hFlt3-L at 50 ng/µL) and 8 µg/mL protamine sulfate
(*see*Note 9). Transfer CD34-VCM solution to CH-296-
coated T-75 flasks. Wash the 50 mL conical tubes with 3 mL
of the appropriate VCM media, transfer to the respective T-75
flask and incubate for 4 h at 37°C in 5% $CO_2$ and 95% air in a
humidified incubator.

2. Following the first transduction, wash the cells and replace
in the appropriate media (Human Dexter or StemSpan
SFEM) for overnight incubation as described in **Subheading
"Gammaretrovirus Transduction (Unconcentrated Virus
Stocks)," step 3**.

3. To begin the second transduction following the overnight incubation remove the cells, rinse the flasks, and replace the media
as described in **Subheading "Gammaretrovirus Transduction
(Unconcentrated Virus Stocks)," step 3**. The only modification is that the media is the IMDM-VCM or StemSpan SFEM
VCM media described in **step 1**. Incubate for 4 h at 37°C in 5%
$CO_2$ and 95% air in a humidified incubator.

**Cell Preparation
for Infusion**

1. After the second 4-h virus exposure, remove the cells from the
flasks to a 50 mL conical tube and wash each flask with ~5 mL
of room temperature Hanks and transfer the Hanks to the cell
suspension previously removed from the cells. Add 3 mL of
Cell Dissociation Buffer to the flask and incubate in a 37°C
humidified incubator for 5–15 min tapping the flasks gently
several times during the procedure. Remove the Cell Dissociation Buffer from the flasks and add to the cell suspension.
Check the flask for efficient removal of adherent cells under a
standard inverted light microscope. Repeat as necessary until
>95% of the cells have been removed.

2. Centrifuge cells as above, resuspend cell pellets in ~3 mL of
1× PBS, combine fractions, and pellet the cells again. Resuspend the entire cell pellets in ~5 mL of Hanks, make 1:10
and 1:100 of cells in trypan blue, and count viable cells on a
hemocytometer. Remove cells for CFU plating (*see* **Subheading "Colony Forming Unit (CFU) Analysis of Transduced
Canine CD34-Selected Cells"**) and liquid cultures (*see* **Subheading 3.1.4**). Bring the volume to 10 mL with 1× PBS and
draw into a 20 mL syringe with a 16-gauge needle, add an
additional 10 mL of 1× PBS to the 50 mL conical, draw that
into the same syringe, and place on ice until infusion into the
conditioned canine recipient.

**Lentivirus and Foamy Virus Transduction**

Using HIV-derived lentivirus or foamy virus vectors for transplantation studies are particularly attractive for several reasons including established self-inactivating (SIN) vector design *(14, 15)*, although substantial advances have been made in gammaretrovirus-SIN vector design *(16)*, and the ability to limit ex vivo culture for lentivirus and foamy virus transduction protocols (<24 h) compared with gammaretrovirus transduction protocols (~3 days) (*see* **Fig. 2**).

1. Pellet CD34-selected cells (*see* **Subheading "Selection of Canine CD34+Cells from Leukapheresis or Bone Marrow Product," step 14**) at 450 g for 5 min and resuspend at a concentration of $1.0 \times 10^6$ cells/mL in IMDM 10/1 FBS or StemSpan SFEM VCM (*see* **Note 10**) supplemented with growth factors (cG-CSF, cSCF, and hFlt3-L at 50 ng/mL) and 8 μg/mL protamine sulfate (do not add protamine sulfate to foamy virus transductions). Add 10 mL ($1 \times 10^7$) to each CH-296-coated T-75 flask and incubate at 37°C in 5% $CO_2$ and 95% air in a humidified incubator overnight (~18 h).

2. After the overnight virus exposure, wash and prepare CD34 cells for infusion into irradiated canine recipients as detailed above (*see* **Subheading "Cell Preparation for Infusion"**).

**3.1.4. Transduction Efficiency**

Following retrovirus transduction, canine CD34+ cells are plated in IMDM 10/1 FBS supplemented with cG-CSF, cSCF, hMGDF, and hFlt3-L at 50 ng/mL at a density of ~$1 \times 10^5$ in a 12-well plate with nontransduced control cells from the same animal plated in a separate well. After 3 and 10 days of expansion (day 0 is the end of transduction), ~30% of the cells are removed from the respective wells for gene marking analysis using flow cytometric and/or PCR-based methods.

**Flow Cytometric Analysis of Transduced Canine CD34-Selected Cells**

1. Transfer the cells removed from the liquid culture (see above) to a FACS tube, bring to ~3 mL with FACS buffer without propidium iodide, and centrifuge at 450 g for 5 min.

2. Aspirate the media and resuspend the cell pellet in ~250 μL of FACS buffer with propidium iodide [1 μg/mL].

3. Analyze flow cytometric quantification of at least 20,000 events as stated before (*see* **Subheading 3.1.2.**, step 4). Gating should include <0.1% nontransduced cells in the relevant region.

**Taqman PCR Analysis of Transduced Canine CD34-Selected Cells**

1. Transfer the cells removed from the liquid culture (see above) to an Eppendorf tube, bring to 1 mL with 1× PBS, and centrifuge at 10,000 g for 2 min.

2. Aspirate the media and add 200 μL Hanks and vortex the sample until complete resuspension of the cell pellet, then add 250 μL of Qiagen Cell Lysis Buffer and 20 μL of Proteinase K [500 μg/mL].

3. Store samples at −4°C until ready to analyze with Taqman PCR. Complete the DNA extraction of the sample following the manufacturer's protocol (QIAamp DNA Blood Mini Kit). Explanation of Taqman PCR for analysis of gene-marked canine samples is detailed in **Subheading 3.2.3.**

Colony Forming Unit (CFU) Analysis of Transduced Canine CD34-Selected Cells

As a complimentary assay to the FACS data described above, a reliable in vitro test for transduction efficiency of primitive cells is analysis of gene-modified CFUs. CD34-enriched transduced and nontransduced cells are plated on day 0 in a double-layer agar culture system and cultured for 14 days at 37°C in 5% $CO_2$ and 95% air in a humidified incubator before enumeration of cultures by standard inverted light microscopy to determine the total number of CFUs in the transduced and nontransduced plates. If the retrovirus vector expresses GFP or YFP fluorescence microscopy can be performed with a FITC filter for GFP+ and YFP+ visualization. Using these data, it is possible to calculate plating and transduction efficiency.

1. Heat 1% (bottom layer) and 0.6% (top layer) SeaPlaque Agarose on hot plate to boiling and cool to ~45–50°C in a 37°C water bath. It is best to keep the 0.6% Agar in a 56°C water bath to avoid solidification. Prior to using the 0.6% agar in **step 5**, make sure to cool in the 37°C water bath as above.

2. Aliquot 2× CFU media to a 15 mL falcon tube (500 μL/plate). Generally plate each condition of CFU in triplicate.

3. Add the appropriate amount of growth factors to the media (standard canine growth factor cocktail: final concentration of entire 1.5 mL of top and bottom layer is 100 ng/mL of cSCF, cG-CSF, cGM-CSF, hMGDF, 4 U/mL erythropoietin). No growth factor is added to the top layer, so the entire amount of growth factor to bring the concentrations noted above needs to be added to the bottom layer mixture. Prewarm the 2× CFU media and growth factor mixture in a 37°C water bath.

4. Once both solutions (Agar and CFU media with growth factors) have come to the required temperature, add an equal volume of 1% agarose to the CFU medium, mix well (without introducing bubbles), and aliquot 1 mL of the mixture to each 35 mm dish. Quickly swirl the plate to coat and leave at room temperature for at least 25 min before plating top layer.

5. Transfer the 0.6% Agar from the 56°C water bath to the 37°C water bath. Aliquot the appropriate amount of 2× CFU media to a 15 mL falcon tube (0.25 mL for each dish). From the canine CD34-selected cells, take enough cells so each plate will have between 1 and $3 \times 10^3$ cells. Place the CFU media/cells in a 37°C water bath.

6. Once both solutions have come to the required temperature, add an equal volume of 0.6% Agar to the CFU media/cell

mixture, mix well and aliquot 500 µL of the mixture to each 35 mm dish. Coat plates without swirling and allow to solidify for at least 20 min.

7. Place the CFU plates in an aerated clean plexiglas box and fill a 35-mm dish with sterile water to provide humidity in the box. Incubate 7–14 days at 37°C in 5% $CO_2$ and 95% air in a humidified incubator.

**Retrovirus-Specific CFU PCR**

To enumerate gene transfer DNA can be extracted from CFUs (*see* **Subheading "Colony Forming Unit (CFU) Analysis of Transduced Canine CD34-Selected Cells"**) and amplified to detect provirus sequences and general DNA. This analysis can be used when the vector does not carry a fluorescent transgene or to determine the number of colonies that have integrated vectors. DNA from CFUs is extracted in a PCR tube by adding 10 µL of a picked colony into 90 µL of sterile $H_2O$ and 500 µg/mL Proteinase K. After thorough vortexing, the sample is placed in a PCR machine running an extraction program (56°C, 120 min; 99°C; 10 min). Approximately 20 µL of the crude DNA extract is analyzed by PCR using retrovirus-specific forward and reverse primers. Additionally, another 20 µL of each colony extract is used to amplify the β-actin gene using specific forward and reverse primers for general DNA detection. PCR conditions include initial denaturation at 94°C for 1 min, 40 cycles of 94°C for 1 min, 65°C for 30 s and 72°C for 1 min, and a final extension at 72°C for 10 min. Negative controls are run without DNA (water) or with DNA extracted from normal canine cells.

*3.1.5. Infusion of Gene-Modified Cells*

As preparation for transplantation, the animals receive a single nonmyeloablative or myeloablative dose (under mild sedation 0.2–0.4 mg/kg butorphanol (IV)) ranging from 100 to 920 cGy total body irradiation (TBI) at 7 cGy/min delivered by a linear accelerator on day-1 *(9)* or day 0 (day 0 designates the day gene-modified cells are infused). Chemical conditioning (i.e., busulfan or a combination of BCNU and temozolomide) can be used as pretransplant conditioning *(17–19)*. Although pharmacokinetics (i.e., $C_{max}$, half-life, area under the curve (AUC), and clearance) in canines is likely different than humans, targeted drug doses can be approximated to mimic clinical settings by evaluating the suppression based on the duration of days with absolute neutrophil counts (ANC) <500/µL and platelet counts (PLT) <20,000/µL. As a general rule reasonable suppression that leads to stable engraftment of gene modified cells in an immune competent animal is ANC <500/µL and PLT <20,000/µL for 3–5 days each (HPK, unpublished results). Alternatively, for some disease models, minimal to no conditioning is required due to the immunocompromised nature of the animal as described by Ravin et al. for treatment of neonatal canine X-SCID *(20)*. Following irradiation, animals are given 500 mL of Ringer's solution subcutaneously. If potentially immunogenic transgenes are present, immunosuppression should be used in addition to irradiation. Immunosuppression consists of cyclosporine (CSP) (15 mg/kg body weight orally, twice daily) beginning on day -1 or -3, and, in the allogeneic setting, mycophenolate mofetil (MMF) (10 mg/kg body weight subcutaneously, once daily) beginning on day 0 at least 1 h after infusion of gene-modified cells. MMF is administered <1 min after subcutaneous injection of lidocaine (2% v/v) to dull the site of MMF injection to limit discomfort. CSP levels should be closely monitored early after transplantation (up to 1 month) because of potential CSP-associated intussusception that if left untreated is fatal. Gene-modified cells are infused 15–30 min after diphenhydramine administration (1.5 mg/kg body weight intravenously). Gene-modified cells are suspended in 18–20 mL of 1× PBS and infused intravenously over 5–10 min while monitoring clinical response of the animal (i.e., respiration, heart rate, and flushed appearance). If negative selection cells are infused, those are administered directly after the gene-modified cells. In general, the target cell number of gene-modified cells infused is at least $2 \times 10^6$/kg, and if this lower limit is not met prolonged neutropenia and thrombocytopenia is likely.

### 3.2. Posttransplant Gene Marking Analysis

The analysis of posttransplant gene marking of canine peripheral blood and bone marrow cells is described in **Subheadings 3.2.1.–3.2.4.** This includes (a) the description of typical engraftment kinetics of hematopoietic gene-modified cells, (b) the description of flow cytometric-based analysis of gene marked hematopoietic cells, (c) the description of PCR-based analysis of gene marked hematopoietic cells, and (d) the description of gene marking analysis in individual CFUs.

### 3.2.1. Engraftment Kinetics of Hematopoietic Gene-Modified Cells

Gene-modified cells generally engraft with similar kinetics as unmodified product given the cellular thresholds described in **Subheading 3.1.5.** About 1 h after infusion of gene-modified cells, cG-CSF (5 μg/kg body weight subcutaneously, twice a day) injections are started to facilitate neutrophil engraftment. cG-CSF is continued until absolute neutrophil count (ANC) is >1,000/μL for three consecutive days. Transfusion support is maintained until platelets hold at levels greater than 10,000/μL without pronounced petechiae. Daily complete blood counts (CBCs) are continued until platelets reach levels >100,000/μL.

### 3.2.2. Flow Cytometric-Based Analysis of Gene-Marked Hematopoietic Cells

In vivo gene marking is carried out using ~1–2 mL of venous peripheral blood collected using a 21¾ gauge butterfly needle and a Vacutainer K2 EDTA 2.0 mL tube (Becton Dickinson, Franklin Lakes, NJ).

1. Remove a whole blood fraction (~100 μL) from the Vacutainer tube to a FACS tube with 800 μL of 1× PBS.

2. Transfer the remainder of the blood sample to a 15 mL conical tube and bring to ~15 mL with hemolytic buffer, incubate for ~5 min and centrifuge cells at 450 g for 5 min. Aspirate the supernatant and repeat the hemolytic lysis until ~95% of the red blood cells are removed.

3. After final red blood cell lysis, resuspend whole blood cell and nucleated white blood cell fractions in 1 mL FACS buffer, transfer to FACS tubes, and just prior to flow-cytometric analysis take 200 μL cell suspension, add to 300 μL DNase buffer with propidium iodide [1 μg/mL].

4. Analyze flow-cytometric quantification of at least 20,000 events as stated before (*see* **Subheading 3.1.2.**, step 4). Gating should exclude lesser than 0.1% control cells in the relevant region for white blood cells. Control cells from a normal canine, matching the cell type assayed, are used to set the gates.

**Gene Marking in Hematopoietic Cell Subsets**

Samples of cells used for general flow-cytometric analysis (see above) can be used to carry out staining in lymphoid (B cells and T cells) and myeloid (granulocytes and monocytes) subsets. For all staining procedures $1 \times 10^6$ cells are used for all samples and that includes isotype controls and normal canine (nontransplanted) controls.

1. Transfer $1 \times 10^6$ total white blood cells to FACS Tubes, bring to 3 mL with DNase buffer, and centrifuge at 450 g for 5 min. Aspirate the supernatant leaving ~100 μL for staining and add 10 μL of primary antibodies CD3 (T cells), CD14 (monocytes), CD21 (B cells), or DM5 (granulocytes) and 5 μL of PE-IgG$_1$ isotype control to transplanted and control animal tubes.

2. Incubate samples at 4°C for 20 min, bring to ~3 mL with DNase buffer, and pellet cells as above.

3. Aspirate supernatant from all samples leaving ~100 μL and, for primary antibodies that are not directly conjugated (CD3 and DM5), add 5 μL of IgG-PE goat anti-mouse.

4. Incubate samples at 4°C for 20 min, bring to ~3 mL with DNase buffer, and pellet cells as above. Aspirate supernatant and add ~300 μL of DNase buffer with propidium iodide [1 μg/mL]. Carry out flow cytometric analysis as described above (*see* **Subheading 3.2.2.**, step 4).

**3.2.3. PCR-Based Analysis of Gene Marked Hematopoietic Cells**

In the event that the retrovirus of choice has no reporter transgene, relative gene marking can be calculated using retrovirus-specific primers and probes using TaqMan PCR technology. Whole peripheral blood collection and red blood cell lysis is carried out as described in **Subheading 3.2.2.** Nucleated white blood cells are resuspended in 200 µL of Hanks and transferred to Eppendorf tubes where DNA is extracted using Qiagen DNA extraction protocol, and samples are stored at –20°C. For the example detailed here, we will detect gene marking with a lentivirus construct. Initially, 300 ng of DNA is amplified in at least duplicate with a lentivirus-specific primer/probe combination (5'-TGA AAG CGA AAG GGA AAC CA-3', 5'-CCG TGC GCG CTT CAG-3'; probe, 5'-FAM-AGC TCT CTC GAC GCA GGA CTC GGC -TAMRA-3') (Synthegen, Houston, TX). In a separate reaction, a canine interleukin 3 (IL-3)-specific primer/probe combination (5'-ATG AGC AGC TTC CCC ATC C-3', 5'-GTC GAA AAA GGC CTC CCC-3'; probe, 5'-FAM-TCC TGC TTG GAT GCC AAG TCC CAC -TAMRA-3') is used to adjust for equal loading of genomic DNA per reaction. Standards consist of dilutions of DNA extracted from cell lines that contain a single provirus copy of a lentivirus vector. Negative controls consist of DNA extracted from peripheral blood mononuclear cells (PBMCs) obtained before transplantation from control animals or water. Reactions are run by means of the ABI master mix (Applied Biosystems, Branchburg, NJ) on the ABI Prism 7,700 sequence detection system (Applied Biosystems) under the following thermal cycling conditions: 50°C for 2 min and 95°C for 10 min, then 40 cycles of 95°C for 15 s and 60°C for 1 min. Standard curves of DNA and retrovirus are generated from serial dilutions from normal canine DNA and lentivirus cell line DNA, respectively. Standard values indicating an accurate amplification are a slope between –3.2 and –3.7 and $R$ values between 0.98 and 0.99.

*3.2.4. Gene Marking in Individual Colony Forming Units (CFUs)*

As described in **Subheading"Colony Forming Unit (CFU) Analysis of Transduced Canine CD34-Selected Cells,"** CFUs can be grown in double-layer agar plates and subsequently analyzed for provirus integration using retrovirus-specific and general DNA PCR primers. To analyze in vivo gene marking small bone marrow aspirates (5–20 mL) are drawn by a qualified animal technician. Generally, for this size of bone marrow aspirate only one humeri is required. Double-layer agar plates are prepared as detailed in **Subheading"Colony Forming Unit (CFU) Analysis of Transduced Canine CD34-Selected Cells,"** and if bulk bone marrow is used $5 \times 10^4$ to $1 \times 10^5$ cells should be added per CFU plate. Alternatively, small-scale CD34 selection can be carried out as described in **Subheading "Selection of Canine CD34⁺Cells from Leukapheresis or Bone Marrow Product,"** and then $1 \times 10^3$ to $3 \times 10^3$ cells should be added per CFU plate. Enumeration and extraction of CFU DNA and CFU PCR is carried out as described in **Subheadings "Colony Forming Unit (CFU) Analysis of Transduced Canine CD34-Selected Cells"** and **"Retrovirus-Specific CFU PCR."** In general, the percentage of positive CFUs corresponds well with the gene marking data collected by flow-cytometric and Taqman PCR analysis.

**3.3. In vivo Selection**

In vivo selection strategies of canine hematopoietic cells gene-modified with drug resistance genes are described in **Subheadings 3.3.1–Subheading 3.3.4**. This includes (a) the description of clinical pretreatment analysis, (b) the description of chemotherapy preparation and administration, (c) the description of clinical posttreatment analysis, and (d) the description of molecular posttreatment analysis.

*3.3.1. Pretreatment Analysis*

This section will cover broad aspects of pretreatment analysis, but for alternative chemotherapy regimens, the investigator should modify the parameters tailored specifically to the drug of choice (i.e., methotrexate-induced mucositis). Prior to treatment, complete blood cell counts (CBC) are obtained with a differential including total white blood cells, neutrophils, lymphocytes, eosinophils, basophils, hematocrit, and platelets. A basic chemistry panel is taken the day of treatment with a variety of parameters, but the most important for the majority of drugs used for in vivo selection are liver-specific values including alanine aminotransferase (ALT), aspartate aminotransferase (AST), alkaline phosphatase (ALP), and bilirubin levels. Importantly, chemistry panels should include indicators of proper kidney function (i.e., blood urea nitrogen [BUN] and creatinine) and pancreatic function (i.e., amylase and lipase). General clinical evaluations of the animals (i.e., weight, hydration, flushed or pale appearance, malaise, or any indication of jaundice) should also be performed prior to treatment.

*3.3.2. Chemotherapy Preparation and Administration*

There are a variety of agents that can be administered in an attempt to select gene-modified cells in canine's posttransplantation, but we will keep the detailed experiments here confined to the drugs most commonly used in the canine, namely, $O^6$-benzylguanine ($O^6BG$), temozolomide (TMZ), and 1,3-bis(2-chloroethyl)-1-nitrosourea (BCNU) for in vivo selection of methylguanine methyltransferase (MGMT) gene-modified cells.

Administration of $O^6$-benzylguanine, Temozolomide, and BCNU

A standard dose of $O^6BG$/TMZ or BCNU with which to begin in vivo selection is 110 mg/m² $O^6BG$ (~5 mg/kg but never more than 50 mg) and 500 mg/m² TMZ or 4.5 mg/m² (~0.2 mg/kg) BCNU. In an unprotected animal or an animal with <3% MGMT(P140 K) gene marking in peripheral blood cells, this would cause marked and pronounced cytopenia, but with standard care the animal should recover normal blood counts in several weeks. Canines are fasted the night and morning prior to treatment with TMZ so that the stomach contents do not affect absorption of the drug. Also, canines receive a ¼ tablet of Zofran ~30 min prior to receiving TMZ to minimize treatment related nausea.

1. Add 50 mg vial of $O^6BG$ (Sigma) to a 50 mL conical tube with 18 mL PBS pH: 7.0 (room temperature) and 12 mL polyethylene glycol 400.

2. Add 1 mL PBS/PEG solution to 1 vial of $O^6BG$ and transfer back to the 50 mL conical tube to transfer residual $O^6BG$ and repeat as necessary until completely transferred.

3. Bring final volume to 50 mL with 1× PBS pH: 7.0 (Final concentration 1 mg/mL $O^6BG$).

4. Float the conical tube in a sonication water bath (Branson 3,510) heated to 37°C–42°C for 30–40 min (*see* **Note 11**).

5. During sonication remove 150 mL from a 250 mL 0.9% saline bag and prewarm the bag at 37°C.

6. Add the appropriate amount of the $O^6BG$ solution based on the weight of the canine to the prewarmed saline bag. Keep solution warmed until administration.

7. Infuse the $O^6BG$ intravenously over 15–20 min (flow rate ~600 mL/h).

   If combining the $O^6BG$ treatment with TMZ (Schering-Plough, Madison, NJ), capsules are administered orally within 5 min after completion of the $O^6BG$ infusion. In addition to antiemetics, canines receive a small amount of wet food (<2 ounces) immediately after TMZ delivery to alleviate nausea. Animals are put back on food ~4 h after the TMZ capsules are given. Generally, TMZ is well tolerated by the animals with very few cases of vomiting. If BCNU is given instead of TMZ, 30–45 min should pass between the end of the $O^6BG$ infusion and administration of BCNU. BCNU is well tolerated in the canine, but canines are about three times more sensitive to BCNU relative to humans (personal communication and observations).

1. Prepare BCNU by resuspending the lyophilized powder in 100% ethanol to a final concentration of 33.0 µg/µL.

2. Slowly draw the appropriate amount of BCNU solution into a 20 mL syringe containing 18 mL of 0.9% saline (*see* **Note 12**).

3. Wrap the syringe in aluminum foil and administer within 30 min.

4. Administer BCNU over 3–4 min intravenously followed by ~10 mL of PBS flush.

*3.3.3. Posttreatment Clinical Analysis*

Animals are regularly examined by a board-certified veterinarian and weighed at least once per week in the 4-week period after drug treatment. Repeated liver, kidney, and pancreas function tests are carried out and compared with pretreatment levels. Treatment-associated abnormalities associated with $O^6BG$ and TMZ or BCNU is transient transaminitis. All animals should maintain their weight within 10% of baseline following drug administrations. The abnormalities should resolve quickly (<2 weeks) with no overt sustained issues.

Cytopenia is the most common and well-described side effect of TMZ and BCNU treatment and, in general, neutrophil nadir is seen 7–10 days after treatment followed by platelet nadir at 12–15 days after treatment. If the absolute neutrophil count (ANC) drops below $1,000/\mu L$, the animals receive baytril (2.2 mg/kg orally, twice a day), and if the neutropenia becomes more severe (ANC $<500/\mu L$), animals discontinue baytril and instead receive amikacin (20 mg/kg intravenously, once a day) and ceftazidime (37.5 mg/kg intravenously, twice a day). Transfusion support is given if platelets are $<10,000/\mu L$, pronounced petechiae, or other clinical evidence of bleeding (i.e., bloody stool).

*3.3.4. Posttreatment Gene Marking Analysis*

Gene marking can be monitored after chemotherapy treatment by flow cytometry as described in **Subheading 3.2.2** or by Taqman PCR as described in **Subheading 3.2.3**. In general for 1–2 weeks following chemotherapy treatment, gene marking values will rise markedly before stabilizing ~4–6 weeks after treatment. It is believed that this is due to the dual toxicity of the chemotherapy agents described here to more differentiated hematopoietic cells as well as hematopoietic stem cells, and the "true" increase in gene marking as a result of gene-modified stem cell selection is the difference in gene marking before treatment and the level of gene marking after ~4–6 weeks following treatment.

## 4. Notes

1. Prolonged leukapheresis (~4 h), not bone marrow aspiration, is suggested for AMD3100 due to the mobilization kinetics.

2. Blood clots are not uncommon during red blood cell lysis and are disrupted as well as possible with a 5 mL pipet.

3. In general, canine cells aggregate more readily compared with human or nonhuman primate cells, using DNase in the buffer aids in breaking up some of the larger cell aggregates.

4. Titering of anti-canine CD34$^+$ antibody (1H6) is done on normal unmobilized canine bone marrow using standard antibody staining procedures, and the concentration of antibody that yields optimum staining of canine CD34$^+$ cells as determined by maximum mean fluorescence intensity (MFI)

using flow cytometry is the concentration used for transplantation studies (usually concentrations range between 0.1 and 0.6 μL for every $1 \times 10^6$ cells from 1 mg/mL antibody stock).

5. All buffer washes and addition of cells to Miltenyi LS columns are done when >90% of the solution has flowed through the column but prior to the column running dry.

6. 1H6 is added during the initial incubation even to the CD34-selected population because this leads to improved staining. Without the addition of 1H6 during this step and proceeding directly to the streptavidin-PE step results in dimmer staining.

7. If serum-free conditions are desired, StemSpan SFEM gives comparable results to both Human Dexter and IMDM 10/1 FBS. The prestimulation and transduction can be carried out in StemSpan SFEM supplemented with growth factors. Importantly, if you choose to use StemSpan SFEM, then the retrovirus stocks need to be resuspended and stored in StemSpan SFEM and not IMDM, which does not affect titers.

8. Unconcentrated gammaretrovirus stocks can be prepared fresh from producing lines or frozen stocks made prior to transduction. The media that the unconcentrated gammaretrovirus stocks are prepared in cannot be removed, so this should be considered when preparing virus stocks. Test media conditions that yield high titer stocks and also allow for efficient transduction. Concentrated growth factor stocks can be added to the media for transduction along with other small molecules (i.e., polybrene and protamine sulfate) that are not required during virus preparation.

9. Generally, the MOI for gammaretrovirus transductions using an RD114-pseudotyped vector is between 0.5 and 2 and VCM does not exceed 20% (≤2 mL) of the total media volume on the cells.

10. Generally, the MOI for lentivirus (VSV-G) and foamy virus (standard envelope) transductions is between 5–20 and 3–8, respectively, and VCM does not exceed 20% (≤2 mL) of the total media volume on the cells.

11. $O^6$-benzylguanine never completely dissolves in the PBS/PEG solution, but upon addition of the dilute solution to the saline bag, no detectable particles persist.

12. The BCNU/ethanol solution has a tendency to precipitate if drawn up too quickly into the syringe.

## Acknowledgments

We thank Michele Spector, DVM, and the technicians and staff in the canine facilities of the Fred Hutchinson Cancer Research Center. We also acknowledge the assistance of Bonnie Larson and Helen Crawford in preparing the manuscript. This work was supported in part by grants HL74162, HL36444, DK56465, and DK47754 from the National Institutes of Health, Bethesda, MD. HPK is a Markey Molecular Medicine Investigator

## References

1. Neff, T., Beard, B. C. and Kiem, H. -P. (2006) Survival of the fittest: in vivo selection and stem cell gene therapy (Review). *Blood* 107, 1751–1760.

2. Kim, D. H., Rossi, J. J. (2007) Strategies for silencing human disease using RNA interference. *Nat Rev Genet* 8, 173–184.

3. Nathwani, A. C., Davidoff, A. M. and Linch, D. C. (2005) A review of gene therapy for haematological disorders (Review). *Br. J. Haematol* 128, 3–17.

4. Persons, D. A., Tisdale, J. F. (2004) Gene therapy for the hemoglobin disorders (Review). *Semin. Hematol.* 41, 279–286.

5. Horn, P. A., Keyser, K. A., Peterson, L. J., Neff, T., Thomasson, B. M., Thompson, J. and Kiem, H. -P. (2004) Efficient lentiviral gene transfer to canine repopulating cells using an overnight transduction protocol. *Blood* 103, 3710–3716.

6. Wu, T., Kim, H. J., Sellers, S. E., Meade, K. E., Agricola, B. A., Metzger, M. E., Kato, I., Donahue, R. E., Dunbar, C. E. and Tisdale, J. F. (2000) Prolonged high-level detection of retrovirally marked hematopoietic cells in nonhuman primates after transduction of CD34+ progenitors using clinically feasible methods. *Mol Ther* 1, 285–293.

7. Dunbar, C. E., Seidel, N. E., Doren, S., Sellers, S., Cline, A. P., Metzger, M. E., Agricola, B. A., Donahue, R. E. and Bodine, D. M. (1996) Improved retroviral gene transfer into murine and rhesus peripheral blood or bone marrow repopulating cells primed in vivo with stem cell factor and granulocyte colony-stimulating factor. *Proc Natl Acad Sci USA* 93, 11871–11876.

8. Neff, T., Peterson, L. J., Morris, J. C., Thompson, J., Zhang, X., Horn, P. A., Thomasson, B. M. and Kiem, H. -P. (2004) Efficient gene transfer to hematopoietic repopulating cells using concentrated RD114-pseudotype vectors produced by human packaging cells (Letter to the Editor). *Mol Ther* 9, 157–159.

9. Licht, T., Haskins, M., Henthorn, P., Kleiman, S. E., Bodine, D. M., Whitwam, T., Puck, J. M., Gottesman, M. M. and Melniczek, J. R. (2002) Drug selection with paclitaxel restores expression of linked IL-2 receptor gamma-chain and multidrug resistance (MDR1) transgenes in canine bone marrow. *Proc Natl Acad Sci USA* 99, 3123–3128.

10. Neff, T., Horn, P. A., Peterson, L. J., Thomasson, B. M., Thompson, J., Williams, D. A., Schmidt, M., Georges, G. E., von Kalle, C. and Kiem, H. -P. (2003) Methylguanine methyltransferase-mediated in vivo selection and chemoprotection of allogeneic stem cells in a large-animal model. *J Clin Invest* 112, 1581–1588.

11. Neff, T., Beard, B. C., Peterson, L. J., Anandakumar, P., Thompson, J. and Kiem, H. -P. (2005) Polyclonal chemoprotection against temozolomide in a large-animal model of drug resistance gene therapy. *Blood* 105, 997–1002.

12. Neff, T., Horn, P. A., Valli, V. E., Gown, A. M., Wardwell, S., Wood, B. L., von Kalle, C., Schmidt, M., Peterson, L. J., Morris, J. C., Richard, R. E., Clackson, T., Kiem, H. -P. and Blau, C. A. (2002) Pharmacologically regulated in vivo selection in a large animal. *Blood* 100, 2026–2031.

13. Whitwam, T., Haskins, M. E., Henthorn, P. S., Kraszewski, J. N., Kleiman, S. E., Seidel, N. E., Bodine, D. M. and Puck, J. M. (1998) Retroviral marking of canine bone marrow: long-term, high-level expression of human interleukin-2 receptor common gamma chain in canine lymphocytes. *Blood* 92, 1565–1575.

14. Trobridge, G., Josephson, N., Vassilopoulos, G., Mac, J. and Russell, D. W. (2002)

Improved foamy virus vectors with minimal viral sequences. *Mol Ther* 6, 321–328.

15. Dull, T., Zufferey, R., Kelly, M., Mandel, R. J., Nguyen, M., Trono, D. and Naldini, L. (1998) A third-generation lentivirus vector with a conditional packaging system. *J Virol* 72, 8463–8471.

16. Schambach, A., Bohne, J., Chandra, S., Will, E., Margison, G. P., Williams, D. A. and Baum, C. (2006) Equal potency of gammaretroviral and lentiviral SIN vectors for expression of O6-methylguanine-DNA methyltransferase in hematopoietic cells. *Mol Ther* 13, 391–400.

17. Bauer, T. R., Jr., Hai, M., Tuschong, L. M., Burkholder, T. H., Gu, Y. C., Sokolic, R. A., Ferguson, C., Dunbar, C. E. and Hickstein, D. D. (2006) Correction of the disease phenotype in canine leukocyte adhesion deficiency using ex vivo hematopoietic stem cell gene therapy. *Blood* 108, 3313–3320.

18. Deeg, H. J., Schuler, U. S., Shulman, H., Ehrsam, M., Renner, U., Yu, C., Storb, R.

and Ehninger, G. (1999) Myeloablation by intravenous busulfan and hematopoietic reconstitution with autologous marrow in a canine model. *Biol Blood Marrow Transplant* 5, 316–321.

19. Sokolic, R. A., Bauer, T. R., Gu, Y. C., Hai, M., Tuschong, L. M., Burkholder, T., Colenda, L., Bacher, J., Starost, M. F. and Hickstein, D. D. (2005) Nonmyeloablative conditioning with busulfan before matched littermate bone marrow transplantation results in reversal of the disease phenotype in canine leukocyte adhesion deficiency. *Biol Blood Marrow Transplant* 11, 755–763.

20. Ting-De Ravin, S. S., Kennedy, D. R., Naumann, N., Kennedy, J. S., Choi, U., Hartnett, B. J., Linton, G. F., Whiting-Theobald, N. L., Moore, P. F., Vernau, W., Malech, H. L. and Felsburg, P. J. (2006) Correction of canine X-linked severe combined immunodeficiency by in vivo retroviral gene therapy. *Blood* 107, 3091–3097.

# Chapter 24

## Detection of Retroviral Integration Sites by Linear Amplification-Mediated PCR and Tracking of Individual Integration Clones in Different Samples

**Manfred Schmidt, Kerstin Schwarzwaelder, Cynthia C. Bartholomae, Hanno Glimm, and Christof von Kalle**

## Summary

In order to restore or to introduce a gene function integrating viral vector systems are used to genetically modify hematopoietic stem cells. The occurrence of immortalized cell clones after transduction in vitro (Blood 106:3932–3939, 2005) and clonal dominance as well as leukemia in preclinical (Nat. Med. 12:401–409, 2006; Blood 106:2530–2533, 2005; Science 308:1171–1174, 2005; Science 296:497, 2002; Blood 107:3865–3867, 2006) and clinical (Nat. Med. 12:401–409, 2006; Science 302:415–419, 2003; J. Clin. Invest. 118:3143–3150, 2008) gene therapy trials revealed that the nondirected integration of a vector may be associated with serious side effects. By means of the linear amplification-mediated PCR (LAM-PCR) (Blood 100:2737–2743, 2002; Nat. Methods 4:1051–1057, 2007) it is possible to identify miscellaneous vector-genome junctions in one sample, each unique for one integration clone down to the single cell level. Thus this method allows to determine the clonality of a genetically modified hematopoietic repopulation as well as to sequence the vector integration sites and therefore to analyze the integration site distribution and the influence of the vector integration site on the cell fate. The recognition of the integration site sequence corresponding to a specific clone allows the tracking of an individual clone in various samples.

**Key words:** Gene therapy, Genetically modified hematopoietic stem cell, LAM-PCR, Hematopoietic repopulation, Clonality, Retroviral integration site, Integration site distribution, Side effect, Tracking.

## 1. Introduction

The use of integrating vector systems in gene therapy allows the long-term expression of a transgene in a target cell. Concomitantly the integration site represents a molecular marker unique for each integration clone *(1, 2)*. Since the transduction efficiencies increased

Christopher Baum (ed.), *Methods in Molecular Biology, Methods and Protocols, vol. 506*
© Humana Press, a part of Springer Science + Business Media, LLC 2009
DOI: 10.1007/978-1-59745-409-4_24

the immortalization of transduced cell culture cells occurred *(3)*, and clonal dominance as well as leukemia were described as side effects which were mainly caused by the integrated vector in preclinical *(4–7)* and clinical *(8, 9, 10)* gene therapy trials. To know the integration preference of different vectors and to understand the biological consequence of the integration at certain genomic loci it is necessary to detect and to characterize the vector–genome junctions derived from the integration clones contained in the analyzed cell population. The linear amplification-mediated polymerase chain reaction (LAM-PCR) allows the simultaneous detection of vector–genome junctions derived from different integration clones inclosed in minimal samples *(11, 12)*. The first step is the preamplification of the vector–genome junctions through a linear PCR with biotinylated primers hybridizing at one end of the integrated vector. The following steps are carried out on a semisolid streptavidin phase in order to capture DNA strands with an incorporated biotinylated vector primer. After the synthesis of double strands, a restriction digest, the ligation of a linker cassette on the genomic end of the fragment, and the denaturation of the unbiotinylated strands from the biotinylated strands, two exponential PCRs with nested arranged vector- and linker cassette primers are carried out in order to amplify the fragments consisting of linker cassette-, genomic-, and vector-sequence. The generated fragments are sequenced and the integration locus determined. The knowledge of the genomic flank sequence composition derived from a clone of interest allows the creation of primers specific for an individual integration clone. PCR using these primers in combination with vector-specific primers enables the *in vivo* monitoring of the specific clone in different samples. Using these techniques we could analyze the integration site distribution *in vitro* in cell culture cells *(13)* and *in vivo* in preclinical *(4, 6, 7)* and clinical *(8, 9, 14, 15)* gene therapy trials. We could also show that the integrated vector may subtly influence the cell fate in clinical gene therapy trials *(14, 15)* and may also lead to clonal dominance as well as leukemia in preclinical *(4, 6, 7)* as well as clinical *(8, 9, 10)* gene therapy trials.

## 2. Materials

### 2.1. Linear PCR

1. Taq DNA Polymerase (Qiagen, Hilden, Germany).
2. dNTPs (Fermentas, St. Leon-Roth, Germany).
3. Aqua ad iniectabilia (Boehringer, Ingelheim, Germany).
4. Human genomic DNA (Roche Diagnostics, Mannheim, Germany).
5. Primer (MWG Biotech, Ebersberg, Germany):

5′ Biotin-modificated primers are marked with (B)

LTR1 (B)5′ > AGCTGTTCCATCTGTTCTTGGCCCT < 3′

**2.2. Magnetic Capture**

1. Magnetic particles: Dynabeads M-280 Streptavidin (Dynal, Oslo, Norway).
2. Magnetic separation unit MPC-E-1 (Dynal, Oslo, Norway).
3. PBS (pH 7.5)/0.1% BSA.
4. Kilobase binder kit (Dynal, Oslo, Norway).
5. Aqua ad iniectabilia (Boehringer, Ingelheim, Germany).

**2.3. Double-Strand Synthesis (Hexanucle-otide Priming)**

1. Klenow Polymerase (Roche Diagnostics, Mannheim, Germany).
2. Hexanucleotide mixture (Roche Diagnostics, Mannheim, Germany).
3. dNTPs (Fermentas, St. Leon-Roth, Germany).
4. Aqua ad iniectabilia (Boehringer, Ingelheim, Germany).

**2.4. Restriction Digest**

1. Enzyme Tsp509 I (New England Biolabs, Frankfurt am Main, Germany).
2. Aqua ad iniectabilia (Boehringer, Ingelheim, Germany).

**2.5. Construction of the Linker Cassette**

1. 250 mM Tris–HCl, pH 7.5.
2. 100 µM MgCl$_2$.
3. Oligonucleotides (MWG Biotech, Ebersberg, Germany): *Oligonucleotide I:*

   5′ > GACCCGGGAGATCTGAATTCAGTGGCACAGCAGT-TAGG < 3′ *Oligonucleotide II:*

   5′ > AATTCCTAACTGCTGTGCCACTGAATTCAGATC < 3′
4. Microcon-30 (Millipore, Bedford, USA).
5. Aqua ad iniectabilia (Boehringer, Ingelheim, Germany).

**2.6. Ligation of the Linker Cassette**

1. Fast-Link DNA Ligation Kit (Epicentre Biotechnologies, Madison, USA).
2. Aqua ad iniectabilia (Boehringer, Ingelheim, Germany).

**2.7. Denaturation**

1. 0.1 N NaOH.

**2.8. Exponential PCRs**

1. Taq DNA Polymerase (Qiagen, Hilden, Germany).
2. dNTPs (Fermentas, St. Leon-Roth, Germany).
3. Aqua ad iniectabilia (Boehringer, Ingelheim, Germany).
4. Primers (MWG Biotech, Ebersberg, Germany):

   5′ Biotin-modificated primers are marked with (B)

*First exponential PCR*

LTR2 (B)5′ > GACCTTGATCTGAACTTCTC < 3′

LC1 5′> GACCCGGGAGATCTGAATTC < 3′

*Second exponential PCR*

LTR3 5′> TCCATGCCTTGCAAAATGGC < 3′

LC2 5′> GATCTGAATTCAGTGGCACAG < 3′

***2.9. Separation of the LAM-PCR Product on a Spreadex High-Resolution Gel***

1. Spreadex high-resolution gel (Elchrom Scientific, Cham, Switzerland).

2. Submerged gel electrophoresis apparatus SAE 2000 (Elchrom Scientific, Cham, Switzerland).

3. TAE buffer (40×, Elchrom Scientific, Cham, Switzerland).

4. Blue run loading buffer (5×): 25 mM Tris–HCl, pH 7.0; 150 mM ethylenediaminetetraacid (EDTA), pH 8.0; 0.05% bromophenol blue, 25% glycerol.

5. Ethidium bromide (Applichem, Darmstadt, Germany).

6. 100-bp ladder (Invitrogen, Carlsbad, USA).

***2.10. SemiQuantitative Analysis of an Individual Integration Clone***

1. Taq DNA Polymerase (Qiagen, Hilden, Germany).

2. dNTPs.

3. Aqua ad iniectabilia (Boehringer, Ingelheim, Germany).

4. Primers.

   5′ Biotin-modified primers are marked with (B):

   LTR4 (B)5′ > CCTTGCAAAATGGCGTTACT < 3′

   LTR5 5′ > CAAACCTACAGGTGGGGTCT < 3′

5. Human genomic DNA (Roche Diagnostics, Mannheim, Germany).

6. Agarose LE (Roche Diagnostics, Mannheim, Germany).

7. TBE buffer (10×, Amresco, Solon, USA).

8. Blue run loading buffer (5×): 25 mM Tris–HCl, pH 7.0; 150 mM ethylenediaminetetraacid (EDTA), pH 8.0; 0.05% bromophenol blue, 25% glycerol.

9. Ethidium bromide (Applichem, Darmstadt, Germany).

10. 100-bp ladder (Invitrogen, Carlsbad, USA).

# 3. Methods

This section describes (1) the preamplification of retroviral vector–genome junctions derived from LN vectors based on the murine leukemia virus (MLV) via linear polymerase chain reaction (PCR) using a biotinylated LTR primer, (2) the magnetic capture of the biotinylated PCR products, (3) the synthesis of DNA double strands, (4) the restriction digest of the genomic region flanking the vector integration site, (5) the construction of the linker cassette, (6) the ligation of the linker cassette to the genomic flank, (7) the denaturation of the particle bound DNA, (8) the amplification of the retroviral LN vector-derived vector genome junctions through exponential PCR, (9) the separation of the LAM-PCR product on a high-resolvent gel, and (10) the semiquantitative analysis of an individual integration clone using LTR primers in combination with clone-specific primers hybridizing at the genomic flank.

### 3.1. Linear PCR

The first step of the LAM-PCR is a linear PCR accomplished with a 5′-biotinylated (B) primer hybridizing at the 5′-end of the LN vector. The primer sequence is given in **Subheading 2.1**.

1. Insert 0.01 ng to 1 µg of transduced DNA together with 1× concentrated PCR buffer, 200 µM dNTPs each, 5-nM primer, and 2.5 U Taq polymerase into the PCR and fill the reaction up with distilled water to a final volume of 50 µl. Use the same amount of untransduced DNA as a negative control. Carry out the initial denaturation for 5 min at 95°C followed by 50 PCR cycles composed of the denaturation for 1 min at 95°C, the annealing for 45 s at 60°C, and the extension for 90 s at 72°C. Conduct the final extension for 10 min at 72°C. After the PCR is completed add the other 2.5 U Taq polymerase to each reaction and repeat the 50-cycle PCR.

### 3.2. Magnetic Capture

1. Expose 20 µl of the magnetic particles (10 µg/µl) to a magnetic field for 60 s and discard the supernatant in the presence of the magnetic field (*see* **Note 1**).

2. Resuspend the particles in 40 µl PBS/0.1% BSA (pH 7.5) and discard the supernatant in the presence of the magnetic field. Repeat this step once.

3. Wash the particles once in 20 µl Binding Solution and resuspend them in 50 µl Binding Solution.

4. Incubate each product of the linear PCR with 50 µl of the magnetic particle solution (*see* **Note 2**) overnight at room temperature (*see* **Note 3**) on a shaker at g-force value: 0,0255.

5. Expose the sample for 60 s to a magnetic field, discard the supernatant in the presence of the magnetic field, and wash the beads once in 100 µl distilled water.

### 3.3. Double-Strand Synthesis (Hexanucleotide Priming)

1. Blend 1× concentrated hexanucleotide mixture, 200 µM dNTPs each, 2 U Klenow-Polymerase and fill the mixture up with distilled water to a final volume of 20 µl.

2. Resuspend the particles with 20 µl of the reaction mixture.

3. Incubate the sample for 1 h at 37°C.

4. Add 80 μl of distilled water and expose the sample to a magnetic field for 60 s.

5. Discard the supernatant in the presence of the magnetic field and wash the particles once with 100 μl of distilled water.

### 3.4. Restriction Digest

1. Blend 1× concentrated restriction buffer, 4 U of the enzyme Tsp509 I and fill the reaction up with distilled water to a final volume of 20 μl.

2. Incubate the DNA–particle complex with 20 μl of the restriction mixture for 1 h at 65°C (*see* **Note 4**).

3. Add 80 μl of distilled water to the reaction, expose the sample for 60 s to a magnetic field, discard the supernatant in the presence of the magnetic field, and wash the DNA–particle complex once with 100 μl of distilled water.

### 3.5. Construction of the Linker Cassette

1. Blend 40 μl of the oligonucleotide I (100 pmol/μl); 40 μl of the oligonucleotide II (100 pmol/μl); 110 μl 250 mM Tris–HCl, pH 7.5; 10 μl 100 mM $MgCl_2$ and incubate the reaction for 5 min at 95°C in a heat block. Switch off the heat block and let the sample cool down slowly overnight within the heat block. The sequences of oligonucleotide I and oligonucleotide II are given in **Subheading 2.5**.

2. Add 300 μl of distilled water and put the sample on a Microcon-30 column.

3. Centrifuge the sample for 12 min at room temperature and $14,000 \times g$.

4. Place the column reversed onto a fresh tube and centrifuge the sample for 3 min at room temperature and $1,000 \times g$.

5. Fill the concentrated sample up with distilled water to a final volume of 80 μl.

6. Aliquot the sample and freeze it at –20°C (*see* **Note 5**).

### 3.6. Ligation of the Linker Cassette

1. Blend 2 μl of the linker cassette, 1× concentrated fast link ligation buffer, 10 mM ATP, 2 U Fast-Link Ligase, and fill the reaction up to 10 μl with distilled water.

2. Add 10 μl of the ligation mixture to the DNA–particle complex and incubate the reaction for 5 min at room temperature.

3. Add 90 μl of distilled water, expose the sample for 60 s to a magnetic field, and discard the supernatant in the presence of the magnetic field.

4. Wash the DNA–particle complex once with 100 μl of distilled water.

**3.7. Denaturation**

1. Incubate the DNA–particle complex with 5 μl of 0.1 N NaOH for 10 min at room temperature and at g-force value: 0,0255 on a shaker.

2. Expose the sample for 60 s to a magnetic field and collect the supernatant which contains the ssDNA in the presence of the magnetic field.

**3.8. Exponential PCRs**

Primer sequences for the first and second exponential PCR are listed in **Subheading 2.8**.

1. Use 2 μl of the denaturation product as template for the first exponential PCR and 0.1–5% of the first exponential PCR product as template for the second exponential PCR. For each exponential PCR blend 1× concentrated PCR Buffer, 200 μM dNTPs each, 0.5 μM of each primer, 5 U Taq polymerase and fill the reaction up with distilled water to a final volume of 50 μl.

2. Carry out an initial denaturation of 95°C for 5 min followed by 35 cycles composed of 95°C for 1 min and 30 s, 60°C for 45 s, and 72°C for 1 min. As final extension choose 72°C for 10 min.

**3.9. Separation of the LAM-PCR product on a Spreadex High-Resolution Gel**

1. Fix a Spreadex gel within the electrophoresis tank using an appropriate catamaran and fill the electrophoresis tank with 1.9 L of 1× concentrated TAE buffer.

2. Load 10 μl of each LAM-PCR product with 2 μl of 5× concentrated blue run loading buffer. Include 1 well for a molecular weight marker.

3. Let the gel run at 10-V/cm electrode gap.

4. After 5 min switch on the buffer pump.

5. When the dye front reached the lower edge of the gel, detach the gel from the plastic matrix using a nylon fiber and stain the gel for 20 min in ethidium bromide solution (0.5 μg ethidium bromide/ml distilled water) on a shaker at g-force value: 0,0007 and room temperature.

**3.10. SemiQuantitative Analysis of an Individual Integration Clone**

The analysis of an individual integration clone in different samples is possible by using primers hybridizing at the vector LTR in combination with primers hybridizing at the genomic flank of the clone of interest. Primer sequences are given in **Subheading 2.10**.

1. Design 2 nested arranged primers hybridizing at the genomic flank of the clone of interest with a GC content not higher than 60%.

2. Use 10–50 ng of sample DNA as PCR template and blend 1× concentrated PCR buffer, 200 µM dNTPs each, 0.5 µM each of the primer LTR4 and the outer genomic flank primer, 5 U Taq polymerase, and fill the reaction up to 50 µl with distilled water. Use the same amount of untransduced DNA as a negative control. Carry out the initial denaturation at 95°C for 5 min and let the PCR run for 35 cycles consisting of 95°C for 1 min, 54–60°C for 45 s, and 72°C for 1 min. Accomplish a final extension for 10 min at 72°C.

3. Insert 2% of the first PCR product into a second PCR (*see* **Note 6**) using the inner genomic flank primer in combination with primer LTR5. Use the amplification conditions described earlier.

4. Prepare a 2% agarose gel using 2 g agarose per 100 ml 1× concentrated TBE buffer.

5. Boil the mixture until the agarose is completely dissolved.

6. Let the gel cool down for 5 min at room temperature and add 0.5 µg/ml ethidium bromide (*see* **Note 7**).

7. Cast the gel in a sledge in which you have inserted a comb (*see* **Note 8**).

8. Once the gel has polymerized remove the comb and transfer the sledge into an electrophoresis tank.

9. Fill the tank with 1× concentrated TBE buffer until the gel is completely covered.

10. Load 10 µl of the PCR product with 2 µl of 5× concentrated blue run loading buffer in each well. Include 1 well for a molecular weight marker.

11. Connect the gel unit to a power supply and run the gel at 10 V/cm electrode gap until the dye front runs 7 cm below the upper edge of the gel.

12. Apply the gel on a gel documentation system.

## 4. Notes

1. The magnetic beads may not become dry or frozen since this could result in reduced performance.

2. The ratio of PCR product and binding solution must always be 1:1.

3. Carry out the capture step for at least 8 h.

4. The restriction time may be extended to several hours.

5. After thawing one aliquot does not refreeze the linker cassette.

6. After the first exponential tracking PCR you may insert a magnetic capture purification step. To accomplish this purification it is necessary that the vector primer carries a biotin modification at the 5′-end. The incubation time should be carried out not shorter than 1 h and the denaturation should be carried out with 10 μl NaOH. Introduce 1 μl of the denaturation product into the second PCR. Optionally the product of the first PCR may be diluted 1:20 with distilled water and 1 μl of the dilution may be inserted into the second PCR.

7. Since ethidium bromide is toxic we accomplish the insertion of the ethidium bromide beneath a flue.

8. Cast the gel carefully into the sledge since air bubbles are undesirable.

# References

1. Scherdin U., Rohdes K., Breindl M. (1990) Transcriptionally active genome regions are preferred targets for retrovirus integration. *J. Virol.* 64, 907–912.

2. Collas P., Husebye H., Alestrom P. (1996) The nuclear localization sequence of the sv40 antigene promotes transgene uptake and expression in zebrafish embryo nuclei. *Transgenic Res.* 5, 451–458.

3. Du Y., Jenkins N. A., Copeland N. G. (2005) Insertional mutagenesis indentifies genes that promote the immortalization of primary bone marrow progenitor cells. *Blood* 106, 3932–3939.

4. Calmels B., Ferguson C., Laukkanen M. O., Adler R., Faulhaber M., Kim H. J., et al.(2005) Recurrent retroviral vector integration at the Mds1/Evi1 locus in nonhuman primate hematopoietic cells. *Blood* 106, 2530–2533.

5. Kustikova O., Fehse B., Modlich U., Yang M., Düllmann J., Kamino K., et al. (2005) Clonal dominance of hematopoietic stem cells triggered by retroviral gene marking. *Science* 308, 1171–1174.

6. Li Z., Düllmann J., Schiedlmeier B., Schmidt M., von Kalle C., Meyer J., et al. (2002) Murine leukemia induced by retroviral gene marking. *Science* 296, 497.

7. Seggewiss R., Pittaluga S., Adler R. L., Guenaga F. J., Ferguson C., Pilz I. H., et al. (2006) Acute myeloid leukemia is associated with retroviral gene transfer to hematopoietic progenitor cells in a rhesus macaque. *Blood* 107, 3865–3867.

8. Ott M., Schmidt M., Schwarzwaelder K., Stein S., Siler U., Koehl U., Glimm H., et al. (2006) Correction of X-linked chronic granulomatous disease by gene therapy, augmented by insertional activation of *MDS1-EVI1, PRDM16* or *SETBP1. Nat. Med.* 12, 401–409.

9. Hacein-Bey-Abina S., von Kalle C, Schmidt M, McCormack M. P, Wulffraat N, Leboulch P, et-al., (2003) LMO2-associated clonal T cell proliferation in two patients after gene therapy for SCID-X1. *Science* 302, 415–419.

10. Howe S. J., Mansour M. R., Schwarzwaelder K., Bartholomae C. C., Hubank M., Kempski H., et al. (2008) Insertional mutagenesis combined with acquired somatic mutations causes leukemogenesis following gene therapy of SCID-X1 patients. *J. Clin. Invest.* 118, 3143–3150.

11. Schmidt M., Zickler P., Hoffmann G., Haas S., Wissler M., Muessig A., et al. (2002) Polyclonal long-term repopulating stem cell clones in a primate model. *Blood* 100, 2737–2743.

12. Schmidt M., Schwarzwaelder K., Bartholomae C. C., Zaoui K., Ball C., Pilz I., et al. (2007) High-resolution insertion-site analysis by linear amplification-mediated PCR (LAM-PCR). *Nat. Methods* 4, 1051–1057.

13. Bartholomae C. C., Arens A., Balaggan K. S., Yáñez-Muñoz R. J., Montini E., Howe S. J., et al. Specific Lentiviral Vector Integration Profiles in Rodent Postmitotic Tissues. In submission.

14. Deichmann A., Hacein-Bey-Abina S., Schmidt M., Garrigue A., Brugman M., Hu J., et al. (2007) Vector integration is nonrandom and clustered and influences the fate of lymphopoiesis in SCID-X1 gene therapy. *J. Clin. Invest.* 117, 2225–2232.

15. Schwarzwaelder K., Howe S. J., Schmidt M., Brugman M., Deichmann A., Glimm H., et al. (2007) Gammaretrovirus-mediated correction of SCID-X1 is associated with skewed vector integration site distribution in vivo. *J. Clin. Invest.* 117, 2241–2239.

# Chapter 25

# Retroviral Insertion Site Analysis in Dominant Haematopoietic Clones

## Olga S. Kustikova, Ute Modlich, and Boris Fehse

## Summary

Identification of retroviral vector insertion sites in single, dominating cell clones has become an important tool for the investigation of cellular signalling pathways involved in clonal expansion and malignant transformation. Also, recent severe adverse events in clinical trials resulting from retroviral vector-mediated insertional mutagenesis underline the need of well-designed safety studies including integration site analyses to estimate cost/benefit ratios in gene therapy. We have recently described a modified ligation-mediated PCR (LM PCR) method allowing preferential retrieval of insertion sites causally linked to clonal dominance of an affected clone. In the first part of the given work we focus on particularities of the LM PCR procedure to be taken into account when working with self-inactivating as compared to 'classical' retrovectors. In the following sections we focus on data acquisition, processing, organisation, and analysis. Thus the protocol presented here should be helpful in establishing and utilising databases of retroviral integration sites.

**Key words:** Gene transfer, Retroviral vector, Insertional mutagenesis, Clonal dominance, LM PCR, Retroviral insertional sites, Common insertion sites.

## 1. Introduction

Based on current technologies stable gene transfer is best achieved using integrating, virus-derived vectors. Retroviral (including lentiviral) vectors are most widely used since they integrate into the host genome by the means of an enzyme-mediated process not causing significant genetic changes (other than the insertion itself) at the integration site. However, since insertions occur in an unpredictable, semi-random fashion, retroviral vector integration (RVI) may result in deregulated expression of genes at the

Christopher Baum (ed.), *Methods in Molecular Biology, Methods and Protocols, vol. 506*
© Humana Press, a part of Springer Science+Business Media, LLC 2009
DOI: 10.1007/978-1-59745-409-4_25

insertion site, e.g. due to the disruption of regulatory or coding sequences. This process termed insertional mutagenesis *(1)* may have variable consequences – from a non-measurable impact up to a survival disadvantage (e.g. apoptosis induction) or, which may be more dramatic in its consequence, growth advantage of an affected clone *(1–3)*.

Based on those initial findings, replication-competent retroviruses (RCR) were deliberately used as mutagens which strongly facilitated the identification of oncogenes *(2, 3)*. The approach was based on the fact that the integrated form of any retrovirus, the provirus, has a well-characterised genomic structure with a known DNA sequence. Based thereon, the provirus may serve as a kind of 'genomic anchor' or starting point for 'walking' into (i.e. identifying) neighbouring gene loci *(4)*.

Later it turned out that 'first generation' *Murine Leukaemia Virus* (MLV-) derived, replication-defective γ-retroviral vectors containing complete long terminal repeats (LTR) potentially mediate the same effects as RCR upon integration. In fact, today the observed consequences of insertional gene dysregulation mediated by γ-retroviral gene transfer vectors range from temporary survival advantages up to benign and even malignant clonal outgrowth *(5–8)*. Based on these data, various alternative vector systems derived from other retroviruses (e.g. immunodeficiency viruses such as HIV) and/or lacking the (duplicated) strong viral enhancer/promoter sequences in the LTRs ('self-inactivating' or SIN vectors) have been proposed as potentially safer substitutes.

Identification of insertion sites/preferences of different vector types in unselected as well as selected cell populations (e.g. dominant clones in vivo) represents one important aspect of comparative biosafety testing performed for various integrating vectors. Consequently, different techniques allowing high-throughput determination of large numbers of insertion sites in unselected samples (to identify possible general preferences of given vector types) *(9–12)* as well as the retrieval of insertions in single clones dominating haematopoiesis are required *(5, 7, 13)*. One crucial condition for the unequivocal allocation of retrovirus-flanking sequences to the genome allowing fast identification of neighbouring genes has been the completion of the human and mouse genome projects.

Adaptation of ligation-mediated PCR (LM PCR), initially described by Riggs and colleagues *(14)*, to the retrieval of retrovirus-flanking sequences by Schmidt et al. *(9)* represented a real breakthrough for the field. We have recently described a slightly modified version of the Schmidt protocol for the identification of flanking regions of γ-retroviral LTR vectors *(13)*. That protocol had been designed to focus on clones which dominate haematopoiesis at the time of analysis (e.g. in murine long-term study) mainly ignoring insertions in single cells *(7)*.

In the given work we concentrate on another very important aspect of insertion site analysis, namely data processing, organisation, and interpretation. Development of various highly efficient tools for insertion sites recovery from retrovirally transduced cells (see above) has allowed accumulating large amounts of data. Thus, the quality of downstream data processing may have a significant impact on the liability of drawn conclusions. To avoid additional work it is important to organise databases from the beginning in a 'future-safe way'. This means that new databases should contain as much potentially significant information as possible to ensure efficient 'post-database' analysis by various bioinformatics tools. Analytical algorithms and statistical approaches should be developed with respect to exactly defined questions.

In the first part of the present chapter we provide a basic overview on LM PCR protocol applicable to the identification of flanking sites of standard *(13)* as well as SIN γ-retroviral vectors with a focus on principal differences in the molecular analysis of integrated LTR- and SIN-vectors. In the remaining sections we capitalise on current methods for data processing, analysis, and systematisation which could be useful for the establishment of insertion site databases as recently described *(15)*. Since the field is developing very fast, the latter part should not be viewed as a ready-to-use protocol but rather an introduction into possibilities of interpreting RVIS databases, together with examples of problem areas.

## 2. Materials

### 2.1. Molecular Analysis of RVISs by LM PCR

#### 2.1.1. LM PCR Strategy and Primer Design

1. Electronic version of the vector sequence.
2. Program for restriction site analysis of the sequence, e.g. DNA Star program (DNASTAR, Inc., Madison, WI, USA).
3. List of the most commonly used four-cutter restriction enzymes, not sensitive to *dam*, *dcm*, or mammalian CpG methylation: Tsp509 I (10 U/µl, New England BioLabs (NEB), Frankfurt a. Main, Germany), Mse I (10 U/µl, NEB), HpyCH4 V (5 U/µl, NEB), Msp I (20 U/µl, NEB).
4. Program for primer design, e.g. http://www.premierbiosoft.com/netprimer/index.html.
5. Examples of primers used for investigation of 5'-/3'-genomic-flanking region of different proviruses (with and without internal control of LM PCR) are summarised in **Table** 1 together with the respective references.

**Table 1**
**Examples of primers used for LM PCR to retrieve flanking sequences of different retroviral vectors**

| Primer | Sequence | Target sequence | Internal control | Reference |
|--------|----------|-----------------|------------------|-----------|
| rvLTR 1 | 5′-Biotin-CTGGGGACCATCTGT-TCTTGGCCTC-3′ | γ-RV (SF-LTR) 5′-flanking region | Yes | *(13)* |
| rvLTR 2 | 5′-GCCCTTGATCTGAACTTCTC-3′ | | | |
| rvLTR 3 | 5′-CCATGCCTTGCAAAATGGC-3′ | | | |
| SINPRE 1 | 5′-Biotin-GCACTGATAATTCCGT-GGTGTTGTC-3′ | γ-RV SIN (containing the wPRE) 3′-flanking region | No | *(16)* |
| SINLTR 2 | 5′-AGCGATATCGAATTCACAACC-3′ | | | |
| SINLTR 3 | 5′-CCCAATAAAGCCTCTTGCTGT-3′ | | | |

*2.1.2. Genomic DNA Preparation*

QIAamp DNA blood kit (QIAGEN, Hilden, Germany).

*2.1.3. Preparation of Asymmetric Polylinker Cassette*

1. Polylinker oligonucleotides:
   - Linker-Oligo 1: 5′-GACCCGGGAGATCTGAATTCAGT-GGCACAGCAGTTAGG-3′, 200 pmol/μl
   - Linker-Oligo 2: 5′-CCTAACTGCTGTGCCACTGAATT CAGATCTCCCG-3′, 200 pmol/μl.
2. 5× annealing buffer: 0.5 M Tris–HCl pH 7.4, 0.35 M $MgCl_2$. Store at –20°C.
3. Water bath set at 70°C.

*2.1.4. Restriction Digest and Precipitation*

1. Tsp509 I (10 U/μl, NEB) or another four-cutter restriction enzyme (*see* **Subheadings 2.1.1, item 3**).
2. 10× restriction buffer (NEB).
3. 3 M NaAc pH 5–6.
4. Glycogen (20 μg/μl, Roche, Penzberg, Germany).
5. Absolute ethanol and 70% ethanol.

*2.1.5. Primer Extension*

1. Native Pfu DNA polymerase (2.5 U/μl, Stratagene, La Jolla, CA, USA).
2. 10× Native Pfu buffer (Stratagene).
3. dNTP mix (25 mM each, Stratagene).
4. 5′-Biotinylated primers rvLTR 1 or SINPRE 1 (*see* **Table 1**), 0.25 pmol/μl.

5. Thermocycler.

6. QIA Quick PCR kit (QIAGEN).

*2.1.6. Biotin-streptavidin Interaction*

1. Dynabeads M-280 Streptavidin (10 µg/µl, Invitrogen, Karlsruhe, Germany).

2. 2× BW buffer: 10 mM Tris–HCl pH 7.5, 1 mM EDTA, 2.0 M NaCl.

   • Store aliquots at –20°C.

3. Dynal Magnetic Particle Concentrator MPC-S (magnetic tube holder) (Invitrogen).

4. Rotator: sample mixer Dynal MX1 (Invitrogen).

*2.1.7. Polylinker Cassette Ligation*

1. T4 DNA ligase (400 U/µl, NEB).

2 10× NEB ligation buffer (NEB).

3. 40 µM asymmetric polylinker mix. Store in single-use aliquots at –20°C.

*2.1.8. First and Nested PCR*

1. 2× Extensor Hi-Fidelity PCR Master mix (ABgene, Hamburg, Germany).

2. Primers OC1, OC2 (13):

   • OC1: 5′-GACCCGGGAGATCTGAATTC-3′, 25 pmol/µl

   • OC2: 5′-AGTGGCACAGCAGTTAGG-3′, 25 pmol/µl.

3. Primers RvLTR 2, RvLTR 3 (for SF vectors) or SINLTR 2, SINLTR 3 (*see* **Table 1**), 25 pmol/µl.

4. Thermocycler.

*2.1.9. Separation of PCR Products in an Agarose Gel*

1. Agarose.

2. 10× TBE buffer.

3. Gel tanks.

4. Power supply.

5. Ethidium bromide.

6. 100-bp ladder (or comparable DNA size marker).

*2.1.10. Purification of Dominant Products and Direct Sequencing*

1. Gel extraction kit (QIAGEN).

2. 3 M NaAc pH 5–6, glycogen (20 µg/µl, Roche) and absolute Ethanol.

3. Big Dye Terminator v1.1 Cycle Sequencing Kit (Applied Biosystems, Foster City, CA, USA) or CEQ Dye Terminator Cycle Sequencing with Quick Start Kit (Beckman Coulter, Fullerton, CA, USA).

4. Sequencing primer: for SF vectors: 5′-CTTGCAAAAT-GGCGTTAC-3′.

**2.2. Post-Molecular Analysis of RVISs Using BLAST or Equivalent Search Databases**

1. http://www.ncbi.nlm.nih.gov/blast
2. http://www.ensembl.org
3. http://www.ncbi.nlm.nih.gov/entrez
4. http://www.genome.ucsc.edu
5. http://www.repeatmasker.org

**2.3. Post-BLAST Database Analysis**

1. http://rtcgd.ncifcrf.gov
2. http://stemcell.princeton.edu
3. http://david.abcc.ncifcrf.gov
4. http://www.ingenuity.com or others

## 3. Methods

**3.1. Molecular Analysis of RVISs by LM PCR**

Insertional site analysis of dominant haematopoietic clones comprises the following major procedures: (1) molecular analysis of RVISs by LM PCR; (2) post-molecular analysis of RVISs using the NCBI (National Centre for Biotechnology Information) database and BLAST (Basic Local Alignment Search Tool) program or equivalent databases/tools, (3) 'post-BLAST' database analysis.

*3.1.1. LM PCR Strategy and Primer Design*

1. Delineate the configuration of the integrated provirus: pay special attention to the rearrangement of LTR modules (U3, U5) during integration, in particular with respect to SIN design, if applicable (*see* **Note 1**).

2. Choose whether the 5′- or 3′-genomic-flanking region of the provirus will be the object of investigation. Based on this decision, design vector-specific primers (*see* **Fig.1a, b**). If it is impossible to locate all three vector-specific primers within the LTR (as exemplified for SIN vectors in **Fig. 2**), there will be no internal (vector-derived) control in the PCR (*see* **Note 2**).

3. Select four-cutter enzyme for digestion of genomic DNA. The chosen enzyme must not be sensitive to any form of methylation and must not cut within the internal control amplicon.

4. If the PCR primers are located within the LTRs, you have to check the size of the internal control which is expected to be within the range of 130–500 bp (*see* **Fig.1a, b**). However, in those cases when primers are located between the LTRs there will be no amplification of vector sequences ('internal control', *see* **Fig.2**) (*see* **Note 2**).

*3.1.2. Preparation of Genomic DNA from Peripheral Blood, Spleen, or Bone Marrow*

To extract genomic DNA from cells of mouse bone marrow, spleen, liver, or peripheral blood, use the QIAamp DNA blood

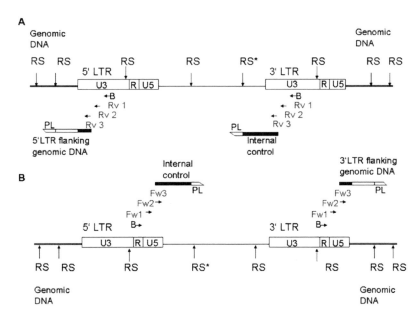

Fig. 1. Strategy of primer design for retrieving 5′LTR (**a**) and 3′LTR (**b**) flanking genomic regions of γ-retroviral (LTR) vectors by LM PCR. Note that the optimal amplification range of LM PCR is within 100–800 bp. The size of the internal control should be 130–500 bp. It can be calculated by adding the distance between the 5′-end of the 3′LTR and the first recognition site of the restriction enzyme within the vector (RS*) to the length of the co-amplified linker and LTR sequences (including the specific primers). *RS* 4-bp-cutter restriction enzyme recognition site; *B* biotin; *Rv 1, Rv 2, Rv 3* reverse primers (LTR specific); *Fw 1, Fw 2, Fw 3* forward primers (LTR specific).

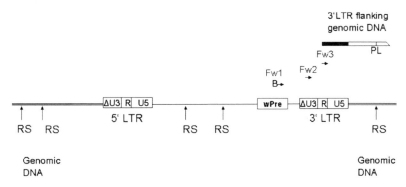

Fig. 2. Schematic representation of primer design for γ-retroviral SIN vectors. For SIN vectors it might be necessary to design alternative primers, because of the almost complete deletion of the U3 region of the LTR. Here, the first (biotinylated) primer binds to a unique sequence of the retroviral vector (e.g. the wPRE element) that is as close as possible to the 3′LTR. The second LTR primer covers the deletion in the U3 region. Note that there will be no internal control detectable when using this strategy. *Fw 1, Fw 2, Fw 3* forward primers.

kit (QIAGEN) or any alternative procedure that yields high-quality genomic DNA. In these procedures, pay attention to avoid potential contamination with plasmids or PCR products to which the PCR primers used later may anneal.

*3.1.3. Preparation of Asymmetric Polylinker Cassette*

1. Mix in an Eppendorf tube 40 µl of $H_2O$, 20 µl of Linker-Oligo 1 (200 pmol/µl), and 20 µl of Linker-Oligo 2 (200 pmol/µl).

2. Keep the mixture for 5 min at 70°C in a water bath. Then add at room temperature (RT) 20 µl of 5× annealing buffer, mix, and incubate for another 5 min at 70°C. Switch off the water bath and leave the tube in the self-cooling water overnight.

3. Next morning prepare RT aliquots of 5–10 µl each, which may be stored at –20°C. Use each aliquot only once.

*3.1.4. Restriction Digestion of Genomic DNA*

1. Restriction digest is carried out in a final volume of 30 µl. Mix in an Eppendorf or PCR tube (*see* **Note 3**): 0.1–1.0 µg of genomic DNA, 3 µl of 10× restriction buffer, 0.5 µl of restriction enzyme Tsp509 I and add $H_2O$ to a final volume 30 µl.

2. Incubate for 2 h at 65°C (or at 37°C depending on the chosen restriction enzyme). Incubation in a thermocycler is recommended.

3. Precipitate overnight at –20°C with 0.3 M NaAc pH 5–6 (final concentration), 2 volumes of absolute Ethanol, and 20 µg of glycogen.

4. Next morning spin down at 16,000 g (Eppendorf centrifuge) for 20 min, carefully discard supernatant, add 200 µl of 70% ethanol, centrifuge again for 5 min, and discard the supernatant.

5. Re-suspend the air-dried pellet in 10 µl of $H_2O$ (*see* **Note 4**).

6. Restricted genomic DNA is best stored at –20°C.

*3.1.5. Primer Extension Step*

1. Mix in a PCR tube (end volume 20 µl) (*see* **Note 3**): 10 µl of DNA (*from* **Subheadings 3.1.4, step 5**), 2 µl of 10× Native Pfu buffer, 0.16 µl of 25 mM dNTP, 1 µl of Primer rvLTR 1 – biotin or SINPRE 1 – biotin (0.25 pmol/µl), 1 µl of Native Pfu DNA polymerase and 5.84 µl of $H_2O$. Using Pfu DNA polymerase results in blunt-ended products, which are ideal for asymmetric linker cassette ligation (*see* **subheadings 3.1.7**).

2. Incubate in a thermocycler with following program configuration: 95°C for 5 min, 64°C for 30 min, 72°C for 15 min.

3. After primer extension (PE) reaction clean the product using QIAQuick PCR kit (QIAGEN), elute in 40 µl of $H_2O$ in an Eppendorf tube.

   The product of primer extension may be stored at –20°C. The protocol may be interrupted at this point before proceeding to the next step.

*3.1.6. Target DNA Enrichment*

1. 200 µg of Dynabeads M-280 Streptavidin (*see* **Note 5**) should be prepared for the enrichment of the primer extension (PE) product. Therefore, transfer 20 µl of standard beads solution (10 µg/µl) into an Eppendorf tube and wash two times with

150 µl of 2× BW buffer as follows: add 2× BW buffer, mix gently by finger knocking, put onto a magnetic tube holder, remove the buffer, take out from holder, and repeat the procedure.

- Re-suspend in 40 µl of 2× BW buffer.

2. Add the washed Dynabeads (200 µg in 40 µl of 2× BW buffer) to 40 µl of purified PE reaction (*see* **subheadings 3.1.5, step 3**) (*see* **Note 6**).

3. Incubate the solutions with gentle rotation of the tubes at RT for 2–8 h. Then wash the mixture twice with 100 µl of $H_2O$ (the same way as with 2× BW buffer, *see* **step 1**). Re-suspend by pipetting in 5 µl of $H_2O$ (*see* **Note 4**).

*3.1.7. Linker Cassette Ligation*

1. Prepare all components for linker cassette ligation on ice: thaw the aliquot of asymmetric polylinker mix (*see* **3.1.3**) on ice; thaw ligation buffer at room temperature, vortex (to dissolve the DTT pellet) and spin down, put on ice.

2. Mix the following components to an end volume of 10 µl in a PCR tube or an Eppendorf tube (*see* **Note 3**): 5 µl of DNA (*from* **3.1.6, step 3**), 1 µl of 10× NEB ligation buffer, 1 µl of asymmetric polylinker mix, 0.2 µl of Ligase, and 2.8 µl of $H_2O$ (4°C). Put the ice-cold ligation reaction into a thermostat (best a thermocycler) and incubate at 16°C overnight.

3. Next morning wash the ligation mix two times with 100 µl of $H_2O$ using the magnetic tube holder (*see* **3.1.6, step 1**), re-suspend in 10 µl of $H_2O$ by pipetting. After washing, the ligation mix may be stored at –20°C for several months (*see* **Note 7**).

*3.1.8. First PCR and Nested PCR to Amplify the Target DNA (Insertion Site)*

1. Program configuration for first and nested PCR: initial denaturation at 94°C for 2 min followed by 30 cycles composed of 94°C for 15 s, 60°C for 30 s, 68°C for 2 min. Carry out the final extension at 68°C for 10 min.

2. First PCR end volume is 25 µl. Mix in a PCR tube (*see* **Note 3**): 1 µl of DNA (*from* **3.1.7, step 3**), 12.5 µl of 2× Extensor Hi-Fidelity PCR Master mix, 1 µl of Primer OC1 (25 pmol/µl), 1 µl of Primer RvLTR 2 or SINLTR 2 (25 pmol/µl), 9.5 µl of $H_2O$.

3. Nested PCR: the same cycling conditions as for the first reaction. To prepare DNA template for nested PCR dilute the first PCR mix 1:500. Mix the following components in a PCR tube (*see* **Note 3**): 1 µl of DNA, 12.5 µl of 2× Extensor Hi-Fidelity PCR Master mix, 1 µl of Primer OC2 (25 pmol/µl), 1 µl of Primer RvLTR 3 or SINLTR 3 (25 pmol/µl), 9.5 µl of $H_2O$ (*see* **Note 8**).

*3.1.9. Insertion Pattern Visualisation by Gel Electrophoresis*

For insertion pattern visualisation it is recommended to use 2% agarose 0.5× TBE gel electrophoresis. Pay attention to the size of internal controls (if applicable) of your probes, which serve as an in-reaction quality measure for your LM PCRs (*see* **Note 2**).

*3.1.10. Purification and Sequencing of Dominant Products*

Cut out dominant bands and extract DNA fragments for further analysis using Gel extraction kit (QIAGEN) or any other procedure for extraction of DNA from agarose gels.

To improve the quality of direct sequencing it is better to precipitate the eluted fragments using 3 M NaAc pH 5–6, Glycogen, and absolute Ethanol. Conditions for sequencing should be established according to the used sequencing reagents and device. In addition to the LTR-specific, a linker-specific primer may be used for sequencing. For those cases when direct sequencing of eluted PCR products is not working well we recommend sequencing after subcloning of eluted PCR fragments.

*3.2. Post-Molecular Analysis of RVISs Using the National Centre for Biotechnology Information Database and Basic Local Alignment Search Tool or Equivalent Search Databases/Tools*

Sequenced PCR products may only be viewed as RVISs if they contain the expected vector-(LTR-)specific sequences. If sequencing spans the whole PCR fragment (that would not always be the case for directly sequenced PCR products but could be expected with cloned fragments), polylinker sequences should also be revealed. Before database search, sequences should be processed manually or using DNA analysis software tools to remove bases that correspond to the retroviral vector and to the polylinker.

*3.2.1. Pre-submission Processing of the PCR Sequences*

*3.2.2. Submission of the Edited Sequences and Primary Analysis of Search Data*

To locate the genomic position of the obtained 3′- or 5′-vector-flanking sequences in the respective genome you should first perform a search of one of the available genome databases. The procedure is exemplarily set out for the mouse genome and the National Centre for Biotechnology Information (NCBI) database (as available in April 2007):

1. Go to National Centre for Biotechnology Information (NCBI) website http://www.ncbi.nlm.nih.gov/.

2. Select Basic Local Alignment Search Tool (BLAST) http://www.ncbi.nlm.nih.gov/blast/.'The program compares nucleotide sequences to sequence databases and calculates the statistical significance of matches' (this and the following direct quotations are from the NCBI website).

3. Go to Genomes/Mouse.

4. Paste your sequence to be analysed (in FASTA or txt format) into the respective field. Select the following parameters *Database*: 'genome (all assemblies)' and *Program*: 'BLASTN: compare nucleotide sequences' before starting your search.

5. The BLAST search may have different outcomes:

   (a) 'No significant similarity found' may indicate that your analysed sequence is too short (*see* **Note 9**) or contains highly repetitive DNA stretches. However, in some cases such result reflects the fact that the given Genome project is just 'almost complete' (*see* **Note 10**). In all those instances it is suggested to repeat LM PCR with an alternative restriction enzyme or at the other end of the vector using another primer set.

   (b) In rare cases you will find only one alignment with 100% or nearly 100% homology representing the flanking region of your vectors. More often a few sequences with different RefIDs do have identical degrees of homology. Usually, those sequences are all located at the same place in the genome, thus mirroring the input of different consortia to the given genome project. If your query sequence aligned within or close to a gene locus this will be indicated under each RefID ('Features in this part of subject sequence:').

   (c) When blasting larger fragments, besides the high-degree homologies you may find additional 'sequences producing significant alignments'. This may be due to repetitive sequences within your analysed fragment or to the presence of related or pseudo genes in the genome. To avoid this situation and mask stretches corresponding to repetitive sequences inside analysed fragments one can use corresponding programs, e.g. http://www.repeatmasker.org/. For further analysis you should focus on the entry revealing the minimal $E$ value (='expected value' gives an indication of the statistical significance of a given alignment and reflects the size of the database and the scoring system used), and the highest bit score (bit score gives an indication of how good the alignment is), maximal identities and minimal gaps, i.e. (close to) 100% homology.

6. By simply clicking onto the linked ID of the chosen sequence you will be directed to a chromosome map indicating location of your query sequence (NCBI Map Viewer). The map will reproduce a 100-kB window of the genomic region surrounding the sequence homolog to the analysed flanking region, i.e. will indicate the genomic location of the given RVIS. On the left hand side of this page you will find an ideogram of the chromosome highlighting the position of the query sequence (=your RVIS) on the respective chromosome.

7. Using the map determine the gene(s) closest to the RVIS (*see* **Note 11** and **subheading 3.3.1, steps 1 and 2**).

8. Determine orientation of the vector with regard to (each of) the nearest gene(s). To do so you best establish orientation as related to the chromosome for both your query sequence and the gene(s) of interest (indicated in the NCBI Map viewer). To define orientation of the integrated vector you have to take into account whether the 5′ or the 3′ genome-flanking regions were amplified by LM PCR and from which side sequencing was accomplished. For instance: in case you have amplified the 3′ genome-flanking region of a provirus and directly sequenced it using an LTR-based primer, orientation of the provirus will be 'forward' regarding the gene of interest, if both, query sequence and gene of interest have *the same* orientation with respect to the chromosome. Vice versa, in case you have retrieved the 5′ genome-flanking region of a provirus and sequenced it using an LTR-based primer, orientation of the provirus will be 'forward' regarding the gene of interest, if query sequence and gene of interest have *different* orientations with respect to the chromosome (*see* **Fig. 3**).

9. To get more information on the gene of interest you have to click onto the gene symbol in the NCBI Map Viewer. The following data will be provided: orientation (as related to the chromosome), all information linked to the given gene symbol (e.g. via Mouse Genome Informatics MGI, including Gene ID; exact location on chromosome (*see* **Note 12**); official gene symbol and others names; bibliography/related articles in PubMed, gene ontology information: function, process, component; conserved domains), Sequence viewer with indication of the transcriptional start site (TSS, according to NCBI definitions not always reflecting the real TSS but conditional 'gene start'), exon/intron structure of the gene, etc.

10. To determine the distance from an RVIS to the TSS of a gene of interest you need to recall exact position of your RVIS as related to the chromosome (*see* **subheading 3.2.2, step 5**). Calculate the difference between position of the RVIS and the indicated coordinates of the TSS for the gene of interest (*see* **3.2.2, step 9**) (*compare* **Fig. 3**).

11. In some cases, other databases like http://www.ensembl. orgor http://www.genome.ucsc.edu can provide the investigator with additional or more detailed information.

12. Different groups are establishing computerised analysis tools for insertion sites which may in the near future help to automate most parts of the afore described analysis (e.g. http://www.gtsg.org).

13. Please pay attention that different groups may use diverse criteria for RVIS analysis and sometimes even different parameter definitions.

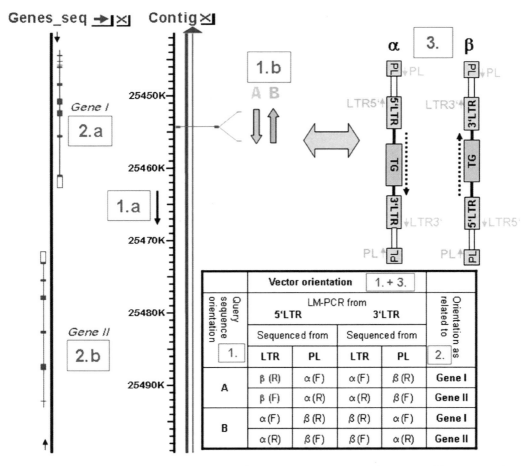

Fig. 3. Schematic representation of identification of provirus orientation regarding a gene of interest based on results of an NCBI genomic blast search. The *left* part of the figure corresponds to the schematic representation of results of a genomic BLAST search as provided by the NCBI database. Under 'Contig', a nucleotide scale (in kb) is provided for the chromosome to which your query sequence was aligned. You can use the following algorithm to define orientation of the integrated vector. [1.] To determine the Query sequence orientation: [1a.] The chromosome's orientation/scale in those schemas is always up-to-down (see the respective *arrow*). [1b.] If nucleotide numbers of your blasted Query sequence are increasing together with those of the chromosome, the aligned sequences are directed in the same direction (**A**) as the chromosome; in the opposite case it is directed in an anti-parallel orientation (**B**). [2.] Gene directions are indicated by arrows besides the 'Genes_seq' bar: [2a] Genes located on the right side of the bar (e.g. *Gene I*) are always directed co-linear with the chromosome, [2b] genes outlined on the left side (e.g. *Gene II*) are always directed contra-linear. [3.] To finally determine the orientation of the inserted vector (*dotted arrow*) as related to the chromosome (α = parallel, β = anti-parallel) and the genes of interest (*F* forward or *R* reverse) you have to take into account the location of the adjoining sequences retrieved by LM PCR (5′ vs. 3′) *and* the direction of your initial sequence reaction (started from the LTR or the polylinker (PL)). All possible variants of sequencing primers are indicated in the *right* part of the figure. The colours of the primers indicate the sequence direction as related to the chromosome (A vs. B). All orientations towards genes and chromosomes are summarised in the Table in the lower part of the picture. For example, after LM PCR you obtained the 5′ adjacent region of your integrated vector and sequenced it using an LTR-specific primer. You first determine the query sequence orientation with respect to the chromosome, which turns out to be co-linear (A). Consequently, your vector is oriented in the opposite direction to the chromosome (β). Finally, as related to *Gene I* the vector is oriented in a reverse (R), as related to *Gene II* in a forward (F) orientation (see table, highlighted in *blue* and *bold*).(*See Color Plates*)

*3.2.3. Summary of BLAST/*
*Pubmed Analysis*

Altogether, the blast search to annotate insertion sites (*see* **sub-heading 3.2.2, steps 1–12**) in conjunction with subsequent PubMed analyses (http://www.ncbi.nlm.nih.gov/entrez) will provide you with all necessary locus-specific information including: exact position of retroviral integration site with respect to the transcriptional start site (TSS) of the nearest and other genes, symbols, IDs, exon/intron structures, and proposed functions of those neighbouring genes as well as orientation of the retroviral vector as related to any of them (forward or reverse).

### 3.3. 'Post-BLAST' Database Analysis

Information obtained up to this point in many cases needs systematisation as well as further processing to make it suitable for subsequent bioinformatics analysis. We recommend considering the following important guidelines for your 'post-BLAST' database analysis *(15)*.

*3.3.1. Develop Your*
*Own Algorithm for Data*
*Processing and Analysis*

The algorithm that one develops for data analysis should take the following points into account during initial data acquisition:

It is desirable to include for 'post-BLAST' analysis a genomic region of 200 kb up- and downstream of the RVIS. This recommendation is based on the described long-range effects of strong (viral) enhancers *(4, 17)* (*see* **Note 11**).

1. Sometimes retroviral insertions occur in regions where many genes are located (so-called gene dense region). Initially it might be suggestive to focus on the gene closest to the insertional site. However, this may result in significant loss of information. Therefore, try to initially extract and store as much information as possible for a given hit.

2. Avoid 'cherry-picking': define criteria in advance and follow them.

*3.3.2. Summary of the*
*RVIS Data Obtained from*
*Long-Term Animal*
*Experiments*

Organise acquired data in accordance with the needs of easy and fast re-analyses by the means of bioinformatics. Establish clear criteria for the functional grouping of genes adjacent to insertion sites. The latter should be based on the information on the gene's function provided in the genome databases as well as additional literature data (e.g. PubMed: http://www.ncbi.nlm.nih.gov/entrez). You could also refer to Gene ontology databases such as www.geneontology.org. A possible example of grouping genes according to their functions into three categories (a) proto-oncogenes, (b) signalling genes, and (c) other/unknown genes has been introduced in a recent publication of ours establishing an *Insertional dominance database* (IDDb) of genes potentially involved in clonal dominance of haematopoietic stem cells in a model of serial bone marrow transplantation *(15)*.

*3.3.3. Database Quality*
*Control*

Make sure that all contributors to your database follow the strict quality criteria you introduced.

*3.3.4. Database Internal Analysis*

Perform regular analyses within your database, e.g. to reveal common insertion sites (CISs).

*3.3.5. Database External Analysis*

We recommend for each novel insertion (RVIS) to include into post-BLAST analysis a cross-check step with already existing databases of retrovirus insertion sites. The broadest database of RCR insertions correlated with malignant transformation is the RTCGD (retrovirus tagged cancer gene database) *(18)*: http://rtcgd.ncifcrf.gov. The IDDb *(15)* comprises hits in haematopoietic stem cells (HSC) representing different levels of clonal dominance. The stem cell database (SCDb) encompasses genes expressed in HSC *(19)*: http://stemcell.princeton.edu. Genes already listed as common integration sites (CISs) in the RTCGD or contained in IDDb as well common CISs found in your own accumulated data may be particularly interesting, even if they are not always the closest genes with respect to a given retroviral insertion *(15)*.

*3.3.6. Bioinformatics Analysis of Data Contained in the Database*

If you have accumulated sufficient data in your own database, it may, possibly in conjunction with data from related databases such as RTCGD or IDDb, be subjected to further bioinformatics studies such as Ingenuity analysis/Gene Ontology analysis: http://www.ingenuity.com/http://david.abcc.ncifcrf.gov (*see* **Note 12**).

# 4. Notes

1. In an infected cell, the two long terminal repeats (LTRs) of an integrated retrovirus (provirus) are identical. They have a defined structure consisting of three elements: *U3-R-U5*. Consequently, an integrated provirus looks as follows: 5′ *U3-R-U5* – genes + cis-acting elements – *U3-R-U5* 3′. In contrast, the retroviral RNA genome has *no LTRs*, but the following architecture: 5′ *R-U5* – genes + *cis*-acting elements – *U3-R* 3′. This genomic structure is easy to remember – the 'U' stands for 'unique', the 'R' for 'redundant'. From a comparison of the RNA genome and the provirus two important conclusions become obvious: (a) The viral genome transcribed from a provirus contains the U5 element from the 5′LTR and the U3 element from the 3′LTR and (b) in an infected cell both U3 regions should be identical and derived from the 3′LTR of the initial provirus, whereas both (identical) U5 regions are derived from the 5′LTR of the initial provirus. (This is being accomplished by complex rearrangements

during reverse transcription of the viral genome – please refer to a *Virology* textbook for further details.)

Since recombinant proviruses bearing two different LTRs could be designed and introduced into retrovirus-producer cells by transfection, these peculiarities of the retroviruses life cycles are being used to deliberately introduce mutations or deletions into LTRs, e.g. to generate self-inactivating (SIN) LTRs by destroying promoter/enhancer elements located in the 3′LTR U3-region.

For LM PCR primer design it is *important to remember* that for an integrated provirus the U3 region of the 5′LTR is identical to the U3 of the 3′LTR in the vector plasmid bearing the initial provirus, and the U5 region of the 3′LTR is identical to the U5 of the 5′LTR in the original recombinant vector.

2. Additional measures for quality control of LM PCR results are important, e.g. checking reproducibility with the same and alternative enzymes, controlling RVISs numbers by Southern blot analysis (particularly in the case of investigating flanking genomic regions of SIN vectors). The source of the genomic DNA together with clonal dominance stage may influence the results. Inclusion of a mock control from the same mouse strain is recommended *(13)*.

3. Preparation of «master mix» starting from water is recommended.

4. Take care to collect all pellets from the tube walls by pipetting, especially when working with dynabeads.

5. Do not vortex or centrifuge Dynabeads M-280 Streptavidin.

6. The product of the PE step simply purified with QIAQuick PCR kit (QIAGEN) (13).

7. We do not perform a NaOH-mediated denaturation step (13).

8. The use of an enzyme mix, e.g. so called Extensor Hi-fidelity PCR Master mix, containing a proof-reading polymerase for amplification steps instead of Taq polymerase is strongly recommended. This master mix contains a mixture of polymerases (with proof-reading activity), reaction buffer, dNTPs, MgCl2, and also gel-electrophoresis loading buffer (13).

9. From a theoretical point of view, given a genome size of approximately $3 \times 10^9$ bp, unknown sequences to be used for homology (BLAST) searches should not be shorter than 16 nucleotides ($4^{16} = 4.29 \times 10^9$) to allow non-ambiguous allocation.

10. If there is no alignment using the NCBI mouse chromosome database you could try searching Nucleotide-nucleotide BLAST (blastn) in order to find significant homologies, for instance with some BAC (Bacterial Artificial Chromosome)

clones. The latter may subsequently be used for chromosome alignments. Alternatively, you may try searching other databases (http://www.ensemble.org) which sometimes contain alternative data sets.

11. A given insertion may have happened into a gene-dense as well as a gene-sparse region. Since enhancer effects from retroviral LTRs have been reported to take place over a distance of 200 kb *(4)*, it makes in any case sense to look at both sides of an insertion for additional genes.

12. Public databases such as NCBI, Ensembl, and RTCGD are frequently updated based on growing data sets. Sometimes this may lead to significant changes, e.g. of the chromosomal organisation. Please keep, therefore, in mind that such parameters as exact location of RVIS on a given chromosome, Gene ID, gene names, RTCGD membership, etc. could be subject to change.

## Acknowledgements

This work was supported by Grants from the Deutsche Forschungsgemeinschaft (DFG SPP1230). The authors wish to thank Martijn Brugman and Hartmut Geiger for critical reading of the manuscript.

## References

1. Jaenisch, R., Harbers, K., Schnieke, A., Löhler, J., Chumakov, I., Jahner, D., Grotkopp, D., and Hoffmann, E. (1983) Germline integration of moloney murine leukemia virus at the Mov13 locus leads to recessive lethal mutation and early embryonic death. *Cell* 32, 209–216.

2. Peters, G., Brookes, S., Smith, R., and Dickson, C. (1983) Tumorigenesis by mouse mammary tumor virus: evidence for a common region for provirus integration in mammary tumors. *Cell* 33, 369–377.

3. Mikkers, H., and Berns, A. (2003) Retroviral insertional mutagenesis: tagging cancer pathways. *Adv Cancer Res* 88, 53–99.

4. Uren, A. G., Kool, J., Berns, A., and van Lohuizen, M. (2005) Retroviral insertional mutagenesis: past, present and future. *Oncogene* 24, 7656–7672.

5. Li, Z., Düllmann, J., Schiedlmeier, B., Schmidt, M., von Kalle, C., Meyer, J., Forster, M., Stocking, C., Wahlers, A., Frank, O., Ostertag, W., Kühlcke, K., Eckert, H. G., Fehse, B., and Baum, C. (2002) Murine leukemia induced by retroviral gene marking. *Science* 296, 497.

6. Hacein-Bey-Abina, S., Von Kalle, C., Schmidt, M., McCormack, M. P., Wulffraat, N., Leboulch, P., Lim, A., Osborne, C. S., Pawliuk, R., Morillon, E., Sorensen, R., Forster, A., Fraser, P., Cohen, J. I., de Saint Basile, G., Alexander, I., Wintergerst, U., Frebourg, T., Aurias, A., Stoppa-Lyonnet, D., Romana, S., Radford-Weiss, I., Gross, F., Valensi, F., Delabesse, E., Macintyre, E., Sigaux, F., Soulier, J., Leiva, L. E., Wissler, M., Prinz, C., Rabbitts, T. H., Le Deist, F., Fischer, A., and Cavazzana-Calvo, M. (2003) LMO2-associated clonal T cell proliferation in two patients after gene therapy for SCID-X1. *Science* 302, 415–419.

7. Kustikova, O., Fehse, B., Modlich, U., Yang, M., Düllmann, J., Kamino, K., von Neuhoff, N.,

Schlegelberger, B., Li, Z., and Baum, C. (2005) Clonal dominance of hematopoietic stem cells triggered by retroviral gene marking. *Science* 308, 1171–1174.

8. Ott, M. G., Schmidt, M., Schwarzwaelder, K., Stein, S., Siler, U., Koehl, U., Glimm, H., Kuhlcke, K., Schilz, A., Kunkel, H., Naundorf, S., Brinkmann, A., Deichmann, A., Fischer, M., Ball, C., Pilz, I., Dunbar, C., Du, Y., Jenkins, N. A., Copeland, N. G., Luthi, U., Hassan, M., Thrasher, A. J., Hoelzer, D., von Kalle, C., Seger, R., and Grez, M. (2006) Correction of X-linked chronic granulomatous disease by gene therapy, augmented by insertional activation of MDS1-EVI1, PRDM16 or SETBP1. *Nat Med* 12, 401–409.

9. Schmidt, M., Hoffmann, G., Wissler, M., Lemke, N., Mussig, A., Glimm, H., Williams, D. A., Ragg, S., Hesemann, C. U., and von Kalle, C. (2001) Detection and direct genomic sequencing of multiple rare unknown flanking DNA in highly complex samples. *Hum Gene Ther* 12, 743–749.

10. Schmidt, M., Zickler, P., Hoffmann, G., Haas, S., Wissler, M., Muessig, A., Tisdale, J. F., Kuramoto, K., Andrews, R. G., Wu, T., Kiem, H. P., Dunbar, C. E., and von Kalle, C. (2002) Polyclonal long-term repopulating stem cell clones in a primate model. *Blood* 100, 2737–2743.

11. Wu, X., Li, Y., Crise, B., and Burgess, S. M. (2003) Transcription start regions in the human genome are favored targets for MLV integration. *Science* 300, 1749–1751.

12. Mitchell, R. S., Beitzel, B. F., Schroder, A. R., Shinn, P., Chen, H., Berry, C. C., Ecker, J. R., and Bushman, F. D. (2004) Retroviral DNA integration: ASLV, HIV, and MLV show distinct target site preferences. PLoS Biol 2, E234.

13. Kustikova, O., Baum, C., and Fehse, B. (2008) Retroviral Integration Site Analysis in Hematopoietic Stem Cells. *Methods in Molecular Biology – Hematopoietic Stem Cell Protocols* (Ed. Bunting, K., Humana Press Inc., Totowa, NJ) 430, 255–267.

14. Pfeifer, G. P., Steigerwald, S. D., Mueller, P. R., Wold, B., and Riggs, A. D. (1989) Genomic sequencing and methylation analysis by ligation mediated PCR. *Science* 246, 810–813.

15. Kustikova, O., Geiger, H., Li, Z., Brugman, M. H., Chambers, S. M., Shaw, C. A., Pike-Overzet, K., de Ridder, D., Staal, F. J. T., von Keudell, G., Cornils, K., Nattamai, K. J., Modlich, U., Wagemaker, G., Goodell, M. A., Fehse, B., and Baum, C. (2007) Retroviral vector insertion sites associated with dominant hematopoietic clones mark "stemness" pathways. *Blood* 109, 1897–1907.

16. Modlich, U., Bohne, J., Schmidt, M., von Kalle, C., Knoss, S., Schambach, A., and Baum, C. (2006) Cell culture assays reveal the importance of retroviral vector design for insertional genotoxicity. *Blood* 108, 2545–2553.

17. West, A. G., and Fraser, P. (2005) Remote control of gene transcription. *Hum Mol Gen* 14, R101–R111.

18. Akagi, K., Suzuki, T., Stephens, R. M., Jenkins, N. A., and Copeland, N. G. (2004) RTCGD: retroviral tagged cancer gene database. *Nucleic Acids Res* 32, D523–D527.

19. Ivanova, N. B., Dimos, J. T., Schaniel, C., Hackney, J. A., Moore, K. A., and Lemischka, I. R. (2002) A stem cell molecular signature. *Science* 298, 601–604.

# Chapter 26

## Tracking Gene-Modified T Cells In Vivo

### Alessandra Recchia and Fulvio Mavilio

### Summary

Identification, monitoring, and analysis of genetically modified cells in the peripheral blood are an important component of the clinical follow-up of patients treated by hematopoietic cell gene therapy. Analysis of gene-marked peripheral blood cells provides crucial information on gene transfer efficiency as well as on the nature and characteristics of the genetically modified cells, and may provide early evidence of the occurrence of potentially detrimental side effects. T lymphocytes are a convenient target for this type of analysis, due to their abundance and their relatively long life span in vivo. Tracking of gene-marked T cells is based on relatively simple, FACS- and PCR-based techniques, which may be applied to monitoring genetically modified T cells as well as T cells derived from transplanted, genetically modified hematopoietic stem cells. This chapter provides a description of these techniques and clues to their rational use in a clinical setting.

**Key words:** Hematopoietic stem cells, Bone marrow transplantation, T cells, T-cell transplantation, Retroviral vectors, Gene transfer, Polymerase chain reaction.

### 1. Introduction

Gene therapy of genetic or acquired blood disorders is currently based on transplantation of autologous, genetically modified hematopoietic stem cells. Genetic modification of stem cells is commonly accomplished by transduction with a replication-defective retroviral vector (RV) encoding the therapeutic gene. The outcome of such transplants is usually monitored by the analysis of peripheral blood mononuclear cells, which provide information on the efficiency of gene transfer (gene marking), on the long-term persistence of the genetic modification, on the expression of the transferred gene, and on the nature and

Christopher Baum (ed.), *Methods in Molecular Biology, Methods and Protocols, vol. 506*
© Humana Press, a part of Springer Science + Business Media, LLC 2009
DOI: 10.1007/978-1-59745-409-4_26

characteristics of the genetically modified cells (e.g., long-term repopulating stem cells, short lived pluripotent progenitors, committed progenitors) *(1–5)*. In addition, peripheral blood cell monitoring may provide early evidence of the occurring of potentially detrimental side effects, or clues to their molecular basis *(6)*. Peripheral blood T cells are an easy target for genetic analysis, due to their abundance and their relatively long life span in vivo. In addition, autologous or allogeneic T cells may be direct targets of genetic manipulation and transplantation, as a cure of a genetic (e.g., SCID) *(7, 8)* or an acquired (e.g., AIDS) *(9, 10)* T-cell immunodeficiency, as anticancer vaccines *(11)*, or as adjunctive treatment to allogeneic bone marrow (BM) transplantation in the therapy of leukemia/lymphoma *(12, 13)*.

The infusion of donor T lymphocytes after allogeneic, T-depleted BM transplantation favors engraftment, provides early immune reconstitution, and prevents or cures relapse by providing additional graft-versus-leukemia activity. Genetic modification by the transfer of a drug-inducible "suicide" gene provides the additional benefit of allowing efficacious control of graft-versus-host disease, a common side effect of T-cell transplantation, particularly in the case of mismatched (haploidentical) BM transplantation. Genetic modification of T cell is also commonly achieved by transduction with an RV encoding the therapeutic/effector gene. The timidine kinase gene of the herpes simplex virus (HSV-TK) is a commonly used suicide gene, which can be activated in vivo by administration of the antiviral drug ganciclovir® *(12, 14, 15)*. In the most clinically advanced protocols, the RV encodes also a cell surface marker, such as the truncated form of the low-affinity receptor for nerve growth factor (ΔLNGFR) *(16)*, which allows purification of T cells before infusion and provides a convenient tracking strategy after administration *(13)*.

In vivo tracking of gene-marked T lymphocytes can be accomplished by relatively simple techniques, which are the same whether T cells derive from transplanted stem cells or are the direct target of the genetic manipulation. These techniques are aimed at detecting the transgene, or other vector-specific sequences, in the DNA extracted from peripheral blood mononuclear cells or purified T cells, and are essentially based on PCR amplification. When T cells are marked by a surface antigen, their abundance or persistence can be easily monitored by cytofluorimetry with a specific antibody. The same antibody can be used to purify T cells from apheresis and analyze them in more detail as a bulk population or after limiting-dilution cloning *(15, 17)*.

## 2. Materials

### 2.1. Purification of T Cells from Peripheral Blood Samples

1. Lymphoprep (Nycomed, Oslo, Norway). Store at 4°C.
2. 1× PBS/2 mM EDTA solution. Store at 4°C.
3. FITC- or PE-conjugated anti-CD8 mAb (PharMingen, San Diego, CA). Store at 4°C.
4. FITC- or PE-conjugated anti-CD4 mAb (PharMingen, San Diego, CA). Store at 4°C.
5. Mouse anti-LNGFR mAb (20.4, from American Type Culter Collection). Store at 4°C.
6. Goat-antimouse-IgG1-coated magnetic beads (Dynabeads M450; Dynal AS Oslo, Norway). Store at 4°C.

### 2.2. T-Cell Culture

1. Medium for T-cell culture: RPMI 1640 (Biowhittaker Europe, Verviers, Belgium). For 500 ml final: add 5 ml of 200 mM glutamine (Gibco, Grand Island, NY), 0.5 ml of 100,000 IU/ml penicillin (Pharmacia, Milan, Italy), 0.5 ml of 100,000 IU/ml streptomycin (Bristol-Meyers Squibb, Sermoneta, LT), 25 ml autologous serum, 1 ml of 50,000 U/ml human recombinant IL-2 (Chiron, Milan, Italy). Store at 4°C.
2. Phytohemagglutinin (PHA) (Boehringer Mannheim-Roche GmbH, Mannheim, Germany). Use at final concentration of 2 μg/ml. Store at –20°C.
3. Anti-CD3 monoclonal antibody (mAb) (OKT3, Orthoclone, Milan, Italy). Use at the final concentration of 30 ng/ml. Store at 4°C.
4. Anti-CD28 mAb (PharMingen, San Diego, CA). Use at the final concentration of 1 μg/ml. Store at 4°C.

### 2.3. Cytofluorimetric Analysis of Purified T Cells

1. Fluorescein isothiocyanate (FITC)- and phycoerythrin (PE)-conjugated antihuman CD2, CD3, CD4, CD8, CD56, CD19 (Becton Dickinson, Mountain View, CA), CD62L, CD28, and CD95 (PharMingen Biosciences, San Diego, CA) antibodies. Store at 4°C.
2. FITC- and PE-conjugated isotype control.
3. FITC- and PE-conjugated antihuman CD25, HLA-DR, CD69, CD45RA, and CD45RO antibodies (Becton Dickinson).
4. Mouse anti-CCR7 mAb, biotinylated anti-mouse IgM mAb and PerCP-conjugated streptavidin. Store at 4°C.

**2.4. Extraction of DNA from Purified T Cells**

1. QIAmp DNA Blood Micro, Mini, Midi, or Maxi kit (Qiagen Inc., Valencia, CA). Store at room tempearture.

**2.5. Quantitative Polymerase Chain Reaction Analysis of Gene-Marked T Cells**

1. Primers and TaqMan® MGB (minor groove binder) probe designed over a specific region of the therapeutic cDNA sequence. Store at −20°C.

2. TaqMan® Fast Universal PCR Master Mix (Applied Biosystem, Foster City, CA). Store at 4°C.

3. Genomic DNA from a human T-cell clone containing a known copy number of integrated retroviral vector. Store at −20°C.

4. 18S ribosomal RNA control (Applied Biosystem, Foster City, CA) for normalization of DNA content in quantitative polymerase chain reaction (Q-PCR). Store at −20°C.

**2.6. Quantitative Tracking of Specific RV Integrants by Real-Time Q-PCR**

1. Custom-synthetized, single-stranded oligonucleotide primer complementary to a unique genomic sequence flanking a specific proviral integration site. Store at −20°C.

2. Retroviral LTR-specific oligonucleotide primer: 5′-GTTT-GCATCCGAATCGTGGT-3′.

3. Retroviral LTR-specific oligonucleotide probe: 6-(carboxyfluorescein)FAM-TCTCCTCTGAGTGATTGACTACCCAC-GACG-MGB. Store at −20°C.

4. TaqMan® Fast Universal PCR Master Mix (Applied Biosystem). Store at 4°C.

5. Bacterial plasmid containing a cloned, specific integration junction, or DNA from serial dilutions of a test population of T cells containing the specific integrant, with untransduced PB lymphocytes (from 1:10 to $1:10^4$). Store at −20°C.

**2.7. Analysis of the T-Cell Repertoire**

1. Mouse antihuman antibodies specific for the T-cell receptor variable (V) regions V2(a), V5(a), V5(b), V5.1, V5.2, V5.3, V8(a), V12.1, V 13.1, V14, V16, V17, V18, and V22 (Immunotech, Marseilles, France). Store at 4°C.

2. FITC-conjugated goat antimouse antibody. Store at 4°C.

3. Isotype-specific control antibody from the same manufacturer. Store at 4°C.

# 3. Methods

**3.1. Purification of T Cells from Peripheral Blood Samples and T-Cell Culture**

Peripheral blood mononuclear cells (PBMCs) are isolated by leukapheresis and Ficoll-Hypaque (Lymphoprep) cushion separation.
1. Dilute heparinized peripheral blood 1:4 with PBS/EDTA solution.

2. Layer the cell suspension on Ficoll-hypaque in a 2:1 ratio, and spin at 1,800 rpm for 30 min at 4°C. Do not use brake.

3. Carefully collect the mononuclear cell layer at the interphase, and wash twice with 10 ml of PBS/EDTA solution (spin for 10 min at 1,500 rpm).

The mononuclear cell fraction can be resuspended in the culture medium described in **Subheading 2.2**.

PBMCs can be used as such, or T lymphocytes purified by fluorescence-activated cell sorting (FACS) with anti-CD3 (pan-lymphocyte) antibodies, and/or fractionated into CD4$^+$ or CD8$^+$ T-cell fractions by FACS with FITC- or PE-conjugated anti-CD8 or anti-CD4 mAbs.

T cells can be expanded in culture from PBMC or purified lymphocyte fractions upon mitogen activation. Different activation signals can be used for this scope:

1. 2 μg/ml PHA.

2. 30 ng/ml anti-CD3 mAb (OKT3).

3. 30 ng/ml anti-CD3 plus 1 μg/ml anti-CD28 mAb.

Activated T cells are cultured at a density of 10$^6$ cells/ml in T-cell culture medium containing 100 U/ml rhIL-2. Medium and IL-2 are replaced every 3–4 days. If necessary, T cells may be restimulated after 3 weeks with CD3/CD28 mAbs.

CD4$^+$ or CD8$^+$ T-cell fractions can be obtained also after expansion in culture.

### 3.2. Cytofluorimetric Analysis of Purified T Cells

Purified T cells may be phenotyped by FACS analysis of the expression of specific surface markers upon single or multiple staining with FITC- and PE-conjugated antibodies. Commonly used markers are CD2, CD3, CD4, CD8, CD56, CD19, CD62L, CD28, and CD95. FITC- and PE-conjugated isotype-matched Abs are always used as controls.

Activation and differentiation of T cells may be evaluated by staining with FITC- and PE-conjugated antibodies against CD25, HLA-DR, CD69, CD45RA, CD45RO, and CCR7. CCR7 is revealed by PerCP-conjugated streptavidin after staining with an anti-CCR7 mAB and biotinylated antimouse IgM mAb. In particular, expression of CD45RA and CCR7 allows to identify the following T-cell phenotypes:

- CD45RA$^+$/CCR7$^+$: naive T cells

- CD45RA$^-$/CCR7$^-$: effector memory T cells

- CD45RA$^-$/CCR7$^+$: central memory T cells

- CD45RA$^+$/CCR7$^-$: terminally differentiated, effector T cells

Characterizing in detail a transduced T-cell population provides important information on their activity, evolution, and effector functions in vivo in a number of clinical applications (15, 18).

If the gene transfer vector carries a cell surface marker gene, such as the truncated form of the low-affinity receptor for nerve growth factor ( LNGFR), gene-marked cells can be purified by FACS or magnetic immunoselection with a specific antibody. Immunoselection allows to obtain a highly enriched bulk population of transduced T cells expressing the gene of interest.

Magnetic immunoselection procedure:

1. Stain the transduced T cell with the anti-LNGFR mAb for 30 min at 4°C.

2. Wash cells twice with PBS.

3. Stain the cells with goat-antimouse-IgG1-coated magnetic beads for 30 min at 4°C.

4. Wash cells twice with PBS.

5. Load the transduced cells on magnetic beads (Dynabeads M450; Dynal AS Oslo, Norway).

6. Eliminate the beads from selected cells by exposure to the magnet.

7. Plate the selected T cells in the fresh culture medium.

8. Analyze the purity of the selected T cells by FACS analysis with the same antibody used for selection (*see* **Note 1**).

**3.3. Extraction of DNA from Purified T Cells**

DNA is extracted from T cells by the QIAmp micro, mini, midi, or maxi column kit, depending on the available cell number. Simply follow the manufacturer's instructions in the kit handbook (*see* **Note 2**). DNA is stored in the final elution buffer at –20°C.

**3.4. Quantitative PCR Analysis of Gene-Marked T Cells**

In the absence of a cell surface marker, gene-marked T cells can be tracked pre- and post-transplantation by real-time Q-PCR by using the Taqman ABI 7900 real-time PCR machine (Applied Biosystem). Genomic DNA is extracted and quantified against a standard genomic DNA amplifying the 18S rRNA gene (provided as a kit by Applied Biosystem). This procedure allows the normalization of the PCR for DNA content.

To define the frequency of gene-marked T cells, or in case of uniformly transduced cell population, to estimate the average copy number of RV integrated in the T-cell genome, two primers and a TaqMan MGB probe (*see* **Note 3**) are designed by using the Primer Express® Software v3.0 (Applied Biosystem) on a unique region of the RV vector (*see* **Note 4**). For a correct estimate, an appropriate standard is represented by serial dilutions of genomic DNA from a T-cell clone containing a known copy number per cell of the same RV. Raw amplification data are analyzed using the Sequence Detection Software (Applied Biosystem) (*see* **Note 5**). The output is the number of copies of the amplified sequence per μg of input DNA in the PCR (*see* **Note 6**). The average number of integrated RV copies per cell can be calculated considering that one diploid cell contains 7.1 pg of DNA (*see* **Note 7**).

**3.5. Quantitative Tracking of Specific RV Integrants by Real-Time Q-PCR**

Several techniques, such as LM- or LAM-PCR, are now available to clone and sequence vector–genome junctions in a transduced T-cell population (*see* Chapter ??). Some of the vector integration sites may have specific characteristics (e.g., they are within or close to a proto-oncogene) *(17)* that makes their specific tracking a necessary or desirable component of a clinical follow-up. To determine the relative contribution of a T clone carrying a specific integrant in peripheral blood over time, real-time Q-PCR is set up to specifically amplify a specific proviral–genome junction *(19)*.

Q-PCR is performed by a Taqman ABI 7900 real-time PCR machine (Applied Biosystem). Genomic DNA is extracted from T cells by the QIAamp DNA Blood kit, as described in **Subheading 3.3** (*see* **Note 8**) Normalization of DNA content is performed as described in **Subheading 3.4**.

To estimate the relative frequency of a specific integrant, primers and TaqMan MGB probe are designed by using the Primer Express® software. In particular, one primer and the probe are designed on the LTR region of the RV, while the other primer is designed on a unique genomic sequence flanking the proviral–genome junction (*see* **Note 9**).

To track a specific integration, the standard in the PCR is represented by serial dilutions of plasmid DNA containing the specific vector–genome junction. This is usually the plasmid coming from the LM- or LAM-PCR mapping of that particular integration site (*see* Chapter ??). As described before, the raw amplification data are analyzed using the Sequence Detection Software, and the frequency of the cells carrying the specific RV integration is estimated as described in **Subheading 3.4** (*see* **Note 10**).

**3.6. Analysis of the T-Cell Repertoire**

The clonal composition of a gene-marked T-cell population can be established also by the analysis of the T-cell receptor (TCR) Vβ-chain repertoire. A polyclonal population is characterized by a wide array of T-cell receptor rearrangements, and by usage of a wide repertoire of Vβ-chains. Oligoclonality is typically accompanied by a reduction of the repertoire. Oligoclonality of a gene-marked T-cell population may be the consequence of different factors, among which is the emergence of dominant T-cell clones caused by an insertion-related expansion *(6)*. Monitoring clonality is therefore a desirable component of a clinical follow-up.

Analysis of the TCR repertoire is performed by staining T cells with a panel of mouse antibodies specific for different Vβ chains, such as (V2(a), V5(a), V5(b), V5.1, V5.2, V5.3, V8(a), V12.1, V 13.1, V14, V16, V17, V18, and V22, followed by FACS detection after binding to a FITC-conjugated goat anti-mouse antibody.

# 4. Notes

1. Enrichment depends on the starting concentration of gene-marked T cells. If this is higher than 5%, enrichment can be as high as 95%.

2. Elute the genomic DNA in the lower volume suggested by the kit handbook, in order to have DNA concentrated as much as possible.

3. An MGB probe is an oligonucleotide with a reporter fluorescent dye (such as FAM) attached to its 5′-end, and a nonfluorescent quencher attached to its 3′-end. The probe is coupled with a minor groove binder (MGB) which increases its melting temperature ($T_m$). The MGB probe should be as shorter as possible, without being shorter than 13 nucleotides. Because of the asymmetric placement of the minor groove binder at the 3′-end, complementary MGB probes do not necessarily have the same Tm as sense probe sequences. It is therefore necessary to test the Tm of the complement MGB probe sequence in the TaqMan MGB probe Test Document section of the Primer Express® software. This kind of TaqMan probe is recommended with respect to the 6-carboxy-4,7,2′,7′-tetrachlorofluorescein (TET)-TAMRA probe. A TAMRA probe is an oligonucleotide with a reporter fluorescent dye (TET, FAM or VIC) attached to its 5′-end and a quencher fluorescent dye (TAMRA) attached to its 3′-end. The presence of fluorescent quencher can interfere with the reporter dye fluorescence that is measured by a fluorescent sequence detector.

4. Amplification primers can be designed on vector-specific sequences or on the cDNA sequence of the transgene. If the transgene is a human cell gene, the region delimited by the two primers should contain at least one intron in the corresponding genomic sequence. This allows avoiding amplification of genomic DNA and the unequivocal identification of a vector-derived amplification signal.

5. $C_t$ (threshold cycle) of the reactions should be between 15 and 30. An optimal slope for the standard curve is −3.3.

6. Fifty nanograms of genomic DNA is an optimal amount in a quantitative PCR.

7. Q-PCR gives an average vector copy number in the input DNA. In the case of a uniformly transduced cell population, this value corresponds to the actual vector copy number per cell. If the percentage of transduced cells in the input sample is lower than 100%, the vector copy number value can be used to estimate it, but cannot give at the same time the

percentage of transduced cells and the average vector copy number per cell. As an example, an average copy number of 0.2 indicates that the starting cell population contains 20% of gene-marked cells assuming one integrated provirus per cell, 10% assuming two integrated proviruses per cells, and so on. In general, T-lymphocytes transduced by an RV at the currently used multiplicity of infection contain an average of one to two copies of integrated provirus per cell.

8. For this specific purpose, a high proportion of transduced cells in the starting sample is highly desirable. Therefore, it is strongly suggested to start with a T-cell pure population (e.g., a CD3$^+$ population), and, whenever possible, to try to enrich for cells that express the transgene.

9. The optimal DNA amplification length is around 100 nucleotides. This factor should be considered in designing the genome-specific primer, which should be at a distance from the integrated provirus that allows amplifying a DNA fragment as close as possible to 100 nucleotides.

10. If a T-cell clone carrying the specific integration is available, an ideal standard for the Q-PCR can be obtained by diluting the clone with untransduced PB lymphocytes in a proportion ranging from 1:10 to 1:10$^{-4}$. With such standard, the frequency of the cells carrying the specific integrant in the test sample can be directly calculated by comparing the amplification signal with that of the standard curve, after normalization for DNA content.

## Acknowledgments

The authors would like to thank Chiara Bonini for her support and critical reading of the manuscript.

## References

1. Cavazzana-Calvo, M., Hacein-Bey, S., de Saint Basile, G., Gross, F., Yvon, E., Nusbaum, P., Selz, F., Hue, C., Certain, S., Casanova, J. L., Bousso, P., Deist, F. L., and Fischer, A. (2000) Gene therapy of human severe combined immunodeficiency (SCID)-X1 disease, *Science* 288, 669–72.

2. Aiuti, A., Vai, S., Mortellaro, A., Casorati, G., Ficara, F., Andolfi, G., Ferrari, G., Tabucchi, A., Carlucci, F., Ochs, H. D., Notarangelo, L. D., Roncarolo, M. G., and Bordignon, C. (2002) Immune reconstitution in ADA-SCID after PBL gene therapy and discontinuation of enzyme replacement, *Nat Med* 8, 423–5.

3. Hacein-Bey-Abina, S., Le Deist, F., Carlier, F., Bouneaud, C., Hue, C., De Villartay, J. P., Thrasher, A. J., Wulffraat, N., Sorensen, R., Dupuis-Girod, S., Fischer, A., Davies, E. G., Kuis, W., Leiva, L., and Cavazzana-Calvo, M. (2002) Sustained correction of X-linked severe combined immunodeficiency by ex vivo gene therapy, *N Engl J Med* 346, 1185–93.

4. Schmidt, M., Carbonaro, D. A., Speckmann, C., Wissler, M., Bohnsack, J., Elder, M., Aronow, B. J., Nolta, J. A., Kohn, D. B., and von Kalle, C. (2003) Clonality analysis after retroviral-mediated gene transfer to CD34+ cells from the cord blood of ADA-deficient SCID neonates, *Nat Med* 9, 463–8.

5. Ott, M. G., Schmidt, M., Schwarzwaelder, K., Stein, S., Siler, U., Koehl, U., Glimm, H., Kuhlcke, K., Schilz, A., Kunkel, H., Naundorf, S., Brinkmann, A., Deichmann, A., Fischer, M., Ball, C., Pilz, I., Dunbar, C., Du, Y., Jenkins, N. A., Copeland, N. G., Luthi, U., Hassan, M., Thrasher, A. J., Hoelzer, D., von Kalle, C., Seger, R., and Grez, M. (2006) Correction of X-linked chronic granulomatous disease by gene therapy, augmented by insertional activation of MDS1-EVI1, PRDM16 or SETBP1, *Nat Med* 12, 401–9.

6. Hacein-Bey-Abina, S., Von Kalle, C., Schmidt, M., McCormack, M. P., Wulffraat, N., Leboulch, P., Lim, A., Osborne, C. S., Pawliuk, R., Morillon, E., Sorensen, R., Forster, A., Fraser, P., Cohen, J. I., De Saint Basile, G., Alexander, I., Wintergerst, U., Frebourg, T., Aurias, A., Stoppa-Lyonnet, D., Romana, S., Radford-Weiss, I., Gross, F., Valensi, F., Delabesse, E., Macintyre, E., Sigaux, F., Soulier, J., Leiva, L. E., Wissler, M., Prinz, C., Rabbitts, T. H., Le Deist, F., Fischer, A., and Cavazzana-Calvo, M. (2003) LMO2-associated clonal T cell proliferation in two patients after gene therapy for SCID-X1, *Science* 302, 415–9.

7. Bordignon, C., Notarangelo, L. D., Nobili, N., Ferrari, G., Casorati, G., Panina, P., Mazzolari, E., Maggioni, D., Rossi, C., Servida, P., Ugazio, A. G., and Mavilio, F. (1995) Gene therapy in peripheral blood lymphocytes and bone marrow for ADA-immunodeficient patients., *Science* 270, 470–75.

8. Blaese, R. M., Culver, K. W., Miller, A. D., Carter, C.S., Fleisher, T., Clerici, M., Shearer, G., Chang, L., Chiang, W., Tolstoshev, P., Greenblatt, J. J., Rosenberg, S. A., Klein, H., Berger, M., Mullen, C. A., Ramsey, W. J., Muul, L., Morgan, R. A., and Anderson, W. F. (1995) T Lymphocyte-directed gene therapy for ADA-SCID: initial trial results after 4 years, *Science* 270, 475–80.

9. Levine, B. L., Humeau, L. M., Boyer, J., MacGregor, R. R., Rebello, T., Lu, X., Binder, G. K., Slepushkin, V., Lemiale, F., Mascola, J. R., Bushman, F. D., Dropulic, B., and June, C. H. (2006) Gene transfer in humans using a conditionally replicating lentiviral vector, *Proc Natl Acad Sci U S A* 103, 17372–7.

10. Dropulic, B., and June, C. H. (2006) Gene-based immunotherapy for human immunodeficiency virus infection and acquired immunodeficiency syndrome, *Hum Gene Ther* 17, 577–88.

11. Russo, V., Tanzarella, S., Dalerba, P., Rigatti, D., Rovere, P., Villa, A., Bordignon, C., and Traversari, C. (2000) Dendritic cells acquire the MAGE-3 human tumor antigen from apoptotic cells and induce a class I-restricted T cell response, *Proc Natl Acad Sci U S A* 97, 2185–90.

12. Bonini, C., Ferrari, G., Verzeletti, S., Servida, P., Zappone, E., Ruggieri, L., Ponzoni, M., Rossini, S., Mavilio, F., Traversari, C., and Bordignon, C. (1997) HSV-TK gene transfer into donor lymphocytes for control of allogeneic graft-versus-leukemia, *Science* 276, 1719–24.

13. Ciceri, F., Bonini, C., Marktel, S., Zappone, E., Servida, P., Bernardi, M., Pescarollo, A., Bondanza, A., Peccatori, J., Rossini, S., Magnani, Z., Salomoni, M., Benati, C., Ponzoni, M., Callegaro, L., Carradini, P., Bregni, M., Traversari, C., and Bordignon, C. (2007) Anti-tumor effects of HSV-TK engineered donor lymphocytes after allogeneic stem cell transplanation, *Blood*.

14. Verzeletti, S., Bonini, C., Marktel, S., Nobili, N., Ciceri, F., Traversari, C., and Bordignon, C. (1998) Herpes simplex virus thymidine kinase gene transfer for controlled graft-versus-host disease and graft-versus-leukemia: clinical follow-up and improved new vectors, *Hum Gene Ther* 9, 2243–51.

15. Marktel, S., Magnani, Z., Ciceri, F., Cazzaniga, S., Riddell, S. R., Traversari, C., Bordignon, C., and Bonini, C. (2003) Immunologic potential of donor lymphocytes expressing a suicide gene for early immune reconstitution after hematopoietic T-cell-depleted stem cell transplantation, *Blood* 101, 1290–8.

16. Mavilio, F., Ferrari, G., Rossini, S., Nobili, N., Bonini, C., Casorati, G., Traversari, C., and Bordignon, C. (1994) Peripheral blood lymphocytes as target cells of retroviral vector-mediated gene transfer, *Blood* 83, 1988–97.

17. Recchia, A., Bonini, C., Magnani, Z., Urbinati, F., Sartori, D., Muraro, S., Tagliafico, E., Bondanza, A., Stanghellini, M. T., Bernardi, M., Pescarollo, A., Ciceri, F., Bordignon, C., and Mavilio, F. (2006) Retroviral vector integration deregulates gene expression but has no consequence on the biology and function of transplanted T cells, *Proc Natl Acad Sci U S A* 103, 1457–62.

18. Bondanza, A., Valtolina, V., Magnani, Z., Ponzoni, M., Fleischhauer, K., Bonyhadi, M., Traversari, C., Sanvito, F., Toma, S., Radrizzani, M., La Seta-Catamancio, S., Ciceri, F., Bordignon, C., and Bonini, C. (2006) Suicide gene therapy of graft-versus-host disease induced by

central memory human T lymphocytes, *Blood* 107, 1828–36.

19. Aiuti, A., Cassani, B., Andolfi, G., Mirolo, M., Biasco, L., Recchia, A., Urbinati, F., Valacca, C., Scaramuzza, S., Aker, M., Slavin, S., Cazzola, M., Sartori, D., Ambrosi, A., Di Serio, C., Roncarolo, M. G., Mavilio, F., and Bordignon, C. (2007) Multilineage hematopoietic reconstitution without clonal selection in ADA-SCID patients treated with stem cell gene therapy, *J Clin Invest* 117, 2233–40.

# Chapter 27

## DNA Microarray Studies of Hematopoietic Subpopulations

Karin Pike-Overzet, Dick de Ridder, Tom Schonewille, and Frank J.T. Staal

### Summary

In recent years, one of the most quickly incorporated methods in biomedical research has been microarray technology. Microarrays have been designed for the purpose of genotyping (e.g., SNP analysis), expression analysis (mRNA, miRNA, exon arrays), chromatin immunoprecipitations (ChIP-on-chip technology), and DNA sequencing. In this chapter we will focus on the application of DNA microarrays in gene expression analysis of mRNA. This technology allows for the simultaneous analysis of transcription patterns of literally thousands of genes, making it a very powerful approach. Hematopoietic subpopulations are relatively easy to obtain by using fluorescence-activated cell sorting or magnetic bead cell separation. Over the last decade, the combination of these purification techniques with microarray analysis has resulted in an enormous gain of knowledge of blood lineages and their development.

**Key words:** Expression analysis, Microarray, RNA quality.

### 1. Introduction

Determining gene expression profiles by using DNA microarrays is applicable to many areas of medicine and biology, such as studying the effect of treatment, disease classification, and identification of developmental stages. DNA microarrays are usually glass or silicone slides that have thousands to millions different DNA clones or oligonucleotide probes adhered to them. Each clone or probe is positioned in a precise spot on the array called a probe cell. A probe cell holds $10^5$–$10^6$ copies of a specified DNA clone or oligonucleotide corresponding to a certain (partial) gene transcript. Some DNA microarrays are enclosed in cartridges for easy handling during shipping and laboratory procedures. Ready-made arrays can be purchased from one of the many microarray technology companies.

Christopher Baum (ed.), *Methods in Molecular Biology, Methods and Protocols*, vol. 506
© Humana Press, a part of Springer Science+Business Media, LLC 2009
DOI: 10.1007/978-1-59745-409-4_27

Alternatively, several companies as well as local array facilities could supply custom-made arrays. Generally two kinds of DNA microarrays can be distinguished: cDNA (spotted) microarrays and oligonucleotide microarrays. Spotted microarrays are usually hybridized with cDNA from two samples to be compared, labeled with two different fluorophores, e.g., Rhodamine (red) and Fluorescein (green). The samples are mixed and hybridized to a single microarray that is then scanned, allowing the visualization of up-regulated and down-regulated genes using a single chip. The drawback of this approach is that the absolute levels of gene expression cannot be measured. Oligonucleotide microarrays give estimates of the absolute value of gene expression. Consequently, the comparison of two conditions requires the use of two separate microarrays. Samples for oligonucleotide microarrays are generally labeled with a single fluorophore.

Even though many platforms for expression analysis exist, in this chapter we will focus on the Affymetrix oligonucleotide microarray, as it is the most widely used platform. This platform yields highly reproducible results due to its standardized production process and experimental equipment. All reagents and methods are described as recommended by Affymetrix.

For expression analysis on Affymetrix microarrays, total RNA has to be isolated from, preferably purified, cell populations. Subsequently, double-stranded cDNA is generated. An in vitro transcription (IVT) reaction is then carried out to produce biotin-labeled cRNA. The cRNA is fragmented before it is hybridized to the array during a 16-h incubation. Directly following hybridization, the array undergoes an automatic washing and staining procedure on a fluidics station. As a final step, the array is scanned and data can be stored and analyzed.

## 2. Materials

### 2.1. General Supplies

1. Sterile, RNase-free, 1.5-mL microcentrifuge vials.
2. Sterile-barrier, RNase-free pipette tips.
3. Water used in the protocols is molecular biology grade, i.e., nuclease free.

### 2.2. RNA Isolation

1 RNeasy Mini Kit, QIAGEN, Hilden, Germany.
2. 14.3 M β-mercaptoethanol.
3. Ethanol, 96–100% (v/v).
4. QIAshredder homogenizer, QIAGEN, Hilden, Germany.
5. DEPC-treated water.

**2.3. Assessment of RNA Quantity and Quality**

1. Spectrophotometer.
2. Agilent 2100 Bioanalyzer.

**2.4 Preparation of Biotin-Labeled cRNA**

1. For preparation of Poly-A RNA controls, Poly-A RNA Control Kit (Affymetrix, Santa Clara, CA, USA).
2. For first-strand cDNA synthesis, one-Cycle cDNA Synthesis Kit (Affymetrix, Santa Clara, CA, USA).
3. For second-strand cDNA synthesis, one-Cycle cDNA Synthesis Kit (Affymetrix, Santa Clara, CA, USA).
4. For cleanup of double-stranded cDNA, Sample Cleanup Module (Affymetrix, Santa Clara, CA, USA) and ethanol, 96–100% (v/v).
5. For synthesis of biotin-labeled cRNA, IVT Labeling Kit (Affymetrix, Santa Clara, CA, USA).
6. For cleanup and quantification of biotin-labeled cRNA, ethanol, 96–100% (v/v); ethanol, 80% (v/v); spectrophotometer; Sample Cleanup Module (Affymetrix, Santa Clara, CA, USA).
7. For *fragmentation of biotin-labeled cRNA*, Sample Cleanup Module (Affymetrix, Santa Clara, CA, USA).

**2.5 Eukaryotic Target Hybridization**

1. Water, Molecular Biology Grade.
2. Bovine Serum Albumin (BSA) solution (50 mg/mL).
3. Herring Sperm DNA.
4. GeneChip Eukaryotic Hybridization Control Kit, Affymetrix, Santa Clara, CA, USA (contains control cRNA and control oligo B2).
5. DMSO.
6. 12× MES stock buffer, filtered: 1.22 M MES, 0.89 M [Na+], pH 6.5–6.7 (Do not autoclave. Store at 2–8°C, and shield from light. Discard solution if yellow).
7. 2× Hybridization buffer: 200 mM MES, 2 M [Na+], 40 mM EDTA, 0.02% Tween-20, store at 2–8°C and shield from light.
8. Microarrays.
9. Hybridization Oven 640, Affymetrix, Santa Clara, CA, USA.

**2.6. Washing, Staining, and Scanning**

1. Water, Molecular Biology Grade.
2. 50 mg/mL Bovine Serum Albumin (BSA) solution.
3. 1 mg/mL R-Phycoerythrin Streptavidin (SAPE), Molecular Probes, Invitrogen Corporation, Carlsbad, CA, USA. Store in the dark at 4°C, either foil-wrapped or kept in an amber tube. Do not freeze SAPE.
4. 20× SSPE: 3 M NaCl, 0.2 M $NaH_2PO_4$, 0.02 M EDTA.
5. Antistreptavidin antibody (goat), biotinylated.

6. Wash buffer A, Nonstringent wash buffer, filtered: 6× SSPE, 0.01% Tween-20.

7. Wash buffer B: stringent wash buffer, filtered: 100 mM MES, 0.1 M [Na⁺], 0.01% Tween-20, store at 2–8°C, and shield from light.

8. 2× Stain buffer: 200 mM MES, 2 M [Na⁺], 0.1% Tween-20.

9. Goat IgG stock: 10 mg/mL, 150 mM NaCl. Store at 4°C.

10. GeneChip® Fluidics Station 450, Affymetrix, Santa Clara, CA, USA.

11. Tough-Spots, Label Dots, USA Scientific, Ocala, FL, USA.

12. GeneArray® Scanner or GeneChip® Scanner 3000, Affymetrix, Santa Clara, CA, USA.

### 2.7. Basic Quality Control

1. Affymetrix GeneChip© Operating Software (GCOS). Affymetrix, Santa Clara, CA, USA.

### 2.8. Data Analysis

1. Affymetrix GeneChip© Operating Software (GCOS).

2. Data analysis tools: Rosetta Resolver (commercial *(1)*), Bioconductor (open source *(2)*), BRB Arraytools (free for nonprofit use *(3)*), Cluster/Treeview (free for nonprofit use *(4)*), Spotfire Decision Site (commercial *(5)*) or other software.

3. Annotation tools: NetAffx (free *(6)*), Ingenuity Pathways Analysis (commercial *(7)*), GenMAPP (open source *(8)*), DAVID (free *(9)*), or other software.

## 3. Methods

### 3.1. RNA Isolation (See Note 1)

The cells of interest are used for isolation of total RNA, using the RNeasy Mini Kit from QIAGEN. Up to $1 \times 10^7$ cells can be used for each purification per tube and set of reagents. It is important to use the right amount of starting material in order to get optimal RNA yield and purity. Depending on the study design, populations need to be purified before RNA is isolated by magnetic bead cell separation or fluorescence-activated cell sorting (*see* **Note 2**). Efficient disruption and homogenization of the starting material is an absolute requirement for all total RNA purification procedures. Homogenized cell lysates (from **step 3**) can be stored at –70°C for several months, although some drop in quality occurs over time.

1. Add 10 µL b-mercaptoethanol per 1 mL buffer RLT needed. Make a fresh solution each day.

2. For pelleted cells, loosen the cell pellet thoroughly by flicking the tube. Disrupt the cells by adding the appropriate volume

of buffer RLT supplemented with β-mercaptoethanol (*see* **Note 3**).

3. Homogenize the lysate by pipetting the lysate directly into a QIAshredder spin column placed in a 2-mL collection tube, and centrifuge for 2 min at full speed. At this point the homogenized lysate can be stored at –70°C for months.

4. Add 1 volume of 70% ethanol to the homogenized lysate, and mix well by pipetting. Do not centrifuge.

5. Transfer up to 700 μL of the sample, including any precipitate that may have formed, to an RNeasy spin column placed in a 2-mL collection tube. Close the lid gently, and centrifuge for 15 s at ≥8,000 × $g$. Discard the flow-through. If the sample volume exceeds 700 μL, centrifuge successive aliquots in the same RNeasy spin column. Discard the flow-through after each centrifugation. Reuse the collection tube in **step 6**.

6. Add 700 μL buffer RW1 to the RNeasy spin column. Close the lid gently. Incubate at room temperature for 5 min and centrifuge for 15 s at ≥38,000 × $g$ to wash the spin column membrane. Discard the flow-through. Reuse the collection tube in **step 7**.

7. Add 500 μL buffer RPE to the RNeasy spin column. Close the lid gently, and centrifuge for 15 s at ≥38,000 × $g$ to wash the spin column membrane. Discard the flow-through. Reuse the collection tube in **step 8**.

8. Add 500 μL buffer RPE to the RNeasy spin column. Close the lid gently, and centrifuge for 2 min at ≥8,000 × $g$ to wash the spin column membrane.

The long centrifugation dries the spin column membrane, ensuring that no ethanol is carried over during RNA elution. Residual ethanol may interfere with downstream reactions. After centrifugation, carefully remove the RNeasy spin column from the collection tube so that the column does not contact the flow-through. Otherwise, carryover of ethanol will occur.

9. Place the RNeasy spin column in a new 2-mL collection tube and discard the old collection tube with the flow-through. Close the lid gently, and centrifuge at full speed for 1 min. Perform this step to eliminate any possible carryover of buffer RPE, or if residual flow-through remains on the outside of the RNeasy spin column after **step 8**.

10. Place the RNeasy spin column in a new 1.5-mL collection tube. Add 30–50 μL RNase-free water directly to the spin column membrane. Close the lid gently, and centrifuge for 1 min at ≥8,000 × $g$ to elute the RNA. When working with RNA from low cell numbers, elute using 25 μL RNase-free water.

### 3.2. Assessment of RNA Quantity and Quality

1. Quantify RNA yield by spectrophotometric analysis using the convention that 1 absorbance unit at 260 nm equals 40 µg/mL RNA. The absorbance should be checked at 260 and 280 nm for determination of sample concentration and purity. The A260/A280 ratio should be close to 2.0 for pure RNA. Ratios between 1.9 and 2.1 are acceptable.

2. Integrity of total RNA samples should also be assessed qualitatively on an Agilent 2100 Bioanalyzer. If the RNA is of poor quality, it should not be used for further DNA microarray analysis (*see* **Note 4**). **Figures 1** and **2** show the analysis of an RNA sample of good and bad quality, respectively.

### 3.3. Preparation of Poly-A RNA Controls

1. The poly-A RNA control stock and poly-A control dilution buffer are provided with the Poly-A RNA Control Kit. Prepare the appropriate serial dilutions based on **Table 1**. The first dilution of the poly-A RNA controls can be stored up to 6 weeks in a nonfrost-free freezer at –20°C and frozen-thawed up to eight times.

### 3.4. First-Strand cDNA Synthesis

Briefly spin down all tubes in the one-Cycle cDNA Synthesis Kit before using the reagents.

1. Place total RNA (1–15 µg) in a 1.5-mL tube.
2. Add 2 µL of the appropriately diluted poly-A RNA controls.
3. Add 2 µL of 50 µM T7-Oligo(dT) Primer.
4. Add RNase-free water to a final volume of 12 µL.
5. Gently flick the tube a few times to mix, and then centrifuge briefly (~5 s) to collect the reaction at the bottom of the tube.
6. Incubate the reaction for 10 min at 70°C.

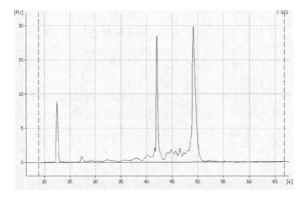

Fig. 1 Electropherogram from the Agilent 2100 Bioanalyzer for total RNA of good quality. For a high-quality total RNA sample, two well-defined peaks corresponding to the 18S (around 42 s) and 28S (around 48 s) ribosomal RNAs should be observed with surface ratios approaching 2:1 for the 28S to 18S bands.

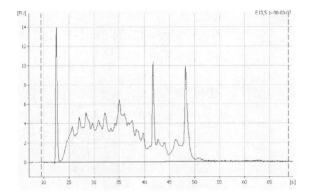

Fig. 2 Electropherogram from the Agilent 2100 Bioanalyzer for total RNA of poor quality. The surface ratio for the 28S to 18S bands is not 2:1 and a lot of breakdown products are observed. When the electropherogram looks like this, the sample should not be used for further analysis.

### Table 1
### Serial dilutions of poly-A RNA controls

| Starting amount | Serial dilutions | | | Spike-in |
|---|---|---|---|---|
| Total RNA (μg) | First | Second | Third | volume (μL) |
| 1 | 1:20 | 1:50 | 1:50 | 2 |
| 1.5 | 1:20 | 1:50 | 1:37.5 | 2 |
| 2 | 1:20 | 1:50 | 1:25 | 2 |
| 5 | 1:20 | 1:50 | 1:10 | 2 |

7. Cool the sample at 4°C for at least 2 min.

8. Centrifuge the tube briefly to collect the sample at the bottom of the tube.

9. In a separate tube, assemble the First-Strand Master Mix. Prepare sufficient First-Strand Master Mix for all the RNA samples. When there are more than four samples, it is wise to include additional material to compensate for potential pipetting inaccuracy or solution lost during the process. The recipe in **Table 2** is for a single reaction.

10. Mix well by flicking the tube a few times. Centrifuge briefly to collect the master mix at the bottom of the tube.

11. Transfer 7 μL of First-Strand Master Mix to each RNA/T7-Oligo(dT) Primer mix for a final volume of 19 μL. Mix thoroughly by flicking the tube a few times.

**Table 2**
**Preparation of the First-Strand Master Mix**
**(for a single reaction)**

| Component | Volume (µL) |
| --- | --- |
| 5× First-Strand Reaction Mix | 4 |
| DTT, 0.1 M | 2 |
| dNTP, 10 mM | 1 |
| Total volume | 7 |

12. Centrifuge briefly to collect the reaction at the bottom of the tube, and immediately place the tubes at 42°C and incubate for 2 min.

13. Add the appropriate amount of SuperScript II to each RNA sample for a final volume of 20 µL. For 1–8 µg of total RNA: 1 µL SuperScript II. For 8.1–15 µg of total RNA: 2 µL SuperScript II.

14. Mix thoroughly by flicking the tube a few times. Centrifuge briefly to collect the reaction at the bottom of the tube, and immediately place the tubes at 42°C. Incubate for 1 h at 42°C, then cool the sample for at least 2 min at 4°C (*see* **Note 5**).

15. After incubation at 4°C, centrifuge the tube briefly to collect the reaction at the bottom of the tube and immediately proceed to steps in **3.5**.

**3.5. Second-Strand cDNA Synthesis**

1. In a separate tube, assemble sufficient Second-Strand Master Mix for all the samples. It is recommended to prepare Second-Strand Master Mix immediately before use. The recipe in **Table 3** is for a single reaction.

2. Mix well by gently flicking the tube a few times. Centrifuge briefly to collect the solution at the bottom of the tube.

3. Add 130 µL of Second-Strand Master Mix to each first-strand synthesis sample for a total volume of 150 µL. Gently flick the tube a few times to mix, and then centrifuge briefly to collect the reaction at the bottom of the tube.

4. Incubate for 2 h at 16°C.

5. Add 2 µL of T4 DNA Polymerase to each sample and incubate for 5 min at 16°C.

6. After incubation with T4 DNA Polymerase add 10 µL of EDTA, 0.5 M, and proceed to steps in **3.6**.

**3.6. Cleanup of Double-Stranded cDNA**

1. Add 600 µL of cDNA-binding buffer to the double-stranded cDNA synthesis preparation. Mix by vortexing for 3 s. Check

**Table 3**
**Preparation of the Second-Strand Master Mix**
**(for a single reaction)**

| Component | Volume (μL) |
|---|---|
| RNase-free water | 91 |
| 5× Second-Strand Reaction Mix | 30 |
| dNTP, 10 mM | 3 |
| E. coli DNA ligase | 1 |
| E. coli DNA Polymerase I | 4 |
| RNase H | 1 |
| Total volume | 130 |

that the color of the mixture is yellow (similar to cDNA-binding buffer without the cDNA synthesis reaction).

2. Apply 500 μL of the sample to the cDNA Cleanup Spin Column sitting in a 2-mL Collection Tube.

3. Centrifuge for 1 min at ≥8,000 × $g$. Discard flow-through.

4. Reload the spin column with the remaining mixture and centrifuge as earlier. Discard flow-through and Collection Tube.

5. Transfer spin column into a new 2-mL Collection Tube. Pipet 750 μL of the cDNA wash buffer onto the spin column.

6. Centrifuge for 1 min at ≥8,000 × $g$. Discard flow-through.

7. At this time, it is recommended to label this collection tube with the sample name. During centrifugation in this step caps may break, resulting in loss of sample information. Open the cap of the spin column and centrifuge for 5 min at maximum speed (≤25,000 × $g$). Place columns into the centrifuge using every second bucket. Position caps over the adjoining bucket so that they are oriented in the opposite direction to the rotation. This avoids damage of the caps. Centrifugation with open caps allows complete drying of the membrane. Discard flow-through and Collection Tube.

8. Transfer the spin column into a 1.5-mL Collection Tube, and pipet 14 μL of cDNA elution buffer directly onto the spin column membrane. Incubate for 1 min at room temperature.

9. Centrifuge 1 min at maximum speed (≤25,000 × $g$) to elute. The average volume of eluate is 12 μL from 14 μL elution buffer.

10. After cleanup, proceed to steps in **3.7**.

*3.7. Synthesis of Biotin-Labeled cRNA*

1. Add reaction components to the template cDNA in the order indicated in **Table 4.** If more than four IVT reactions are to be performed, a master mix can be prepared by multiplying the reagent volumes by the number of reactions. Do not assemble the reaction on ice, since spermidine in the 10× IVT labeling buffer can lead to precipitation of the template cDNA.

2. Carefully mix the reagents and collect the mixture at the bottom of the tube by brief centrifugation.

3. Incubate at 37°C for 16 h (*see* **Note 6**).

4. Store labeled cRNA at –20 or –70°C if not purifying immediately. Alternatively, proceed to steps in **3.8.**

*3.8. Cleanup and Quantification of Biotin-Labeled cRNA*

1. Add 60 μL of RNase-free water to the IVT reaction and mix by vortexing for 3 s.

2. Add 350 μL IVT cRNA-binding buffer to the sample and mix by vortexing for 3 s.

3. Add 250 μL ethanol (96–100%) to the sample, and mix well by pipetting. Do not centrifuge.

4. Apply sample (700 μL) to the IVT cRNA Cleanup Spin Column sitting in a 2-mL Collection Tube. Centrifuge for 15 s at ≥8,000 × *g*. Discard flow-through and Collection Tube.

5. Transfer the spin column into a new 2-mL Collection Tube. Pipet 500 μL IVT cRNA wash buffer onto the spin column.

6. Centrifuge for 15 s at ≥8,000 × *g* to wash. Discard flow-through.

7. Pipet 500 μL 80% (v/v) ethanol onto the spin column.

8. Centrifuge for 15 s at ≥8,000 × *g*. Discard flow-through.

## Table 4
## IVT reaction (for a single reaction)

| Component | Volume |
| --- | --- |
| Template cDNA | All (±13 μL) |
| RNase-free water | variable (±7 μL) |
| 10× IVT labeling buffer | 4 μL |
| IVT labeling NTP mix | 12 μL |
| IVT labeling enzyme mix | 4 μL |
| Total volume | 40 μL |

9. At this time, it is recommended to label this collection tube with the sample name. During centrifugation in this step, caps may break, resulting in loss of sample information. Open the cap of the spin column and centrifuge for 5 min at maximum speed ($\leq$25,000 × $g$). Place columns into the centrifuge using every second bucket. Position caps over the adjoining bucket so that they are oriented in the opposite direction to the rotation. This avoids damage of the caps.

10. Transfer spin column into a new 1.5-mL Collection Tube and pipet 11 µL of RNase-free water directly onto the spin column membrane.

11. Centrifuge 1 min at maximum speed ($\leq$25,000 × $g$) to elute.

12. Pipet 10 µL of RNase-free water directly onto the spin column membrane.

13. Centrifuge 1 min at maximum speed ($\leq$25,000 × $g$) to elute.

14. For subsequent photometric quantification of the purified cRNA, it is recommended to use the Nanodrop. Store cRNA at –20°C, or –70°C if not quantified immediately. Alternatively, proceed to **step 15**.

15. Quantify cRNA yield by spectrophotometric analysis using the convention that 1 absorbance unit at 260 nm equals 40 µg/mL RNA. The absorbance should be checked at 260 and 280 nm for determination of sample concentration and purity. The A260/A280 ratio should be close to 2.0 for pure RNA. Ratios between 1.9 and 2.1 are acceptable.

16. For quantification of cRNA when using total RNA as starting material, an adjusted cRNA yield must be calculated to reflect carryover of unlabeled total RNA. Using an estimate of 100% carryover, use the formula given later to determine adjusted cRNA yield:
    Adjusted cRNA yield = RNAm – (total RNAi) ($y$)
    RNAm = amount of cRNA measured after IVT (µg)
    Total RNAi = starting amount of total RNA (µg)
    $y$ = fraction of cDNA reaction used in IVT
    Example: Starting with 10 µg total RNA, 50% of the cDNA reaction is added to the IVT, giving a yield of 50 µg cRNA. Therefore, adjusted cRNA yield = 50 µg cRNA – (10 µg total RNA) (0.5 cDNA reaction) = 45.0 µg.

17. Use adjusted yield in **3.9**. Store cRNA at –20°C short-term, or –70°C long-term.

***3.9. Fragmentation of Biotin-Labeled cRNA***

1. Add sufficient fragmentation buffer to all the cRNA samples from **3.8** according to **Table 5**. Refer to specific probe array package insert for information on array format.

**Table 5**
**Sample fragmentation reaction by array format**

| Component | 49/64 Format (Standard) | 100 Format (Mini) |
|---|---|---|
| cRNA | 20 μg (1–21 μL) | 15 μg (1–21 μL) |
| 5× Fragmentation buffer | 8 μL | 6 μL |
| RNase-free water | To 40 μL final volume | To 30 μL final volume |
| Total volume | 40 μL | 30 μL |

2. Incubate at 94°C for 35 min. Put on ice following the incubation. The standard fragmentation procedure should produce a distribution of RNA fragment sizes from approximately 35 to 200 bases. Store undiluted, fragmented sample cRNA at –20°C (or –70°C for longer term storage) until ready to perform the hybridization.

***3.10. Eukaryotic Target Hybridization***

1. Please refer to **Table 7** for the necessary amount of cRNA required for a specific probe array format. These recipes take into account that it is necessary to make extra hybridization cocktail due to a small loss of volume (10–20 μL) during each hybridization.

2. Equilibrate probe array to room temperature immediately before use.

3. Wet the array by filling it with 1× hybridization buffer (Total fill volume, **Table 6**) through one of the septa with a volume that is appropriate for the array format using a micropipettor and appropriate tips.

4. Incubate the probe array filled with 1× hybridization buffer in the Hybridization Oven, set to 45°C, for at least 10 min, with rotation.

5. Mix the hybridization cocktail according to **Table 7** for each target, scaling up volumes for hybridization to multiple probe arrays.

6. Heat the hybridization cocktail to 99°C for 5 min in a heat block.

7. Transfer the hybridization cocktail to a 45°C heat block for 5 min.

8. Centrifuge hybridization cocktail at maximum speed for 5 min to remove any insoluble material from the hybridization mixture.

9. Remove the buffer solution from the probe array cartridge and fill with appropriate hybridization volume **Table (6)** of the clarified hybridization cocktail, avoiding any insoluble matter at the bottom of the tube.

**Table 6**
**Probe array cartridge volumes**

| Array | Hybridization volume (μL) | Total fill volume (μL) |
|---|---|---|
| 49 Format (Standard) | 200 | 250 |
| 64 Format | 200 | 250 |
| 100 Format (Midi) | 130 | 160 |
| 169 Format (Mini) | 80 | 100 |
| 400 Format (Micro) | 80 | 100 |

**Table 7**
**Hybridization cocktail for single probe array**

| Component | 49 Format (Standard) | 100 Format (Midi) | 169 Format (Mini) 400 Format (Micro) | Final concentration |
|---|---|---|---|---|
| Fragmented cRNA (adjusted) | 15 μg | 10 μg | 5 μg | 0.05 μg/μL |
| Control oligo B2 (3 nM) | 5 μL | 3.3 μL | 1.7 μL | 50 pM |
| 20× Eukaryotic hybridization; controls (*bioB*, *bioC*, *bioD*, *cre*) | 15 μL | 10 μL | 5 μL | 1.5, 5, 25, and 100 pM respectively |
| Herring Sperm DNA (10 mg/mL) | 3 μL | 2 μL | 1 μL | 0.1 mg/mL |
| BSA (50 mg/mL) | 3 μL | 2 μL | 1 μL | 0.5 mg/mL |
| 2× Hybridization buffer | 150 μL | 100 μL | 50 μL | 1× |
| DMSO | 30 μL | 20 μL | 10 μL | 10% |
| $H_2O$ | To final volume of 300 μL | To final volume of 200 μL | To final volume of 100 μL | |
| Final volume | 300 μL | 200 μL | 100 μL | |

10. Place probe array into the Hybridization Oven, set to 45°C. To avoid stress to the motor, load probe arrays in a balanced configuration around the axis. Rotate at 60 rpm.

11. Hybridize for 16 h.

*3.11. Washing, Staining, and Scanning*

1. Prepare the Staining Reagents according to **Tables 8** and **9**. Volumes given are sufficient for one probe array. Remove SAPE from the refrigerator and tap the tube to mix well before preparing stain solution. Always prepare the SAPE stain solution fresh, on the day of use.

**Table 8**
**SAPE solution mix**

| Components | Volume | Final concentration |
|---|---|---|
| 2× Stain buffer | 600.0 μL | 1× |
| 50 mg/mL BSA | 48.0 μL | 2 mg/mL |
| 1 mg/mL Streptavidin Phycoerythrin (SAPE) | 12.0 μL | 10 μg/mL |
| DI H$_2$O | 540.0 μL | – |
| Total volume | 1,200 μL | |

**Table 9**
**Antibody solution mix**

| Components | Volume (μL) | Final concentration |
|---|---|---|
| 2× Stain buffer | 300.0 | 1× |
| 50 mg/mL BSA | 24.0 | 2 mg/mL |
| 10 mg/mL Goat IgG Stock | 6.0 | 0.1 mg/mL |
| 0.5 mg/mL biotinylated antibody | 3.6 | 3 μg/mL |
| DI H$_2$O | 266.4 | – |
| Total volume | 600 | |

2. Mix the SAPE solution well and divide into two aliquots of 600 μL each to be used for stainings 1 and 3 during the washing and staining procedure.

3. Mix the Antibody Solution well and use it for staining 2 during the washing and staining procedure.

4. To wash, stain, and scan a probe array, an experiment must first be registered in GCOS. Please follow the instructions detailed in the "Setting Up an Experiment" section of the appropriate GCOS User's Guide.

5. The Fluidics Station 450 is used to wash and stain the probe arrays. It is operated using GCOS. Prime the Fluidics Station to ensure that the lines are filled with the appropriate buffers (nonstringent wash buffer A and stringent wash buffer B) and the fluidics station is ready for running fluidics station protocols.

6. After 16 h of hybridization, remove the hybridization cocktail from the probe array and fill the probe array completely with the appropriate volume of nonstringent wash buffer A,

as given in **Table 6.** If necessary, at this point, the probe array can be stored at 4°C for up to 3 h before proceeding with washing and staining. Equilibrate the probe array to room temperature before washing and staining.

7. Insert the appropriate probe array into the designated module of the fluidics station.

8. Using the GCOS software, run a washing and staining protocol that is appropriate for the probe array format used.

9. Remove any microcentrifuge vial remaining in the sample holder of the fluidics station module(s) being used.

10. If prompted to "Load Vials 1-2-3," place the three staining reagent vials into the sample holders 1, 2, and 3 on the fluidics station, which correspond to stains 1, 2, and 3.

11. At the end of the run, or at the appropriate prompt, remove the microcentrifuge vials and replace with three empty microcentrifuge vials.

12. Remove the probe arrays from the fluidics station modules.

13. Check the probe array window for large bubbles or air pockets. If bubbles are present return the probe array to the module and engage it. Alternatively, try to remove bubbles manually by removing and refilling the probe array with non-stringent wash buffer A using a micropipettor and appropriate tips.

14. If you do not scan the arrays right away, keep the probe arrays at 4°C and in the dark until ready for scanning. If there are no more samples to hybridize, shut down the fluidics station following the "shutdown" procedure.

15. GCOS software also controls the scanner. The probe array is scanned after the wash protocols are complete. Make sure the laser is warmed up prior to scanning (10–15 min). If the probe array was stored at 4°C, warm to room temperature before scanning.

16. If necessary, clean the glass surface of the probe array with a non-abrasive towel or tissue before scanning. Do not use alcohol to clean glass. On the back of the probe array cartridge, clean excess fluid from around septa. Before scanning the probe array cartridge, apply Tough-Spots to each of the two septa on the probe array cartridge to prevent the leaking of fluids from the cartridge during scanning.

17. Scan the probe arrays.

18. After the scan, inspect for the presence of image artifacts (i.e., high/low intensity spots, scratches, high regional, or overall background, etc.) on the array. The boundaries of the probe area (viewed upon opening the .DAT or .CEL file) are easily identified by the hybridization of the B2 oligo, which

is spiked into each hybridization cocktail. Hybridization of B2 is highlighted on the image by the alternating pattern of intensities on the border, the checkerboard pattern at each corner and the array name, located in the upper left or upper middle of the array. B2 Oligo serves as a positive hybridization control and is used by the software to place a grid over the image (*see* **Note 7**).

19. Make a .CHP file by selecting the .DAT file and running an analysis using MAS or GCOS software.

### 3.12. Basic Quality Control

1. Make an expression report (.RPT) file from the .CHP file. The report will give additional information about quality control, such as the average background and noise values. There are no official guidelines regarding background, but typical average background values range from 20 to 100 for arrays scanned with the GeneChip® Scanner 3000 when average intensity is set at 500 or 1,000. Arrays being compared should ideally have comparable background values.

2. Poly-A RNA controls can be used to monitor the entire target labeling process. All the Poly-A controls should be called "Present" with increasing signal values in the order of lys, phe, thr, dap.

3. The GeneChip Eukaryotic Hybridization Control Kit contains 20× eukaryotic hybridization controls that are composed of a mixture of biotin-labeled cRNA transcripts of bioB, bioC, bioD, and cre prepared in staggered concentrations. This allows for evaluation of the sample hybridization efficiency on eukaryotic gene expression arrays. BioB is at the level of assay sensitivity (1:100,000 complexity ratio) and should be called "Present" at least 50% of the time. BioC, bioD, and cre should always be called "Present" with increasing Signal values, reflecting their relative concentrations.

4. For the majority of GeneChip expression arrays, β-actin and GAPDH are used to evaluate RNA sample and assay quality. Particularly, the signal values of the 3′ probe sets for actin and GAPDH are compared to the Signal values of the corresponding 5′ probe sets. The ratio of the 3′ probe set to the 5′ probe set is generally no more than 3. A high 3′–5′ ratio may indicate degraded RNA or inefficient transcription of ds cDNA or biotinylated cRNA.

5. The number of probe sets called "Present" relative to the total number of probe sets on the array is displayed as a percentage in the Expression Report (.RPT) file. Percent Present (%P) values depend on multiple factors including cell/tissue type, biological or environmental stimuli, probe array type, and overall quality of RNA. Replicate samples

should have similar %P values. Extremely low %P values are a possible indication of poor sample quality. However, the use of this parameter must be evaluated with care and in combination with the other sample and assay quality metrics.

6. For replicates and comparisons involving a relatively small number of changes, the scaling/normalization factors should be comparable among arrays. Larger discrepancies among scaling/normalization factors (e.g., threefold or greater) may indicate significant assay variability or sample degradation leading to noisier data.

**3.13. Data Analysis**     Expression can be calculated in GCOS, using the MAS algorithm *(10)*, or in Bioconductor. The three basic steps involved in expression calculation are: background removal, (multichip) normalization, and summarization of individual probe measurements by a probe set signal. Bioconductor offers a wide variety of algorithms to do so: RMA *(11)* and GCRMA *(12)* (model-based background removal, quantile normalization, and summarization by the robust median polish procedure), dChip *(13)* and VSN *(14)* are widely used. For small sample analyses, which the studies of hematopoietic subpopulations typically are, normalization is an extremely important step to reduce the influence of outlier arrays.

Differential expression between two single arrays can be assessed using GCOS, resulting in a *p*-value for each probeset. The use of simple fold changes (i.e., calling a probeset differentially expressed when the fold change exceeds 2) is deprecated. For multiple arrays, differential expression is typically assessed using a statistical test such as the *t*-test or Wilcoxon test *(15, 16)*. However, since hematopoietic subpopulations are generally studied using only few samples (2–3) per subpopulation, it may be better to use the individual probe measurements to directly arrive at a *p*-value for differential expression without first summarizing as probeset value, e.g., using a two-way analysis of variance *(17)*. The resulting *p*-values should always be adjusted for multiple testing, since many false positives are expected when a large number of probesets (20,000–55,000 depending on platform and array type) are tested at an individual *p*-value cut-off of 5%. Multiple testing correction can be performed to control the family-wise error rate (the probability of incorrectly calling at least one probeset differentially expressed) using, e.g., Bonferroni or Šidák step-down correction *(18)*, or to control the false discovery rate (FDR), the probability of each probeset called differentially expressed being incorrect *(19)*. The latter can be obtained using a test such as Significance Analysis of Microarrays (SAM *(20)*), producing *q*-values corresponding to false discovery rates, or by converting *p*-values to *q*-values *(21)*. The differentially expressed probesets (or microarrays) can next be clustered, i.e., groups of probesets (or microarrays) having similar expression profiles can be identified. There are numerous software packages available to perform cluster analyses, e.g., the freely available Cluster/ Treeview or Spotfire Decision site.

Individual differentially expressed probesets or clusters of these can be annotated using the Affymetrix NetAffx website, which contains a number of probe set annotations as well as hyperlinks to databases describing the corresponding genes or ESTs. Clusters can also be investigated for over-representation of certain (functional) annotations or involvement in pathways, using Ingenuity Pathways Analysis, GenMAPP, or DAVID to arrive at conclusions regarding expression of functionally related genes.

## 4. Notes

1. It is important to wear powder-free gloves throughout the procedure to minimize introduction of powder particles into sample or probe array cartridges. Take steps to minimize the introduction of exogenous nucleases; use RNase-free reagents and materials. Change gloves often.

2. It is highly recommended to purify cell populations before isolating RNA and doing microarray analysis, as using heterogeneous populations will cloud your analysis.

3. Incomplete loosening of the cell pellet may lead to inefficient lysis and reduced RNA yields. Therefore, it is very important to make sure the cells are loosened properly before adding the lysis solution.

4. One of the crucial parameters in microarray analysis is the RNA quality. If the electropherogram from Agilent 2100 Bioanalyzer analysis shows a 28S to 18S peak surface ratio of about 2:1 and no breakdown products can be observed, the sample can be processed for microarray analysis. If the electropherogram shows RNA of bad quality it is strongly discouraged to use the sample. Amplification will be far from optimal, and because of breakdown the expression profile does not properly represent the population.

5. After first-strand cDNA synthesis, cooling the samples at 4°C is required before proceeding to the next step. Adding the Second-Strand master mix directly to solutions that are at 42°C will compromise enzyme activity.

6. To prevent condensation that may result from water bath-style incubators, incubation of the IVT reaction is best performed in oven incubators for even temperature distribution, or in a thermal cycler.

7. Variation in B2 hybridization intensities across the array is normal and does not indicate variation in hybridization efficiency. If the B2 intensities at the checkerboard corners are either too low or high, or are skewed due to image artifacts, the grid will not align automatically. The grid can be aligned manually using the mouse to click and drag each grid corner to its appropriate checkerboard corner.

## References

1. http://www.rosettabio.com/.
2. http://www.bioconductor.org/.
3. http://linus.nci.nih.gov/brb/.
4. http://rana.lbl.gov/.
5. http://www.spotfire.com/.
6. http://www.affymetrix.com/analysis/.
7. http://www.ingenuity.com/.
8. http://www.genmapp.org/.
9. http://david.abcc.ncifcrf.gov/.
10. Affymetrix. (2001) *Microarray suite user guide*, Vol. 5, Affymetrix, Santa Clara, CA USA.
11. Irizarry RA, Hobbs B, Collin F, Beazer-Barclay YD, Antonellis KJ, Scherf U, Speed TP. (2003) Exploration, normalization, and summaries of high density oligonucleotide array probe level data. *Biostatistics* 2, 249–264.
12. Wu Z, Irizarry RA, Gentleman R, Murillo F, Spencer F. (2004) A model based background adjustment for oligonucleotide expression arrays. Technical Report, John Hopkins University, Baltimore, MD.
13. Li C, Wong WH. (2001) Model-based analysis of oligonucleotide arrays: expression index computation and outlier detection. *Proc. Natl Acad. Sci. U. S. A.* 1, 31–36.
14. Huber W, von Heydebreck A, Sultmann H, Poustka A, Vingron M. (2002) Variance stabilization applied to microarray data calibration and to the quantification of differential expression. *Bioinformatics* 18, Suppl 1:S96–S104.
15. Affymetrix. (2004.) Data analysis fundamentals. http://www.affymetrix.com/Auth/support/downloads/manuals/data_analysis_fundamentals_manual.pdf.
16. Sheskin DJ. (2004) *Handbook of parametric and nonparametric statistical procedures*, 3rd edition, Chapman & Hall/CRC, Boca Raton, FL.
17. Dik WA, Pike-Overzet K, Weerkamp F, de Ridder D, de Haas EF, Baert MR, van der Spek P, Koster EE, Reinders MJ, van Dongen JJ, Langerak AW, Staal FJ. (2005) New insights on human T cell development by quantitative T cell receptor gene rearrangement studies and gene expression profiling. *J. Exp. Med.* 11, 1715–1723.
18. Dudoit S, Yang Y, Callow M, Speed T. (2002) Statistical methods for identifying differentially expressed genes in replicated cDNA microarray experiments. *Stat. Sin.* 12, 111–139.
19. Benjamini Y, Hochberg Y. (1995) Controlling the false discovery rate: a practical and powerful approach to multiple testing. *J. R. Stat. Soc. Ser. B* 57, 289–300.
20. Tusher V, Tibshirani R, Chu C. (2001) Significance analysis of microarrays applied to transcriptional responses to ionizing radiation. *Proc. Natl Acad. Sci. U. S. A.* 98. 5116–5121. http://www-stat.stanford.edu/~tibs/SAM/
21. Storey JD. (2002) A direct approach to false discovery rates. *J. R. Stat. Soc. Ser. B* 64, 479–498. http://faculty.washington.edu/jstorey/qvalue/

# Chapter 28

# Quantification of Genomic Mutations in Murine Hematopoietic Cells

## Hartmut Geiger, David Schleimer, Kalpana J. Nattamai, and Jan Vijg

## Summary

Maintaining the stability of the genome is critical to cell survival and normal cell growth. Genetic modification of hematopoietic cells might bear an inherent increased risk for the accumulation of DNA mutations. It frequently requires cultivation of the cells under super-physiological oxygen levels, which can result in increased oxidative damage, as well as under super-physiological concentrations of cytokines, which might interfere with DNA-damage checkpoint activation and by this means might result in an increased mutational load. We describe here a protocol for monitoring the frequency of DNA mutations in bone marrow cells post transduction or upon selection either in vitro or in vivo based on the lacZ-plasmid (pUR288) transgenic mouse (small blue mouse) mutation indicator strain.

**Key words:** Hematopoietic cells, Gene transfer, Mutation, DNA damage, Transgenic, pUR288.

## 1. Introduction

The lacZ-plasmid (pUR288) transgenic mouse (small blue mouse) has been generated to serve as an in vivo indicator for mutagenic events in the somatic genome (1). In this model, nontranslated concatameric plasmids that are integrated into the murine genome can be analyzed for mutations in the lacZ part of the plasmid. For the quantification of somatic mutations in this model, the plasmids are excised from the genomic DNA, enriched with magnetic beads, ligated and electroporated into a distinct *E. coli* strain that allows for metabolic selection of plasmids that bear mutations in the lacZ gene. The type of mutation is then determined by a PCR in combination with restriction digestion.

Christopher Baum (ed.), *Methods in Molecular Biology, Methods and Protocals, vol. 506*
© Humana Press, a part of Springer Science+Business Media, LLC 2009
DOI: 10.1007/978-1-59745-409_28

The use of this animal model for the quantitative determination of the frequency and nature of somatic mutations due to for example organismal aging in various tissues has been extensively characterized *(2–11)*. Here we present a protocol specifically adapted for lacZ mutation analysis in murine hematopoietic cells *(8)*.

## 2. Materials

### 2.1. Mice

C57BL/6-Tg(LacZpl)60Vij/J (little blue mice) can be obtained from The Jackson laboratory (http://jaxmice.jax.org), stock number 002754.

### 2.2. DNA Isolation

QIAamp Blood Midi Kit (Qiagen, Valencia, CA).

### 2.3. Competent Cells

1. LB medium: 5g tryptone (Difco, Sparks, MD), 2.5g yeast extract (Difco), 5g NaCl, 500mL water, autoclave for 20min (*see* **Notes 1–3**).
2. puC19 or pUR288 DNA is dissolved at 0.04 or 0.4pg/μL in water, store at –80°C at 5μL single-use aliquots.
3. TB medium: 0.9g tryptone, 1.18g yeast extract, 0.47g $K_2HPO_4$, 0.11g $KH_2PO_4$ is dissolved in 50mL water, autoclave for 20min.
4. Top agar: 1.0g tryptone, 0.5g yeast extract, 0.025g NaCl, 0.35g agar (Difco) is dissolved in 100mL water, autoclave for 20min, keep top agar at 41°C.

### 2.4. Preparation of lacZ/lacI Fusion Protein

1. Streak lacZ–LacI fusion protein producing strain on M9CA plates (Difco, add 0.35g agar/100mL), grow overnight at 30°C.
2. Select at least five colonies, inoculate in 1.5mL M9CA medium, streak 100μL of medium after 4h on TY X-gal plates (to 100mL TY medium (Difco) add 0.35g agar, autoclave for 20min, add X-gal (to 75μg/mL)), incubate the plates and the cultures overnight at 30°C.
3. Select the colony with the highest beta-galactosidase activity (bluest plate), inoculate overnight culture in 50mL M9CA medium, culture again overnight at 30°C.
4. Transfer 30mL of this culture to 8L of TY medium, incubate at 30°C to $OD_{600}$ of 1.5 (late log phase), monitor growth every 30min after first 3h.
5. Weigh centrifuge beakers. Concentrate cells by centrifugation at 4,000×*g* for 20min.
6. Resuspend pellet carefully in 1mL 50mM Tris–HCl (pH 7.5), 10% (w/v) sucrose per 1g of cell pellet.

7. Drip cell slurry slowly into liquid nitrogen (will generate popcorn like structures), harvest popcorn and store at –80°C.

8. Dissolve two vials of protease inhibitor cocktail VI (A.G Scientific) in 75mL of 50mM Tris–HCl, pH 7.5/10% (w/v) sucrose and prewarm to 37°C. Take 60g of frozen cell slurry and add it quickly to this solution. Stir carefully, add 7.5mL lysis solution (2M NaCl, 7.6% (w/v) spermidine·HCl, 10% (w/v) sucrose), add 30mg lysozyme, mix carefully, and leave on ice for 1h.

9. Transfer the suspension to a 37°C water bath for 4min, swirl every 30s, then put suspension back on ice.

10. Centrifuge for 1h at 23,000 × $g$ at 4°C. Pour supernatant off, measure volume, and put it on ice.

11. While stirring, add ammonium sulfate to the supernatant to 40% saturation (24g/100 mL supernatant), stir on ice for another 2 h.

12. Centrifuge at 18,000×$g$ for 15min at 4°C. Discard supernatant.

13. Resuspend pellet in 6mL of storage buffer (25mM Tris–HCl (pH 8.0), 150mM NaCl, 1mM EDTA, 5mM MgCl$_2$, 0.05% (v/v) Tween 20, 25% (v/v) glycerol, 14.3mM beta-mercaptoethanol), store at single-use aliquots (10μL) at –80°C.

### 2.5. Preparation of lacI/lacZ Coated Magnetic Beads

1. Take 1mL of Dynal magnetic beads coated with sheep anti-mouse IgG (4×10$^8$ magnetic beads/mL) and pellet the beads on a magnetic stand. Wash the beads 3 times in 1mL PBS.

2. Resuspend the magnetic beads in 950μL PBS and 50μL anti-beta-galactosidase antibody (2mg/mL, Promega, Madison, WI)). Incubate at 37°C for 1h while rotating. Wash the beads 3 times in 1mL PBS as earlier.

3. Resuspend beads in 990μL of PBS and 10μL of lacZ/lacI fusion protein. Incubate at 37°C for 2h while rotating. Wash the beads 3 times in 1mL PBS as earlier. Resuspend the beads in 1mL PBS and store at 4°C. Coated beads are good for at least 6 months.

### 2.6. Plasmid Enrichment and Electroporation Procedure

1. 5× binding buffer: 50mM Tris–HCl (pH 7.5), 50mM MgCl$_2$, 25% (w/v) glycerol, adjust to pH 6.8 with HCl, sterilize with 0.22-μm filter. Store at room temperature. To obtain 1× buffer, dilute with water.

2. IPTG stock solution: dissolve IPTG (isopropyl-beta-d-thiogalactopyranoside) at 25mg/mL in water, sterilize with 0.22-μm filter, store in single-use aliquots (60μL) at –20°C.

3. IPTG elution buffer: 10mM Tris–HCl (pH 7.5), 1mM EDTA, 125mM NaCl, sterilize with 0.22-μm filter, store at room temperature.

4. ATP: dissolve at 10mM in water, sterilize with 0.22-μm filter, store in single-use aliquots (10μL) at –80°C.

5. 3M Sodium acetate (pH 4.9), sterilize with 0.22-μm filter.

6. X-gal stock solution: dissolve 50mg X-gal (5-bromo-4-chloro-3-indolyl-beta-D-galactopyranoside) in dimethylformamide, store protected from light at 4°C.

7. TB/glycerol medium: 0.9g tryptone, 1.18g yeast extract, 0.47g $K_2HPO_4$, 0.11g $KH_2PO_4$ is dissolved in 50mL water, autoclave for 20min. Add 400μL of 50% (w/v) glycerol to the medium and mix. Prepare always fresh.

8. Top-agar: 1.0g tryptone, 0.5g yeast extract, 0.025g NaCl, 0.35g agar is dissolved in 100mL $dH_2O$, autoclave for 20min, keep top agar at 41°C. Prepare 2 times 100mL. To prepare X-gal top agar, add to 100mL top agar 50μg/mL 2,3,5-triphenyl-2H-tetrazolium chloride, 25μg/mL kanamycin, 150μg/mL ampicillin and 37.5μg/mL X-gal. To prepare p-gal top agar, add to 100mL top agar 50μg/mL 2,3,5-triphenyl-2H-tetrazolium chloride, 25μg/mL kanamycin, 150μg/mL ampicillin and 0.3% (w/v) p-gal (phenyl-beta-D-galactopyranoside).

***2.7. Analysis of Mutant Clones***

1. LB medium (*see* **Subheading 2.3**).

2. X-gal plates for galactose-insensitive screening: to 500mL water, add 20 capsules LB Agar, autoclave for 20min. Then add kanamycin (to 25μg/mL), Ampicillin (to 150μg/mL), X-gal (to 75μg/mL), pour 35mL agar per 15-cm plate and dry in dark. Store plates at 4°C.

3. PCR Master mix (per sample): 11.5μL water, 0.5μL 12.5μM pUR4923-F(5′-TGG AGC GAA CGA CCT ACA CCG AAC TGA GAT-3′), binds to the ORI of pUR288, 0.5μL 12.5μM pUR3829-R (5′-ATA GTG TAT GCG ACC GAG TTG CTC TTG-3′), binds to the Ampicillin resistance gene locus in pUR288, 12.5μL Qiagen HotStarTaq Master Mix.

4. PCR program pUR288CO: Step 1: 95°C, 10min; Step 2: 95°C, 20s; Step 3: 68°C, 8min; Step 4: go to step 2 for 34 more times; Step 5: 68°C, 10min; Step 6: 4°C, ∞.

5. 1% TBE gel: 1% agarose (w/v) in 1× TBE-buffer, add 2μL ethidium bromide solution per 100mL gel (from stock at 10mg/mL).

6. 6× Gel loading buffer: 0.25% (w/v) Bromophenol Blue, 0.25% (w/v) xylene cyanol FF, 30% glycerol (w/v).

# 3. Methods

There are now multiple types of transgenic mice available for mutagenicity testing as well as mice that utilize endogenous reporter genes as readout targets. The mutation assay based on the pUR288 transgenic plasmid systems allows qualitative as well as quantitative comparisons of the mutation frequency in almost any given tissue/cell type in vivo as well as in vitro, as it does not, as many other models, require in vitro proliferation/selection of primary single cell clones (like for example the HPRT or APRT mutation indicator systems). The pUR288 system detects point mutations and other small intragenic lesions as well as large deletion and nonhomologous recombination events. However, it cannot detect other large chromosomal events such as mitotic recombination (LOH) or nondysjunction events. Data generated in our laboratory indicate that with a sample size of ten biological repeats, the assay has the power to assign a twofold difference in the total mutation frequency among experimental groups with statistical significance of $p<0.05$.

### 3.1. Preparation of Electrocompetent Cells

1. Prepare 1L sterile ice-cold water and 125mL sterile ice-cold 10% glycerol in water.

2. Add 50μL of an *E. coli* C (ΔlacZ/galE⁻) glycerol stock and 5μL of 50mg/mL kanamycin to 10mL LB medium. Incubate overnight at 37°C at 250rpm. We always start two cultures, in case one does not grow.

3. Next morning: Make sure the rotor for the 50-mL round-bottom tubes is refrigerated.

4. Resuspend the overnight cultures carefully, add 1.5mL of the overnight culture to each of two Erlenmeyers with 500-mL LB medium which contains no kanamycin (*see* **Notes 2** and **3**). Grow the cells to an $OD_{600}$ of 0.45 at 31.5°C and 250 rpm. Take the cell density every 15min after the first 3h. Cells should follow the growth curves depicted in **Fig. 1a** and **Fig. 1b**.Discard flask that do not follow these curves (e.g., cells growing faster or slower).

5. Cool the cells by placing the Erlenmeyers on ice for 30 min on a shaker. Use styrofoam boxes filled with ice to hold Erlenmeyers as you will work always on ice from now on.

6. Divide the cell cultures over two 500-mL centrifuge bottles, centrifuge the cells at $3,800 \times g$ for 15min at 4°C.

7. Very gently resuspend each pellet in 250mL ice-cold water, shake slowly on ice on rotary shaker until pellet is dissolved. Never pour water over pellet. Centrifuge the cells again at $3,800 \times g$ for 15min at 4°C. Take the supernatant off by vacuum, but never touch the pellet directly.

8. Repeat **step** 7.

9. Resuspend the cell pellets in 50mL 10% (w/v) ice-cold glycerol (in water) by shaking tubes on ice on rotary shaker and divide the two suspensions over four 50-mL round-bottom high-speed tubes. Centrifuge cells at $3,800 \times g$ for 15min at 4°C. Take the supernatant off by vacuum.

a

**Growth curve flask 1 and flask 2**

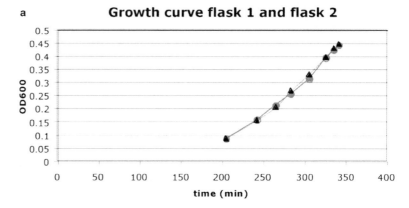

b

**Growth curve flask1 and flask 2**

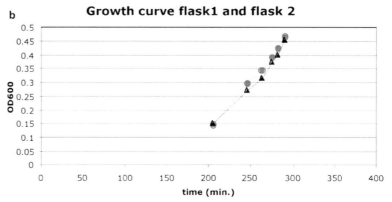

Fig. 1. (**a** and **b**) Representative growth curves for E. coli C (ΔlacZ/galE⁻) strains for the preparation of electrocompetent cells from two independent experiments.

10. Resuspend each pellet in 0.5mL 10% (w/v) ice-cold glycerol (shake on ice on rotary shaker) and combine the pellets. The $OD_{600}$ should be the following: 10µL cells in 3mL LB should have an $OD_{600}$ of at least 0.34 up to 0.41; adjust density of the suspension if necessary with 10% (w/v) glycerol (linear math if density is too high). Keep cells on ice.

11. Aliquot the cell suspension into microcentrifuge tubes at 255µL per tube and freeze tubes immediately in a dry ice/ethanol bath. Add number of vials generated, date of production, and the information on growth curve into the database. Store cells at –80°C (*see* **Note 4**).

12. Test competency of the cells the next day: Take 5µL of pUR288 at 0.4pg/µL at a total of 2pg (or a total of 0.2pg pUC19) and electroporate competent cells according to a standard protocol (*see* **Subheading 2.6**), and plate on X-gal top agar (*see* **Subheading 2.6**). Count colonies the next day. Be sure to always include as a control electro-competent cells from a previous cell preparation with a known efficiency.

13. Calculate the efficiency:

$$\frac{\text{Number of colonies}}{2 \text{ (pg) of DNA}} \times \frac{1 \times 10^6 \text{ pg}}{\mu g} \times 100 (\text{dilution factor})$$

14. We expect to obtain at least around $1.5 \times 10^{10}$ colonies/μg plasmid when cells are tested with pUC19 and tenfold less when tested with pUR288.

**3.2. DNA Purification from Bone Marrow Cells**

1. Bone marrow cells are flushed out of the femur and/or tibia with a syringe and a 20–22G needle in a volume of at least 2mL of PBS. Take the cell count, and aliquot at least $1 \times 10^7$ cells into a microcentrifuge tube (make aliquots in case there are more than $1 \times 10^7$ cells).

2. Cells will be centrifuged at $1,500 \times g$ at 4°C for 5min. The supernatant will be discarded and the cells resuspended in 1.5mL of PBS. Cells will be again centrifuged at $1,500 \times g$ at 4°C for 5min and the supernatant will be discarded. Quick freeze cell pellet in ethanol/dry ice bath. Store cell pellet at –80°C for further use.

3. At least $1 \times 10^7$ cells should be used for the DNA isolation to obtain enough DNA for the assay. Thaw cell pellet and follow precisely instructions provided with the QIAamp Blood Midi Kit.

**3.3. Determination of the Frequency of Mutated Plasmids via the Plasmid Enrichment and Electroporation Procedure**

1. Adjust 10–30μg genomic DNA to a 150-μL volume with water (we always use 30μg in case the samples were heterozygous for the lacZ transgene). Add 30μL of 5× binding buffer and 60U of HindIII. Mix gently and incubate mixture for 1h at 37°C in water bath. The time can be extended to up to 2h. As a control, always perform the assay on DNA from a non-pUR288 transgenic tissue at the same time (*see* **Note 5**).

2. Pellet 60μL of lacZ/lacI fusion protein-coated beads using the magnetic stand and remove the PBS storage buffer. Resuspend the beads in the DNA/HindIII/binding buffer mixture. Incubate for 1h at 37°C while rotating.

3. Wash the beads 3 times with 250μL 1× binding buffer and resuspend them in 75μL IPTG-elution buffer, 5μL IPTG stock solution, 100μL water, and 20μL NEBuffer #2. Vortex softly and incubate for 30min at 37°C while rotating.

4. To further support dissociation of the plasmid from the beads as well as inhibition of the restriction enzyme, place mixture at 65°C for 20min. Allow cooling to room temperature and spin shortly to get down condensation drops. Add freshly thawed ATP to a final concentration of 0.1mM (2μL of a

10mM stock) and 0.1U T4 DNA ligase (1μL of a 10× dilution of 1U/μL T4 DNA ligase in 1× T4 DNA ligase buffer). Vortex softly and incubate at room temperature for 1h.

5. Resuspend beads by light vortexing (for better suspension) and pellet them on the magnetic stand and transfer the solution (without beads) to a clean tube. Repeat this two more times to remove all remaining beads in the solution.

6. Precipitate DNA with 30μg glycogen, 0.1volume 3M sodium acetate, pH 4.9 (approximately 22μL) and mix before ethanol is added. Add 2.5volumes of 95% ethanol (approximately 560μL). Vortex and place mixture at –80°C for at least 30min for precipitation. A southern blot tracking the plasmid along these steps is depicted in **Fig. 2** (excision from genomic DNA, enrichment, ligation, precipitation).

7. While the DNA is precipitating at –80°C, prepare X-gal and p-gal top agar.

8. Spin down the precipitated DNA for 30min at 14,000×$g$ at 4°C. Wash the pellet once with 250μL 70% (v/v) ethanol, vortex, and spin again at 14,000× $g$ for 5min. Carefully remove all remaining ethanol using a fine tip pipette and allow the DNA to dry for at least 10min at room temperature. Resuspend the DNA in 5μL water.

9. While the DNA is drying, place electroporation cuvettes (0.1-cm gap width) and electrocompetent cells (directly out of –80°C) on ice for a minimum of 20min.

10. Set Gene Pulser to 25mF and 1.8kV. Set Pulse Controller to 200W (*see* **Note 6**).

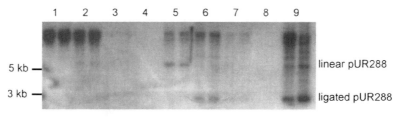

1. genomic DNA
2. digested genomic DNA
3. supernatent first wash with 1X binding buffer
4. supernatent second wash with 1X binding buffer
5. digested plasmid on beads
6. ligated plasmid in supernatant
7. beads after plasmid elution
8. plasmid precipitation supernatant
9. plasmid resuspended in water

Fig. 2. Southern blot depicting the whereabouts of the plasmid pUR288 along the enrichment procedure. The blot demonstrates that the plasmid is highly enriched in the final precipitated product by the procedure, with only minor plasmid loss in any of the supernatants analyzed.

11. Pipet 60µL of electrocompetent cell slurry (*E. coli* lacZ/galE⁻) carefully to the 5µL plasmid solution. Mix cells with the plasmid carefully by swirling the mixture with pipette tip (do not pipet up and down). Pipet cells/plasmid mixture into electroporation cuvette and electroporate.

12. After the electroporation, immediately add 1mL of ice-cold TB medium to the cuvette.

13. Transfer the cells to a 15-mL culture tube containing an additional 1mL TB medium and incubate at 37°C for 1h at 225rpm in a shaker/incubator.

14. Dilute either 2 or 4µL of the cell suspension in 2mL TB medium which is mixed with 13mL X-gal top agar and plated. We target this dilution to obtain later at least 100 colonies on the X-gal plate. The remaining cells (1.998 or 1.995mL) are plated using 13mL p-gal top agar. Wait until top agar is solidified (usually takes 30min).

15. All plates are incubated for 15h at 37°C.

16. Count colonies on both p-gal and X-gal top agar plates. The raw mutation frequency is determined by dividing the total number of colonies on the selective plate (p-gal) by the number of colonies on the titer plate (X-gal) and the dilution factor (500× or 1,000×). The final mutation frequency will be determined after the analysis of the mutant clones.

*3.4. Analysis of Mutant Clones*

1. To determine the type of mutation of the clone, the following PCR/restriction digest procedure is performed. The PCR amplifies the part of the plasmid containing the lacZ gene.

2. Transfer individual mutant colonies from selective plates to individual wells of 96-well round-bottom-plates in an LB medium containing ampicillin (150µg/mL) and kanamycin (25µg/mL). Works best with 2-µL tips.

3. Cover and tape plate shut. Grow overnight in a shaker/incubator at 37°C and 200rpm.

4. Examine overnight cultures for growth. Note any cultures that did not grow on a spreadsheet (follow 12 by 8 matrix of the 96-well plate).

5. Transfer a sample of each culture with a Boekel replicator to an X-gal plate for galactose sensitivity screening. Mark the right orientation on the bottom plate and incubate upside down at 37°C. Examine plate after 1–2h for blue staining cultures; these are galactose-insensitive host cells containing wild-type plasmids (false mutants, *see* **Fig. 3**). Circle blue colonies and note them on spreadsheet.

6. Incubate plate overnight: Note any colonies that did not grow on spreadsheet (false mutants, *see* **Fig. 4**).

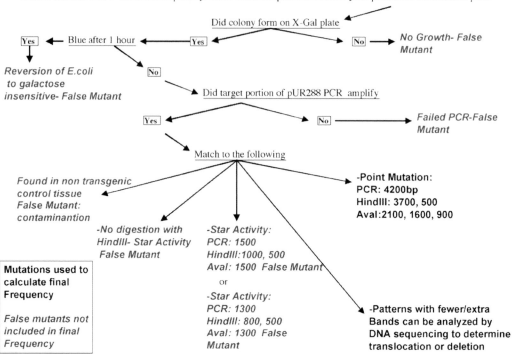

Pick 20 colonies from P-Gal selective plate → Grow o/n in Amp/Kan LB media → Re-plate on X-Gal selective plate

Did colony form on X-Gal plate

Yes ← Blue after 1 hour ← Yes ← → No → No Growth- False Mutant

Reversion of E.coli to galactose insensitive- False Mutant

No

Did target portion of pUR288 PCR amplify

Yes ← → No → Failed PCR-False Mutant

Match to the following

Found in non transgenic control tissue False Mutant: contaminantion

-Point Mutation: PCR: 4200bp HindIII: 3700, 500 AvaI:2100, 1600, 900

-No digestion with HindIII- Star Activity False Mutant

-Star Activity: PCR: 1500 HindIII:1000, 500 AvaI: 1500 False Mutant

or

-Star Activity: PCR: 1300 HindIII: 800, 500 AvaI: 1300 False Mutant

-Patterns with fewer/extra Bands can be analyzed by DNA sequencing to determine translocation or deletion

Mutations used to calculate final Frequency

False mutants not included in final Frequency

Fig. 3. Decision tree for determining the final mutation frequency from the raw mutation frequency in combination with the analysis of the mutant clones picked from the p-gal top agar plate. Any mutant clone that falls in a "black" category will be counted as a mutant clone and will be used to determine the final mutation frequency according to the formula in Subheading 3. Any clone that falls into a "lighter gray" area will be regarded as false positive and discarded. Up to 30% of all clones might fall into that group.

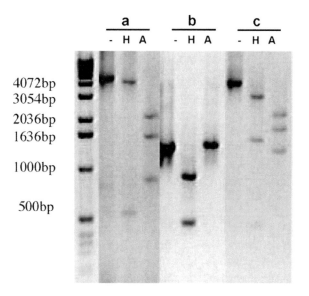

a   b   c
- H A   - H A   - H A

4072bp
3054bp
2036bp
1636bp
1000bp
500bp

Fig. 4. Typical restriction patterns for mutant clones, depicting (**a**) a pattern resembling a point mutation, (**b**) a pattern resembling a star activity, and (**c**) a pattern resembling either a deletion or a translocation (size-change mutation). *H* digested by HindIII; *A* digested by AvaI.

7. Transfer 2μL of each culture with a multichannel pipetter to a 96-well plate containing the PCR Master mix. Run the PCR.

8. Prepare Mastermix for PCR product digestions as follows:

| HindIII (per sample) | AvaI (per sample) |
|---|---|
| 10.25μL water | 10μL water |
| 1.5μL NEBuffer# 2 | 1.5μL NEBuffer# 4 |
| 0.25μL HindIII (5U) | 0.5μL AvaI (5U) |

9. Load 12μL of each master mix into an individual well of a 96-well plate, add 3μL of the PCR product to the digestion master mix and incubate at 37°C for 1h.

10. Once digestion is complete, load samples onto a 1% TBE gel the following way: Aliquot 5μL of undigested PCR product, add 5μL 6× loading buffer. Mix and load 8μL to gel. Add 5μL 6× loading buffer directly to each digested sample and load 8μL to gel.

11. Samples should be loaded in the following order: PCR product undigested, HindIII digest, AvaI digest. Load 5μL 1-kb DNA size marker to flank each row of samples.

12. Run gel for 2h at 100V. Take gel picture.

**3.5. Scoring of Mutant Clones**

1. Depending on the length of the PCR product and the type of the pattern of the digested clones, three distinct types of mutations can be identified after gel electrophoresis (**Fig. 4**).

   • Nonsize-change mutations (point mutations). The PCR amplifies a 4.2-kb part of pUR288, containing the lacZ part of the gene. A clone that shows the size and restriction pattern identical to the PCR product generated from the parental pUR288 plasmid (undigested: 4.2kb; HindIII: 3.7, 0.5kb; AvaI: 2.1, 1.6, 0.9kb) is regarded to bear a point mutation that renders the lacZ gene inactive. Such clones are scored as point mutations (**Fig. 4**, pattern a).

   • Mutant clones due to star activity: Under certain extreme conditions, restriction endonucleases are capable of cleaving at sequences which are similar but not identical to their defined recognition sequence (star activity), which will usually not result in regeneration of the consensus HindIII restriction recognition site upon religation. So any PCR product that is not recut by HindIII is regarded to be such a product (which does not count as a mutation conferred upon in the animal and thus is regarded as a false mutant) and will not be scored and thus subtracted from the number of mutant clones. Secondly, as confirmed by DNA sequencing, two of the star activity sites in PCR product will regenerate a HindIII restriction site upon religation, but these clones show a very characteristic gel pattern: The PCR product is either 1.5 or 1.3kb, while the HindIII sizes are 1.0 and 0.5 or 0.8 and 0.5kb and the product is uncut by AvaI (**Fig. 4**, pattern b). These clones will not be scored as mutations and thus subtracted from the number of mutant clones.

- Size-change mutations (deletions, translocations, and inversions). Any clone that deviates from the point mutation pattern as described in (a) and does not register as a star activity mutation will be scored as a mutation that is either a deletion or a translocation (**Fig. 3**, pattern c). A single sequencing reaction though can be used to identify translocations among these types of mutations (*see* **Note 8**).

- Contaminations. Any clone that is lacZ negative, but resistant to kanamycin and ampicillin will read out as a mutant clone in this assay. Thus even a very minor contamination with such a plasmid can result in false positive mutant clones in the p-gal plate. As we carry DNA from a nontransgenic animal alongside with test-DNA as a control in this assay, any mutant clone that is found in the nontransgenic sample is regarded as a contamination. This also means that any clone in a test sample with an identical restriction pattern has to be subtracted from the number of mutant clones in this sample (*see* **Note 8**).

2. Finally, any clone that did not amplify by PCR is regarded a false positive and will not be scored as a mutation.

3. In summary, this results in the "decision tree" diagram shown in **Fig. 4.** Follow that diagram to correctly subtract false mutants from the raw mutation frequency and to then determine the final mutation frequency of the sample with the formula given in **Subheading 3.1** and the percen-tage of point mutations and large deletions/translocations (*see* **Notes 7** and **8**).

## 4. Notes

1. We always use either ddH$_2$O or ultrapure 18.2MΩcm resistivity water when we refer to water.

2. We found that the source of the water used for making the electrocompetent cells does influence the efficiency for electroporation of the product. So it might be a good idea

to initially test water from different purification systems/ resources in the facility.

3. Do not use pellets to prepare the LB for making electrocompetent cells; always prepare from powder and always prepare medium fresh.

4. Do not store electrocompetent cells longer than 3 months. They may lose competency.

5. As any lacZ-kan+amp+plasmid will read out as a mutation in this assay, it is central for the assay to always carry along a nontransgenic tissue throughout the analysis, as well as to have dedicated pipettes, reagents, and equipment that are not shared with any other kind of plasmid work (like minipreps for example).

6. The settings for the electroporation might have to be tested/ optimized on your specific electroporator. We use a product from BioRad (Hercules, CA).

7. To unequivocally identify the type of mutation, the whole PCR product has to be sequenced. As in almost all instances though a point mutation pattern by restriction digestion was verified as a point mutation by sequencing, the current approach is a straightforward and relatively inexpensive but reliable way to determine the type of mutation.

8. If it is not clear whether the clones are contaminations, sequencing is a very straightforward way to identify whether the clone is derived from pUR288 (most likely mutation) or not (contamination). We use the following sequencing primer pUR3839.R (5′-ATA GTG TAT GCG GCG ACC GAG TTG CTC TTG-3′). The elongation of this primer spans the transition from the plasmid backbone over the HindIII restriction site into the lacZ part of pUR288. Also based on this one primer-sequencing reaction, it is possible to distinguish between deletions and translocations (plasmid contains sequences from murine genomic DNA, see examples in **Table 1**.

## Table 1
## Mutation frequency (×10$^{-5}$)

| Bone marrow: 6.0±3.5 | | | Small intestine: 12.8±7.9 | | |
|---|---|---|---|---|---|
| Point mutations | Translocations | Deletions | Point mutations | Translocations | Deletions |
| 20% | 59% | 21% | 61% | 11% | 28% |

Bone marrow shows only half the mutational load in young animals compared to small intestine. Interestingly though, bone marrow has with 60% of all mutant clones being translocations the highest spontaneous rate of translocations for any tissue reported so far

## Acknowledgments

The author would like to thank Martijn Dollé and Rita Busuttil for their support in learning how to perform the assay and Sina Albert for support in establishing the assay in his laboratory. This work was supported in part by NIH grant R01 HL076604 as well as a New Scholar in Aging grant from The Ellison Medical Foundation.

## References

1. Dolle, M.E., et al., *Evaluation of a plasmid-based transgenic mouse model for detecting in vivo mutations.* Mutagenesis, 1996. **11**(1): 111–8.

2. Dolle, M.E., et al., *Rapid accumulation of genome rearrangements in liver but not in brain of old mice [see comments].* Nat Genet, 1997. **17**(4): 431–4.

3. Dolle, M.E., et al., *Distinct spectra of somatic mutations accumulated with age in mouse heart and small intestine.* Proc Natl Acad Sci U S A, 2000. **97**(15): 8403–8.

4. Giese, H., et al., *Age-related mutation accumulation at a lacZ reporter locus in normal and tumor tissues of Trp53-deficient mice.* Mutat Res, 2002. **514**(1–2): 153–63.

5. Vijg, J. and M.E. Dolle, *Large genome rearrangements as a primary cause of aging.* Mech Ageing Dev, 2002. **123**(8): 907–15.

6. Dolle, M.E. and J. Vijg, *Genome dynamics in aging mice.* Genome Res, 2002. **12**(11): 1732–8.

7. Tutt, A.N., et al., *Disruption of Brca2 increases the spontaneous mutation rate in vivo: synergism with ionizing radiation.* EMBO Rep, 2002. **3**(3): 255–60.

8. Geiger, H., et al., *Mutagenic potential of temozolomide in bone marrow cells in vivo.* Blood, 2006. **107**(7): 3010–1.

9. Vijg, J., et al., *Transgenic mouse models for studying mutations in vivo: applications in aging research [corrected and republished article originally printed in* Mech Ageing Dev, 1997. **98**(3): 189–202]. Mech Ageing Dev, 1997. **99**(3): 257–71.

10. Dolle, M.E., et al., *Characterization of color mutants in lacZ plasmid-based transgenic mice, as detected by positive selection.* Mutagenesis, 1999. **14**(3): 287–93.

11. Dolle, M.E., et al., *Background mutations and polymorphisms in lacZ-plasmid transgenic mice.* Environ Mol Mutagen, 1999. **34**(2–3): 112–20.

# Chapter 29

# Proteomics Studies After Hematopoietic Stem Cell Transplantation

## Eva M. Weissinger, Petra Zürbig, and Arnold Ganser

## Summary

Complex biological samples hold significant information on the health status and on development of disease. Approximately 35,000 human genes give rise to more than 1,000,000 functional entities at the protein level. Thus, the proteome provides a much richer source of information than the genome for describing the state of health or disease of humans. The composition body fluids comprise a rich source of information on changes of protein and peptide expression. Here we describe the application of capillary electrophoresis (CE) coupled online to an electrospray-ionization time-of-flight mass spectrometer (ESI-TOF-MS) to analyze human urine for the identification of biomarkers specific for complications after allogeneic hematopoietic stem cell transplantation (HSCT). Sequencing of native proteins/peptides is necessary for the identification of possible new therapeutic targets.

**Key words:** Proteomics, Hematopoietic stem cell transplantation, Capillary electrophoresis, Mass spectrometry, Polypeptide, Clinical diagnosis.

## 1. Introduction

Proteome analysis is now emerging as key technology for deciphering biological processes and the discovery of biomarkers for diseases from tissues and/or body fluids in clinical research and diagnostics. The complexity and wide dynamic range of protein expression poses an enormous challenge to both separation technologies and subsequent detection of the molecules. Recent advances in mass spectrometry (MS) enable characterization of mass or mass/charge ($m/z$) of the peptides and proteins with high resolution and speed. The combination of the high-resolution separation properties of capillary electrophoresis (CE) with

Christopher Baum (ed.), *Methods in Molecular Biology, Methods and Protocols, vol. 506*
© Humana Press, a part of Springer Science+Business Media, LLC 2009
DOI: 10.1007/978-1-59745-409-4_29

the sensitive and accurate mass identification of MS as well as approaches for CE-MS coupling to date *(1, 2)* lead to robust CE-MS applications with sensitivity in the high attomol range, comparable to nanoliquid chromatography systems *(3)*. Several applications have been published *(4–6)*. Thus, CE-MS offers a fast, accurate, and reproducible system for the analysis of clinical samples. More than 2,000 individual proteins and peptides can be detected in a single CE-MS run within 45–60 min, thus fulfilling the requirements for analysis in a clinical setting *(7)*. Allogeneic hematopoietic stem cell transplantation (HSCT) offers the only permanent cure for many patients with hematological malignancies or marrow dysfunction syndromes. Complications after HSCT, such as concurrent infection in the time of aplasia, reactivation of viruses, such as cytomegalovirus (CMV), or the development of immunological complications, such as graft-versus-host-disease (GvHD), limit the application of HSCT. Here we describe the application of screening for complications after HSCT with CE-MS analysis of urine collected from patients after allogeneic HSCT ( *(7)*, Weissinger et al., 2007), thus giving an example for clinical decision making based on proteomic patterns. In addition sequencing of peptides forming patterns specific for disease is discussed.

Current literature indicates that CE-MS is a powerful tool allowing fast and reliable analysis of polypeptides from several types of highly complex biological samples, such as urine, blood, or cerebrospinal fluid. Information on several hundred polypeptides from an individual sample can be obtained quickly *(4–7)*. Although these polypeptides can serve as excellent biomarkers for diagnostic purposes, their potential physiological role remains unknown as long as their identity defined by their amino acid sequence is not determined. The identification of the defined biomarkers presents some unique challenges. The biomarkers cannot be easily isolated; the sequence analysis has to be performed from a complex mixture, and potential biomarkers are frequently post-translationally modified. Potential biomarkers detected by CE-MS are likely to be small fragments of larger proteins. Thus, to identify a 2–10 kDa (modified) portion of a protein with a possible molecular weight greater than 60 kDa requires extensive de novo sequencing.

For this purpose, CE can be interfaced online with MS/MS instruments. Neususs et al. *(3)* describe a capillary electrophoresis-tandem mass spectrometry (CE-MS/MS) approach for routine application in proteomic studies. Stable coupling is achieved by using a standard coaxial sheath-flow sprayer. The applied sheath flow is reduced to 1–2 µL/min in order to increase sensitivity. Detection limits are as low as 500 attomol. Low femtomole amounts are required for unequivocal identification by MS/MS experiments in the used ion trap and subsequent database search.

Alternatively, the entire CE-MS run can be spotted off-line onto a MALDI target plate, and subsequently the polypeptides of interest can be analyzed using MALDI-TOF-TOF *(2, 8)*. This method has the advantage that the signal of interest can be located in MS mode and optimal fragmentation conditions can be determined without repeated separation. However, sequencing with MALDI-TOF-TOF generally does not result in data of sufficient quality from urinary peptides with molecular weights above approximately 3 kDa. Several biomarker candidate peptides were identified using MALDI-MS/MS, as shown for GvHD *(9)*, diabetic nephropathy *(10)*, dialysis fluid *(11)*, or bladder cancer *(12)*.

Fourier-transform ion cyclotron resonance mass spectrometry (FT-ICR MS) instruments facilitate the identification of urinary polypeptides even larger than 8 kDa *(13)*. The authors described CE off-line coupled FT-ICR MS to identify polypeptides in the urine from patients with focal segmental glomerulosclerosis (FSGS), membranous nephritis (MN), minimal change disease (MCD), Immunoglobulin A nephropathy (IgAN), and diabetic nephropathy DN and validated multiple biomarkers for the control and each of the diseases.

A comparison of the different CE-MS/MS options was recently reported by Zürbig et al. *(14)*. Compared to other high-performance separation methods coupled either online or off-line to MS devices, CE-MS provides a unique advantage: the number of basic amino acids correlates at pH 2 with the polypeptide migration time. This unique feature facilitates independent entry of different sequencing platforms for peptide sequencing of CE-MS-defined biomarkers from highly complex mixtures.

However, it is important to note that sequencing should not be restricted to the primary amino acids, but should subsequently include the deciphering of any post-translational modification (PTM). If the biomarker identification workflow involves the separation of intact proteins, enzymatic digestions, and MS analysis of the digestion products, standard methods for MS/MS sequencing can be used. However, this approach generally does not result in the identification of PTM.

Many of the higher-throughput methods directly detect native peptides and proteins present in tissues or body fluids. As these do not have boundaries generated by defined enzymatic cleavage, sequencing based on standard MS/MS is not possible. Standard MS/MS sequencing methods are based on comparing observed peptide fragmentation patterns with calculated expected patterns resulting from "in silico" digestion and fragmentation of a protein database. Thus, the result is a statistical probability that the peptide sequence has been identified correctly. However, if the peptide boundaries are not defined, the search space explodes and the sequence has to be deduced by "de novo" sequencing, which poses major challenges that are beyond the scope of discussion for this chapter. In addition, many biomarker peptides are extensively modified by PTM, making identification even more problematic. Furthermore, some PTMs may be disease specific and can serve as biomarkers. In combination, these issues are a large source for errors and ambiguities in sequence assignments, and sequencing biomarkers remains a formidable challenge to date.

While sequencing of biomarkers and their post-translational modifications is a desirable goal for the future but currently not always feasible, at least a clear definition that reliably allows to detect the potential biomarker with high confidence in other samples is essential (e.g., by affinity, isoelectric point, migration characteristics, mass, etc.). This requires a high-quality separation and subsequent MS analysis to reproducibly assign relative quantities, migration in the separation dimension, and mass, all with reasonably high accuracy and resolution.

## 2. Materials

1. Urine samples collected cold and immediately frozen and stored at –20°C.
2. Buffer for sample dilution: 4 M Urea, 20 mM $NH_4OH$ containing 0.2% SDS.
3. Amicon Ultra centrifugal filter device (30 kDa; Millipore, Bedford, USA).
4. C2-column (Pharmacia, Uppsala, Sweden).
5. Elution buffer: 50% acetonitrile (ACN) in HPLC-grade $H_2O$ containing 0.5% formic acid.
6. Lyophilization Christ Speed-Vac RVC 2-18/Alpha 1-2 (Christ, Osterode am Harz, Germany).
7. Capillary Electrophoresis P/ACE MDQ system (Beckman Coulter, Fullerton, USA).
8. Capillary: 90 cm, 50 μm I.D. fused silica capillary.
9. Rinse between runs: 0.1 M NaOH.
10. CE-Running buffer: 20% ACN, 0.25 M formic acid.
11. Sheath liquid 30% (v/v) iso-propanol (Sigma-Aldrich, Germany) and 0.4% (v/v) formic acid in HPLC-grade water (flow rate: 2 μl/min).
12. ESI-TOF sprayer-kit (Agilent technologies, Palo Alto, CA, USA).

13. Micro-TOF-MS (ESI-TOF; Bruker Daltonik, Bremen, Germany).

14. MALDI-TOF-MS-MS (Bruker Daltonik, Bremen, Germany).

15. Software: Mosaiques Visu: for peak detection (biomosaiques software, Hannover, Germany).

16. MosaCluster: SVM-based program for grouping according to multiple parameters. (biomosaiques software, Hannover, Germany).

## 3. Methods

The methods described below outline (1) the sample preparation, (2) the CE-MS analysis, (3) data processing, analysis to evaluate the abundant information obtained after a single run in the CE-MS for clinical application to obtain particular patterns typical for different diseases or complications, and (4) de novo sequencing of possible biomarkers.

### 3.1. Sample Preparation

Body fluids are complex mixtures of molecules with a wide range of polarity, hydrophobicity, and size. When analyzing complex biological samples, major concerns are loss of analytes (here: polypeptides) during the sample preparation as well as reproducibility of the data generated. Ideally, a crude unprocessed sample should be analyzed. This would avoid all artifacts, losses, or biases arising from sample preparation. To date the presence of large molecules, e.g., albumin, immunoglobulin, and others, hampers this direct approach, since these molecules bare little information in the clinical setting but will interfere with the detection of smaller, less abundant proteins and peptides. Thus a manipulation of the samples is necessary, but can be limited to a few steps.

#### 3.1.1. Collection and Storage of the Samples

Urine samples are obtained from patients at different time points before and after HSCT, starting prior to conditioning and weekly thereafter for the time on the transplantation ward. The samples are typically taken as the second spot urine, voiding the first urine of the day. Proper handling of the samples is of outmost importance; the urine should be aliquoted and frozen immediately after collection and stored at –20°C until analysis (*see* **Note 1**).

#### 3.1.2. Sample Preparation

Shortly before analysis, 1 ml aliquots are thawed and diluted with 1 ml of 4 M Urea, 20 mM $NH_4OH$ containing 0.2% SDS. This is followed by an ultrafiltration step using the Amicon Ultra centrifugal filter device (30 kDa; Millipore, Bedford, USA). Samples are spun at 3,000 × $g$ until 1.5 ml of filtrate has passed through the filter, thus removing the high molecular weight proteins such as albumin, transferrin, and others (*see* **Note 2**). The filtrate is then desalted and applied onto a reversed-phase C2-column to remove urea, salts, and other confounding materials. Polypeptides are eluted with 50% acetonitrile (ACN) in HPLC-grade $H_2O$ with 0.5% formic acid. Next, samples are lyophilized overnight and resuspended in 50 μl of HPLC-grade $H_2O$ shortly before injection.

### 3.2. CE-MS Analysis of the Samples

#### 3.2.1. Capillary Electrophoresisa

The samples are transferred to appropriate vials and stored in the CE-auto sampler section at 5°C. For capillary electrophoresis a P/ACE MDQ (Beckman Coulter, Fullerton, USA) system is used equipped with a 90-cm, 50-μm inner diameter, bare-fused silica capillary. The use of coated capillaries appears not to be beneficial (*see* **Note 3**). The capillary is first rinsed with running buffer (20% acetonitrile, 0.5% formic acid, 79.5% HPLC-grade water) for 3 min prior to sample injection. The sample is injected for 99 s with 1–6 psi, resulting in the injection of 60–300 nL sample. Separation is performed with +30 kV at the injection side, and the capillary temperature is set to 35°C for the entire length of the capillary up to the ESI interface. After each run, the capillary is rinsed for 5 min with 0.1 M NaOH, to remove protein build up that accumulates, if uncoated capillaries are used (*see* **Note 3**). This is followed by a 5-min rinse with HPLC-grade $H_2O$ and subsequently with running buffer. The CE-MS set up is depicted in Fig. 1.

#### 3.2.2. CE-MS Interface and Analysis

The MS analysis is performed in positive electrospray mode with an ESI-TOF sprayer kit (Agilent technologies, Palo Alto, CA, USA) using a Micro-TOF-MS (Bruker Daltonik, Bremen, Germany). The ESI sprayer is grounded and the ionspray interface potential is set between –3,700 and –4,100 V. The sheath liquid is applied coaxially, consisting of 30% (v/v) iso-propanol (Sigma-Aldrich) and 0.4% (v/v) formic acid in HPLC-grade water, at a flow rate of 1–2 μl/min. These conditions result in a detection limit of about 1 fmol of different standard peptides (7). MS spectra are accumulated every 3 s, over a mass-to-charge range from 400 to 2,500 or 3,000 $m/z$ for about 45–60 min.

### 3.3. Data Processing and Statistical Analysis

#### 3.3.1. Data Processing: Peak Annotation

$m/z$ Values of MS peaks are deconvoluted into mass and combined if they represented identical molecules at different charge states using MosaiquesVisu (15). MosaiquesVisu (accessible at http://www.proteomiques.com) employs a probabilistic clustering algorithm and uses both isotopic distribution and conjugated masses for charge-state determination of polypeptides. In a first electronic analysis, the software identifies all peaks within each single spectrum. This usually results in more than 100,000 peaks from a single sample. Since true analytes must appear in several,

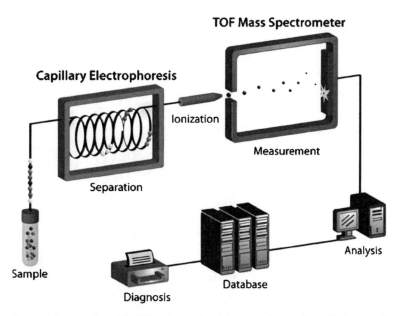

Fig. 1. Online coupling of CE-MS and set up for data processing: a schematic drawing of the online coupling of CE-MS beginning at the preparation of the sample preparation to the final data processing and identification of the pattern is shown.

e.g., at least three successive spectra, signals from individual peptides present in consecutive spectra are collected, evaluated with respect to charge state, and combined in the next step. This reduces the list to 3,000–7,000 "CE/MS" peaks. These data are termed "peak list" of an individual sample. The peak list can be converted into a 3-dimensional plot, $m/z$ on the $y$-axis plotted against the migration time on the $x$-axis and the signal intensity color coded (Fig. 2). Signals have to meet certain criteria, like: (a) the signal intensity must be greater than the threshold (usually S/N > 7); (b) single charged peaks are discarded; and/or (c) peaks width must be below the threshold of 2 min. Molecules that do not fit into the aforementioned criteria are not proteins and are removed from the list.

*3.3.2. Data Processing: Peak Deconvolution, Protein Contour Plots*

In the next step of the data processing for peaks representing identical molecules with different charge states are identified. These are deconvoluted into a single mass. The actual mass of the polypeptides plotted against the migration time results in the "protein-plot" (Fig. 2). This theoretical CE/MS spectrum that now contains the information on migration time, mass, and signal intensity for each individual polypeptide generally consists of between 1,000 and 2,500 individual polypeptides per sample. To allow comparison and search for conformity and differences between the samples obtained at different time points after

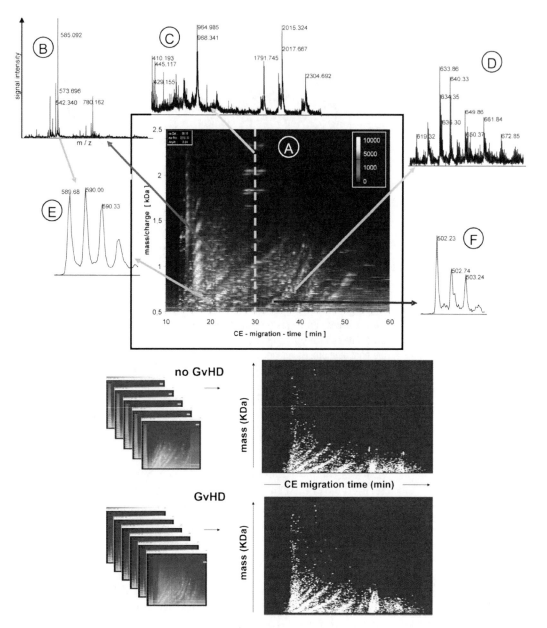

Fig. 2. Data generation and processing. (**a**) The CE-MS ionogramm is shown in the *upper part*, the resulting 3-dimensional raw data plot (**a**) obtained from urine of a patient after HSCT with no problems is shown in the *center*. The *m/z* (*y*-axis) is plotted against the migration time (*x*-axis) in the CE, while the signal amplitude is depicted as a color code ranging from black to white with increasing intensity of the signal (*right side*). Each *dash* shown in this contour plot corresponds to a single polypeptide analyzed in the CE-MS run. To display the information content in each of the about 1,000 single spectra contained in one CE-MS analysis, a magnified view of the MS data of three individual spectra is shown (**b–d**). In the spectra the signal intensity (amplitude) on the *y*-axis is plotted onto the mass/charge on the *x*-axis, as indicated in (**b**). The cross section on the raw data plot is shown as *arrows* in (**b**), (**c**), or (**d**). An even further magnification of MS spectra from two individual polypeptides reveals the high resolution of the technology used (**e**, $z = 3$; and **f**, $z = 2$). Reprinted with permission from *(9)*. (**b**) Digital data compilation. Individual datasets from CE-MS analysis of human urine samples were calibrated using internal standards. The *left panel* displays these data from different patients (no GvHD *upper panel*, GvHD *lower panel*) in a 3-dimensional contour plot: mass (in kDa on a logarithmic scale) plotted against normalized migration time (min). The MS signal intensity is represented by the peak height as well as color. The data were digitally compiled to a group-specific polypeptide pattern, shown in the *right panel* (no GvHD *upper panel*, GvHD *lower panel*).

HSCT and from different individuals, CE-migration times have to be normalized using ca 200 polypeptides generally present in a urine sample that serve as internal standards *(16)*. The signal intensity is normalized to the total ion current of the utilized signals (TIC). Polypeptides within different samples are considered identical, if the mass deviation is less than 50 ppm and the CE migration-time deviation is less than 2 min.

Intra- and interassay variability are ascertained by repeated analysis of one sample and by analysis of samples obtained at different time points from the same patient under comparable conditions, respectively.

*3.3.3. Polypeptide Patterns for Prediction of Complications After Stem Cell Transplantation*

Complications after stem cell transplantation can be predicted with high significance by screening patient's urine routinely with CE-MS. Urine from 50 patients after hematopoietic stem cell transplantation (HSCT; 45 after allogeneic and 5 after autologous HSCT) and 8 with sepsis were collected up to day +365 after HSCT. Screening led to the generation of polypeptide patterns yielding an early recognition of changes, if complications occurred during the observation period. Twenty patients developed GvHD after allo-HSCT. The polypeptide patterns yielded differentially excreted, statistically relevant polypeptides (Fig. 3) forming a pattern specific for recognition of GvHD. Comparison with patients with sepsis allowed to distinguish sepsis from GvHD with a specificity 97% and a sensitivity of 98% based on an inclusion of sepsis-specific polypeptides. Thus the application of CE-MS to the screening of patients after HSCT can help to reduce transplantation related morbidity and mortality significantly.

*3.4. Mass Spectrometry for Polypeptide Identification*

The so-called bottom-up approach (2D-gel electrophoresis and multidimensional protein identification technology) of analytical protein mass spectrometry utilizes an initial treatment with a protease (usually trypsin) for protein identification to break proteins into relatively small peptides of which the $m/z$ can be accurately determined by mass spectrometry. This so-called peptide mass fingerprinting allows reliable identification (see for example *(17–22)*).

In biomarker discovery experiments a different approach has become popular: A so-called top-down approach, in which polypeptides are delivered to the mass spectrometer without prior trypsinization. Depending on the mass accuracy of the mass spectrometer being used, peptides and small proteins up to 10–20 kDa can often be recognized. Top-down strategies have been exploited using CE-MS (capillary electrophoresis-mass spectrometry) *(9, 10, 23–26)*, LC-MS (liquid chromatography-mass spectrometry) *(27–31)*, and SELDI-TOF-MS (surface-enhanced

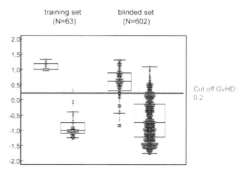

aGvHD in prospective cohort: 71 total, 33 grade I, 20 grade II, 9 grade III, 9 grade IV

# Training set: 33 patients

Disease:         13 AML, 6sAML, 5 ALL, 2 MM, 2 NHL, 2 SAA,
                 1 MDS, 1 CLL, 1 HD
HSCT:            28 PBSCT, 4 BM
Donor:           20 MUD, 13 MRD

| Patient ID | Age | days after HSCT | grade | manifesaton | SVM score |
|------------|-----|-----------------|-------|-------------|-----------|
| 2791 | 54 | 29 | II  | skin             | 1.000 |
| 2719 | 40 | 43 | II  | skin             | 1.000 |
| 2890 | 66 | 13 | II  | skin             | 1.237 |
| 2725 | 21 | 33 | III | skin             | 1.000 |
| 3064 | 50 | 19 | II  | skin, intestine  | 1.000 |
| 2787 | 48 | 7  | IV  | intestine, liver | 1.124 |
| 2800 | 51 | 19 | IV  | intestine        | 1.251 |
| 3197 | 20 | 23 | III | intestine        | 1.136 |
| 3049 | 18 | 28 | IV  | intestine        | 1.330 |
| 2249 | 34 | 34 | II  | intestine        | 1.000 |

Fig. 3. Prospective evaluation of proteomic pattern. The training set consisting of 13 samples from 10 patients with aGvHD and 50 samples from 23 patients without this complication yielded the aGvHD proteomic pattern was used for scoring with support vector machines yielding (SVM). Samples with a classification factor (CF) above the cut off (cut off: +0.2) scored positive for aGvHD and below the cut off indicated no aGvHD developing in these patients. Classification of the blinded test set resulted in a sensitivity of 83% and a specificity of 75%. *Abbreviations AML* acute myeloid leukemia; *sAML* secondary acute myeloid leukemia; *ALL* acute lymphoid leukemia; *MM* multiple myeloma; *NHL* non-Hodgkin lymphoma; *SAA* severe and very severe aplastic anemia; *MDS* myelodysplastic syndromes; *CLL* chronic lymphocytic leukemia; *HD* Hodgkin disease; *PBSCT* peripheral blood stem cell transplant; *BM* bone marrow; *MUD* matched unrelated donors; *MRD* matched related donors.

laser desorption/ionization-time of flight-mass spectrometry) *(32–36)* systems (Fig. 4).

Absolute identification of proteins is more difficult with top-down approaches than with bottom-up approaches ($m/z$ ratios are often reported rather than protein identifications). For the bottom-up analysis the first provision is the mass fingerprint for the correct identification of a protein. To confirm the identity of the protein, sequence analysis of a few tryptic peptides can be performed. The sequence analysis of the peptides resulting from the top-down approach is often a make-or-break outcome. Due to the fact that the fragmentation of some peptides results in water elimination of an amino acid (asparagines, aspartic acid, glutamine, glutamic acid) *(37, 38)* or proline residues caused in a partial fragment spectrum, the identification of the full sequence becomes uncertain. In this case, only $MS^n$ approaches with the

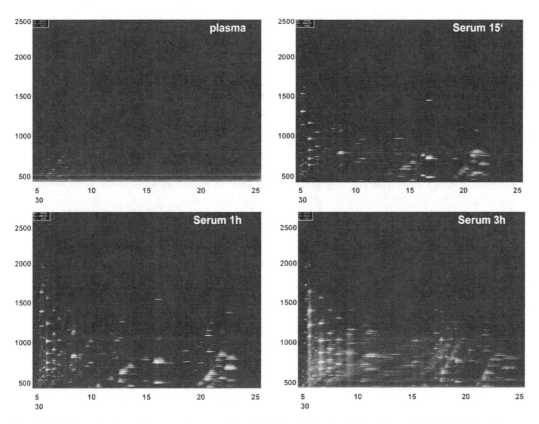

Fig. 4. CE-MS spectra from plasma and serum after clotting for different periods of time (as indicated). Mass/charge ratio is indicated on the *left*, CE-migration time (in minutes) at the *bottom*, signal intensity is *color coded*. All samples were taken from a healthy volunteer at the same time. Equal amounts (150 nl, corresponding to 3 μl of crude serum or plasma) of sample were prepared and injected into the CE. While in plasma only a small number of polypeptides are visible, an increasing amount of obviously similar polypeptides (yet clearly different from plasma) can be observed in serum depending on the clotting time. Reprinted with permission from *(2)*.

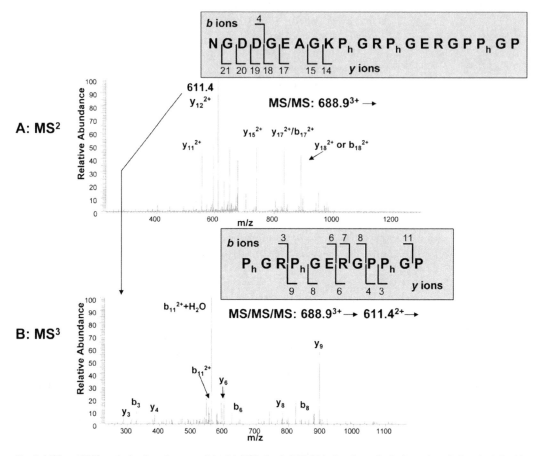

Fig. 5. MS² and MS³ analysis of a urinary peptide. (**a**) MS² of *m/z* 688.931 showing a limited number of abundant doubly charged *y*-ions. (**b**) MS^3 of the *m/z* 611.421 fragment ion from CID of 688.931 showing multiple fragment ions that allows complete sequence identification. Reprinted with permission from *(14)*.

use of an ion trap device lead to satisfactory results. The following example illustrates the demanding challenges of the top-down compared to bottom-up approaches: MS/MS of the triply charged ion (688.931³⁺) corresponding to the peptide (Fig. 5a) does not yield sufficient fragment ions to enable unambiguous sequence identification. The MS² spectrum (Fig. 3a) consists primarily of a few abundant ions representing only larger portions of the peptide. This results in uncertain full sequence identification, especially the locations of the hydroxyproline residues. Using MS³, the initial precursor ion (688.931³⁺) was mass selected and subjected to collision-induced dissociation (CID). One of the resulting fragment ions (611.421²⁺) was then mass selected for further CID, and the resulting product ions were measured (Fig. 5b). This MS³ analysis yielded fragment ions, which are not easily detected in the MS² spectrum and enables a complete sequence determination.

## 4. Notes

1. The samples must be collected under proper conditions especially when highly sensitive analysis like MS is applied. While in common proteomic techniques, such as western blotting, ELISA, or even 2D-gel electrophoresis minor degradation of the proteins generally does not cause problems, CE-MS or actually any MS-based proteomic analysis will be severely hampered by degradation. The samples require storage at least at −20°C immediately after collection, in order to prevent degradation of proteins, which would result in additional fragments of proteins that are irrelevant to the underlying disease. Immediate cooling of the samples is another way to preserve the stability. Urine is a very stable body fluid, compared to serum or plasma, thus in our hands is preferable over the other fluids.

2. Initial experiments and also our previous data *(39)* revealed that higher molecular weight proteins in the urine sample generally appear not to contain significant information, but cause severe problems during the CE-MS runs, like precipitation, clogging of the capillary, and overloading. Smaller molecules may be more interesting, but can be lost due to the large amounts of molecules like albumin or transferrin or others. Therefore, an optimized protocol to remove the confounding large molecules using ultrafiltration has been developed. The most abundant higher molecular weight protein is albumin, a carrier protein that binds a substantial fraction of other proteins and peptides. In order to prevent binding of small fragments to albumin, we use a chaotropic agent, urea, in combination with a detergent, SDS, for the dilution of the samples. Consistently good and reproducible results and recovery of >80% of added standard polypeptides were observed in the presence of 2 M urea and 0.1% SDS *(12)*.

3. Under neutral or basic conditions the silanol groups of the capillary form a negatively charged surface which interacts with the positive charge of the proteins. This leads to peak broadening especially of highly charged molecules. To circumvent these problems, several different coatings and coating protocols have been described *(40–42)*. Hence, we have examined several mostly silyl-based coatings *(2)*.Unfortunately, most of the coating procedures described in the literature are quite tedious and time consuming. Coated capillaries gave good results when using standard polypeptides. However, coatings that were examined in our laboratory were not satisfactory with "real" clinical samples like urine, since these appear to deposit in part on the coating material leading to peak

broadening and increased migration times. In a very recent manuscript Ullsten et al. *(43)* describe a new coating that seems to represent a major improvement. This polyamine coating (PolyE-323) was reported to show good stability even at high pH, involves no complicated chemical reactions, and can also be applied onto the capillary within minutes. It still remains to be seen whether this type of coating is compatible with clinical samples, but it appears to be a very promising approach.

Consequently, our experience using different types of coating so far was quite unsatisfactory, and the optimal approach appears to be the use of uncoated capillaries. Decreasing the pH of the background electrolyte (BGE) reduces the negative charge of the surface, thus reducing capillary-protein interaction. Additionally, hydrophobic interactions can be reduced by adding organic solvent to the BGE. Therefore, best results are obtained at low pH in the presence of acetonitrile.

Last but not least, it is important to point out that most coatings result in a positively charged capillary wall and cause an inverse electro-osmotic flow; consequently, the electrical field must be reversed.

## References

1. Moini M. Capillary electrophoresis mass spectrometry and its application to the analysis of biological mixtures. Anal Bioanal Chem. 2002;373:466–480.

2. Kolch W, Neususs C, Pelzing M, Mischak H. Capillary electrophoresis-mass spectrometry as a powerful tool in clinical diagnosis and biomarker discovery. Mass Spectrom Rev. 2005;24:959–977.

3. Neususs C, Pelzing M, Macht M. A robust approach for the analysis of peptides in the low femtomole range by capillary electrophoresis-tandem mass spectrometry. Electrophoresis. 2002;23:3149–3159.

4. Manabe T. Capillary electrophoresis of proteins for proteomic studies. Electrophoresis. 1999;20:3116–3121.

5. Oda RP, Clark R, Katzmann JA, Landers JP. Capillary electrophoresis as a clinical tool for the analysis of protein in serum and other body fluids. Electrophoresis. 1997;18:1715–1723.

6. Jellum E, Dollekamp H, Blessum C. Capillary electrophoresis for clinical problem solving: analysis of urinary diagnostic metabolites and serum proteins. J Chromatogr B Biomed Appl. 1996;683:55–65.

7. Kaiser T, Wittke S, Just I, et al. Capillary electrophoresis coupled to mass spectrometer for automated and robust polypeptide determination in body fluids for clinical use. Electrophoresis. 2004;25:2044–2055.

8. Rejtar T, Hu P, Juhasz P, et al. Off-line coupling of high-resolution capillary electrophoresis to MALDI-TOF and TOF/TOF MS. J Proteome Res. 2002;1:171–179.

9. Kaiser T, Kamal H, Rank A, et al. Proteomics applied to the clinical follow-up of patients after allogeneic hematopoietic stem cell transplantation. Blood. 2004;104:340–349.

10. Mischak H, Kaiser T, Walden M, et al. Proteomic analysis for the assessment of diabetic renal damage in humans. Clin Sci (Lond). 2004;107:485–495.

11. Weissinger EM, Kaiser T, Meert N, et al. Proteomics: a novel tool to unravel the patho-physiology of uraemia. Nephrol Dial Transplant. 2004;19:3068–3077.

12. Theodorescu D, Wittke S, Ross MM, et al. Discovery and validation of new protein biomarkers for urothelial cancer: a prospective analysis. Lancet Oncol. 2006;7:230–240.

13. Chalmers MJ, Mackay CL, Hendrickson CL,et al. Combined top-down and bottom-up mass spectrometric approach to characterization of biomarkers for renal disease. Anal Chem. 2005;77:7163–7171.

14. Zurbig P, Renfrow MB, Schiffer E,et al. Biomarker discovery by CE-MS enables sequence analysis via MS/MS with platform-independent separation. Electrophoresis. 2006;27:2111–2125.

15. Neuhoff N, Kaiser T, Wittke S,et al. Mass spectrometry for the detection of differentially expressed proteins: a comparison of surface-enhanced laser desorption/ionization and capillary electrophoresis/mass spectrometry. Rapid Commun Mass Spectrom. 2004;18:149–156.

16. Wittke S, Mischak H, Walden M,et al. Discovery of biomarkers in human urine and cerebrospinal fluid by capillary electrophoresis coupled to mass spectrometry: towards new diagnostic and therapeutic approaches. Electrophoresis. 2005;26:1476–1487.

17. Biron DG, Joly C, Marche L,et al. First analysis of the proteome in two nematomorph species, *Paragordius tricuspidatus* (Chordodidae) and *Spinochordodes tellinii* (Spinochordodidae). Infect Genet Evol. 2005;5:167–175.

18. Gagnaire V, Piot M, Camier B,et al. Survey of bacterial proteins released in cheese: a proteomic approach. Int J Food Microbiol. 2004;94:185–201.

19. Gras R, Muller M. Computational aspects of protein identification by mass spectrometry. Curr Opin Mol Ther. 2001;3:526–532.

20. Pang JX, Ginanni N, Dongre AR, Hefta SA, Opitek GJ. Biomarker discovery in urine by proteomics. J Proteome Res. 2002;1:161–169.

21. Raharjo TJ, Widjaja I, Roytrakul S, Verpoorte R. Comparative proteomics of *Cannabis sativa* plant tissues. J Biomol Tech. 2004;15:97–106.

22. Thongboonkerd V, McLeish KR, Arthur JM, Klein JB. Proteomic analysis of normal human urinary proteins isolated by acetone precipitation or ultracentrifugation. Kidney Int. 2002;62:1461–1469.

23. Meier M, Kaiser T, Herrmann A,et al. Identification of urinary protein pattern in type 1 diabetic adolescents with early diabetic nephropathy by a novel combined proteome analysis. J Diabetes Complications. 2005;19:223–232.

24. Rossing K, Mischak H, Parving HH,et al. Impact of diabetic nephropathy and angiotensin II receptor blockade on urinary polypeptide patterns. Kidney Int. 2005;68:193–205.

25. Weissinger EM, Wittke S, Kaiser T, et al. Proteomic patterns established with capillary electrophoresis and mass spectrometry for diagnostic purposes. Kidney Int. 2004;65:2426–2434.

26. Wittke S, Haubitz M, Walden M,et al. Detection of acute tubulointerstitial rejection by proteomic analysis of urinary samples in renal transplant recipients. Am J Transplant. 2005;5:2479–2488.

27. Jurgens M, Appel A, Heine G,et al. Towards characterization of the human urinary peptidome. Comb Chem High Throughput Screen. 2005;8:757–765.

28. Schrader M, Schulz-Knappe P. Peptidomics technologies for human body fluids. Trends Biotechnol. 2001;19:S55–S60.

29. Selle H, Lamerz J, Buerger K,et al. Identification of novel biomarker candidates by differential peptidomics analysis of cerebrospinal fluid in Alzheimer's disease. Comb Chem High Throughput Screen. 2005;8:801–806.

30. Svensson M, Skold K, Svenningsson P, Andren PE. Peptidomics-based discovery of novel neuropeptides. J Proteome Res. 2003;2:213–219.

31. Tammen H, Schorn K, Selle H,et al. Identification of Peptide tumor markers in a tumor graft model in immunodeficient mice. Comb Chem High Throughput Screen. 2005;8:783–788.

32. Clarke W. Proteomic research in renal transplantation. Ther Drug Monit. 2006;28:19–22.

33. Forde CE, Gonzales AD, Smessaert JM,et al. A rapid method to capture and screen for transcription factors by SELDI mass spectrometry. Biochem Biophys Res Commun. 2002;290:1328–1335.

34. Tan CS, Ploner A, Quandt A, Lehtio J, Pawitan Y. Finding regions of significance in SELDI measurements for identifying protein biomarkers. Bioinformatics. 2006;22:1515–1523.

35. Tang N, Tornatore P, Weinberger SR. Current developments in SELDI affinity technology. Mass Spectrom Rev. 2004;23:34–44.

36. Yip TT, Lomas L. SELDI ProteinChip array in oncoproteomic research. Technol Cancer Res Treat. 2002;1:273–280.

37. Geiger T, Clarke S. Deamidation, isomerization, and racemization at asparaginyl and aspartyl residues in peptides. Succinimide-linked reactions that contribute to protein degradation. J Biol Chem. 1987;262:785–794.

38. Stephenson RC, Clarke S. Succinimide formation from aspartyl and asparaginyl

peptides as a model for the spontaneous degradation of proteins. J Biol Chem. 1989;264:6164–6170.

39. Wittke S, Fliser D, Haubitz M,et al. Determination of peptides and proteins in human urine with capillary electrophoresis-mass spectrometry, a suitable tool for the establishment of new diagnostic markers. J Chromatogr A. 2003;1013:173–181.

40. Belder D, Deege A, Husmann H, Kohler F, Ludwig M. Cross-linked poly(vinyl alcohol) as permanent hydrophilic column coating for capillary electrophoresis. Electrophoresis. 2001;22:3813–3818.

41. Johannesson N, Wetterhall M, Markides KE, Bergquist J. Monomer surface modifications for rapid peptide analysis by capillary electrophoresis and capillary electrochromatography coupled to electrospray ionization-mass spectrometry. Electrophoresis. 2004;25 :809–816.

42. Liu CY. Stationary phases for capillary electrophoresis and capillary electrochromatography. Electrophoresis. 2001;22: 612–628.

43. Ullsten S, Zuberovic A, Wetterhall M,et al. A polyamine coating for enhanced capillary electrophoresis-electrospray ionization-mass spectrometry of proteins and peptides. Electrophoresis. 2004;25: 2090–2099.

44. Weissinger EM, Schiffer E, Hertenstein B et al. Proteomic patterns predict acute graft-versus-host disease after allogeneic hematopoletic stem cell transplantation. Blood. 2007;109:5511–5519.

# Chapter 30

# Spectral Karyotyping and Fluorescence In Situ Hybridization of Murine Cells

## Cornelia Rudolph and Brigitte Schlegelberger

## Summary

Cytogenetic characterization of murine chromosomes using banding techniques like R- or G-banding is technically demanding due to the similar size and the acrocentric structure of all chromosomes. The molecular cytogenetic technique of spectral karyotyping (SKY) overcomes that difficulty by karyotyping metaphase chromosomes after different and simultaneous fluorescence labeling of the whole genome. SKY allows the detection and identification of numerical as well as structural chromosome aberrations with a resolution of approximately 2 Mb. The technique is applicable to all fast-proliferating cells, e.g., cells of the hematopoietic system like stem cells or T- and B-lymphocytes. It is also applicable to murine embryonic fibroblast or cells isolated from tissues with increased proliferation – especially tumor tissues. Furthermore, SKY is recommended for the cytogenetic characterization of newly established cell lines.

**Key words:** Spectral karyotyping, Fluorescence in situ hybridization, Chromosomal instability, In vitro and in vivo models,

## 1. Introduction

Spectral karyotyping was first published in 1996 by Schröck et al. for human chromosomes *(1)* and a few months later for murine chromosomes *(2)*. The method is based on the combination of five fluorochromes in only one hybridization and acquisition. Thus, SKY enables the detection of discrete aberrations and complex rearrangements as well as the identification of marker chromosomes in a simple manner.

Spectral karyotyping requires special technical equipment that combines Fourier spectroscopy, CCD imaging, and optical

Christopher Baum (ed.), *Methods in Molecular Biology, Methods and Protocols, vol. 506*
© Humana Press, a part of Springer Science+Business Media, LLC 2009
DOI: 10.1007/978-1-59745-409-4_30

microscopy as well as a specialized software for analysis. All components are available from Applied Spectral Imaging (ASI, Migdal Ha`Emek, Israel). An alternative method that is based on the use of different fluorescent filter sets compared to only one custom-designed filter in SKY is the multiplex-fluorescence in situ hybridization (M-FISH) *(3)*. Referring to the results, SKY and M-FISH are comparable. In contrast to murine cells, human chromosome band-specific probes are available and can be used for M-banding analyses *(4)*.

In the last few years, the investigation of genetic alterations, in particular the detection and characterization of chromosomal instability (CIN), has become increasingly important in the scientific research of genetically modified cells, e.g., transgenic in vivo models *(5)* **Fig. (1)** or hematopoietic stem cell transplantation models after retroviral gene transfer. It has been shown that hematopoietic cells can be forced to undergo malignant transformation by insertional mutagenesis and by induction of CIN *(6, 7)*. Reproducible leukemias developed as a result of multiple insertions of retroviral vectors inducing the activation of oncogenes and/or signaling genes. The mechanisms responsible for CIN are largely unknown as yet, although recent data suggest an important role of epigenetic modifications. For example, reduced expression of the methyltransferase 1 (*Dnmt1*) promotes

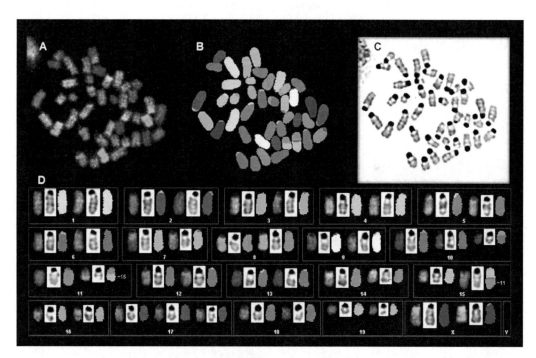

Fig. 1 SKY analysis of metaphase chromosomes prepared from a murine B-cell lymphoma. Detection of a balanced translocation T(11B1;15D2–D3), an additional chromosome 10 with terminal deletion, and a trisomy 17. ((**a**) RGB (red-green-blue) image; (**b**) RGB image after classification; (**c**) inverted DAPI image; (**d**) karyotype). (*See Color Plates*)

the development of aggressive T-cell lymphomas with increased genomic instability *(8)*.

The murine karyotype comprises 40 chromosomes. Synteny maps enable the assignment of murine genomic regions to human genomic regions and vice versa. Thus, chromosomal alterations of the murine genome detected by SKY can be "translated" into the human genome (http://www.ensembl.org). It is important to note that precise description of breakpoints and the identification of genes involved in structural chromosomal aberrations require additional molecular cytogenetic techniques like fluorescence in situ hybridization (FISH) or array-based comparative genomic hybridization (array-CGH). Array-CGH allows the detection of genomic imbalances with a resolution up to the single gene level and is recommended for the validation of unbalanced translocations, deletions, or insertions, for example. Copy number changes, e.g., a complete or partial chromosomal deletion, can result in the monoallelic loss of a tumor suppressor gene. If chromosome aberrations lead to the generation of fusion genes, long-distance PCR, 5′ and 3′ RACE or FISH can help to identify the respective genes. If chromosomal alterations are expected to induce changes in gene expression, Western blot analyses, quantitative *real-time* PCR, or microarray analyses are appropriate methods to investigate these findings. In most instances, validation of alterations detected by SKY is performed by FISH. Therefore, the technical principles of FISH will be also described here.

## 2. Materials

### 2.1. Chromosome Preparation

1. Colcemid (Roche, Mannheim, Germany) 10 µg/ml. Store at 4°C.
2. KCl. Dissolve at 75 mM in water. Store at room temperature. Warm up to 37°C for use.
3. Carnoy?s fixative: methanol and acetic acid. Prepare a mixture of methanol:acetic acid 3:1. Store at room temperature.
4. Ethanol series (70, 90, 100%). Store at room temperature.
5. SuperFrost®Plus microscope slides (Menzel, Braunschweig, Germany).

### 2.2. Spectral Karyotyping

#### 2.2.1. Pretreatment

1. Pepsin (Sigma, Seelze, Germany). Dissolve 100 mg in 1 ml water. Store aliquots at –20°C.
2. HCl 1 M. Store at room temperature. Prepare 100 ml of a 10 mM HCl solution for use and heat to 37°C in a water bath.
3. Phosphate-buffered saline (1× PBS). Store at room temperature.

*2.2.2. Denaturation and Hybridization*

1. 20× SSC: Prepare 20× SSC stock solution by dissolving 132 g SSC in 400 ml water. Adjust pH to 5.3 with HCl and fill up with water to 500 ml. Store at room temperature and use within 6 months.

2. Deionized formamide. Store at 4°C.

3. Denaturation solution: Mix 35 ml formamide, 10 ml water, and 5 ml 20× SSC (final concentration is 70% formamide in 2× SSC), adjust pH to 7.0. Heat in a glass coplin jar in a water bath to 72–74°C (*see* **Note 1**).

4. Ethanol series (70, 90, 100%). Store at –20°C.

5. SKY-Paint-Kit DNA-M10 (ASI, Applied Spectral Imaging, Migdal Ha`Emek, Israel). Store light protected at 4°C or –20°C.

6. Cover slips Ø 12 mm and 18 × 18 mm (Menzel, Germany).

7. Fixogum (Marabuwerke, Tamm, Germany). Store at room temperature.

8. Coplin jars (Fisher Scientific, Schwerte, Germany) (*see* **Note 2**).

9. Humidified chamber.

*2.2.3. Detection*

1. 20× SSC stock solution. *see* **Subheading 2.2.2, item 1**.

2. Tween 20 (Merck, Darmstadt, Germany). Store at room temperature.

3. Deionized formamide. Store at 4°C.

4. Washing solution A: add 75 ml formamide, 60 ml water, and 15 ml 20× SSC (final concentration is 50% formamide in 2× SSC). Adjust pH to 7. Distribute solution to 3 × 50 ml in glass coplin jars and heat to 45°C in a water bath (*see* **Note 1**).

5. Washing solution B: add 12.5 ml 20× SSC and 237.5 ml water. Adjust pH to 7 and heat to 45°C in a water bath.

6. Washing solution C: add 100 ml 20× SSC and 400 ml water. Adjust pH to 7 and add 500 µl Tween 20. Heat to 45°C in a water bath.

7. Blocking buffer: 3% (w/v) bovine serum albumin (BSA) in washing solution C. Store aliquots at –20°C.

8. Antibodies: unconjugated mouse monoclonal antidigoxin (Sigma), Cy5-conjugated streptavidin (Amersham Bioscience, Freiburg, Germany), Cy5.5-conjugated sheep antimouse (Sigma). Prepare stock solution by diluting 1 mg antibody in 1 ml water. Store aliquots at –20°C.

9. Nuclear staining: DAPI (4,6-diamidino-2-phenylindole). Prepare stock solution by dissolving 1 mg DAPI in 5 ml water. Store aliquots at –20°C. Working solution: 10 µl stock solution diluted in 50 ml 2× SSC, pH 7; light protected stable up to 2 weeks at 4°C.

10. Mounting medium: Vectashield (Vector Laboratories, Burlingame, CA, USA).

11. Cover slips 24 × 60 mm (Menzel).

12. Coplin jars (Fisher Scientific) (*see* **Note 2**).

13. Humidified chamber.

### 2.3. Validation of Chromosomal Alterations by Fluorescence In Situ Hybridization

#### 2.3.1. Generation of Biotinylated DNA Probes Using Random Priming

1. DNA isolated from BAC clones (CHORI, Children's Hospital Oakland Research Institute via bacpacorders@chori.org) encoding the gene of interest (http://www.ensembl.org. Store at –20°C (*see* **Note 1**). Human or murine inserts of BAC clones comprise DNA fragments of 100–300 kb in size. DNA fragments generated by random priming are approximately 100- to 500-bp long and cover the whole DNA insert.

2. BioPrime® DNA Labeling System (Invitrogen, Karlsruhe, Germany). Store at –20°C.

3. Washing buffer: 10 mM Tris–HCl, 1 mM EDTA, pH 8.0. Store at room temperature.

4. Microcon YM-30 Centrifugal Filter Units (Millipore, Schwalbach, Germany).

5. Murine Cot-1 DNA (Invitrogen) 1 mg/ml. Store at –20°C.

6. Salmon sperm DNA solution (Invitrogen) 10 mg/ml. Store at –20°C.

7. 20× SSC stock solution. *see* **Subheading 2.2.2.** Prepare a 2× SSC solution (pH 7) for use. Store at room temperature.

8. Hybridization buffer: 50% deionized formamide and 10% (w/v) dextran sulfate in 2× SSC. Store at –20°C.

#### 2.3.2. Fluorescence In Situ Hybridization on Metaphase Chromosomes and Interphase Nuclei

1. Ready-to-use DNA probe.

2. Tween 20 (Merck). Store at room temperature.

3. 20× SSC, *see* **Subheading 2.2.2, item 1**. Prepare 100 ml 0.4× SSC/0.5% Tween 20 pH 7 and 500 ml 2× SSC/0.5% Tween 20. Store at room temperature.

4. Fixogum (Marabuwerke). Store at room temperature.

5. Blocking buffer: 3% (w/v) BSA (bovine serum albumin) in 2× SSC/0.5% Tween 20. Store aliquots at –20°C.

6. Fluorescein Avidin DCS (Vector Laboratories). Store at 4°C.

7. Biotinylated anti-Avidin D (Vector Laboratories). Store at 4°C.

8. Nuclear staining: DAPI (4,6-diamidino-2-phenylindole). Prepare a solution according to **Subheading 2.2.3**.

9. Mounting medium: Vectashield (Vector Laboratories). Store at 4°C.

10. Cover slips, Ø 10 mm (Menzel).

11. Coplin jars (Fisher Scientific).

12. Humidified chamber.

13. Slide warmer.

## 3. Methods

The methods described later outline (1) the chromosome preparation, (2) the spectral karyotyping, and (3) the fluorescence in situ hybridization for validation of structural chromosomal aberrations detected by SKY analysis.

The SKY technique requires metaphase chromosomes of high quality and quantity. Therefore, a critical and careful assessment of the quality of metaphase chromosomes and protease pretreatment, if cytoplasm and debris are visible **Fig. (2)**, strongly influences the success of SKY hybridization and analysis.

Characterization of structural chromosomal alterations detected by SKY analysis, in particular balanced and unbalanced translocations, terminal or interstitial deletions, insertions, duplications, or amplifications, requires a precise description of the chromosomal breakpoints, i.e., altered chromosomal bands. The inverted DAPI image that is achieved in combination with the spectral image for SKY analysis and banding schemes of murine chromosomes provided by the http://www.ensembl.orgdatabase of the Wellcome Trust Sanger Institute facilitates not only the identification and description of the chromosomal breakpoints

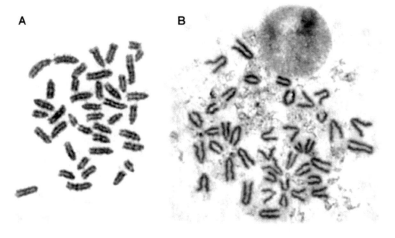

Fig. 2. Giemsa-stained chromosomes. (**a**) The high quality of the metaphase spread requires no pretreatment before SKY hybridization. (**b**) Pepsin treatment is recommended to reduce cytoplasm and debris, thus increasing the success of the SKY hybridization and detection..

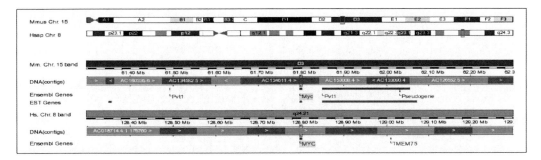

Fig. 3. The murine chromosome 15 shows synteny to the human chromosomes 5, 8, 12, and 22. (*See Color Plates*)

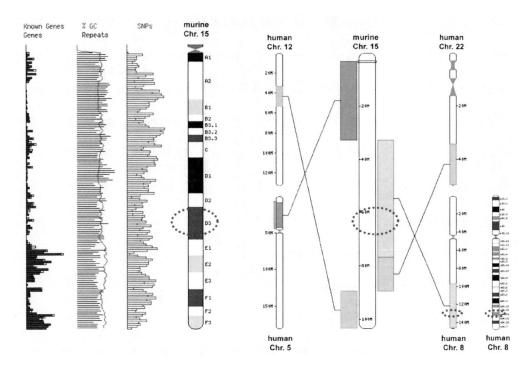

Fig. 4. Synteny of the c-Myc locus (http://www.ensembl.org(. In mice, the proto-oncogene c-myc is located in chromosomal region 15D3. The syntenic human region is 8q24.21. (*See Color Plates*)

but also the assignment of the syntenic human region. For example, the murine chromosome 15 shows synteny to the human chromosomes 5, 8, 12, and 22 **Figs. (3** and **4)**. Thus, a structural chromosomal alteration detected in chromosomal band 15D3 of the murine genome – syntenic to the human band 8q24 – indicates a possible rearrangement of the proto-oncogene c-myc **Fig. (1)**. It is important to note that there are different rules for the description of structural chromosomal alterations of the murine and the human genome divergence (Rules for Nomenclature of Chromosome Aberrations (http://www.informatics.jax.org/mgihome/nomen/anomalies.shtml)).

Translocations can be confirmed by FISH in different ways, e.g., by the generation of a single-color **Fig. (5)** or a dual-color break-apart probe. Generation of labeled DNA probes can be performed by different approaches like random priming, nick translation *(9)*, or DOP-PCR *(10)*. For labeling DNA probes 100–300 kb in size, these methods give similar results. Random-primed labeling uses oligonucleotides that bind approximately all 80–100 bases to the DNA and represent the start site for the Klenow fragment of the DNA polymerase. The polymerase isolated from *E. coli* incorporates labeled and unlabeled nucleotides to synthesize a new DNA strand. To amplify the fluorescent signal of the probe, the DNA must be indirectly labeled, e.g., using the BioPrime® DNA Labeling System protocol for the generation of biotinylated probes that will be described here. Generating a dual-color break-apart probe requires two directly but differently labeled DNA probes, e.g., SpectrumGreen and SpectrumOrange. The generation of a FISH probe to validate a numerical aberration like a monosomy or a trisomy is normally less difficult but not less important, since copy number changes, e.g., a complete or partial chromosomal deletion, can result in the monoallelic

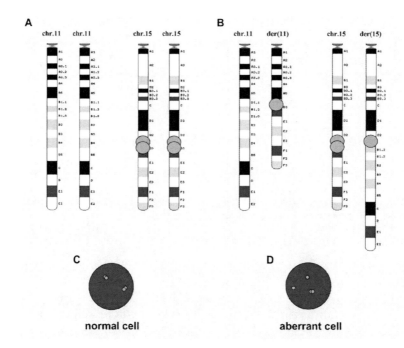

Fig. 5. Presentation of the ideograms of murine chromosomes 11 and 15 and confirmation of a balanced translocation T(11B1;15D3) by FISH using a single-color break-apart probe. One probe is generated to hybridize centromerically close to the breakpoint in 15D3, the other probe to hybridize telomerically close to the breakpoint in 15D3. Thus, two normal chromosomes 15 show 2 fusion signals (**a**). A rearrangement results in the split of one fusion signal into two single signals, one on Der(11) and the other on Der(15) (**b**). (**c**) and (**d**) show the signal constellation as expected in interphase nuclei in a normal cell and in a cell with T(11B1;15D3). (*See Color Plates*)

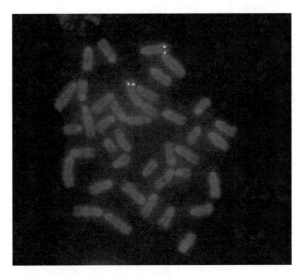

Fig. 6. FISH on murine metaphase chromosomes. The SpectrumGreen labeled DNA probe has been generated by random priming using the BAC clone RP23-382J5. The probe hybridizes to region XF2 and shows two signals representing the expected signal constellation of a normal cell. (*See Color Plates*)

loss of a tumor suppressor gene. In **Fig. 6** metaphase FISH using a directly labeled probe generated by random priming and hybridizing in region XF2 of the murine genome is shown. Additional helpful remarks to perform FISH are available at http://info.med. yale.edu//1span>genetics/ward/tavi/FISHguide. html.

### 3.1. Chromosome Preparation

1. To arrest metaphase chromosomes, add colcemid at a final concentration of 0.035 µg/ml to the culture medium. Incubate cells under cell culture conditions for 6–8 h (*see* **Note 4**).

2. Transfer cells to a 15-ml tube and centrifuge cell suspension at $256 \times g$ for 10 min at room temperature.

3. Discard the supernatant remaining 2 ml above the cell pellet. Vortex to resuspend the cells and slowly add prewarmed KCl during vortexing up to 12–13 ml. Close the tube and shake gently 2–3 times. Incubate the cells for 20 min at 37°C (*see* **Note 5**).

4. Add approximately 2 ml of freshly prepared Carnoy's fixative to prefix the cells. Close the tube and shake gently 2–3 times. Centrifuge cell suspension at $256 \times g$ for 10 min at room temperature and discard the supernatant remaining 2 ml above the cell pellet (*see* **Note 6**).

5. Vortex to resuspend the cells and slowly add freshly prepared Carnoy's fixative during vortexing up to 12–13 ml. Close the tube and shake gently 2–3 times. Centrifuge cell suspension at $256 \times g$ for 10 min at room temperature and discard the supernatant up to 2 ml.

6. Repeat washing and centrifugation steps 4–5 times.

7. Discard the supernatant carefully up to 1–1.5 ml and resuspend cells (*see* **Note 7**).

8. Place one drop of the cell suspension onto a glass slide that is precleaned with Carnoy?s fixative. Control the quality and quantity of metaphases using a phase-contrast microscope (*see* **Note8**).

9. Dehydrate the slide in an ethanol series (70, 90, and 100%) 3 min each, air-dry the slide, and incubate it overnight at 37°C for aging (*see* **Note 9**).

### 3.2. Spectral Karyotyping

#### 3.2.1 Pretreatment (See Note 10)

1. Apply 10 µl of pepsin to an empty beaker, add 100 ml prewarmed HCl, mix well and transfer to a coplin jar.

2. Start pepsin treatment immediately for 5 min, wash briefly in PBS, and control the quality of the metaphases using a phase-contrast microscope (*see* **Note 11**).

3. Dehydrate slide in ethanol series (70, 90, and 100%) at room temperature 3 min each, air-dry, and age the slide overnight at 37°C.

#### 3.2.2. Denaturation and Hybridization

1. Mark region to hybridize (*see* **Note 12**).

2. Denature metaphase chromosomes by incubating 1.5 min in denaturation solution (*see***Note 13**).

3. Fix the denatured DNA by placing the slide immediately in ice-cold ethanol, 70, 90, and 100% 3 min each. Air-dry the slide.

4. Denature 5 or 10 µl of the SKY probe by incubating 7 min at 80°C and preanneal 1 h at 37°C.

5. Place denatured SKY probe on the region to hybridize. Cover with cover slip (Ø 12 mm using 5 µl SKY probe and 18 × 18 mm using 10 µl SKY probe). Seal with Fixogum.

6. Hybridize for 36–48 h at 37°C in a humidified chamber (*see***Note 14**).

#### 3.2.3. Detection

1. Carefully remove Fixogum and cover slip and wash the slide 3 × 5 min in washing solution A, 3 × 5 min in washing solution B and briefly in washing solution C. Allow fluid to drain (*see***Note 15**).

2. Apply 120 µl blocking solution, cover with cover slip 24 × 60 mm and incubate for 30 min at 37°C in a humidified chamber.

3. Wash slide briefly in washing solution C and allow fluid to drain.

4. Prepare antibody solution A by diluting mouse antidigoxin and Cy5-conjugated streptavidin 1:200 in washing solution C. Apply 120 µl diluted antibodies onto the slide, cover with

cover slip 24 × 60 mm, and incubate 45 min at 37°C in a humidified chamber (*see* **Note 16**).

5. Wash slide 3 × 5 min in washing solution C.

6. Prepare antibody solution B by diluting Cy5.5 sheep antimouse 1:200 in washing solution C. Apply 120 µl diluted antibody onto the slide, cover with cover slip 24 × 60 mm, and incubate 45 min at 37°C in a humidified chamber (*see* **Note 16**).

7. Wash the slide 3 × 5 min in washing solution C, incubate 10 min in DAPI solution at room temperature, and wash 5 min in water at room temperature. Tilt slide to allow fluid to drain, apply two drops Vectashield. and cover with cover slip 24 × 60 mm.

8. Slide is prepared for acquisition and analysis.

### *3.3. Validation of Chromosomal Alterations by Fluorescence In Situ Hybridization*

*3.3.1. Generation of Biotin-Labeled DNA Probes by Random Priming*

1. Label DNA isolated from BAC clone according to the BioPrime® DNA Labeling System protocol.

2. Purify the labeled DNA probe following the ethanol precipitation of the BioPrime® DNA Labeling System protocol or alternatively using Microcon columns YM 30. Therefore, apply the total volume of the labeled probe at the column, add 380 µl washing buffer, centrifuge 12 min at 10,000 rpm, and discard the flow. Wash with 380 µl washing buffer, centrifuge 8 min at 13,000 rpm, and discard the flow. Wash with 450 µl washing buffer, centrifuge 9.5 min at 13,000 rpm, and discard the flow. Turn the column, place it in a new tube, and centrifuge 2 min at 10,000 rpm. Dilute the flow up to 50 µl with water. Store the probe stock solution light-protected at –20°C.

3. For preparing the ready-to-use probe, add 5 µl of the probe stock solution with 2 µl murine Cot-1 DNA, 1 µl salmon sperm, and 17 µl hybridization buffer. Store at –20°C.

*3.3.2. Fluorescence In Situ Hybridization Using Indirect Fluorescent Labeled Probes*

1. Prepare slides for hybridization according to **Subheading 3.1**.

2. Place 3 µl of the ready-to-use DNA probe to the region to hybridize, cover with a cover slip, and seal with Fixogum.

3. Incubate the slide for DNA denaturation 10 min on a slide warmer at 80°C.

4. Hybridize at 37°C overnight in a humidified chamber.

5. Carefully remove Fixogum and cover slip and wash the slide 2 × 2 min in 0.4× SSC/0.5% Tween 20 at 70°C (*see***Note 17**).

6. Apply 120 µl blocking buffer, cover with cover slip 24 × 60 mm, and incubate 30 min at 37°C in a humidified chamber.

7. Wash briefly in 2× SSC/0.5% Tween 20 at room temperature.

8. Dilute Fluorescein Avidin DCS 1:1,000 in 2× SSC/0.5% Tween 20. Apply 120 µl of antibody solution on the slide and

cover with cover slip 24 × 60 mm. Incubate 30 min at 37°C in a humidified chamber.

9. Wash 3 × 3 min with 2× SSC/0.5% Tween 20 at room temperature.

10. Dilute biotinylated anti-Avidin D 1:100 in 2× SSC/0.5% Tween 20. Apply 120 µl of antibody solution on the slide and cover with cover slip 24 × 60 mm. Incubate 30 min at 37°C in a humidified chamber.

11. Dilute Fluorescein Avidin DCS 1:1,000 in 2× SSC/0.5% Tween 20. Apply 120 µl of antibody solution on the slide and cover with cover slip 24 × 60 mm. Incubate 30 min at 37°C in a humidified chamber (*see* **Note 18**).

12. Wash 3 × 3 min with 2× SSC/0.5% Tween 20 at room temperature.

13. Incubate the slide for 10 min in DAPI solution at room temperature and wash 5 min in water at room temperature. Tilt slide to allow fluid to drain, apply two drops Vectashield, and cover with cover slip 24 × 60 mm.

14. The slide is prepared for acquisition and analysis. It can be stored for several weeks at 4°C.

## 4. Notes

1. Formamide is a toxic agent. Heating should be done in a water bath under a flow. Dispose solution separately in a container for toxic agents.

2. It is recommended to use light-protected coplin jars.

3. DNA from BAC clones can be isolated, e.g., using the Perfectprep® Plasmid Maxi protocol (Eppendorf, Wesseling-Berzdorf, Germany).

4. To increase the quantity of metaphase chromosomes, the incubation time with colcemid can be increased up to 24 h. Colcemid is toxic. Note safety instructions. Following the protocol of chromosome preparation and spectral karyotyping, there are very important steps that can be varied to increase the quality as well as the quantity of metaphase chromosomes. The most important steps are the incubation time with colcemid and KCl, as well as the pretreatment with proteases.

5. Spreading of chromosomes within the metaphase can be improved by variation of the incubation time with KCl. In some cases, a prolonged incubation time up to 30 min is required, e.g., for cells with tetraploid chromosome complements.

6. Methanol is toxic. Note safety instructions and dispose of Carnoy?s fixative separately in a container for toxic agents.

7. Cell suspension can be stored for several years at –20°C.

8. Dropping the cell suspension onto the slide using a climate chamber with constant temperature (22°C) and constant humidity (48%) increases the quality of the metaphase spreads.

9. Slides should be aged for at least 24 h at 37°C before SKY hybridization. Long-term storage of dehydrated slides is recommended under humidity protection in slide boxes at –80°C. Slide boxes should be prewarmed for 30 min at 37°C before taking the slides out of the boxes.

10. Pretreatment is only recommended if cytoplasm and debris are visible. Continue with denaturation and hybridization steps if no cytoplasm is recognizable.

11. Pepsin treatment can be repeated for 5–10 min if necessary. For the assessment of the success of the pretreatment by phase-contrast microscopy cover with a cover slip 24 × 60 mm. Slides should not dry during the procedure. Alternatively, slides can be pretreated with trypsin *(11)*.

12. Amount of SKY probe used for hybridization depends on the quantity of metaphases. Hybridized region should contain more than 30 analyzable metaphases.

13. Do not overdenature chromosomes due to prolonged denaturation time.

14. Hybridization time can be increased up to 72 h.

15. To facilitate the removal of the cover slip, dip the slide briefly in washing solution A. All following steps should be done under light protection of the slide as far as possible. Take care that the slide is always wet during the following procedures. For washing steps, a rocking platform is recommended.

16. Do not store diluted antibodies.

17. Optimize posthybridization temperature and SSC concentration for each probe.

18. For signal amplification repeat **steps 8–10**. Note that each signal amplification also increases the background.

## Acknowledgments

The authors wish to thank Michael Köhler (ASI GmbH, Germany) and Evelin Schröck for their support and guidance in SKY technology and Gillian Teicke for carefully reading the manuscript.

## References

1. Schrock, E., du Manoir, S., Veldman, T., Schoell, B., Wienberg, J., Ferguson-Smith, M.A., Ning, Y., Ledbetter, D.H., Bar-Am, I., Soenksen, D., Garini, Y., Ried, T. (1996) Multicolor spectral karyotyping of human chromosomes. *Science* **273**, 494–497.

2. Liyanage, M., Coleman, A., du Manoir, S., Veldman, T., McCormack, S., Dickson, R.B., Barlow, C., Wynshaw-Boris, A., Janz, S., Wienberg, J., Ferguson-Smith, M.A., Schrock, E., Ried, T. (1996) Multicolour spectral karyotyping of mouse chromosomes. *Nat Genet* **14**, 312–315.

3. Speicher, M.R., Gwyn Ballard, S., Ward, D.C. (1996) Karyotyping human chromosomes by combinatorial multi-fluor FISH. *Nat Genet* **12**, 368–375.

4. Chudoba, I., Plesch, A., Lorch, T., Lemke, J., Claussen, U., Senger, G. (1999) High resolution multicolor-banding: a new technique for refined FISH analysis of human chromosomes. *Cytogenet Cell Genet* **84**, 156–160.

5. Braig, M., Lee, S., Loddenkemper, C., Rudolph, C., Peters, A.H.F.M., Schlegelberger, B., Stein, H., Dörken, B., Jenuwein, T., Schmitt, C.A. (2005) Inactivation of the Suv39h1 histone methyltransferase promotes Ras-driven lymphomagenesis by disabling cellular senescence. *Nature* **436**, 660–665.

6. Li, Z., Dullmann, J., Schiedlmeier, B., Schmidt, M., von Kalle, C., Meyer, J., Forster, M., Stocking, C., Wahlers, A., Frank, O., Ostertag, W., Kuhlcke, K., Eckert, H.G., Fehse, B., Baum, C. (2002) Murine leukemia induced by retroviral gene marking. *Science* **296**, 497.

7. Modlich, U., Kustikova, O.S., Schmidt, M., Rudolph, C., Meyer, J., Li, Z., Kamino, K., von Neuhoff, N., Schlegelberger, B., Kuehlcke, K., Bunting, K.D., Schmidt, S., Deichmann, A., von Kalle, C., Fehse, B., Baum, C. (2005) Leukemias following retroviral transfer of multidrug resistance 1 (MDR1) are driven by combinatorial insertional mutagenesis. *Blood* **105**, 4235–4246.

8. Gaudet, F., Hodgson, J.G., Eden, A., Jackson-Grusby, L., Dausman, J., Gray, J.W., Leonhardt, H., Jaenisch, R. (2003) Induction of tumors in mice by genomic hypomethylation. *Science* **300**, 489–492.

9. Zhang, Y., Schlegelberger, B. (2002) Simultaneous fluorescence immunophenotyping and FISH on tumor cells. *Methods Mol Biol* **204**, 379–390.

10. Fiegler, H., Carr, P., Douglas, E.J., Burford, D.C., Hunt, S., Scott, C.E., Smith, J., Vetrie, D., Gorman, P., Tomlinson, I.P., Carter, N.P. (2003) DNA microarrays for comparative genomic hybridization based on DOP-PCR amplification of BAC and PAC clones. *Genes Chromosomes Cancer* **36**, 361–374.

11. Loja, T., Kuglik, P., Oltova, A., Smuharova, P., Zitterbart, K., Bajciova, V., Veselska, R. (2007) The optimization of sample treatment for spectral karyotyping with applications for human tumour cells. *Cytogenet Genome Res* **116**, 186–193.

## Databases

http://www.ensembl.org

http://www.informatics.jax.org/mgihome/nomen/anomalies.shtml

http://info.med.yale.edu/genetics/ward/tavi/FISHguide.html

# Chapter 31

## Database Setup for Preclinical Studies of Gene-Modified Hematopoiesis

### Brenden Balcik, Elke Grassman, and Lilith Reeves

## Summary

Murine safety studies are routinely used for gathering preclinical safety and efficacy data and, for Phase I studies, Good Laboratory Practice (GLP) compliance is not mandated. However, extensive amounts of data must be gathered and analyzed. An inter-relational database is an effective tool for storing, sorting, and reviewing data.

**Key words:** Murine preclinical safety studies, Retroviral gene transfer preclinical studies, Preclinical safety study database.

## 1. Introduction

While (the) retroviral human gene transfer trials have been conducted since the late 1980s and, to date, there have been more than 200 gene transfer trials reviewed by the National Institutes of Health (NIH) Office of Biotechnology Activities (OBA) Recombinant DNA Advisory Committee (RAC), these trials continue to be early-phase studies (1, 2). Unlike later phase trials, Phase I trials do not generally require animal pharmacology and toxicology studies to have been performed according to strict Good Laboratory Practices (GLP) as defined in 21 The Code of Federal Regulations (CFR) Part 58; however, these studies must be conducted and recorded using sound scientific proof of principle and must address the potential risks to multiple organs and systems.

Small animal murine models are frequently used for preclinical safety studies in hematopoietic gene transfer trials. It is widely

Christopher Baum (ed.), *Methods in Molecular Biology, Methods and Protocols, vol. 506*
© Humana Press, a part of Springer Science+Business Media, LLC 2009
DOI: 10.1007/978-1-59745-409-4_31

accepted that large animal models better mimic the human system; however, the need for statistically relevant numbers of test and control animals and the extended time from treatment to demonstration of toxicity make this model impractical for many Phase I studies. A well-designed murine model allows for a statistically relevant number of animals to be studied and offers an accelerated process of toxicities associated with the treatment *(3)*. Critical parameters for these studies as related to gene transfer in hematopoietic progenitor cells include adequate numbers of test and control animals, standardized animal conditioning regimens, ability to distinguish host and donor cells, establishment of appropriate levels of gene transfer, and detailed examination of the animals at relevant time points in the study.

The ability to draw valid conclusions from the studies is contingent upon adequate recording and retrieval of the data. A relational database system such as FileMaker Pro (5.0) provides an organized system for managing the vast data obtained from the studies. The system must accommodate all details of the safety and efficacy data according to the study design and accommodate retrieval of details as well as summary data. A well-designed and managed system facilitates thorough evaluation and reporting of the data from a large number of experimental animals. This chapter defines the basic database knowledge and tools required and a stepwise approach for designing and creating a relational database. It also provides simple directions for accessing, exporting, and printing the compiled data. **Subheading 31.4** provides hints for novice database users and offers suggestions to further tailor a database after becoming familiar with the basic relational database methods described in this text.

## 2. Materials

1. FileMaker Pro (5.0) database.
2. Visual QuickStart Guide: FileMaker Pro 6 for Windows and Macintosh *(4)* (*see* **Note 1**).

## 3. Methods

Prepare a separate database for each category of information pertinent to each study to include all cohorts in the study.

### 3.1. Learn the Fundamentals of a Database System

FileMaker Pro is conducive to use by scientists who are not experienced in designing databases; however, a basic understanding of database units and how they interconnect and build upon one another to form the relational database system is essential to developing a database to serve the needs of the study.

1. *Fields* are the fundamental units of data storage. Typical "fields" in a safety study database include: Study identification, specimen identification, dates, data points, etc.

2. *Records* are groups of fields (containing data; i.e., a database) pertaining to a specific topic.

3. *Databases* comprise groups of related records. These can be divided into two types of databases: A Main Database serves as the "hub" that connects the multiple categories (Accessory or Portal Databases) of information to be recorded, and an Accessory or Portal Database includes the data for a given category of information such as the data for all complete blood counts (CBCs), etc.

4. *Layouts* are the organizational designs for each record and can be tailored to be study-specific by the user to provide quick and efficient data analysis. Layouts may be modified as needed over the course of the study.

5. *The Database System (Relational Database)* is a group of databases containing groups of similar information (Records), each linked to one another via information pathways (*Portals*).

### 3.2. Define the Information to be Included in Database System

1. Define the information to be included in the *Main (Hub) Database* (*see* **Note 2** for an example of mapping a database hub and **Note 3** for additional considerations relevant to the Main Database).
   - Study identification
   - Vector identification
   - Animal identification and number
   - Date of initiation
   - Date of death
   - Additional fields for each unique characteristic of the specific study

2. Define the *Accessory or Portal Database* categories.
   - Vector information
   - Transduced cell information
   - Animal blood counts (CBC)
   - Animal clonogenic data
   - Animal immunophenotyping

- Animal molecular analyses – qPCR
- Insertion site analyses-LM or LAM
- Histopathology data
- Necropsy data
- Functional assays as relevant to the study

3. Define the information and time points (*Fields*) critical for interpreting and relating the results to the study for each *Record* within each *Portal Database*. Though it is helpful to input all information in the initial design, FileMaker Pro allows for on-going modifications of the fields (*see* **Note 4**).

- Vector information: Vector design, titer, safety testing results, method of preparation.
- Transduced cell information: Method of transduction, cytokines, percent transduction, cell number, immunophenotyping.
- Animal blood counts (CBC): White blood cell count (WBC), hemoglobin concentration, platelet count, differential blood cell count.
- Animal clonogenic data: CFU quantitation and qPCR copy number.
- Animal immunophenotyping: percentage of each lineage contribution in each tissue, raw data scatter plots, interpretation.
- Animal molecular analyses – qPCR: Primer sequences, target gene, copy number per cell.
- Insertion site analyses: LM and/or LAM PCR gel pictures and interpretation.
- Histopathology data: Pathology report and photomicrographs.
- Necropsy data: organ weights, observations during necropsy.
- Functional assays as relevant to the study: Study specific assay information.

***3.3. Create the Database in Filemaker Pro***

1. Open FileMaker Pro(5.0).
2. Select "Create a new empty file".
3. Define file name and location for storing the database.
4. Enter the field names and type of information to be input (text, number, calculation, date, or container – to be used for graphical data) for the first "Main Database Record".
   - Study identification (Text)
   - Date of initiation (Date)
   - Vector identification (Text)
   - Animal identification (Text and/or Number)

- Additional fields for each Portal Database Category as applicable (*see* **Subheading 31.3.2, step 2**).
  - Vector information
  - Transduced cell information
  - Animal blood counts (CBC)
  - Animal clonogenic data
  - Animal immunophenotyping
  - Animal molecular analyses – qPCR
  - Insertion site analyses
  - Histopathology data
  - Necropsy data
  - Functional assays as relevant to the study

5. Select "Done".

6. Create a new "Database" for each of the Database Categories for which data will be entered by repeating steps in **Subheading 31.3.3, steps 2–5**.

7. Create a unique database identifier for each mouse.
   - Define a single field as Mouse Database Number (Mouse DB #).
   - The Mouse DB # should be independent of the study specimen ID (*see* **Note 5**).
   - The Mouse DB # will also be used for sorting and searching through records.

8. Group the unique identifiers according to each study (*see* **Note 1**).

9. Select "View" and "Layout Mode" to provide the tools for designing the layout. Design for ease and efficiency for entering and viewing the information to be relevant to each category.

10. Organize the fields within each database in as many different layouts as necessary insuring adequate data review.
    - Create a CBC layout to view CBC data over time for a specific mouse to include all data as defined for the category.
    - Format the layout so that, when printed, it appears on a single page (*see* **Note 7**).
      - Select "View" and "Layout Mode".
      - From the "File" drop-down menu choose "Print Setup".
      - Choose paper "Orientation" (Portrait versus Landscape) to best accommodate the data to be shown.
      - Arrange fields to fit the orientation of the layout.
    - Create a CBC layout to view CBC data for an entire mouse cohort.

11. Continue creating layouts in each database until all data defined as relevant support information for each *Portal Database* are easily viewed and interpretable.

12. Create a study summary layout in the Main Data database for each study.
    - Include information pertinent to the study not found elsewhere in the database.
    - Include any development work that may have been performed prior to study initiation and that is relevant for study interpretation.
    - Include summarized data as the study progresses.

13. Create navigational buttons within each database.
    - Open the Layout view.
        - From the "Insert" drop-down menu select "Button".
        - Specify the action to be performed by the button.
        - Select button style.
        - Choose "OK".
    - Move the button to its assigned space (while in the Layout view).
    - Label the button (according to its action) by using the text tool.
    - Define and create buttons to do the following:
        - Sort records (by user-defined criteria).
        - Scroll through records.
        - Choose, search and find specific records.

14. Define the Relationships for each Record.
    - Access the Main "Hub" Database.
    - From the "File" drop-down menu select "Define Relationships"
    - Select "New" and locate the first file (Database) specific to the category of information and select to link.
    - Choose the second file (Database) to which the first file (Database) will be linked.
    - Define each relationship using the Mouse DB # field as the link between each database (*see* **Note 8**).
    - Name the relationship between the two databases.
    - Create portals in the Main Data database so that key data from every other database can be viewed while reviewing in the Main Data database.
    - Include, in each portal, only the fields necessary for quick data review. More complete data review can be performed while in the specific database.

- Define the number of records to be viewed in each portal.
- Repeat for all data.

**3.4. Enter Data**

1. Choose the "View Format" to enter data.
   - To type enter data by directly typing into the "Record", select the "View" drop down menu and select "View as Form". Type data directly into each field for each record.
   - To type enter data as a table into the "Record", select the "View" drop down menu and select "View as Form". Copy data that have been entered into a standard spreadsheet and paste into the appropriate record field.
   - Enter data in the "List" view by typing data directly into each field and scrolling through each record rather than changing screens.
2. Create links to large data files, specifically Flow Cytometry, scatter plots, and pathology pictures.
   - Left click in field in which to place the link.
   - Right click and choose "Insert Object".
   - Select "Create from File".
   - Locate the file in the directory.
   - Select "Link" and "Display as Icon".
   - Choose "OK".
3. *See* **Note 9** for additional tips regarding efficient data entry.

**3.5. Report Data**

1. Report data directly from the database.
   - Sort records by criteria specified by the particular study aims.
   - Print sorted records.
2. Export data to another program (*see* **Note 10**).
   - Sort records by criteria specified by the particular study aims.
   - Export sorted records.
     - From the "File" drop-down menu choose "Export Records".
     - Define the file name, file type, and destination.
     - Designate and order the fields to be exported using "Specify Order" window.
     - Define the field labels to be exported (if any).
     - Format data in the destination file as necessary.

**3.6. Create Passwords**

1. Create passwords for user access privileges. From the "File" drop-down menu choose "Access Privileges".
   - Select "Passwords".
   - Define the password.

- Select privileges granted by this password by activating the check boxes corresponding to each privilege.
- Choose "Done".

2. *See* **Note 11** for more information about passwords.

---

## 4. Notes

The following information is included to assist the novice database designer in creating a functional database with minimal database program knowledge and to provide helpful suggestions for improving and tailoring a database as the user becomes more proficient.

1. A knowledgeable database consultant available to field queries while designing the database and reports, may be helpful in streamlining the process but is not essential for developing an effective tool.

2. A "Main Data Hub" serves as a central unifying database **Fig. (1)**. It is the cover record viewed when the program is opened and is the "window" and "launch point" to access the other databases.

3. It is essential to have a specific location defined to store general study information such as study identification, vector identification, transduction parameters, cohorts, relevant dates, recipient and donor mouse data, transplant data, animal death date and cause, and fields to calculate time between study points and total time on study. General comments and

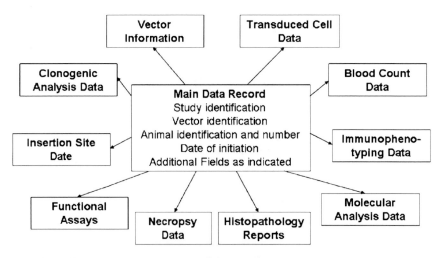

Fig. 1. Preclinical Safety Study Inter-relational Database Organizational Map.

observations do not fit within the confines of more data-specific databases, but the design and the central location of the Main Data database work well for this information.

4. As the database system develops, it is inevitable that changes will be made. The system is extremely flexible in that fields and layouts can be added, deleted, and edited at any point during the system design and construction process. It is important to not be preoccupied with formatting, font size, color, and other noncritical attributes. The system should be designed to be dynamic and with expectations for multiple changes as it is used and evaluated.

5. In choosing a database identifier, it is important to understand that FileMaker Pro only recognizes numbers and not separators that may be used in identifying mice. Decimals and dashes are not recognized by the database, so mouse 1.1 or 1-1 is actually 11. Similarly, mouse 1A and 1B are recognized as the mouse number "1." This limitation must be honored in order to effectively sort data. A simple means of accomplishing this is to assign each mouse a unique numeric ID (Mouse Database Number (Mouse DB #)) independent of its study ID. Each new mouse entered into the system receives the next numeric ID (in sequence) and no Mouse DB # is duplicated in the database. For example, two different cohorts may each have a control mouse 2C. By using the sequential Mouse DB numbering system, each mouse is recognized as a unique specimen when sorting and reviewing the database. The only time this Mouse DB # is used is in the database and when data are presented, the database ID is removed and the study ID is substituted for clear reporting.

6. Due to the large numbers of mice needed for these studies, it is necessary to group the mouse cohorts by study. For example, reserve Mouse DB numbers 1-1000 for mice used in Study #1, Mouse DB #s 1001-1999 for mice used in Study #2, etc.

7. For easier data reporting, it is beneficial to configure as many layouts as possible to print on a single page. This allows for data to be generated directly from the database, for the purpose of reporting or archiving, without having to export records to another program or file type.

8. When creating relationships between the databases, use a field that will be present in all the databases, such as the Mouse DB #. This allows the user to move between the databases and view data from the same mouse within each database.

9. When massive amounts of data are entered, some of the data entries are repetitive. When this is the case, "Value Lists" should be created to lessen the amount of time spent typing

during data entry. Value Lists are drop-down menus that can be tailored for each field. For example, when entering pathology information, rather than typing the same tissue type or preparation with each record, a Value List can be created to choose the type of preparation that corresponds to those records so that the same information is not typed into the field with each record.

10. While the database system is an efficient storage system, it is not well equipped for preparing data for presentation. To generate tables, charts, and graphs, the data may be exported to another software program with presentation capabilities. In addition, like many software programs, the database system can only be viewed from a computer with the software installed. Since these studies are conducted by multiple investigators, it is recommended to export data to more commonly used file types to share the data with investigators.

11. An effective data system for multi-investigator studies must allow for access by multiple users; however, in order to maintain data integrity, it is necessary to set access privileges for each user. Passwords may be assigned with varying degrees of access privileges. For example, one password allows data viewing only. A second password allows for data viewing and data entry. A third password allows for data viewing, entry and deleting. A fourth or master password is issued to a very limited number of users and allows the user to view, add, and delete data as well as create fields, layouts, databases, and relationships.

## References

1. Rosenberg, S. A., Aebersold, P. M., Cornetta, K., Kasid, A., Morgan, R. A., Moen, R., Karson, E. M., Lotze, M. T., Yang, J. C., Topalian, S. L., Merino, M. H., Culver, K., Miller, A. D., Blaese, M. D., and Anderson, W. F. (1990) Gene transfer into humans-immuno-therapy of patients with advanced melanoma, using tumor infiltrating lymphocytes modified by retroviral gene transduction. *N Engl J Med.* 323, 570–578.

2. http://www.gemcrics.od.nih.gov/

3. Will, E., Bailey, J., Schuesler, T., Modlich, U., Balcik, B., Burzynski, B., Witte, D., Layh-Schmitt, G., Rudolph, C., Schlegelberger, B., von Kalle, C., Baum, C., Sorrentino, B. P., Wagner, L. M., Kelly, P., Reeves, L., and Williams, D. A. (2007) Importance of murine study design for testing toxicity of retroviral vectors in support of phase I trials. *Mol Ther.* 15, 782–791.

4. Hester, Nolan. *Visual QuickStart Guide: FileMaker Pro 6 for Windows and Macintosh.* California: PeachPit, 2003.

# Chapter 32

## The US and EU Regulatory Perspectives on the Clinical Use of Hematopoietic Stem/Progenitor Cells Genetically Modified Ex Vivo by Retroviral Vectors

### Carolyn A. Wilson and Klaus Cichutek

## Summary

A primary safety issue presented by human hematopoietic stem cells/progenitor cells (HS/PC) genetically modified by gammaretroviral or lentiviral vectors is the risk of oncogenesis. This risk is a potential consequence of either of the following events: (a) the possible unintended generation of replication-competent vector-derived viruses (replication-competent retrovirus, RCR; replication-competent lentivirus, RCL) leading to neoplasia due to RCR/RCL infection of target and nontarget cells in vivo, or (b) intended vector integration in the chromosomal DNA of the target somatic cells leading to neoplasia due to insertional mutagenesis. These risks should be addressed in nonclinical and clinical studies. In the US and the EU, a combination of regulations and guidance documents are available to investigators and sponsors of gene therapy clinical trials. Guidance documents provide a facile way to adapt regulatory recommendations, in line with the changing state of the art in medical science. In the field of retroviral vectors, a number of innovations are being tested in nonclinical or clinical investigations, and each of these will raise their own regulatory issues. Some recent examples of these types of innovations include development of novel vector structures to minimize risks associated with vector integration, such as lentiviral vectors currently used in clinical trials for HS/PC modification that have been designed with deletions of the strong retroviral enhancer associated with oncogenesis.

Key words: Hematopoietic stem/progenitor cells, Gene therapy, Gene transfer, Retroviral vector, Genotoxicity, Inadvertent germline transmission, Integration, Insertional mutagenesis, Oncogenesis.

## 1. Introduction

Hematopoietic stem/progenitor cells (HS/PC) possess three properties that make these cells attractive targets for the therapeutic application of gene therapy to the treatment of disorders

Christopher Baum (ed.), *Methods in Molecular Biology, Methods and Protocols, vol. 506*
© Humana Press, a part of Springer Science + Business Media, LLC 2009
DOI: 10.1007/978-1-59745-409-4_32

affecting cells of the hematopoietic lineage: (1) the ability to give rise to all cells of the hematopoietic lineage; (2) the maintenance and self-renewal of pluripotent stem cells to support ongoing hematopoiesis; (3) the availability in sufficient amounts for ex vivo manipulation, either from bone marrow or from peripheral blood, in conjunction with reinfusion as a simple method of administration to patients. Since the inception of the clinical use of gene therapy, the US has received over 50 requests to perform investigational clinical trials using genetically modified HS/PC (defined here as cells that are characterized as CD34 or CD133 positive). These products have been clinically evaluated in one of four settings: cancer, genetic disorders, HIV, or for tracking of cells in vivo. All these clinical trials, except two, used gammaretroviral vectors to ex vivo modify the HS/PC. In the European Union (EU), the EudraCT clinical trial database, into which newly initiated clinical trials have been introduced since mid-2004, showed 58 gene therapy clinical trial notifications as of May 2007. Not captured in the EudraCT are four gene therapy clinical trials in the EU using HS/PC that are ex vivo transduced with gammaretroviral vectors to treat inherited immunodeficiency diseases in newborns and adults (for details, see later).

In light of this information, the scope of this chapter will be confined to the discussion of regulatory issues relevant to the use of gamma-retroviral (for example, murine leukemia virus-derived) and lentiviral vectors (for example, human immunodeficiency virus-derived) used to genetically modify HSC/P ex vivo. Retroviral vectors have been the primary choice for the genetic modification of HSC/P because of their characteristic ability to integrate the retroviral expression vector into the host cell genome, potentially resulting in long-term gene expression in multiple hematopoietic lineages. This same property, the ability to integrate into the target cell genome, is the basis for the major risk associated with retroviral vectors: insertional mutagenesis. Depending upon where the retroviral DNA inserts into the host cell genome, there is the potential of gammaretroviral vectors with wild-type promoter/enhancer sequences for causing insertional mutagenesis. Insertional mutagenesis may lead to dysregulated gene expression because of distal gene activation via the strong retroviral enhancer, read-through transcription from the second copy of the enhancer, or depending on the site of integration, disrupted gene expression. In rare cases, the vector insertion may result in malignancy depending upon the nature of the genes affected by the insertion.

In recent years, the tumorigenic risk of using retroviral vectors has been transformed from a theoretical risk to an actual risk. In two European clinical trials in children suffering from X-SCID, at least 5 of the successfully treated children have developed leukemia subsequent to gene therapy. No leukemia has been observed in any other clinical trials using retrovirally modified HS/PC. The observed activation of the known human oncogene LMO-2 in some of these cases (1, 2) has led the scientific and medical communities to conclude that the vector integration directly contributed to the leukemogenic events. Additional contributing factors in discussion include the strong proliferation advantage of the modified cells in context with the disease and the therapeutic gene encoding the gamma c-chain of the IL-2 receptor (Report from the Ad hoc Meeting of CHMP Gene Therapy Expert Group 23–24 January 2003; Report from the Ad hoc Meeting of CHMP Gene Therapy Expert Group 26–27 June 2003; http://www.emea.europa.eu/htms/human/genetherapy/genetherapy.htm).

While the risk of tumorigenesis due to vector insertion may seem novel to the gene therapy field, it is noteworthy that there is an international regulatory framework that addresses this class of risks, broadly known as "genotoxicity." The International Conference on Harmonisation of Technical Requirements for the Registration of Pharmaceuticals for Human Use (ICH) Guidelines and Considerations allows for a mutual understanding of requirements for chemical (drugs) and biological medicinal products (biologicals) between Japan, the USA, Canada, EFTA countries, and the EU. For example, the step 5 version of the ICH S2B-related "Note for guidance on genotoxicity: A standard battery for genotoxicity testing of pharmaceuticals" defines genotoxicity as the fixation of damage to DNA "generally considered to be essential for heritable effects and (essential) in the multi-step process of malignancy." With respect to clinical gene therapy in general, it is important to differ between genotoxicity of the active ingredient and other components of the product. Only genotoxicity of the active ingredient of retroviral vector preparations will be considered here.

As discussed, to date, genotoxicity of gene transfer/therapy medicinal products with the outcome of malignancy has only been observed in patients enrolled in clinical trials aimed at the development of a gene therapy medicinal product for the treatment of X-SCID. However, this is a general consideration for any gene therapy product that carries the potential for "fixation of damage to DNA." In the context of retroviral vector-mediated gene therapy, the genotoxicity is mainly in the form of integration of the gene transfer vector or parts of it. Genotoxicity is a general consideration because new technology may lead to increased risk. For example, the following alterations to vector transduction methods may increase the risk of genotoxicity: (1) improved transduction efficiency, either by better cell isolation methods, more efficient maintenance of "true" hematopoietic stem cells, or modifying the growth factors and cytokines during cell culture, any of which may result in higher vector copy number per cell or an increased outgrowth of preneoplastic cells; and (2) new treatment strategies (such as preconditioning of patients by myeloablative drugs before the infusion of genetically modified cells) may enhance engraftment of genetically modified cells, potentially increasing the risk of engrafting with a vector integrated into a potentially tumorigenic genetic locus. Besides insertional mutagenesis in somatic cells, genotoxicity of gene transfer/therapy products inducing inherited effects relates to inadvertent germline integration and transmission of vectors or vector parts to progeny. In the case of ex vivo modification of HS/PC, the risk of inadvertent germline integration and transmission of vectors to progeny would be relevant only in those rare cases where there is potential for vector mobilization to occur. As an ICH Considerations paper titled "General Principles to Address the Risk of Inadvertent Germline Transmission of Gene Therapy Vectors" is available (http://www.ich.org), the risk of inadvertent germline transmission of vectors will not be further discussed here. Rather, genotoxicity of vectors in somatic cells with the possible outcome of malignancy in the treated subject or patient will be described in the context of the regulatory frameworks that have been developed in the US and the EU to address these risks to clinical trial participants.

### 1.1. Scope of This Chapter

The information presented in this chapter is confined to the discussion of regulatory issues that pertain to one class of product: HS/PC that have been transduced ex vivo with retroviral vectors. However, we will not review the general regulatory issues associated with the use of cell therapies. It is important to note that these issues are equally critical to consider when preparing a regulatory submission, so we refer the reader to other resources available to address the important issues associated with cellular therapies, such as donor screening, freedom from adventitious agents, and product characterization (relevant US guidance documents may be found: http://www.fda.gov/cber/genetherapy/gtpubs.htm;relevant EU legislation and guidance documents may be found under http://ec.europa.eu/enterprise/pharmaceuticals/eudralex/index.htm).

### 2. Regulatory Issues Associated with the Use of Retroviral Vectors Used to Modify Hematopoietic Stem Cells/Progenitors Ex Vivo

Retroviral vectors are replication defective by design. This results from the incorporation of cis-acting elements required for packaging, reverse transcription, and transcription into the retroviral expression vector, while expression of the structural and enzymatic proteins is provided in trans from independent expression

## 2.1. Addressing the Risks Associated with Replication-Competent Gamma-Retrovirus or Lentivirus

cassettes either by transient transfection or use of so-called vector packaging cells to result in release of infectious, but replication-defective, vector particles. Hence, the production method yields retroviral vectors that are physically identical to a retrovirus particle, while genetically devoid of the coding regions for retroviral proteins necessary for generation of progeny virions in transduced target cells. In rare instances, depending upon the design of the cis- and trans-acting expression cassettes, recombination events between the elements present in the expression cassettes or endogenous retroviral sequences in the vector producer cell (vpc), or, more theoretical, in the transduced cells, may occur during retroviral vector manufacture or in the transduced cells. When such events occur, they may result in generation of a new retroviral genome that is restored in its ability to replicate, thereby producing replication-competent retrovirus (RCR) or replication-competent lentivirus (RCL). Sponsors in the EU and the US are encouraged to consider changes to the design of their vector-manufacturing schemes in order to reduce the risk of RCR/RCL generation through one or more of the following strategies: (1) limiting homologous sequences available for recombination, both by reducing the overlapping sequences between the cis- and trans-acting vector expression plasmids and by use of cells with nonhomologous endogenous retroviruses; (2) separating trans-acting sequences onto more than one expression cassette; and (3) introducing stop codons within the overlapping regions in the vector genome to prevent expression of protein, in the event of recombination.

The contamination of retroviral vectors for clinical use with RCR/RCL translates into risks to patient safety. The following preclinical experiment in nonhuman primates provides evidence for the safety concern with RCR contamination of vectors. In a nonclinical experiment three out of ten immune suppressed monkeys developed lymphomas after infusion with bone marrow cells transduced with a preparation of RCR-positive gammaretroviral vector (not known at the time to be RCR-positive) (3). The subsequent observation that recombinants between components of the retroviral vector producer cells were found in the lymphomas provided suggestive evidence that the RCR contamination contributed to the development of lymphomas in these animals (4, 5). The more recent clinical use in the US and the EU of lentiviral vectors based on human immunodeficiency virus, a virus known to cause a fatal infection in humans, underscores the need to rigorously test for RCR/RCL contamination in all retroviral vector lots manufactured for clinical use. It would be an obvious undesirable outcome to introduce either an MLV-derived or an HIV- or other lentivirus-derived pathogen into patients or subjects enrolled in a clinical trial. The regulatory frameworks in the US and the EU for testing of retroviral vectors and clinical trials participants are described as follows.

*2.1.1. US and EU*
*Regulatory Framework*

In the US, sponsors of clinical trials using retroviral vectors (either ex vivo or in vivo) are referred to the Guidance to Industry: Supplemental Guidance on Testing for Replication-Competent Retrovirus in Retroviral Vector-Based Gene Therapy Products and During Follow-up of Patients in Clinical Trials Using Retroviral Vectors (hereby referred to as the RCR Guidance) (note that this guidance was initially issued in 2000 and was recently reissued with minor changes in 2006). The RCR Guidance document provides detailed recommendations regarding when RCR testing should be performed during retroviral vector manufacture. Due to the stochastic nature of the recombination events that give rise to RCR, the RCR Guidance recommends testing for RCR at multiple stages during vector manufacture, beginning with the Master Cell Bank for the vector production cells (if using a stable packaging cell line) through the end of production cells and final product. In addition, the RCR Guidance provides detailed recommendations regarding the quantities of materials to test, and, when possible to use an existing RCR standard, available through the American Type Culture Collection (ATCC, VR-1450 for the virus stock or VR-1448 for the virus producer cell line), how to apply a statistical analysis to the assay method in order to allow for reduced volumes for testing. When the retroviral vector product is used for ex vivo transductions, further RCR testing is recommended if the cells are cultured for a period of greater than or equal to 4 days post the initial exposure to vector. If the culture period is less than 4 days, archiving of transduced cells is sufficient.

In the EU, the legally binding technical requirements in Part IV, Annex I to Directive 2001/83/EC as amended by Directive 2003/63/EC (http://ec.europa.eu/enterprise/pharmaceuticals/eudralex/vol-1/dir_2003_63/dir_2003_63_en.pdf) call attention to the fact that certain gene therapy medicinal products may contain replication-competent vector particles and that patients may have to be monitored during clinical development and postmarketing for the development of possible infections and/or their pathological sequelae.

The European Pharmakopoeia (http://www.edqm.eu/site/page_628.php) General Chapter 5.1.4 titled "Gene transfer medicinal products for human use" gives recommendations for the characterization of recombinant vectors and genetically modified cells. For ex vivo genetically modified cells, it is recommended to test the final cell lot for replication-competent vector generation, which applies if cells are genetically modified with replication-incompetent retroviral vectors. It is, however, allowed to release cells before testing is completed, if necessary. The EU Note for guidance on the quality, preclinical, and clinical aspects of gene transfer medicinal products (CPMP/BWP/3088/99; http://www.emea.europa.eu/htms/human/itf/itfguide.htm) indicates in Chapter 3 (titled "viral vector") safety concerns that should be considered with the use of retroviral vectors such as activation of cellular oncogenes or inactivation of tumor suppressor genes due to proviral integration. An increase of probability of these risks is seen in the presence of RCR. It is therefore recommended to use packaging cell lines with minimal risk of recombination, such as those that independently express the structural and functional genes. In order to detect RCR, cell culture supernatant should be tested after vector production by passage(s) on a cell line permissive for RCR, followed by a suitable detection assay, possibly also allowing RCR quantification. The assay should include an appropriate control in order to evaluate the sensitivity, specificity, and reproducibility of the assay. Supporting data demonstrating the assay performance characteristics should be provided. Similarly, current recommendations for RCL testing are part of the CPMP Position Paper on the quality and safety of lentiviral vectors (CPMP/BWP/2458/03).

The RCR Guidance available in the US also provides recommendations as to the clinical follow-up of clinical trial participants for evidence of RCR transmission. While the details pertaining to the RCR testing schedule and archiving still apply, additional guidance regarding clinical follow-up is now provided in the Guidance for Industry: Gene Therapy Clinical Trials – Observing Participants for Delayed Adverse Events (hereby referred to as GT-Delayed AE Guidance) (described in more detail in **Subheading 32.2.2.1**). In the EU, a Concept Paper on clinical monitoring and follow-up of patients who have been enrolled in certain gene therapy clinical trials or have been treated with marketed gene therapy medicinal products will be published in 2007. Part IV, Annex I to Directive 2001/83/EC also asks for considering long-term follow-up, depending on the circumstances.

It is important to note that the RCR guidance in the US was written mainly with gamma-retroviral-based retroviral vectors in mind, and therefore, not all of this guidance may pertain to RCL testing for lentivirus vector production schemes. Therefore, the US FDA recommends sponsors of US clinical trials to contact FDA for specific guidance on testing for replication-competent lentivirus (RCL), although the reader is also referred to a report from the Lentivirus Vector Working Group that includes a discussion of issues to consider in developing assays for detection of RCL (6). In the EU, the CHMP Position Paper on Quality and Safety of Lentiviral Vectors (CPMP/BWP/2458/03) describes the general quality testing recommended for lentiviral vectors.

*2.2. Addressing the Risks Associated with Insertional Mutagenesis*

Vectors with the ability to integrate into the target cell genome carry the risk of insertional mutagenesis, depending upon where in the genome the integration occurs. Vector insertion is necessary, but not sufficient, to result in a malignant event in the clinical trial participant due to the multistep nature of tumorigenesis. In other words, while vector insertion may induce or deregulate expression of a known human oncogene, additional steps are required before the cells with this insertion event are transformed to malignant cells and a clinically diagnosed malignancy occurs. For these reasons, the malignancies resulting from vector insertion are delayed with unknown periods of latencies. In the only human cases of malignancy resulting from vector-induced insertional mutagenesis in X-SCID-affected children treated with gene therapy, the lag time between exposure and malignancy has been 3–5 years, but theoretically, this period of latency could be even longer. Indeed, if we refer to other well-established genotoxicity-inducing agents or treatments, such as chemotherapy or radiation treatment used in cancer therapy, the latencies in these cases have been reported to occur up to several decades past treatment. For example, in one study following 397 patients with Ewing's sarcoma, 6.5% developed secondary malignancies. In patients treated with chemotherapy the latencies ranged from 1.7 to 12.9 years, and in those treated with radiation therapy, the latencies ranged from 1.5 to 32.5 years (7). Therefore, a combination of careful risk-benefit analysis prior to initiating a gene therapy clinical trial using integrating vectors and long-term observations for signs of malignancies provides tools for managing the risks to these study participants. The specific US and EU recommendations regarding these general approaches are described as follows.

*2.2.1. US Regulatory Framework*

The US FDA has issued Guidance for Industry: Gene Therapy Clinical Trials – Observing Participants for Delayed Adverse Events (hereby referred to as GT-Delayed AE Guidance). The GT-Delayed AE Guidance provides general guidance regarding the need for long-term observations and the nature of those observations to sponsors of gene therapy clinical trials. In addition, specific recommendations are provided for sponsors of clinical trials using integrating vectors. The GT-Delayed AE Guidance recommends that all clinical trial participants should be observed for at least 15 years. Prior to initiating the gene therapy clinical trial, sponsors should collect systematic case histories of the clinical trial participants, in order to have baseline information for comparison, as well as educate the clinical trial participant about the importance of long-term observations, in order to elicit better participation in this aspect of the clinical trial. The baseline history should include information regarding prior exposures to other genotoxic agents (i.e., chemotherapy or radiation therapy) and other comorbidities. After the gene transfer procedure has been performed, the GT-Delayed AE Guidance recommends annual visits to a health care provider for the first 5 years, where new malignancies, neurologic disorders, rheumatologic or autoimmune disorders, or hematologic disorders should be recorded and reported. In the absence of any serious adverse events, the information from these visits can be tabulated in summary form from the sponsor's annual report. After the first 5 years, the GT-Delayed AE Guidance recommends annual contact with specific screening, if indicated by a change in the clinical condition for years 6–15.

In addition to these general recommendations for clinical observations of gene therapy clinical trial participants, the GT-Delayed AE Guidance has specific recommendations for laboratory testing of participants of clinical trials that involve use of cells with high replicative capacity and long survival that have been transduced with integrating gene therapy vectors. As such, the ex vivo modification of HS/PC with retroviral vectors would fall under this category for additional clinical and laboratory follow-up. In these cases, the GT-Delayed AE Guidance recommends that a surrogate of the genetically modified cells (in the case of HS/PC, peripheral blood would provide this surrogate) should be sampled every 6 months for the first 5 years and yearly over the next 10 years for evidence of vector sequences. If vector is detectable in at least 1% of the surrogate cells, then the GT-Delayed AE Guidance recommends that the sponsor should perform an analysis that will indicate the pattern of vector integration sites so that trends toward oligo- or mono-clonality can be identified. Should evidence of oligo- or mono-clonality arise, combined with evidence of clonal expansion or vector integration within or near a locus with known oncogenic activity, the GT-Delayed AE Guidance recommends additional clinical monitoring for evidence of malignancy. Finally, the GT-Delayed AE Guidance provides detailed recommendations regarding specific information that should be provided in the Informed Consent Document to accurately convey the risk of cancer from exposure to products that are genetically modified by retroviral vectors.

*2.2.2. EU Regulatory Framework*

In Chapter 6.4 ("Safety") of the EU Note for guidance on the quality, preclinical, and clinical aspects of gene transfer medicinal products (CPMP/BWP/3088/99), the necessity to include in the viral safety monitoring program, testing for presence or appearance of RCR, when nonreplication-competent retroviral vector is used, is pointed out. In this context, the potential for genomic integration and thus, the necessity of long-term follow-up is stated. In Chapter 6.5 ("Patients monitoring") long-term follow-up is recommended to include assessment for gene expression and stability of the gene transfer procedure, as well as safety assessment, with particular emphasis on potential long-term clinical consequences of vector integration, nonspecific transduction of nontarget cells, or prolonged expression from the vector.

For additional EU guidance on this topic, the reader is referred to Chapter 5.1 ("Unintended and unexpected consequences of gene transfer"), where insertional mutagenesis, induced cellular changes, and vector mobilization are mentioned as quality considerations, and Chapter 5.4.6 ("Genotoxicity/carcinogenicity studies") where the issue of the potential of retroviral vectors for insertional mutagenesis is elucidated.

More detailed EU guidance on clinical monitoring and long-term follow-up has been drafted to include recommendations regarding long-term follow-up after completion of a clinical trial and postmarketing. Specific EU guidance on development and manufacture of lentiviral vectors is also available ("Guideline on Development and Manufacture of Lentiviral Vectors" (CHMP/ BWP/2458/03)), which draws attention to insertional addition of proviral DNA in or close to active genes, which may trigger tumor initiation or promotion, as a major safety concern for replication-incompetent lentiviral vectors. The risk of insertional mutagenesis associated with increased vector copy number per cellular genome is described in the latter guideline in Chapter 6 ("Oncogenesis"). The integration capacity of lentivector lots may be assessed in suitable cell lines by nucleic acid amplification technology.

Following the diagnosis of the first leukemia in one of the children treated in one of the X-SCID trials by administration of retrovirally modified bone marrow cells, the CHMP Gene Therapy Working Party met with international experts and published a report. Here, and in the following report, it was recommended to restrict retroviral vectors to patients suffering from life-threatening disease. Reasons for an overt risk of insertional oncogenesis were estimated to be related to the special physiological condition presented by the disease, the young age of treated children and the modification of progenitor, rather than fully differentiated cells such as lymphocytes.

**2.3. Addressing the Risks Associated with Insertional Mutagenesis by Improved Vector Design and Vector Tracking**

As a response to the occurrence of leukemias in one of the X-SCID gene therapy clinical trials (as of June 2007, the incidence was four out of eight of the children with evidence of engraftment), the CHMP Gene Therapy Working Party has proposed to address the risks of insertional mutagenesis/oncogenesis by platform studies using validated methods of vector tracking, in order to identify the safety profiles of different vectors (Report from the Ad hoc Meeting of CHMP Gene Therapy Expert Group 23–24 January 2003; Report from the Ad hoc Meeting of CHMP Gene Therapy Expert Group 26–27 June 2003; http://www.emea.europa.eu/ htms/human/genetherapy/genetherapy.htm). Scientific studies have been initiated (in the EU?) aimed at comparing the risk of specific gamma-retroviral and lentiviral vectors to induce insertional oncogenesis in rodents in vivo and rodent cells in vitro. In the US, a large study, performed in compliance with Good Laboratory Practices, will be performed to assess the validity of a murine model on the risk of insertional oncogenesis of gamma-retroviral vectors under the auspices of the National Toxicology Program (for information about this program, see http://ntp.niehs.nih.gov/). The primary aims of this study are to identify the limit of sensitivity and general utility of a serial bone marrow transplant model *(8)* to assess the risk of insertional mutagenesis. Additional types of retroviral vectors may be included in the study over time.

Lentiviral vectors currently used in clinical trials are believed to have less risk for insertional oncogenesis than gamma-retroviral vectors currently used in clinical trials. The reduced risk reasoned to increase lentiviral vector safety is based on the use of self-inactivating Long Terminal Repeats (SIN-LTRs) at the 3′-end of the vector carrying the therapeutic gene. The 3′ SIN-LTR is duplicated into the 5′ LTR during reverse transcription of the vector RNA resulting in an integrated vector containing no retroviral promoter or enhancer sequences. Theoretically, a SIN vector no longer deregulates or activates the transcription of neighboring genes. However, it should be noted that the internal promoter/enhancer used to allow transcription of the therapeutic gene may still possess the ability to trans-activate gene expression. Such a design is also chosen for newly developed gamma-retroviral vectors. For both vectors, less risk of insertional oncogenesis in mice seems to result when a weak cellular promoter/enhancer sequence is used internal to the deleted LTRs to direct transgene transcription within the vector (C. Baum, personal communication).

## 3. Conclusions

The main regulatory issues presented by HS/PC genetically modified by gamma-retroviral or lentiviral vectors are (a) the possible generation of replication-competent vector-derived viruses (RCR or RCL), (b) the risk of insertional oncogenesis as a result of insertional mutagenesis due to vector integration into the chromosomal DNA of the transduced cells, and (c) the risk of inadvertent germline integration of vector DNA into germline cells, the latter also presenting possible risks of insertional mutagenesis and adverse effects on children of respective patients. While the risk of inadvertent germline integration is not relevant to ex vivo modified cells, it is important to note that this concern should be addressed for in vivo use of vectors by nonclinical studies on biodistribution and vector persistence (ICH Considerations titled "General Principles to Address the Risk of Inadvertent Germline Transmission of Gene Therapy Vectors"; http://www.ich.org). The absence of RCR or RCL in vector preparations should be achieved by appropriate vector design and rigorous testing of vector producer cells and supernatants, if possible prior to administration of ex vivo modified cells to patients, and testing of patients for possible infections. Long-term follow-up should be performed in clinical trial participants exposed to products made with retroviral vectors because of the potential for delayed adverse effects such as insertional oncogenesis. In the US and the EU, appropriate guidelines are in place.

## Acknowledgments

We thank the ICH Gene Therapy Discussion Group, the EMEA/CHMP Gene Therapy Working Party, the CBER/FDA Office of Cellular, Tissues, and Gene Therapy and the Division of Medical Biotechnology of the Paul-Ehrlich-Institut in Langen, Germany, for helpful discussion and continuous inspiration on regulatory work in gene and cell therapy.

## References

1. Hacein-Bey-Abina, S., et al. (2003) A serious adverse event after successful gene therapy for X-linked severe combined immunodeficiency. *N. Engl. J. Med.* 348(3): 255–256.

2. Hacein-Bey-Abina, S., et al. (2003) LMO2-associated clonal T cell proliferation in two patients after gene therapy for SCID-X1. *Science* 302(5644): 415–419.

3. Donahue, R.E., et al. (1992) Helper virus induced T cell lymphoma in nonhuman primates after retroviral mediated gene transfer. *J. Exp. Med.* 176: 1125–1135.

4. Vanin, E.F., et al. (1994) Characterization of replication-competent retroviruses from non-human primates with virus-induced lymphomas and observations regarding the mechanism of oncogenesis. *J. Virol.* 68(7): 4241–4250.

5. Purcell, D.F.J., et al. (1996) An array of murine leukemia virus-related elements is transmitted and expressed in a primate recipient of retroviral gene transfer. *J. Virol.* 70(2): 887–897.

6. Kiermer, V., et al. (2005) Report from the Lentivirus Vector Working Group: issues for developing assays and reference materials for detecting replication-competent lentivirus in production lots of lentivirus vectors. *BioProcessing* 4(2): 39–42.

7. Fuchs, B., et al. (2003) Ewing's sarcoma and the development of secondary malignancies. *Clin. Orthop. Relat. Res.* 415: 82–89.

8. Li, Z., et al. (2002) Murine leukemia induced by retroviral gene marking. *Science* 296: 497.

# INDEX